Theoretical and Computational Chemistry

MOLECULAR MODELING OF THE SENSITIVITIES OF ENERGETIC MATERIALS

VOLUME 22

THEORETICAL AND COMPUTATIONAL CHEMISTRY

SERIES EDITORS

Professor P. Politzer
Department of Chemistry
University of New Orleans
New Orleans, LA 70148, U.S.A.

Jane S. Murray
Department of Chemistry
University of New Orleans
New Orleans, LA 701487, U.S.A

VOLUME 1
Quantitative Treatments of Solute/Solvent Interactions
P. Politzer and J.S. Murray (Editors)

VOLUME 2
Modern Density Functional Theory: A Tool for Chemistry
J.M. Seminario and P. Politzer (Editors)

VOLUME 3
Molecular Electrostatic Potentials: Concepts and Applications
J.S. Murray and K. Sen (Editors)

VOLUME 4
Recent Developments and Applications of Modern Density Functional Theory
J.M. Seminario (Editor)

VOLUME 5
Theoretical Organic Chemistry
C. Párkányi (Editor)

VOLUME 6
Pauling's Legacy: Modern Modelling of the Chemical Bond
Z.B. Maksic and W.J. Orville-Thomas (Editors)

VOLUME 7
Molecular Dynamics: From Classical to Quantum Methods
P.B. Balbuena and J.M. Seminario (Editors)

VOLUME 8
Computational Molecular Biology
J. Leszczynski (Editor)

VOLUME 9
Theoretical Biochemistry: Processes and Properties of Biological Systems
L.A. Eriksson (Editor)

VOLUME 10
Valence Bond Theory
D.L. Cooper (Editor)

VOLUME 11
Relativistic Electronic Structure Theory, Part 1. Fundamentals
P. Schwerdtfeger (Editor)

VOLUME 12
Energetic Materials, Part 1. Decomposition, Crystal and Molecular Properties
P. Politzer and J.S. Murray (Editors)

VOLUME 13
Energetic Materials, Part 2. Detonation, Combustion
P. Politzer and J.S. Murray (Editors)

VOLUME 14
Relativistic Electronic Structure Theory, Part 2. Applications
P. Schwerdtfeger (Editor)

VOLUME 15
Computational Materials Science
J. Leszczynski (Editor)

VOLUME 16
Computational Photochemistry
M. Olivucci (Editor)

VOLUME 17
Molecular and Nano Electronics: Analysis, Design and Simulation
J.M. Seminario (Editor)

VOLUME 18
Nanomaterials: Design and Simulation
P.B. Balbuena and J.M. Seminario (Editors)

VOLUME 19
Theoretical Aspects of Chemical Reactivity
A. Toro-Labbe (Editor)

VOLUME 20
The Crystalline States of Organic Compounds
A. Gavezzotti

VOLUME 21
Properties and Functionalization of Graphene: A Computational Chemistry Approach
D. Tandabany and F. Hagelberg (Editors)

Theoretical and Computational Chemistry

MOLECULAR MODELING OF THE SENSITIVITIES OF ENERGETIC MATERIALS

VOLUME 22

Edited by

DIDIER MATHIEU
French Alternative Energies and Atomic Energy Commission (CEA), Le Ripault, France

Elsevier
Radarweg 29, PO Box 211, 1000 AE Amsterdam, Netherlands
The Boulevard, Langford Lane, Kidlington, Oxford OX5 1GB, United Kingdom
50 Hampshire Street, 5th Floor, Cambridge, MA 02139, United States

ISBN: 978-0-12-822971-2
ISSN: 1380-7323

For information on all Elsevier publications
visit our website at https://www.elsevier.com/books-and-journals

Publisher: Susan Dennis
Acquisitions Editor: Kathryn Eryilmaz
Editorial Project Manager: Emerald Li
Production Project Manager: Bharatwaj Varatharajan
Cover Designer: Alan Studholme

Typeset by STRAIVE, India

Contents

7. Thermal initiation and propagation of the decomposition process

Romain Claveau, Julien Glorian, and Didier Mathieu

Part III

Relationships involving the crystal structure

8. Some molecular and crystalline factors that affect the sensitivities of explosives

Peter Politzer and Jane S. Murray

9. Interplay between chemical and mechanical factors

Sergey V. Bondarchuk

10. From lattice vibrations to molecular dissociation

Adam A.L. Michalchuk and Carole A. Morrison

11. Role of electronic excited states in the initiation of explosives

Didier Mathieu, Romain Claveau, and Julien Glorian

Part IV

Insight from numerical simulations

12. Molecular dynamics simulation of hot spot formation and chemical reactions

Didier Mathieu and Itamar Borges, Jr

13. Quantum chemical investigations of reaction mechanism

Weihua Zhu

14. Ranking explosive sensitivity with chemical kinetics derived from molecular dynamics simulations

M.J. Cawkwell, S.R. Ferreira, N. Lease, and V.W. Manner

15. Chemical kinetics and the decomposition of secondary explosives

B.F. Henson and L. Smilowitz

Part V

Applications to the design of new materials

16. Implementation of predictive models: Practical aspects

Didier Mathieu, Romain Claveau, and Julien Glorian

17. Molecular and crystal insights into the structural design of low-sensitivity energetic materials

Yi Wang, Siwei Song, and Qinghua Zhang

Contributors

Brian C. Barnes U.S. Army Combat Capabilities Development Command (DEVCOM) Army Research Laboratory, Aberdeen Proving Ground, MD, United States

Sergey V. Bondarchuk Department of Chemistry and Nanomaterials Science, The Bohdan Khmelnytsky National University of Cherkasy, Cherkasy, Ukraine

Itamar Borges, Jr Departamento de Química, Instituto Militar de Engenharia (IME), Rio de Janeiro, RJ, Brazil

Edward F.C. Byrd U.S. Army Combat Capabilities Development Command (DEVCOM) Army Research Laboratory, Aberdeen Proving Ground, MD, United States

Laurent Catoire ENSTA Paris, Institut Polytechnique de Paris, Palaiseau, France

M.J. Cawkwell Los Alamos National Laboratory, Los Alamos, NM, United States

Romain Claveau CEA, DAM, Le Ripault, Monts, France; French-German Research Institute of Saint-Louis, Saint-Louis, France

C. Collet MSIAC Munitions Safety Information Analysis Center, NATO HQ, Belgium

Aurélien Demenay ENSTA Paris, Institut Polytechnique de Paris, Palaiseau, France

S.R. Ferreira Los Alamos National Laboratory, Los Alamos, NM, United States

Julien Glorian French-German Research Institute of Saint-Louis, Saint-Louis, France

B.F. Henson Chemistry Division, Los Alamos National Laboratory, Los Alamos, NM, United States

V. Le Gallo CEA, DAM, Gramat, Gramat, France

N. Lease Los Alamos National Laboratory, Los Alamos, NM, United States

A. Lefrancois CEA, DAM, Gramat, Gramat, France

V.W. Manner Los Alamos National Laboratory, Los Alamos, NM, United States

Didier Mathieu CEA, DAM, Le Ripault, Monts, France

Adam A.L. Michalchuk Federal Institute for Materials Research and Testing (BAM), Berlin, Germany

Jason A. Morrill Department of Chemistry and Biochemistry, William Jewell College, Liberty, MO, United States

Carole A. Morrison The School of Chemistry, University of Edinburgh, Edinburgh, United Kingdom

Jane S. Murray Department of Chemistry, University of New Orleans, New Orleans, LA, United States

Marco Aurélio Souza Oliveira Departamento de Química, Instituto Militar de Engenharia (IME), Rio de Janeiro, RJ, Brazil

Roberta Siqueira Soldaini Oliveira Departamento de Química, Instituto Militar de Engenharia (IME), Rio de Janeiro, RJ, Brazil

Antoine Osmont CEA, DAM, Gramat, Gramat, France

Peter Politzer Department of Chemistry, University of New Orleans, New Orleans, LA, United States

Betsy M. Rice U.S. Army Combat Capabilities Development Command (DEVCOM) Army Research Laboratory, Aberdeen Proving Ground, MD, United States

L. Smilowitz Chemistry Division, Los Alamos National Laboratory, Los Alamos, NM, United States

Siwei Song Institute of Chemical Materials, China Academy of Engineering Physics, Mianyang, China

M. Vaullerin CEA, DAM, Gramat, Gramat, France

Yi Wang Institute of Chemical Materials, China Academy of Engineering Physics, Mianyang, China

Qinghua Zhang Institute of Chemical Materials, China Academy of Engineering Physics, Mianyang, China

Weihua Zhu Institute for Computation in Molecular and Materials Science, School of Chemistry and Chemical Engineering, Nanjing University of Science and Technology, Nanjing, China

Preface

At the end of 2019, as Prof. Peter Politzer invited me to edit a volume in the Elsevier series *Theoretical and Computational Chemistry* devoted to the molecular modeling of the sensitivities of energetic materials, I did not hesitate long before accepting. I had previously experienced the lack of a comprehensive overview introducing newcomers to the different approaches to this topic. The focus of this book is on modeling how molecular structure of energetic materials affects their responses under various kinds of loads. This insight is of primary significance in view of their potential use as explosives, propellants or pyrotechnic compounds.

This book follows Volumes 12 and 13 of the same series edited by Peter Politzer and Jane S. Murray in 2003, and Volume 69 of the series *Advances in Quantum Chemistry* edited by J.R. Sabin in 2014. Previous reviews of the application of molecular modeling techniques to energetic materials mainly address the evaluation of their performances, which are determined primarily by the nature of the constitutive species. This is less the case for the sensitivities that can depend significantly on many additional factors, including the structure of the material at different supramolecular scales, but also some details of the experimental protocols and the environment in which the measurements are performed. In this context, the very existence of a direct relationship between molecular structure and measured sensitivities is questionable. As a matter of fact, molecular modeling of sensitivities has long been limited to empirical correlations linking experimental data to molecular features in a very approximate way, as is still the case for many kinds of insults.

However, significant progress has been made in the last decade regarding impact sensitivity. Enhanced computing capabilities, more efficient reactive potentials, and new physics-based models have led to deeper insight into the microscopic mechanisms involved and to the emergence of increasingly more convincing correlations between drop weight impact test data and theoretical indicators. This supports the assumption that the chemical nature of the material plays a determining role. Although recent approaches are at an early stage of development and still require confirmation and generalization, they provide important motivation to continue this work and address other sensitivity indicators, including decomposition temperature as well as friction or spark sensitivities.

This book aims at providing a comprehensive overview of these recent advances. For a more concise summary of the current state of the art, I would recommend the introduction of the PhD thesis "Mechanochemical Processes in Energetic Materials" published in 2020 by A.L. Michalchuk. Given the real-world significance of sensitivities, and although the present series covers a theoretical discipline, the focus of this volume is on approaches that could lead to practical tools for assessing the stability of potentially high-energy compounds from their structural formula.

Chapter 1 presents a broad overview of energetic materials, their chemical composition, and corresponding applications.

Chapter 2 describes the main safety tests available to quantify the sensitivities of energetic materials toward the various types of loads to which they can be subjected. Although molecular modeling efforts have so far focused on drop weight impact test data, a broad range of testing procedures are described, in the hope that forthcoming studies will address the corresponding sensitivity indicators.

Chapter 3 provides a historical perspective on early attempts to estimate the stability of energetic compounds from simple relationships primarily based on physical insight and involving oxygen balance or bond dissociation energies.

Chapter 4 reviews alternative relationships based on the molecular charge distribution through the use of either local or global descriptors.

Chapter 5 considers the various sensitivity indices in turn and presents some predictive models for each. The emphasis is on empirical correlations as a detailed physical understanding is still lacking for complex properties like friction or spark sensitivities. Therefore, such models may be used as a last resort in spite of their questionable reliability.

Chapter 6 reviews the application of modern quantitative structure–property relationship methodologies to sensitivities of energetic materials, i.e., the search for empirical correlations using statistical and machine learning techniques.

Chapter 7 discusses the initiation of energetic materials as a two-step process, where the energy provided to the loaded sample is first converted into heat that activates endothermic bond cleavages, while in the second step, exothermic recombinations release further energy, leading to local heating that activates subsequent bond cleavages, thus triggering a self-sustaining decomposition process.

Chapter 8 discusses the combined influence of molecular and crystal factors. The former play a primary role and were considered first, while the latter cannot be neglected when designing reduced sensitivity materials and have been the focus of recent attention.

Chapter 9 illustrates the interest in crystal features with an overview of recent studies considering the interplay between chemical and mechanical factors.

Chapter 10 presents the most advanced version of a nonempirical approach to impact sensitivity based on the rate at which mechanical impact energy is converted to intramolecular vibrations.

Chapter 11 addresses the role of electronically excited states, which have long been considered a potentially important factor in understanding and predicting sensitivities.

Chapter 12 reports on the extensive physical insight gained from the molecular simulation of materials subjected to various kinds of loads.

Chapter 13 reviews quantum chemical investigations of the reactions mechanisms involved in the decomposition of energetic materials, which could play a significant role in the observed sensitivities.

Chapter 14 describes one of the most promising approaches currently available to quantitatively estimate impact sensitivity, namely its correlation with chemical decomposition rates derived at the microscale from molecular dynamics simulations.

Likewise, Chapter 15 describes a remarkably successful model describing the decomposition of explosives over the entire temperature range from thermal ignition by direct heating to detonation, using only independently measured kinetic parameters for known processes in the decomposition and combustion of solid explosives.

Chapter 16 provides some practical guidelines to readers interested in implementing predictive models for sensitivities, based on first-principles calculations carried out with standard computer programs or on semiempirical expressions involving descriptors easily calculated using chemoinformatics toolkits.

Finally, Chapter 17 shows how to take advantage of simple structure–sensitivity relationships, in addition to structure–performance relationships, in the design of new high energy density materials.

I am pleased that most leading experts invited to contribute to this book accepted and wrote excellent contributions, despite the constraints arising from the COVID-19 pandemic. By the end of 2020, the whole book was completed except for a couple of chapters for which my search for expert contributors had proved unsuccessful. I set out to fill this gap myself. However, it took more time than expected. I am sorry for the delay this has caused, which could have been even longer without the help of Prof. Itamar Borges in critically reading the first draft of Chapter 11 and writing Chapter 12. Finally, I thank all contributors for their work. I am also very grateful to the editors of this series, Prof. Peter Politzer and Prof. Jane S. Murray, for the trust they placed in me. I am confident that this book will be useful not only to newcomers to the field of sensitivity modeling but also to the broader community of researchers and engineers concerned with the properties of energetic materials.

Didier Mathieu
Hadol, Vosges
July 2021

PART I

Experimental aspects

CHAPTER

1

Overview of energetic materials

Antoine Osmont and A. Lefrancois*
CEA, DAM, Gramat, Gramat, France
*Corresponding author: E-mail: antoine.osmont@cea.fr

1 General principles

1.1 Energetic materials chemistry

Energetic materials are formulations based on a mixture of chemicals that can undergo an exothermic oxydo-reduction reaction to release energy. This energy is used to generate different effects. It can be heat, light, smoke, sound, delay, thrust, and shock.

An oxydo-reduction reaction occurs between reducing and oxidizing agents. Even if a wide variety of reducing agents exists, the number of oxidizing agents is smaller. These agents are based on specific molecular structures.

The more common oxidizing agent is oxygen. In this molecule, the degree of oxidation of the oxygen atom is 0. The more stable form for oxygen atom is at the degree −2 and an exothermic oxidation reaction is linked with an oxidation degree decrease. Despite the usual formula given for this molecule, with a double bond O=O, the real formula is different. Two electrons are involved in one bond. The two other electrons are at the same level of energy and must occupy two different orbitals of the same energy. As a consequence, one finds one electron alone in each orbital, making oxygen a bi-radical, thus explaining its high oxidizing potential. On the opposite, nitrogen, with a real triple bond is totally inert and exists at 78% in the air without any reaction.

If we consider an oxygen atom in an alcohol group, C—O—H, its degree of oxidation is −2. As the atom is already surrounded by eight electrons, it cannot be implied in oxidizing reactions. The only way to make oxygen atoms available for oxidizing reactions is to place them in bonds with hypervalent atoms, such as nitrogen, phosphorous, sulfur, or chlorine. It is possible to build structures in which atoms of oxygen are linked with a lone pair of electrons. For nitrogen, for example, nitrous acid obeys the octet rule, with the formula H—O—N=O. For nitric acid, on the contrary, the formula is $H{-}O{-}N(=O)_2$. The existence of this molecule is allowed by the conjugated system sharing the electrons between the oxygen and nitrogen

Theoretical and Computational Chemistry, Volume 22
https://doi.org/10.1016/B978-0-12-822971-2.00011-5

atom. However, the molecule is not the fundamental state, letting oxygen atoms be available for oxidation reactions. In this molecule, the evaluation of the degree of oxidation is more complex. If we consider that the oxygen atom has a classical environment with two bonds, so a level −2, it implies that the nitrogen atom has a level +5, instead of the classical +3 like in nitrous acid. It reveals that nitrogen is not in its fundamental state and recovering this state frees oxygen radicals in the media, allowing oxidation reactions.

The oxidizing agents used in energetic materials are based on nitrogen and chlorine atoms. For nitrogen, it is nitro groups, $-NO_2$, which can be linked with carbon, nitrogen, or oxygen atoms, in mineral or organic substances. For chlorine, it is chlorate ClO_3^- or perchlorate ClO_4^- mineral salts. We shall only encounter different structures in a very particular application: primary explosives.

1.2 How to obtain energetic materials effects?

All energetic materials are based on a mixture of oxidizing and reducing chemicals. This mixture enables the reaction of the product on itself, without the need for oxygen from the air, making the reaction faster than the combustion of a hydrocarbon in the air for example. When oxidizing and reducing agents are grains of powder mixed together, time is required to transport the radicals from different grains in contact, which slows the combustion. Sometimes, both oxidizing and reducing products are groups present in the same molecule. This enables acceleration of the combustion, which can reach supersonic velocity. Under these conditions, the vicinity does not have enough time to adapt to the evolution of energetic medium, leading to the apparition of shocks. Depending on the velocity, a deflagration or a detonation is obtained.

Depending on the effect searched, the mixtures and the velocity of reaction differ. This difference corresponds to different families of compositions. Energetic materials are divided between four families. The first one is gun propellants, designed to accelerate a projectile in a gun. This family is based on nitrocellulose and nitroglycerine. The second one is propellants, used for the propulsion of rockets. These propellants are generally based on a mixture of ammonium perchlorate, a reducing agent, and a binder or on nitrocellulose/nitroglycerine. The third one is high explosives whose decomposition is a detonation, with shock effects. This family is based on explosive molecules or mixtures with ammonium nitrate. The last family is pyrotechnics. This family is the widest one, with various compositions of oxidizing and reducing agents, enabling different effects.

In the following pages, we shall present these four families, the type of molecules or compositions used, and the process used to manufacture them.

2 Gun propellants

Gun propellants are compositions based on mixtures generating gas in a breach, giving the thrust on a projectile in a gun. This activity is the first application of pyrotechnics for military applications, with black powder. This composition had been used for centuries until the invention of smokeless powder by Vieille in 1884. Since this invention, black powder rapidly disappeared for this application. Nowadays, black powder is only used for pyrotechnics applications. It will be presented in this dedicated section. An enhanced smokeless powder was proposed by Nobel in 1888, with addition of nitroglycerine [1].

2.1 Molecules

The main molecule used for gun propellants is nitrocellulose. It was discovered by Schoenbein in 1845. The combustion of nitrocellulose is smokeless, which is totally different from black powder. Its use as a gun propellant was tested immediately. In France, the pyroxyl commission concluded in 1849 that nitrocellulose enables a reduction of two-third of the charge mass compared with black powder. However, sometimes, an explosion of the gun occurred, without any explanation, which hindered the use of nitrocellulose.

Vieille discovered the possibility to gelatinize the fibrous matter with an ether alcohol mixture. The compact paste obtained is extruded with a press to obtain pieces of regular shape. This regular shape enables the regulation of the combustion, at a constant velocity, and prevents transitions between regimes of decomposition. On the contrary, this transition between regimes is possible with the fibers of nitrocellulose, leading to the gun's explosions encountered by the pyroxyl commission.

A few years after, A. Nobel proposed to add nitroglycerine to nitrocellulose, to increase the power of the powder. It is called double base propellant, by opposition to the single base propellant, containing only nitrocellulose.

Cellulose is a polymer of glucose six-rings, with three alcohol groups. The nitration of these groups with a mixture of sulfuric and nitric acids, called sulfo-nitric acid, leads to a polymer substituted by three nitrate groups by glucose ring (Fig. 1.1). The rate of nitration of cellulose is evaluated by the nitrogen rate in the molecule. Depending on the number of alcohol groups nitrated, 1, 2, or 3, the nitrogen rate is respectively 6.76%, 11.12%, or 14.14%. The military-grade is superior to 12.3% [2].

The chemical properties of nitrocellulose depend on the nitration rate. At 12.6 %N, it is fully soluble in a mixture of diethyl ether and ethanol (2:1). At 13.4 %N, the solubility decreases to 10% in this solvent whereas the pure trinitrate is insoluble in this solvent.

The nitration of cellulose is a reversible reaction. The forward reaction releases nitric acid in the powder, which increases the sensitivity of the powder. To prevent this release, stabilizers are added when manufacturing the powder. Diphenyl amine or centralites are often used to neutralize the acid. The rate of stabilizer should be checked every 5 years to ensure the usability of a powder. Accidents have already happened in storages of energetic materials, due to old powders that ignited spontaneously.

Other products can be added, to reduce the glowing at the muzzle of the gun. It can be potassium nitrate or sulfate.

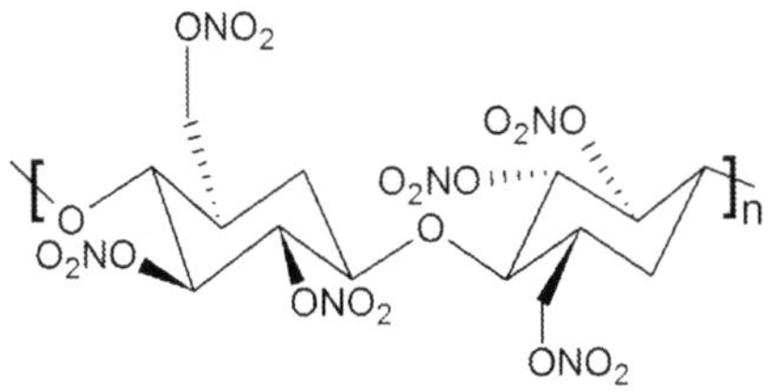

FIG. 1.1 Nitrocellulose molecule.

2.2 Shape of gun propellant grains

Nitrocellulose is gelatinized by an ether alcohol mixture or by nitroglycerine, to obtain a paste. This paste is transformed to obtain objects of regular shape. This shape enables the regulation of the combustion velocity of the propellant.

Several shapes exist: flakes, spheres, crushed spheres, grains, sticks, cords, pipes, multi-hole cylinders (7 or 19), slotted cylinders (Fig. 1.2). Depending on the shape, the laminar combustion leads to an evolution of the burning surface. It decreases for a sphere, is constant for a pipe, and increases for 7 or 19 holes. The appropriate powder for a gun is chosen in function of the interior ballistic searched for the projectile. For small calibers weapons, one finds flakes, spheres, crushed spheres, grains, or sticks. Powders with 7 or 19 holes are used for larger calibers.

2.3 Process

We shall now describe the process used to prepare these powders. The first part is the production of nitrocellulose. A second process is used to transform the nitrocellulose into gun powder.

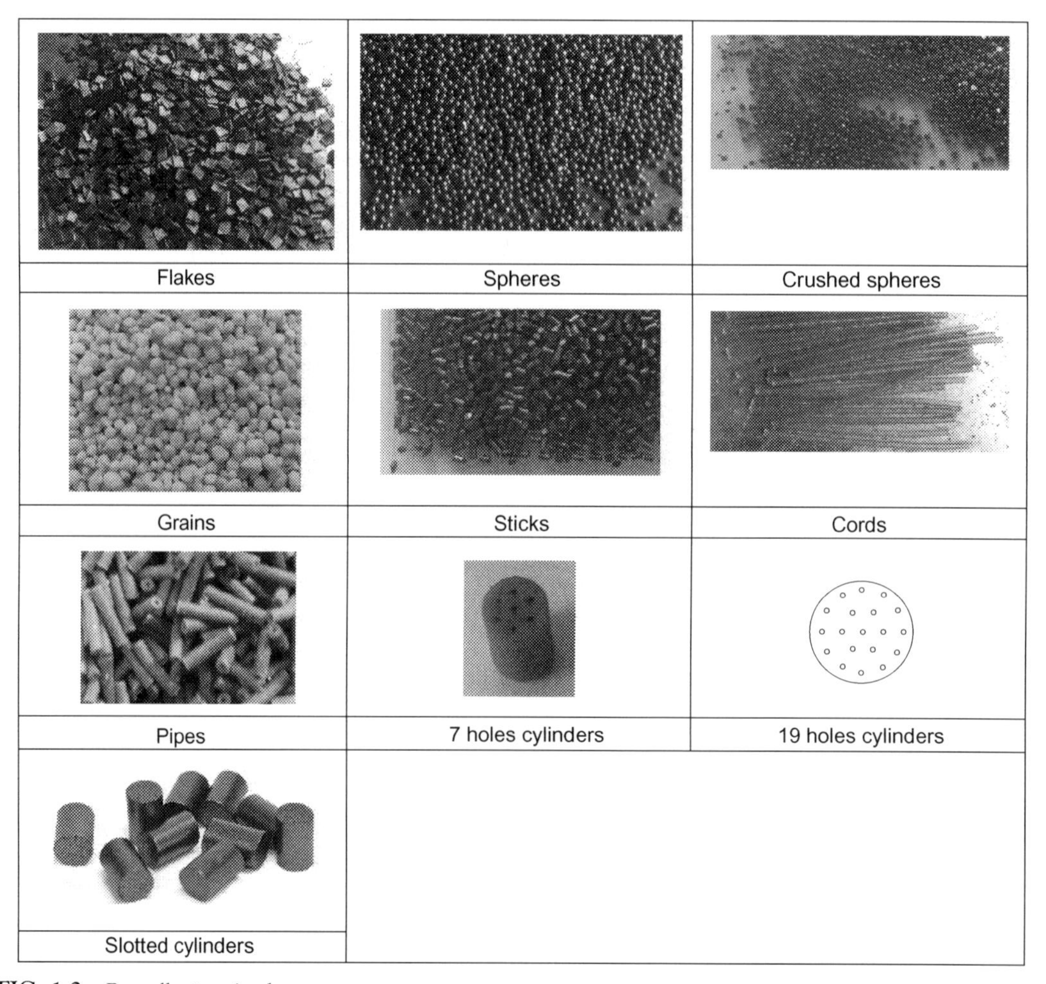

FIG. 1.2 Propellant grain shapes.

2.3.1 Nitrocellulose

The nitration of dried cellulose by nitric acid concentrated above 60% is possible. However, it leads to hydrated nitrocellulose. The maximum nitration rate is 13.2 %N, and the product is heterogeneous.

To solve this problem, the use of a sulfo-nitric medium is necessary. Sulfuric acid is used to fix the water generated by the nitration, letting the nitric acid concentrate enough to proceed up to the cellulose trinitrate. The reaction is very exothermic, implying a thermal regulation of the reactor. The reaction lasts 30 mn at 30°C.

Once the nitration is finished, the nitric bath is removed and the nitrocellulose wringed out. As soon as the wringing is finished, the nitrocellulose is drowned in water so as to avoid spontaneous inflammation. This operation is the more dangerous in the process. The first washing is carried out with cold water.

The product obtained still contains sulfuric acid and sulfate anions trapped in the fibers, causing instabilities that can lead to spontaneous inflammation. To remove these salts, a second washing is needed in boiling water or at a higher temperature in an autoclave for 4 days. To shorten this delay, mechanical agitation can be used.

The next washing is performed with basic water due to the addition of calcium or sodium carbonate. Carbon dioxide is generated by the neutralization of acids, forming calcium or sodium hydroxide that is removed during the last washing. The product is dried by wringing and the process parameters are adjusted to keep enough water in the nitrocellulose to stabilize it.

Dry nitrocellulose is an explosive as sensitive as pentrite. A 13.45 %N nitrocellulose detonates at 7300 m/s [3] and the critical diameter of a 13.3 %N nitrocellulose is 20 mm [4]. Nitrocellulose containing 30% of water is considered only as a flammable compound.

2.3.2 Single base propellants

The first step is the drying of nitrocellulose [5]. To avoid working on dry nitrocellulose, this drying is carried out with pure ethanol in a press or a wringer. The nitrocellulose used is a mixture of two types of different nitration rates: 12.6 and 13.4.

The nitrocellulose with ethanol is poured into a mixer. Once the matter is homogeneous, diethyl ether is added to obtain the solvent ether alcohol 65:35. Diphenylamine is added at the same time. Several hours of mixing give a paste.

This paste is poured into a device where it is pressed to remove air. The paste is added in different samples, pressed after each addition, so as to avoid inflammation. When the matter is homogeneous enough, it is extruded through nozzles to form blocks of defined shape, as described above. The extruded paste is cut at the desired length. Different designs of nozzles exist, to obtain the different shapes of grains as presented in Fig. 1.2.

The grains of powder is dried with a circulation of air and heated slowly up to 60°C. This operation lasts up to 3 days for the larger grains. It is also possible to proceed under a vacuum.

The next phase is washing in a bath of water at 55°C to extract the last trace of solvent. This operation may last up to 15 to 20 days. Then, the powder is dried with air at 55°C.

For small caliber powders, a combustion moderator is added to the surface to slow the initial combustion velocity. The powder is added in an emulsion of centrality, for example.

The last operation, for small and medium calibers only consists in adding graphite to decrease the sensitivity of the propellant to electrostatic discharges.

2.3.3 Double base propellants

The first step is the preparation of a paste of nitroglycerine added to nitrocellulose. Nitroglycerine in emulsion in water is poured in a suspension of nitrocellulose in water. After 20 min of agitation, water is removed by pressing. Nitroglycerine is adsorbed on nitrocellulose, and the paste containing 35% percent of water is allowed for public transport. The maximum rate of nitroglycerine is 50%.

The paste is wringed up to 20% of water. A mixer is first used to homogenize the pasted, then to add side products. The paste is gelatinized by mechanical treatment with rolling mills. Water is removed operation after operation. Sheets are obtained at this step.

If flakes propellant is manufactured, the sheets are simply cut with a guillotine. To obtain other forms, a press with muzzles is needed. The sheets are rolled together and introduced in the chamber of the press. The pressing is carried out under vacuum on a heated matter, at pressures of several hundred bars. The strands of powder obtained are cut with a guillotine.

Flake powders are coated with graphite to decrease the sensitivity to electrostatic discharges.

3 Primary explosives

Primary explosives are substances whose decomposition regime is only detonation. This is due to very particular molecular structures. These substances are used in ignition devices. Detonation of grains is possible at a scale of 30 mg [6,7].

3.1 Molecules

The first primary explosives were discovered in the 16th century. They were the fulminates of gold, silver, and mercury. Their high sensitivity hindered their use and they were only considered as laboratory curiosities. The first use of mercury fulminate was proposed by Howard in 1806 to ignite the black powder in a gun. Nobel proposed in 1867 a detonator containing 1 g of fulminate to ignite the dynamite he had invented. This detonator was dangerous to use due to the high primary loading. In 1877, Nobel proposed a double stage detonator, with a smaller amount of fulminate, and a second charge of picric acid. In 1900, new devices were proposed, based on other secondary loadings, tetryl or TNT. The toxicity concerns linked to the use of mercury are the cause of the abandonment of this product.

Lead azide was discovered by Curtius in 1891. The first detonator based on lead azide was patented in Germany in 1908, enabling its use in this country during the First World War, whereas England and France used only mercury fulminate. In the USA, the production by Dupont de Nemours began in 1932.

Lead styphnate or lead trinitroresorcinate was discovered by Griess in 1874 in Germany and patented for use in a detonator by Von Herz in 1914. It reached the industrial scale between the two world wars.

Mercury fulminate and lead azide have structures that do not respect the octet rule, their existence is possible thanks to conjugated systems (Figs. 1.3 and 1.4). The formula of fulminate is CNO^-. The carbon atom has a lone pair of electrons, which is unusual for this atom. Azide is an anion of formula N_3^-. This anion is linear, the lone pair of electrons of the central nitrogen atom is involved in bonds. The global structure is conjugated, which stabilizes the anion.

The main structures used today are based on salts of heavy metals. The high density of the cations increases the density of the crystals, which increases the energy density. Sodium azide and magnesium styphnate are not explosive substances whereas lead styphnate is a primer (Fig. 1.5).

No metallic primary explosives have been produced. Tetrazene is produced industrially since the 1960s (Fig. 1.6). Its efficiency is not sufficient to be used pure. It is an additive to lead styphnate or lead azide.

The presence of heavy metals in primary explosives is an environmental concern. Studies are carried out to find more environment-friendly "greener" products. The use of silver instead of lead has been proposed. For example, silver azide is a primary explosive.

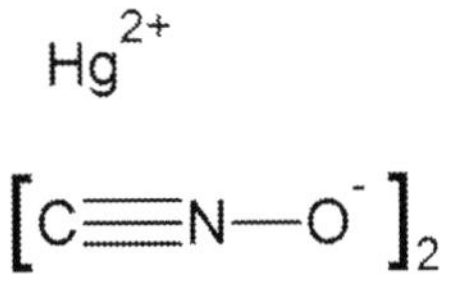

Hg^{2+}

$[C{\equiv}N{-}O^-]_2$

FIG. 1.3 Mercury fulminate.

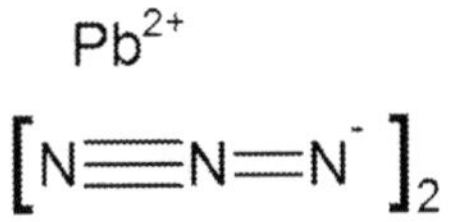

Pb^{2+}

$[N{\equiv}N{=}N^-]_2$

FIG. 1.4 Lead azide.

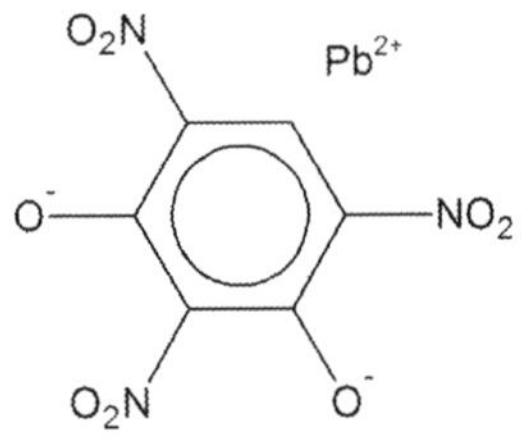

FIG. 1.5 Lead styphnate.

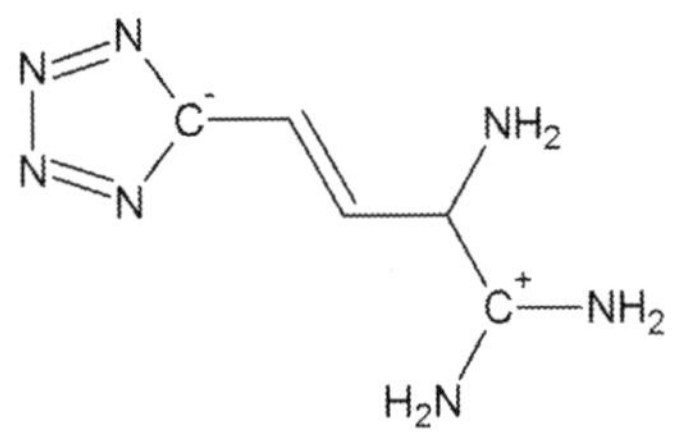

FIG. 1.6 Tetrazene.

TABLE 1.1 Sensitivity of primary explosives

	Mercury fulminate	Lead azide	Lead styphnate	Tetrazene	Pentrite
Shock (J)	0.2–2	0.6–4	2.2–5	0.3–1.6	4
Friction (N)	6.5	0.5	1.5	8.6	67
Electrostatic discharge (mJ)	0.5	8.8	0.1	2.7	450

R. Matyas, J. Pachman, Primary Explosives, Springer, 2013.

Primary explosives have high sensitivities to stimulus. Table 1.1 gives the sensitivities of these explosives to classical stimuli [8] which will be detailed in the next chapter. The sensitivity of PETN, the more sensitive secondary explosive is given as a comparison.

3.2 Process

Mercury fulminate is produced by the addition of ethanol to a nitric solution of mercuric nitrate and catalyzed by copper oxychloride ($CuCl_2$ CuO). Mercury fulminate precipitates. It is filtered and washed up to neutrality. It is kept wet until use.

Lead azide is obtained industrially by batch by precipitation of lead nitrate with sodium azide at a scale of several kilograms. The precipitate is washed with water, fractionated in boxes of 2 kg, and dried. Lead azide is stored dry because it is less stable when wet. All the process is remote-controlled.

Lead styphnate is synthesized in three steps. First, resorcine is nitrated to give trinitroresorcine or styphnic acid. Then a treatment with magnesium oxide leads to magnesium styphnate. Finally, magnesium styphnate is added to a solution of lead nitrate, leading to the precipitation of lead styphnate. Wet styphnate is less sensitive and as chemically stable as the dry form. That is why it is kept wet until its use to load detonators.

Primary explosives are used in pyrotechnic detonators, low-energy electrical detonators, and in inflammation devices (percussion or electrical caps, squibs).

4 Secondary explosives

Secondary explosives are based on molecules or mixtures which can detonate but also burn slowly. The constraints on civil and military explosives are in total opposition, giving very different compositions. Civil explosives must be as cheap as possible. Low efficiency is compensated by a larger quantity, as the product is cheap. The behavior must be globally reproducible and the lifetime is short a few years. On the contrary, military explosives must be as powerful as possible, with a very reproducible behavior, even after a 25-year storage, to assure the weapon's initial performances. The price is not a major constraint, especially on missiles, where the guiding system is very expensive.

4.1 Molecules

The molecules used to prepare secondary explosives exhibit the nitro group, as explained in the introduction. This group can be linked to carbon, nitrogen, or oxygen atoms. For carbon, it can be an aromatic or aliphatic one. For nitrogen, the molecules are called nitramines. For oxygen, the group is a nitrate.

There are dozens of molecules with explosive properties, however, only a small number is used. Examples of molecules of these categories of particular interest and effective use are given here. A larger panel of molecules is considered in [9].

The more frequent nitroaromatic molecules are trinitrotoluene (TNT), triaminotrinitrobenzene (TATB) or hexanitrostilbene (HNS). For nitroaliphatic molecules, one finds compounds like oxynitrotriazole (NTO) or diaminodinitroethylene (DADE or FOX-7), recently synthetized. One molecule belongs to both categories, tetryl.

The main nitramines are hexogen (RDX), octogen (HMX) and hexanitro hexaaza isowurtsitane (HNIW or CL-20). Organic nitrates are found in molecules like pentrite (PETN) or nitroglycerine (NG) and nitroglycol (EGDN) (Table 1.2).

Nitroglycerine was discovered in 1846 by A. Sobrero and identified immediately as an explosive. It was first called pyroglycerine. Its high sensitivity, especially to friction, limited its use. The first industrial process was proposed by Nobel in 1860. The use remained limited until the invention of dynamite 7 years after, the oily explosive being adsorbed on Kieselguhr. Dynamite was used for public works, but not in the army.

Picric acid was discovered by Woulfe in 1771 and rediscovered by Laurent in 1843. It was used as a yellow dye. Turpin patented its loading in molten phase in metallic casings. The French army prosecuted him, leading to the publication in newspapers of the patent by Turpin himself to protect its industrial property. This new loading process was homologated for use in Germany in 1888. Called melinite in France and lyddite in the UK, melinite was largely used during the First World War. After that war, it was replaced by other compositions.

TNT was discovered by Wilbrand in 1863. The synthesis was ameliorated by Hepp in 1880. Its explosive property was identified by Haeussermann in Germany in 1891, leading to its military use in 1901 in this country. TNT remains today largely used as a component of melt-cast compositions.

Tetryl was first synthesized by Michler and Meyer in 1879. It was studied by Van Romburgh in 1883 and Mertens in 1886. The first military application was in Germany in 1894 and it was used to load shells in 1902. Tetryl is a sensitive high explosive, hence its use in initiation devices. It was used to load shells during the First World War and in melt-cast compositions like tetrytol during the Second World War.

PETN was discovered by Tollens and Wigand in 1891 and patented in 1894. It was used in initiation devices in the 1920s and in melt-cast compositions like pentolite since 1926.

RDX was synthetized in 1899 by Henning and used for its medical properties. It was identified as a high explosive in 1920 and studied in the Militaerversuchsamt in Germany. The study in the UK in Woolwich arsenal began in the thirties. The meaning of RDX is research development explosives. It was used during the Second World War to improve TNT in melt-cast comp B and in plastic explosives like C4.

TABLE 1.2 Topological formulas of common explosive molecules

Nitro aromatic and nitroaliphatic carbon		
O_2N, NO_2, O_2N	O_2N, H_2N, H_2N, NO_2, O_2N, NH_2	O_2N, NO_2, O_2N, NO_2, NO_2, O_2N
TNT	TATB	HNS
H, N, O, NO_2, NH—N	O_2N, N, NO_2, O_2N, O_2N	
NTO	Tetryl	
Nitramines		
O_2N, N, N—NO_2, N, O_2N	O_2N, N, N, NO_2, N, N, O_2N, NO_2	O_2N, O_2N, N, N, N, NO_2, NO_2, O_2N, N, N, NO_2
RDX	HMX	CL-20

Nitrate		
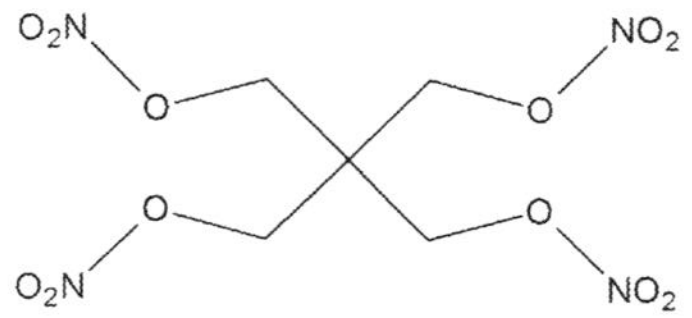 O_2N O O NO_2 O_2N O O NO_2	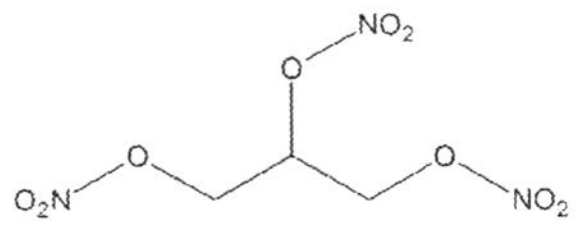 NO_2 O O_2N O O NO_2	 O O_2N O NO_2
PETN	NG	EGDN

HMX was discovered by Bachman in 1940 as he tried, to identify a high melting point impurity in RDX, hence his name, high melting explosive. It was agreed for use in military explosives in 1956 in the US. It is one of the most powerful high explosives in current use.

TATB was first synthetized by Jackson and Wing in 1888 for its medical properties. It was identified as a high explosive after the Second World War and studied for its insensitivity for nuclear applications in both the US and the USSR.

NTO was first synthetized by Manchot and Noll in 1905. It was discovered as an insensitive explosive in 1979 by Becuwe. Today, it is used to obtain insensitive melt-cast or pressed compositions.

CL-20 was synthetized in China Lake in 1987 by Nielsen. It is the most powerful explosive. Its sensitivity and cost limit its use in compositions.

New molecules have been synthetized in the last two decades and their evaluation in compositions is under process. They are based on nitrated aromatic rings, generally with hetero atoms like nitrogen (Table. 1.3). One find BIDN, BITN, DNP, N-Me-TNP, TNAZ and DNAN. A simpler molecule, DADE (or FOX-7) has been proposed also. Their use will be discussed in the section dedicated to military explosives.

A new trend to find higher energetic molecules with sensitivity lower to the one of CL-20 is to precipitate energetic crystals like ADN (ammonium dinitramide, NH4N(NO2)2), GUDN, TKX-50, and DETRA-D.

Last but not least, ammonium nitrate (AN) is of wide use in civil applications and ammonium perchlorate is used as an additive, especially for underwater compositions.

4.2 Process for civil explosives

Explosives for civil applications are based on three molecules, ammonium nitrate, nitroglycerine, and nitroglycol. Nitroglycerine mixed with nitroglycol is used in dynamites, with the addition of ammonium nitrate. Ammonium nitrate is also used pure, mixed with fuel.

The first dynamite was made of nitroglycerine adsorbed on a porous matter: Kieselguhr. To increase the power of its dynamite, A. Nobel replaced the inert Kieselguhr with nitrocellulose at 12.2 %N in 1875. Today, a mixture of nitroglycerine and nitroglycol called nitroglyceroglycol is used, to increase resistance to cold temperatures. This mixture is obtained by the nitration of a mixture of glycerine and glycol. Ammonium nitrate has been added to decrease the cost and the sensitivity of the product. Typical composition is NG-EGDN/nitrocellulose/AN 35:5:60. Dynamite is loaded in plastic cartridges, to ease the use for mining. The low critical diameter enables the loading of cartridges of small diameter, down to 25 mm.

Crystals of ammonium nitrate can be mixed with fuel to obtain ANFO (ammonium nitrate fuel oil). A particular shape of ammonium nitrate crystals is required to have a good ability to detonate. A porosity between 6% and 10% is needed to be sensitive to ignition. Fuel oil is spread on the crystals to obtain the composition, that is loaded in bags. It can be poured directly into mining holes. Typical composition is AN/fuel 96:4. Aluminum can be added to increase efficiency. The ratio is then AN/fuel/Al 91:4:5. A new trend is the use of mobile fabrication units, so as to avoid the transport and storage of high explosives. A premixed is prepared, with AN crystals deprived of porosities. In the mobile unit, a treatment is carried out on the premix to create porosities, making it an explosive. ANFO has a larger critical diameter than dynamite.

TABLE 1.3 Topological formulas of new explosive molecules

New molecules		
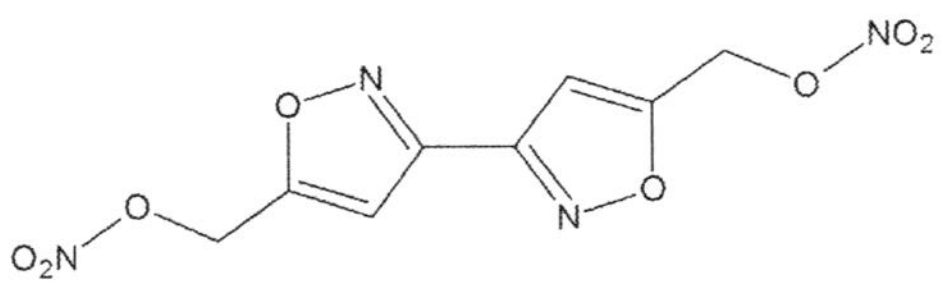 O_2N–O, O–N, N–O, O–NO_2	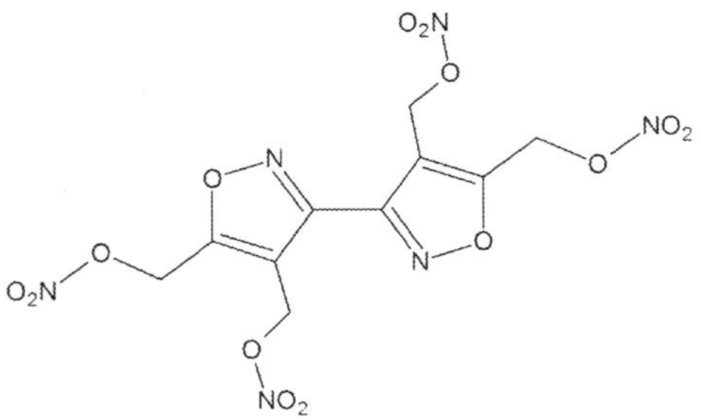 O_2N, O, NO_2, O, O–N, N–O, O_2N–O, O, NO_2	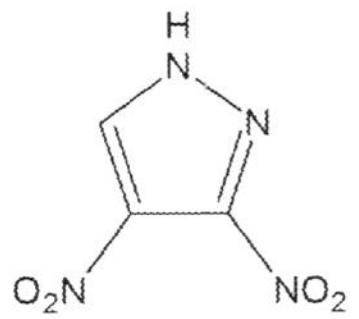 H, N, N, O_2N, NO_2
BIDN	BITN	DNP
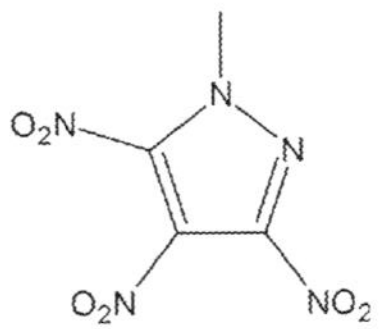 N, N, O_2N, O_2N, NO_2	 NO_2, NO_2, O_2N–N	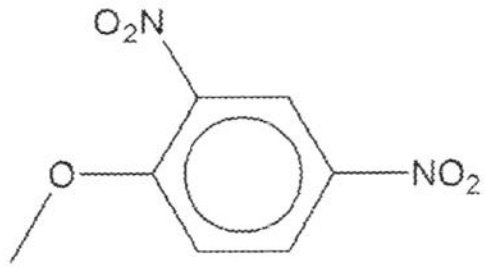 O_2N, O, NO_2
N-Me-TNP	TNAZ	DNAN
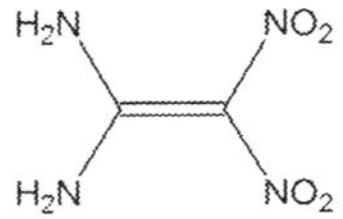 H_2N, NO_2, H_2N, NO_2		
DADE		

Continued

TABLE 1.3 Topological formulas of new explosive molecules—Cont'd

Energetic salts		
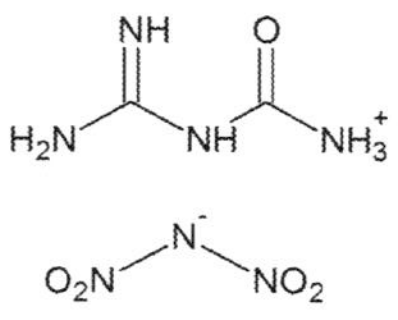 H_2N–C(=NH)–NH–C(=O)–NH_3^+ ; O_2N–N^-–NO_2	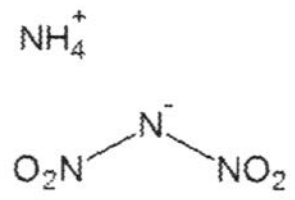 NH_4^+ ; O_2N–N^-–NO_2	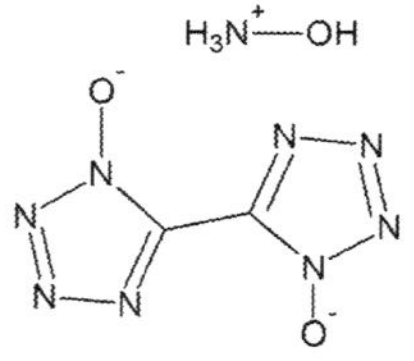 H_3N^+–OH ; N, N, N, N, N–O^- ; N, N, N, N, N–O^-
GUDN	ADN	TKX-50
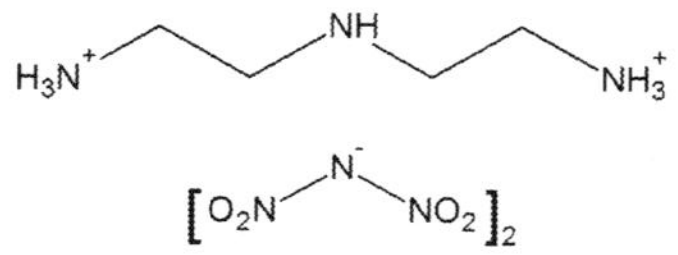 H_3N^+–NH–NH_3^+ ; $[O_2N$–N^-–$NO_2]_2$		
DETRA-D		

TABLE 1.4 Sensitivity of civil explosives

Type	Commercial denomination	Impact sensitivity BAM	Drop weight impact test (30 kg hammer)	Friction sensitivity (N)
ANFO	Anfotite 1+	3/30@50 J		0/30@353 N
Slurry	Emulstar 3000		greater than 4.0 m	0/30@353 N
Dynamite	Eurodyn 2000	12 J	1.75 m	203 N

A drawback of ANFO is its granular shape. This has led to the development of slurries. It is an emulsion between a saturated AN solution and fuel oil. The high solubility of AN, 1180 g/L at 0°C, enables a low quantity of water in the composition, making it detonable. The composition is a paste, which can be loaded into cartridges. Typical diameters are between 25 and 90 mm.

The sensitivities of these compositions are given in Table 1.4. Dynamite is rather sensitive to impact, whereas ammonium nitrate based compositions are very insensitive.

4.3 Process for military explosives

High explosives molecules are crystals, that cannot be used purely to fill weapons, due to the possible motion of grains inside the case, which could lead to a reaction. A shell, for example, is accelerated to 20,000 g in a gun. To prevent reactions, a binder must be used to obtain blocks of explosives and ensure the homogeneity of the loading. Different methods can be used. The oldest is to use a molecule with a low melting point, such as TNT, to fill the cases. It was patented in 1885. Plastic explosives were invented during the Second World War, for demolition and sabotage purposes. High explosive crystals are mixed with a malleable binder. To develop nuclear weapons, high explosives with the highest energetic content were needed. To do this, crystals of high explosives were coated with a few percent of binder and pressed to obtain blocks. The last technique is based on the propellant process. The idea is to use a cast-cured binder like HTPB to link high explosive crystals.

The different families will be presented in the next pages.

4.3.1 *Melt-cast HE*

The melt-cast compositions are prepared with a high explosive with a low melting point between 70°C and 120°C, such as for example often TNT-based, or DNAN (2,4-dinitroanisole), TNAZ, DETRA-D (Diethylene triamine bis-dinitramide), DNP (3(5),4 dinitropyrazole), *N*-methyl TNP (*N*-methyl 3, 4, 5 trinitropyrazole), BIDN (3,3′-biisoxazole-5,5′-bis-methylene dinitrate) or BITN (3,3′-Bis-isoxazole-4,4′,5,5′-tetramethylene tetranitrate). Solid explosive grains such as for examples RDX, HMX, and/or NTO (oxynirotriazole) and inert additives such as aluminum could be added depending on the sensitivity properties and on the required detonation characteristics. DNAN is a good candidate for environmental concerns, but has a lower density and detonation velocity than TNT [10,11]. TNAZ is not suitable for a TNT substitute, because of its high vapor pressure after melting leading to an unacceptable evaporation rate [12]. DETRA-D is sensitive to impact and friction, and is incompatible with a lot of energetic materials and metals, except PETN and GUDN (Fox-12, N-guanylurea-dinitramide). It remains sensitive to impact even in composition with GUDN, see Table 1.5 [13]. The 3(5),4 DNP has a melting point of 89°C. The detonation characteristics are increased in compositions with RDX, HMX, or Fox-7 compared to comp B with similar sensitivity for DNP/RDX 40:60, DNP/HMX 40/60, and reduced sensitivity for DNP/Fox-7 45:55 [14]. The *N*-methyl TNP has a melting point of 91°C and higher detonation performance than comp B with TNT without increasing sensitivity [15]. BIDN has a melting temperature of 88.9°C, a medium sensitivity to impact and electrostatic discharge, and low sensitivity to friction. BITN has a melting temperature of 121.9°C, high sensitivity to impact, friction, and electrostatic discharge see Table 1.5 [11].

The first patent for the melt-cast process is from E.Turpin in 1885. The mixing reactor is heated above the fusion temperature of the fusible explosive, which is introduced first. The explosive grains, additives are added afterward. The mixing phase is performed using a vacuum. The viscous liquid is mix to reach homogeneity and then cast in the mold, which is generally heightened, in order to avoid cracks during the cooling phase. The process is easily castable due to the high content of the fusible explosive, up to 40% with solid explosive grains compositions. Insensible formulations are developed especially with NTO explosives. Settling of the solid explosives and additives could occur during the process, a surface-active agent is intended to solve this issue. The porosity is often very low for these melt-cast compositions. Impurities, air inclusion checks are part of the control process.

The explosives nomenclature depends on each country, an example is given by Anderson [16]. Some melt-cast compositions are listed here [17,18]: comp B (RDX-TNT-wax 64.5:34.5:1), Octol (HMX-TNT 70:30), Pentolite (PETN-TNT 50:50), Tritonal (TNT-Al 80:20), AFX-645 (TNT-NTO-Al-Additives 32:48:12:8), XF11585 (TNT-NTO-RDX-Al-Wax 31:21:27:13.5:7.5), MCX-6100 (DNAN-NTO-RDX 32:53:15). Wax is often used as an inert desensitizer. Ampleman et al. [19] developed a new Greener Insensitive Melt-cast recyclable explosive (GIM), mixing Energetic Thermoplastic Elastomers (ETPE) at 9.5% with melted Octol. ETPE was prepared by reacting Glycidyl Azide Polymers (GAP) as macromonomers with phenylisocyanate. The vulnerability (for example EIDS explosives, Extremely Insensitive Detonating Substances) and toxicity concerns are also taken into account [20,21].

Coulouarn et al. [18] compared the sensitivity of two aluminized melt-cast compositions based on TNT or DNAN: XF11585 and MCX with similar densities, respectively 1.73 and 1.75 g/cc. The sensitivity results are presented in Table 1.5. The two compositions have similar thermal sensitivity; XF11585 is less sensitive for impact, friction, and electrostatic discharge.

4.3.2 *Pressed explosives*

The pressed formulations are prepared with a molding powder, based on coated explosive grains with 3% to 10% by wt. binder with or without plasticizer (viton, estane, wax, etc), less 1% wt. antistatic and lubricant agents (graphite, stearic acid, etc.) and less than 15% wt. inert additives to enhance blast or metal plate acceleration effects. The density quality and homogeneity rely on the chosen uniaxial or isostatic pressing apparatus. The porosity of pressed pellets is often rather high, above 2%.

Here are some pressed composition examples: comp A3 (RDX-wax 91:9), PBXN5 (HMX-Viton 95:5), XP3264 (NTO-RDX-Al-Wax-Graphite 49:33:14:4:0.5), XP3264 Fox-7

TABLE 1.5 Sensitivity results for melt-cast compositions [11,13,18]

Melt-cast composition or matrix	Impact sensitivity BAM (J)	Friction sensitivity (N)	Electrostatic discharges (J)	Temperature of auto-initiation (°C)
MCX 6100	30	117	0.51	206
XF11585	50	248	0.73	203
TNT	15	353 (10%)	0.46	306
DNAN	greater than 24	160	4.5	347
DNP	20	360		
N-methyl TNP	greater than 50.1	greater than 353	—	190
Detra-D/GUDN 70:30	6.1 (same for Detra-D)	60 (same for Detra-D)	—	160 (thermal onset for Detra-D)
Detra-D/GUDN 40:60	7.8	96	—	—
Detra-D/GUDN 30:70	7.8	280	—	—
BIDN	18.1	greater than 360	0.25	185.9
BITN	3	60	0.0625	193.7

TABLE 1.6 Sensitivity results for pressed compositions based on RDX or Fox-7 [22,23]

Melt-cast composition	Impact sensitivity BAM (J)	Friction sensitivity (N)	Electrostatic discharges (J)
XP3264	50.1 (30%)	353 (30%)	0.367 (50%)
XP3264 Fox-7	50.1 (100%)	353 (43%)	0.632 (50%)
HMX/CP52 94:6	7.5	240	—

(NTO-Fox-7-Al-Wax-Graphite 49:33:14:4:0.5). The Plastic Bonded eXplosives (PBX) could be pressed with 4 to 8% binder and plasticizer content or cast-cured. Hahma et al. [22] tested pressed plastic bonded shock insensitive explosives using chlorinated paraffin CP52 without DOA (plasticizers (for example DOA Di (2-ethylhexyl) adipate) to avoid sedimentation, diffusion, and compatibility issues. The impact and the friction sensitivities are listed in Table 1.6.

Coulouarn et al. [23] compared the sensitivity of two aluminized pressed compositions based on NTO, RDX, or DADNE (Fox-7): XP3264 and XP3264 Fox-7. The sensitivity results are presented Table 1.6. XP3264 Fox-7 is slightly less sensitive.

4.3.3 Cast-cured HE

The cast-cured formulations are prepared with high explosive grains between 20% to 92% by wt., HTPB binder 10% to 25% by wt., inert additives up to 40% by wt., and oxidizer content less than 45% by wt. After mixing, the injection could be an option depending on the quantity and the mold volume. The different steps of the batch process and at the workshop are

TABLE 1.7 Sensitivity results for cast-cured compositions [25]

Melt-cast composition	Impact sensitivity BAM (J)	Friction sensitivity (N)	Self ignition temperature (°C)
B2238A	50 J (32%)	240	210
B2263A	24	353 (30%)	209
B2276A	30	353 (30%)	210

presented by Mahé and Nouguez [24]: the shell body preparation, the casting phase with the vertical mixer, the shell finishing, and final inspection, and the traceability. The influence of the particle size distribution, the influence of the RDX content, the influence of the plasticizer content, the influence of the HTPB prepolymer are presented by Mahé [25] to design the cast-cured formulation B2276A with the highest RDX charge content with 91% by wt., reaching high-performance detonation properties as PBXN110, which contain 88% wt. HMX.

Cast-cured explosive nomenclatures are presented by Anderson [16], Mahé [25], and Cooper [17]. Here are some examples of cast-cured formulations: PBXN109 (RDX or Insensitive-RDX/Al/HTPB 64:20:16), B2276A (RDX/HTPB 91:9), PBXN110 (HMX/HTPB 88:12), B2238A (RDX/HTPB 85:15), B2263A (Insensitive-RDX/HTPB 88:12). The impact, friction, and thermal sensitivity of B2238A, B2263A, and B2276A are listed in Table 1.7. The classical vertical mixer replacement by the Resonant Acoustic Mixing (RAM) is studied by many authors [26].

5 Propellants

Propellants are used for autopropulsion devices that deliver a thrust all along the trajectory of a rocket or a missile. Generally, the term missile is used for self-guided devices, when rockets are IR or wire-guided. However, counterexamples exist, like the guided infantry missile Milan [27].

The first rockets using black powder appeared in China in the 8th century, with only fireworks applications. Military rockets were discovered in Europe during the conquest of India by the UK at the end of the 18th century. The first military use in Europe is during the siege of Copenhagen in 1807, with 30,000 rockets launched. The progress of rifled artillery inhibited the development of black powder rockets, considered less safe. Specific works on autopropulsion began during the Second World War like jets assisted take-off rockets (JATO), with a mixture of potassium nitrate and coal pitch. The first propellant was developed by Parson in 1942 in the US. It was a mixture of ammonium perchlorate and roofing asphalt. Rockets with solid propellants have known intensive developments during the cold war, to build nuclear deterrence in different countries, whereas civilian rockets developments were more focused on liquid propulsion. Solid propulsion is more complex but offers better propulsive capacity and better availability.

Today, compositions based on ammonium perchlorate are used for the propulsion of long-range missiles, between 4000 and 8000 km, with an exoatmospheric trajectory. These compositions are also used for boosters of space launchers to deliver the impulse required for take-off. The mass of the solid rocket motors ranges from tons for missiles to 230 t on

Ariane 5 or 500 t on the former US space shuttle. Liquid propulsion is based on a couple of ergols like kerosene, UDMH, or liquid hydrogen with liquid oxygen. Researches are carried out to replace liquid hydrogen with liquid methane. For short-range missiles, under 1000 km, compositions based on nitrocellulose and nitroglycerine are used. However, liquid propulsion is out of the scope of this book.

The decomposition of ammonium perchlorate generates hydrogen chloride, which raises environmental concerns. Replacing it with ammonium nitrate has been studied. However, ammonium nitrate has two major drawbacks. The first one is a smaller specific impulse, a third less, than ammonium perchlorate. The second one is a crystallographic phase transition from a-rhombic to b-rhombic at 32°C, with a 3.6% difference in density, which causes cracks in propellant blocks. These cracks increase the burning surface, which may modify the thrust of the booster and alter the trajectory of the rocket, or worse, lead to an explosion of the booster. The incorporation of metal halides as stabilizers is studied [28].

On smaller missiles, blocks based on nitrocellulose and nitroglycerine are used. These compositions are derived from double-based gun propellants.

The two kinds of solid compositions will be presented in the next sections.

5.1 Nitrocellulose nitroglycerine based propellants

Two processes exist to produce blocks of propellant: extrusion or molding. Extrusion is interesting to produce small or medium rockets. The combustion duration is generally short. The drawback of this technique is the simplicity of the possible forms. To allow more complex blocks, the molding technique has been invented.

Extrusion uses the same process as double base propellants. A paste of nitroglycerine adsorbed on nitrocellulose is laminated to obtain sheets that are rolled to introduce them in an extrusion press. The extruded matter is cut at the desired length with a guillotine and the blocks are stored for 2 or 3 weeks to solidify properly. A machining phase is then required to obtain the appropriate geometry. The larger blocks weigh 20 kg with a 200 mm diameter.

The molding technique is different. Nitrocellulose fibers are cut in small 1 mm sticks. These sticks are poured in a metallic mold, with a central kern to form the central canal, generally with a star shape. A solvent containing nitroglycerine, a plasticizer, triacetin, and a stabilizer, centralite or nitro-2-diphenylamine 74.5:24.5:1 is prepared. This solvent is poured into the mold. The mold is moderately heated for 24 h. The nitrocellulose sticks inflate by adsorbtion of the solvent, which welds them together, forming an isotrop block. The residual solvent is removed and the block is heated for 4 days more. Then, the mold is removed, the kern first, the peripheral casing after. The blocks are cut with a band saw underwater cooling to obtain the desired length. It is possible to obtain several blocks with one draft. Machining operations may be required, depending on the specifications of the block. One safety concern is the risk of exudation of nitroglycerine. As nitroglycerine is an electron acceptor and nitrocellulose electron donor, bonds are created. The nitrocellulose has a nitration rate between 11.5 and 12.5 % N. A higher rate, like 13.5 %N decreases the electron donor behavior of nitrocellulose, facilitating exudation.

5.2 Ammonium perchlorate based propellants

In these propellants, the oxidizer is ammonium perchlorate. An organic binder is used to form blocks. This binder is the reducing agent of the mixture. Aluminum is added to increase the gas temperature by 1000 K, which increases the thrust and the specific impulse. The order of magnitude of specific impulse is between 240 and 280 s.

Several binders have been used. The first is roofing asphalt. Then polymers have been chosen. A prepolymer is used to prepare the composition which is poured into the solid motor. Then the prepolymer reticulation leads to the block. The first polymer is PVC, followed by polyester in the fifties. In the sixties, polyurethane was used. Polybutadiene is used since the seventies. Typical composition is ammonium perchlorate / aluminum / HTPB 68:20:12. To increase the specific impulse, attempts in the 80s, were carried out to add nitramines like RDX or HMX or CL-20, and to use an energetic binder based on nitrate esters like nitroglycerine. These propellants have a higher specific impulse up to 310 s, but were also highly sensitive, mainly to bullet impact. The possibility of transition to detonation has limited their use in ballistic missiles in most countries.

The preparation process is the same as cast-cured high explosives. The ingredients are poured into a vertical mixer. The mixing process lasts several hours under vacuum. The composition is then injected into the motor casing under vacuum. The block is heated at 60°C for one week to reticulate the composition and obtain a block.

The capacity of the mixer is between 3 and 30 t to load long-range motors. When several loading operations are required, especially for long-range missiles, it is possible to prepare during several weeks mixers vessels containing a premix of the composition, without the polymerization precursors. When all vessels are prepared, the polymerization precursors are added to the first vessel, mixed for a few hours, before being poured into the motor casing. During this filling operation, a second vessel is prepared, enabling a continuous filling.

A solid motor has a central canal with a complex form, to keep a constant burning surface during combustion at a constant surface regression velocity, giving a constant thrust. A metallic kern is added to the motor casing before loading. This kern is constituted of several pieces coated with Teflon to permit the extraction of the block after polymerization.

6 Pyrotechnics

Pyrotechnics is the family with the larger range of effects: heat, light, smoke, delay, sound, thrust. This family is the only one to use the oldest pyrotechnic composition: black powder. Different devices are used, to produce these different effects, which will be presented [29].

6.1 Devices producing heat

Compositions called thermites are used to produce localized high thermal effects. These compositions are used in thermal grenades principally for destruction purposes. The first use is during the Second World War, with an application in B17 flying fortress to destroy the Norden bombsight if the plane got lost in an enemy country. It was also intended to be used in Pointe du Hoc during the D-Day to destroy the German guns. Today, thermal grenades are used to destroy tank motors, tank guns, or helicopters.

The classical thermite composition is a mixture of aluminum and ferric oxide in stoichiometry. The reaction generates alumina and molten iron. It can be used for railway welding. In other industrial welding, it is possible to use copper oxide instead of ferric oxide, to produce molten copper.

One US composition is Thermate TH3, based on classical thermite, with the addition of barium nitrate, sulfur, and a binder like PBAN (68.7:29.0:2.0:0.3) [30]. Barium nitrate is used to increase the thermal effect and reduce the ignition temperature.

6.2 Device producing sound

Sounds are produced by the fast combustion of compositions called flash powder. The wider spread composition is a mixture of potassium perchlorate and aluminum 66:34. Aluminum is a very fine powder, above 10 μm. It is used from firecrackers for children to larger fireworks bombs as bursting charges. Other compositions are possible like a black powder with charcoal substituted by aluminum (potassium nitrate, aluminum, sulfur 59:32:9).

Some devices have a whistling composition. This is due to the vivacity of the combustion itself. Some examples of compositions are given in Table 1.8.

6.3 Device producing light

Several classes of devices are used to produce light. Flare can illuminate a scene for several tens of seconds. Fireworks bombs generate colored particles for a shorter time.

Flare is used as a maritime distress signal, during 40 s to 1 mn. Military use consists in lighting a scene during 4 mn with a charge burning under a parachute. Another specific use is countermeasures against IR-guided missiles. The flare is designed to generate multiple luminous pints, hotter than the plane engines, with IR spectra similar to one of kerosene combustion products.

The light produced by a flare is generated by the combustion of the pyrotechnic compositions. Generally, the oxidizer is strontium nitrate, potassium nitrate, or potassium perchlorate. The reducer is charcoal, sulfur, sawdust, aluminum, magnesium, or a polymeric resin. Metallic colorants can be added. To illuminate underwater objects, flares containing calcium are used.

TABLE 1.8 Whistling compositions

Ingredients	Proportion
Potassium chlorate, sodium salicylate, Vaseline oil, iron oxide	73:20:6:1
Potassium perchlorate, gallic oxide,	75:25
Potassium nitrate, potassium picrate	63:37
Potassium perchlorate, potassium benzoate	70:30

D. Brunel. Le grand livre des feux d'artifice, CNRS editions, 2004.

TABLE 1.9 Colored pyrotechnic compositions

Color	Ingredients	Proportions
Red	Potassium perchlorate, resin, lampblack, strontium carbonate, PVC,starch	66:13:2:12:2:5
	Potassium perchlorate, strontium nitrate, magnesium, PVC, lampblack	30:20:30:18:2
Green	Barium nitrate, PVC, magnesium	59:22:19
Yellow	Potassium perchlorate, sodium oxalate, dextrin, shellac, resin	70:14:4:6:6
	Ammonium perchlorate, sodium oxalate, hexamine, iditol	55:19.5:15:10.5
Blue	Ammonium perchlorate, copper acetoarsenate, dextrin, colophon	68:22:4:6
Violet	Ammonium perchlorate, hexamine, calcium carbonate, copper sulfide, iditol	55:20:15:5:5
Rose	Ammonium perchlorate, strontium nitrate, magnesium, polyester	20:30:40:10:

D. Brunel. Le grand livre des feux d'artifice, CNRS editions, 2004.

Fireworks bombs contain a bursting black powder charge with spheres of pyrotechnic composition. Called stars, these spheres have a diameter between 4 and several tens of millimeters. Ignited by the deflagration of black powder, the stars burn for several seconds.

These stars contain metals, generating the lights, whose wavelength is linked to electronic transitions in these atoms. Red, green, yellow, blue, and white are easy to obtain with strontium, barium, sodium, copper, and magnesium. Violet or rose are more difficult to obtain, by a combination of two colors. For example, violet is obtained with a mix of blue and red, so, copper plus strontium. Rose is a lighter red, with strontium enlightened by the combustion of magnesium. Table 1.9 gives examples of compositions used to manufacture colored pyrotechnic compositions.

6.4 Device producing smoke

Smoke is a suspension of solid particles in the air. Dedicated compositions are proposed to burn chemicals with potassium chlorate at a low temperature, above 600°C, and generate dust containing the colored dye. The composition is put inside a cylindrical metallic vessel with a hole on the top to release the smoke. Usually, a smoke grenade lasts for several minutes.

Examples of smoke compositions a given in Table. 1.10.

6.5 Device producing thrust

Fireworks bombs are launched with mortars. Various calibers exist, generally from 50 to 300 mm. However, larger calibers might be encountered. In 1985, the Katakai Japanese corporation launched a 1200 mm caliber bomb. These bombs weigh 300 kg and explode at 850 m from the ground.

In these mortars, the thrust is obtained with a black powder loading.

TABLE 1.10 Colored smoke compositions

Color	Ingredients	Proportion
Yellow	Potassium chlorate, lactose, auramine, chrysoidine	30:25:40:10
Red	Potassium chlorate, lactose, rhodamine	35:25:40
Blue	Potassium chlorate, lactose, methylene blue	35:5:60
Orange	Potassium chlorate, sodium bicarbonate, sulfur, auramine, aminoanthraquinone	26:23:10:16:25
White	Potassium chlorate, lactose, neoprene, cobalt	23:23:47:2
Black	Potassium chlorate, anthracene	55:45
Violet	Potassium chlorate, sodium bicarbonate, sulfur, 1-methylanthraquinone, 1,4-diamino-2,3-dihydro-anthraquinone	30:14:12:18:26

D. Brunel. Le grand livre des feux d'artifice, CNRS editions, 2004.

6.6 Device producing delay

When a fireworks bomb is launched by a mortar, it is ignited by the launch charge but explodes several seconds after. When a pyrotechnic detonator is used, a delay between manual ignition and the detonation of the charge is required. That is why dedicated devices are used to transfer order with a known delay. This is based on the use of fuses.

The first fuses were fine black powder in a paper tube. Today, various compositions exist:

- potassium perchlorate, barium chromate, tungsten;
- lead oxide, silicium;
- lead oxide, barium chromate, boron;
- lead oxide, potassium permanganate, iron, boron.

These compositions are loaded in paper tubes for firework applications. Metallic casing is possible, to obtain devices that can be used in pyrotechnic logic. The fuse can be introduced inside detonators, to obtain various delays. This can be used for destruction, to fire several charges at the same time, without summing the aerial blast effect thanks to different explosion times.

6.7 Black powder manufacture

Black powder is a mixture of potassium nitrate, charcoal, and sulfur 75:15:10. Until the beginning of the 19th century, charcoal was produced from wood and sulfur came from Etna volcano. Potassium nitrate forms on the walls of farms due to the decomposition of fecal matter with urine by bacteria. Walls were cleaned to recover this white powder called saltpeter. At the beginning of nineteenth century, dedicated units were created to produce saltpeter from the same ingredients. This ended with the discovery of Chile guano in the middle of this century. Chile nitrate was the only source of nitrate during the First World War for Europe, except Germany with their new Haber-Bosch process, synthesizing ammonia from air nitrogen and hydrogen, then oxidizing it into nitric acid.

Black powder is produced with grinding wheels. First, wet potassium nitrate and charcoal are ground. Then wet potassium nitrate and sulfur are ground also. The two mixtures are finely ground together to obtain a wet powder. Today, in the US, the mixing is performed with a powerful fan. This powder is transformed to produce grains of defined granulometry, before drying the grains.

References

[1] A. Osmont, Bref historique de la pyrotechnie, formation pyrotechnique, CEA/DAM/ISENDE, 2016.
[2] J. Quinchon, J. Tranchant, Les poudres, propergols et explosifs, tome 2, les nitrocelluloses, Technique et documentation, Lavoisier, 1984.
[3] S. Fedoroff, Encyclopedia of Explosives and Related Items, PATR 2700, Picatinny Arsenal, 1962.
[4] R. Meyer, Explosives, 2nd edition, Verlag Chemie Weinheim, 1981.
[5] J. Quinchon, et al., Les poudres, propergols et explosifs, tome 1, les explosifs. Technique et documentation, second ed., Lavoisier, 1987.
[6] J.M. Darrigrand, Les explosifs primaires, perfectionnement en pyrotechnie, Adera Formation, 2009.
[7] J. Quinchon, J. Tranchant, M. Nicolas, Les poudres propergols, explosifs, tome 3, les poudres pour armes, Techniques et documentation, Lavoisier, 1986.
[8] R. Matyas, J. Pachman, Primary Explosives, Springer, 2013.
[9] A. Osmont, L. Catoire, I. Gökalp, V. Yang, Ab initio quantum chemical predictions of enthalpies of formation, heat capacities, and entropies of gas-phase energetic compounds, Combust. Flame 151 (1–2) (2007) 262–273. Elsevier.
[10] S. Nicolich, J. Niles, P. Ferlazzo, D. Doll, P. Braithwaite, N. Rausmussen, M. Ray, M. Gunger, A. Spencer, Recent developments in reduced sensitivity melt pour explosives, in: 34th International Annual Conference of ICT, 2003.
[11] L.A. Wingard, E. Pablo, P.E. Guzman, E.C. Johnson, J.J. Sabatini, G.W. Drake, Synthesis of novel energetic materials, in: 16th International Detonation Symposium, 2018.
[12] D.S. Watt, M.D. Cliff, Evaluation of 1, 3, 3 Tri Nitro Azetidine (TNAZ)- a High Performance Melt-Castable Explosive. DSTO Technical Report, 2000 (DSTO-TR-1000, Australia).
[13] N.V. Latypov, C. Oscarson, M. Liljedahl, P. Goede, Evaluation of DETRA-D as a melt-cast matrix, in: 44th International Annual Conference of the Fraunhofer ICT, 2013.
[14] J. Ritums, C. Oscarson, M. Liljedahl, P. Goede, K. Dudek, U. Heiche, in: Evaluation of 3(5),4-Dinitropyrazole (Dnp) as a New Melt Cast Matrix, 45th International Annual Conference of the Fraunhofer ICT, 2014.
[15] S. Comte, Nitropyrazoles, new molecules for low sensitive energetic materials, in: 13rd Research Conference, Le Bouchet, 30 January, 2014.
[16] E. Anderson, Chapter 2 explosives, Prog. Astronaut. Aeronaut. vol. 155 (1993) 81–163.
[17] P.W. Cooper, Explosives Engineering, Wiley-VCH, 1996.
[18] C. Coulouarn, R. Aumasson, P. Lamy-Bracq, S. Bulot, Evaluation of melt-cast explosive compositions based on TNT and DNAN, in: 45th International Annual Conference of the Fraunhofer ICT, 2014.
[19] G. Ampleman, A. Marois, P. Brousseau, S. Thiboutot, S. Trudel, P. Béland, Preparation of energetic thermoplastic elastomers and their incorporation into greener insensitive melt cast explosives, in: 43rd International Annual Conference of the Fraunhofer ICT, 2012.
[20] J.A. Chemelle, C. Alliod, R. Terreux, G. Jacob, F. Stankiewicz, Study of prediction of the toxicity of high-energy molecules, in: 46th International Annual Conference of the Fraunhofer ICT, 2015.
[21] ONU, Inscription, Classement et Emballage, Proposition de révision des épreuves de la série 7, Communication de l'Expert du Royaume-Uni, NATIONS UNIES, ST/SG/AC.10/C.3/2007/30, 2007.
[22] A. Hahma, J. Licha, F. Kaiser, O. Pham, D. Clement, Plastic bonded shock insensitive explosives without plasticizers, in: 45th International Annual Conference of the Fraunhofer ICT, 2014.
[23] C. Coulouarn, R. Aumasson, P. Lamy-Bracq, D. Barres, New energetic materials – explosive composition based on FOX-7, in: 46th International Annual Conference of the Fraunhofer ICT, 2015.
[24] B. Mahé, B. Nouguez, A new process to produce cat PBX filled shells at minimized cost, in: 43rd International Annual Conference of the Fraunhofer ICT, 2012.

[25] B. Mahé, A new cast cured PBX formulation B2276A, in: 43rd International Annual Conference of the Fraunhofer ICT, 2012.
[26] M. Andrews, C. Collet, A. Wolff, C. Hollands, Resonant acoustic mixing (RAM): processing and safety, in: 50th International Annual Conference of the Fraunhofer ICT, 2019.
[27] J. Quinchon, et al., Les poudres, propergols et explosifs, tome 4, les propergols. Technique et documentation, second ed., Lavoisier, 1991.
[28] P. Kumar, Advances in phase stabilization techniques of AN using KDN and other chemical compounds for preparing green oxidizers, Def. Technol. 15 (6) (2019) 949–957.
[29] D. Brunel, Le grand livre des feux d'artifice, CNRS editions, 2004.
[30] US Patent 5698812, Song, Eugene, "Thermite Destructive Device", Issued 1997, Assigned to United States Secretary of the Army.

CHAPTER

2

Characterizing responses to insults from energetic materials

A. Lefrancois[a,], Antoine Osmont[a], C. Collet[b], M. Vaullerin[a], and V. Le Gallo[a]*

[a]CEA, DAM, Gramat, Gramat, France [b]MSIAC Munitions Safety Information Analysis Center, NATO HQ, Belgium

*Corresponding author: E-mail: alexandre.lefrancois@cea.fr

1 Impact sensitivity tests

The determination of impact sensitivity is the most commonly used way of evaluating an explosive hazard associated with handling and falling. The sensitivity to impact is the ability of this energetic substance to react under the effect of a sufficient amount of impact energy transmitted to the sample. Almost all substances can react under the action of more or less impact energy.

The impact sensitivity depends on the energetic sample state (granulometry, phase, for example) and the experimental conditions (humidity, temperature, etc.). Indeed the sensitivity to impact decreases with humidity and increases with heat. The drop-weight test data are useful to compare the sensitivity of explosives between them.

Because of the diversity of experimental conditions, many experimental methods for measuring impact sensitivity have been developed like the BAM impactor test [1–5], the impactor test with cap Bourges [6], Sorgues pendular impactor test [7], the 11 kg drop-weight test [8], the 30 kg fall-hammer test [9] and the fall-hammer test for sensitive explosives [10]. Most of those used today are the BAM impactor test [4], the ERL/Bruceton Machine with Type 12 Tools [4,5], and the Rotter test [11].

These sensitivities to impact are determined with an impactor machine named fall-hammer or drop-hammer. The principle of these tests is to drop weights of different sizes from a given height. The parameters to be determined are the height or the energy of the

Theoretical and Computational Chemistry, Volume 22
https://doi.org/10.1016/B978-0-12-822971-2.00006-1

drop-weight associated with a 50% probability of decomposition or no reaction, using the "Bruceton" method.

The impact sensitivity test for liquid or gelatinous explosives is run in the same way as for solids. For primary explosives, the sensitivity is determined with a smaller drop-hammer [10].

For the Rotter test, there is a motion of the sample around the edge thanks to a deliberate clearance in the test assembly consisting of a brass cap and steel anvil. As the other impact tests a mass, 5 kg for solid samples, is dropped onto the sample cell. The gas volume change is monitored and the generation of 1 cm^3 or more of gas and/or smoke in the housing is the go/no go criterion for an initiation. An opening of the housing with smoke can also be used to identify an event for nongassy mixtures [4,11].

The two other main tests for solid high explosives described in detail are the following:

- BAM impactor test
- ERL/Bruceton Machine with Type 12 Tools test

1.1 BAM impactor test

The impact apparatus used is a drop-hammer, developed by the German Bundesanstalt für Materialprüfung (BAM) [2] and manufactured by the Julius Peters company (Fig. 2.1). Depending on the version of the impact apparatus, the weight slides between two vertical

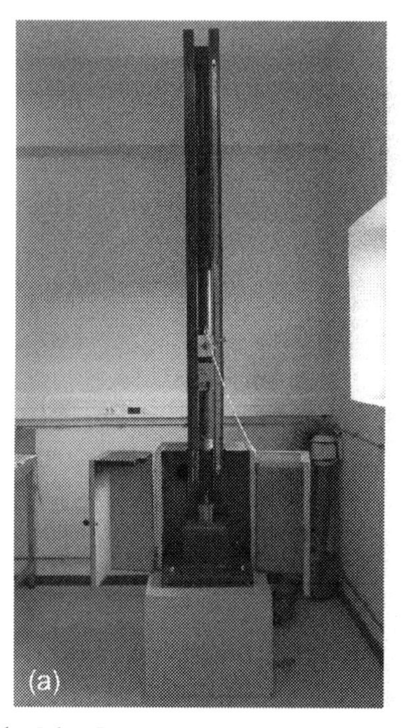

FIG. 2.1 Julius Peters or BAM impact apparatus and 2 kg drop-weight. *Credit: Author.*

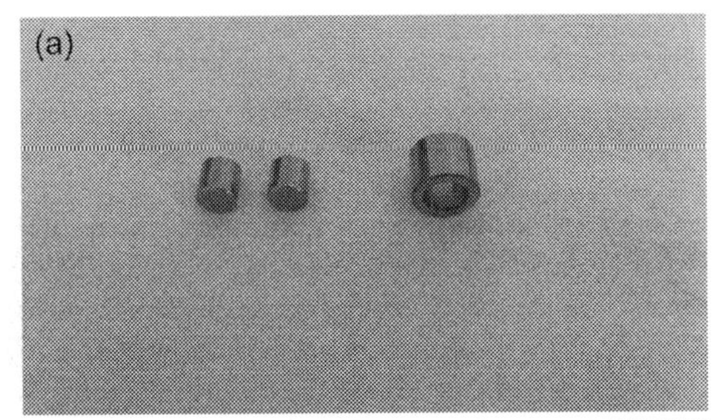

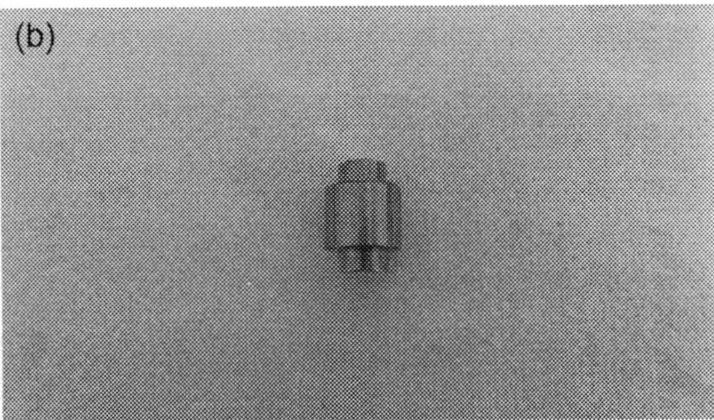

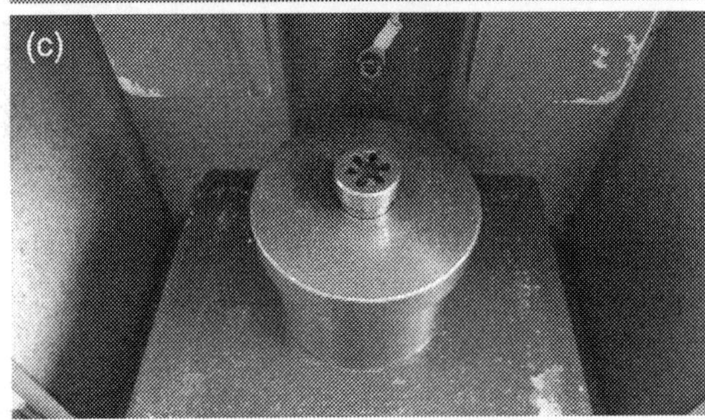

FIG. 2.2 Parts of the impact test device for solids HE, rams and ring, confinement device, and lower part of BAM impact machine. *Credit: Author.*

rails. In this test, a solid energetic material sample (20 mm^3 dried and meshed below 1 mm powder [5], or 40 mm^3 powder with a particle size of 0.5–1.0 mm [4], disc or chip with a volume of about 40 mm^3 [5] is subject to the action of a drop-weight (0.25, 1, 2, 5 or 10 kg) leading to a range of energy from 0.25 to 100 J. The substance is placed in a confinement device (Fig. 2.2), made up of two rams (steel coaxial cylinders: 10 mm height and 10 mm in diameter), placed one on top of the other, and guided by a steel ring. The rams and the ring are replaced after each test. Compared values of impact sensitivity with BAM impactor machine are presented in Table 2.1.

For a series of tests, we determine the energy of the weight which leads to a 50% probability of decomposition. Decomposition can be recognized by noise, flame, smoke, odor, generation of gases [4,13], or by inspection of the impact device for colored and burning marks after the upper ram has been removed.

TABLE 2.1 Values of impact sensitivity with BAM impactor machine.

	Impact sensitivity (J)	
	STANAG 4489 [4]	**Köhler and Ide (1993) Vaullerin et al. [12]**
PETN	5	3
RDX	8	7.5
HMX	9	7.4
Tetryl	17	3
TNT	30	15
TATB	—	50

Credit: Author.

After a preset height (6 tests for [5] or 10 for Stanag 4489 [4]), a started level is chosen and the sensitivity criteria are determined by a probabilistic method, named "Bruceton" or "up-and-down" method [4,5]. When the latest test is positive the following is done with less energy and when the test is negative we do it at a higher height. The sensitivity index is usually calculated with the average of the 30 first results minimum or up to 50 [5]. The number of tests is an even number determined with a difference of one pitch between the levels of the first and final tests. The last result shall be positive if its level is higher than the level of the first test and negative if it is below. In the past, the number of tests was fixed at 50. Of course, the precision of the sensitivity index depends on the number of tests.

For impact energy comparison, the drop height (m) for 50% positive reactions and the standard deviation (m) is multiplied by the mass of the drop-weight used (kg) and by the gravitational constant. In general, values of sensitivity energy index below 10 J usually show high sensitivity to impact. Values between 10 and 30 J indicate materiel of moderate sensitivity that possibly can be handled in accordance with standard procedures. The maximum energy determined by the BAM impactor is usually 50 J and indicates relative insensitivity to impact. Some examples are given in Table 2.1 and [12].

Tables 2.1 and 2.2 show that the differences in the results are dependent on experimental conditions (state of the explosives, humidity, temperature, etc.).

For Stepanov [16], the nanocrystalline Types A and B RDX show less impact sensitivity than the reference 4.8 μm RDX Table 2.2 for uncoated and even coated samples [14]. Type B is less sensitive than type A, despite the mean size distribution.

For Berthe [17], the Spray Flash Evaporation (SFE) RDX nanocrystals with or without ADN are more impact sensitive than the μm size reference in Table 2.3, except for the slow crystallization process between ADN and RDX. The SFE HMX nanocrystals with or without ADN are slightly more sensitive. The nano and micron size CL20 is very sensitive. The nanosize ADN is slightly more sensitive than the micron size ADN.

TABLE 2.2 ESD, impact and shock sensitivity for μm and nano RDX [14].

[15] Uncoated α phase	Picatinny ESD n°1032	Impact sensitivity H50(cm) n°1012	Shock sensitivity (kbar) n°1042
RDX 4.8 μm	0/20 at 0.25 J	32.2	19.6
RESS Type A nano RDX 150 nm	0/20 at 0.25 J	57.3	20.6
RESS Type B nano RDX 500 nm	0/20 at 0.25 J	73.5	26.1

Credit: Author.

TABLE 2.3 ESD, impact and friction sensitivity for μm and nano energetics [17].

Tests Sample, $Ø_{mean}$	ESD OZM (mJ)	Impact BAM sensitivity (J)	Friction BAM sensitivity (N)
α RDX ref. 3.69 μm	199	4	>360
SFE α RDX 0.32 μm	261	<1.6	>360
SFE α ADN/RDX 5 μm	477	<1.6	>360
α ADN/RDX slow tens μm	957	4.5	>360
α ADN ref. μm size	1496	3	>360
α Nano ADN 30 nm	996	4.5	>360
β HMX ref. 0.7 μm	733	5	>360
SFE γ to α, γ HMX 0.15 μm	1212	3	>360
SFE α ADN/HMX few μm	957	2.1	>360
ADN/HMX slow 100 μm	268	1.6	128
ε CL20 ref. 98 μm	157	<1.6	128
SFE β CL20 0.32 μm	186	<1.6	112
SFE α DNA/CL20 few μm	238	<1.6	96
α ADN/CL20 slow 40 μm	488	<1.6	72

Credit: Author.

1.2 ERL/Bruceton machine test

In the past, this test uses a machine designed at the explosives division, Atomic Weapons Research Establishment (AWE).

The ERL machine test principle is the same as the BAM impactor test: a 2.5 kg or 5 kg weight is dropped by means of a hand windlass attached to an electro-magnet (Figs. 2.3–2.5) onto a small sample (about 35 mg powder passing through a 0.50 mm sieve or 1 or 2 mm thick

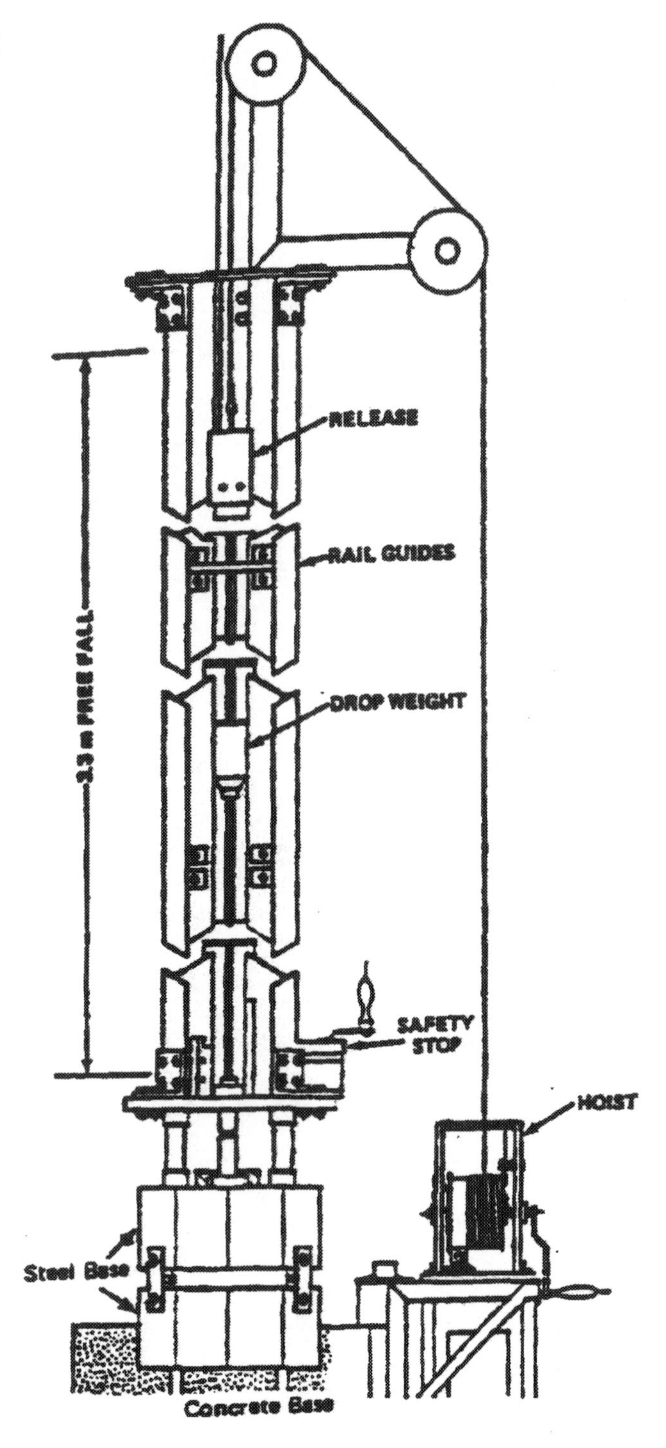

FIG. 2.3 Drop-weight impact apparatus, ERL model, type 12 tools. *Credit: NATO Stanag 4489.*

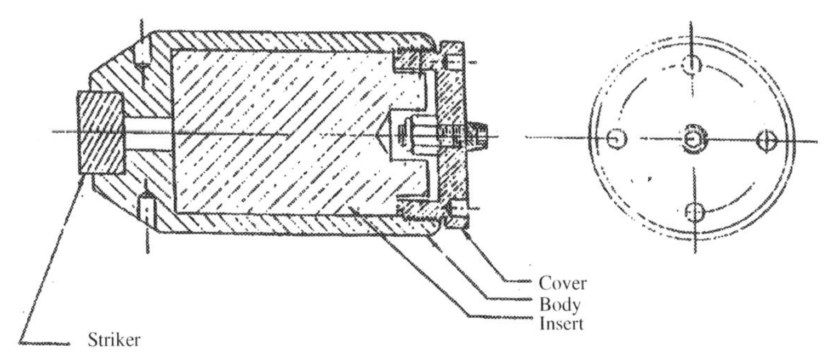

FIG. 2.4 Drop-weight assembly. *Credit: NATO Stanag 4489.*

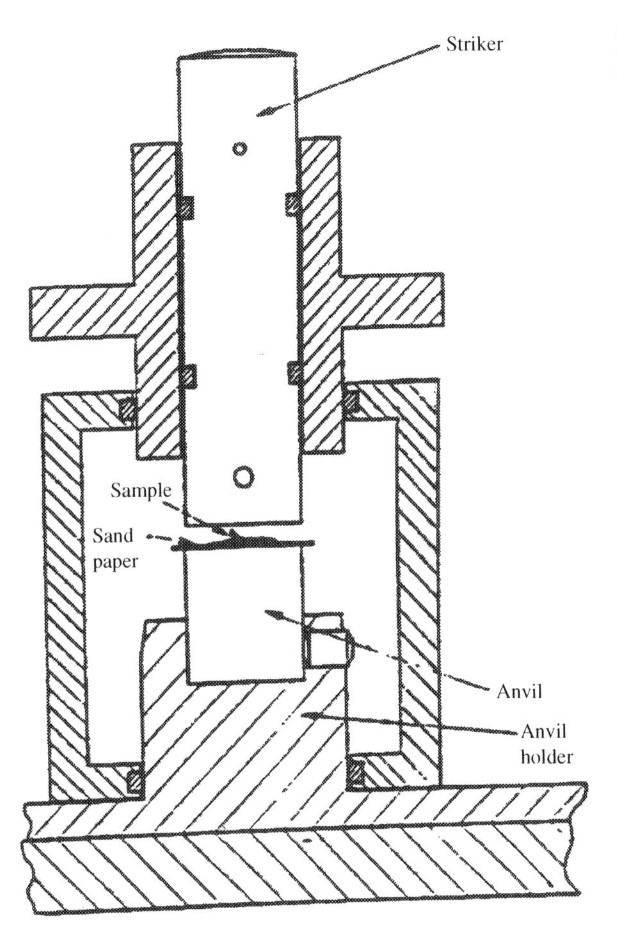

FIG. 2.5 Anvil striker arrangement, ERL machine. *Credit: Stanag 4489.*

TABLE 2.4 Values of impact sensitivity with ERL/Bruceton machine with Type 12 Tools STANAG 4489 [4].

	H_{50} (cm) 2.5 kg weight
PETN	13–17
RDX *Class 5*	18–22
HMX *Class 5*	20–26
Tetryl	32
TNT	154

Credit: Author.

for solid sample) placed in the center of a one-inch square piece of sandpaper. The two surfaces are the sandpaper (type 12 tool) described in NATO STANAG 4489 [4] and roughened steel (type 12B tool). A series of drops are made from different heights with the Bruceton method. The criterion for "explosion" is an arbitrarily set level of sound measured with an electronic noise meter or color, smoke, mark on the material, or some evidence of reaction under the sandpaper produced by the explosive on impact.

The test result is the height H_{50} in meter (some examples are given in Table 2.4) at which the probability of explosion is 50% or the height at which there is no reaction. The results are dependent on the anvil surface.

In general, H_{50} values below 0.25 m usually indicate high sensitivity to impact.

Values between 0.25 and 0.70 m are a sign of material with moderate sensitivity.

Values above 0.70 m usually show relative insensitivity to impact. The maximum drop-heights of the Lawrence Livermore National Laboratory (LLNL) and Los Alamos National Laboratory (LANL) drop-weight apparatus are, respectively, 1.77 and 3.20 m.

2 Friction sensitivity test

After sensitivity to impact, sensitivity to friction is considered part of the set of basic information about the small-scale sensitivity of energetic materials. The friction test is intended to reproduce the possible fretting of energetic material in a casing or the effects of a tool scraping against an energetic material during explosives processing.

Friction tests are well suited for energetic materials in any compact, solid, or divided state. For liquids and pasty substances (up to a certain viscosity), some friction sensitivity tests, like the BAM friction test, may not be appropriate. This is due to their lubricating tendencies and the resulting low heat development that is usually not sufficient to cause a reaction [18].

2.1 Experimental set-up

A variety of different test machines exist, amongst which some that are not considered to deliver a pure friction stimulus but rather a mix of impact and friction, like the UK mallet friction test, described in the UK EMTAP [19] Manual of Tests, or the "skid" test described in Walker [20]. These latter tests will not be described in this section, to the benefit of more common—if not standardized—friction test apparatus preferred by the international community.

In UN MoTaC [21], four friction test methods are described:

- The BAM friction test [Test 3 (b) (i)] originates from Germany. The BAM friction test apparatus is also referred to as the Julius Peters apparatus [22]. It is the recommended test for friction in the UN MoTaC. This test is further described later in this section;
- The UK rotary friction test [Test 3 (b) (ii)]. This test is further described later in this section;
- The Russian friction sensitivity test [Test 3 (b) (iii)]. In this apparatus, the sample is pressed between two rollers, up to the prescribed pressure. The movement of the upper roller along the substance is carried out using impact from a pendulum weight;
- The US ABL friction machine test [Test 3 (b) (iv)], ABL standing for Allegany Ballistics Laboratory. In this apparatus, the sample is placed between a stationary wheel and a moving steel anvil plate. The plate velocity is fixed at 8 ft./s (2.4 m/s), whereas the force applied to the wheel varies within a range of 10–1800 lb. (4.5–816 kg). As for the BAM and rotary friction apparatus, the ABL machine produces a sliding friction effort in the sample.

In the US MIL-STD-1751A [22], the series 1020 describes four friction sensitivity test methods:

- The method 1021 for the Friction Sensitivity of the ABL Sliding Friction Test [same apparatus as in the UN Test 3 (b) (iv) but the assessment method differs];
- The method 1022 for the Friction, Steel/Fiber Shoe;
- The method 1023 for the Roto-Friction Test;
- The method 1024 for the BAM Friction Test [same apparatus and method as in the UN Test 3 (b) (i)].

The NATO STANAG 4487 [18] describes two test methods:

- the BAM Friction Test [same apparatus and method as the UN Test 3 (b) (i)]
- the Rotary Friction Test [same apparatus and method as the UN Test 3 (b) (ii)]. The Rotary Friction Test is also described in EMTAP [23].

The results coming from these two test methods are the most accepted ones among NATO nations in the qualification process of energetic materials NATO AOP-7 [24]. This is the reason why these two methods will be further detailed in the next sections.

All friction tests mentioned above are applicable to all kinds of energetic material (primary explosives, booster explosives, main charge high explosives, solid rocket propellants, solid gun propellants, and pyrotechnics). However, dedicated friction test methods exist for highly friction-sensitive energetic materials such as primary explosives and, to a less extent, pyrotechnics. A small version of the BAM apparatus is used in the Slovak Republic and France, with applied friction loads ranging from 0.1 to 10 N. This test method is described in the French norm NF T 70-509 [25]. The UK uses the Emery paper friction test which is described in UK EMTAP [26].

2.1.1 The BAM friction test

As previously said, the BAM friction test is commonly used and accepted in both the NATO and UN communities. The BAM friction apparatus is shown in Fig. 2.6. The material under test is placed between a fixed porcelain peg and a moving porcelain plate, as shown in Fig. 2.7. A movement consists of a forward and a backward motion of the plate under the porcelain peg of 10 mm in each direction, which results in a sliding friction effort applied to the sample. The total (peak-to-peak) displacement of the horizontal motion is 10 mm with the full cycle taking about 1 s to complete. A variable load is applied to the holder of the porcelain peg carrier, thanks to different weights and positions on the load arm. Loads varying from 5 to 360 Newton can thus be realized.

2.1.2 The rotary friction test

The rotary friction test is also commonly used and accepted in a number of nations. The test apparatus and method originate from the UK. The sample is placed between a mild steel rotating wheel and a mild steel stationary block, as shown in Fig. 2.8. Pressure loading on the sample is achieved using a pneumatically operated ram and adjusted using an air regulator. The friction effort transmitted to the sample is achieved when the wheel rotates through 120° around its center. As a result, and contrary to what its name may suggest, the rotary friction apparatus produces a sliding friction effort which is applied to the sample under test.

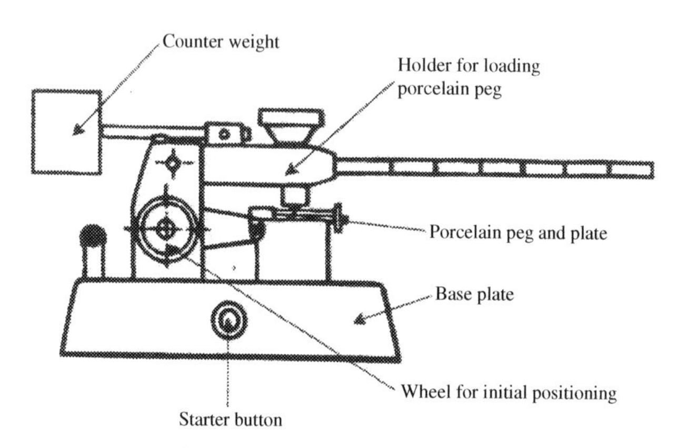

FIG. 2.6 The BAM friction test apparatus NATO STANAG 4487 [18]. *Credit: NATO.*

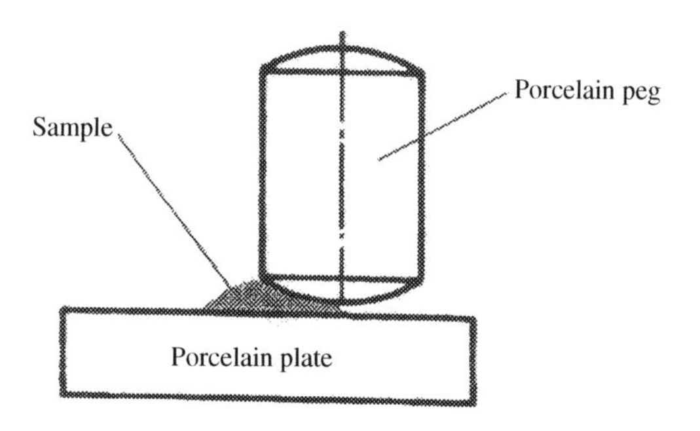

FIG. 2.7 Porcelain plate and peg NATO STANAG 4487 [18]. *Credit: NATO.*

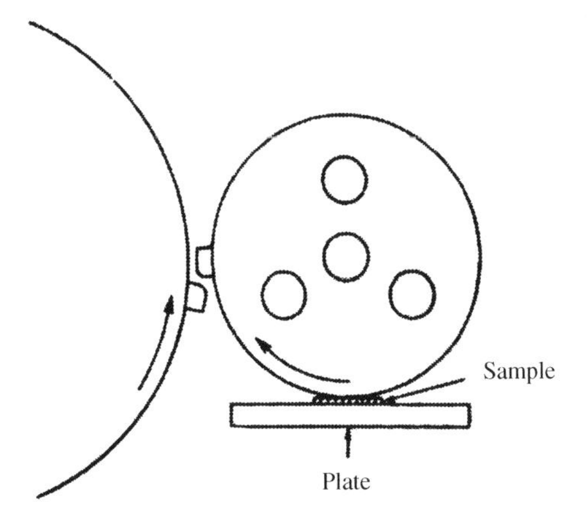

FIG. 2.8 Action of rotary friction sensitivity test NATO STANAG 4487 [18]. *Credit: NATO.*

The angular velocity of the wheel is used as the variable parameter and is controlled by varying the speed of the motor which drives the flywheel. During the test, angular velocities ranging from 40 to 316 rpm, according to UK EMTAP [27] or 398 rpm, according to NATO STANAG 4487 [18], can be realized under a loading pressure of 276 kPa. For very sensitive explosives, it may be necessary to use a smaller load.

2.1.3 *Comparison of friction test machines*

Table 2.5 gathers the main experimental set-up characteristics for the eight friction test machines mentioned above: the BAM, Small BAM, Rotary, ABL, Steel/Fiber Shoe, Roto-Friction, Russian friction test, and Emery paper friction test apparatus. The information contained in this table is extracted from the respective test descriptions listed above, hence, they do not take into account the possible variations described elsewhere.

From Table 2.5, it can be seen that all friction tests operate on similar principles—a sample is held between two friction surfaces and, while a force is applied to the friction surfaces, one of the surfaces is moved (in a sliding or rotating mode) generating dynamic friction between the surfaces and the sample. However, if the test machines operate on similar principles, the type, range, and duration of load applied to the sample differ from one test apparatus to another. The test procedure and the assessment of the sensitivity may also strongly differ, this is discussed in the next section about experimental results.

2.2 Experimental results

The friction test methods require a small amount of material but the number of trials is usually ranging between 20 and 50. This is to allow a statistically representative result [28] specially in the cases where a Bruceton method is used [29]. Each trial is assessed as positive or negative. The criteria for determining a positive trial may not be worded exactly the same in all test procedures but they all include the usual evidence of partial or total decomposition of the sample under test: odor, change of color, audible sound, burn traces, scorching, crackling, sparkling, and flame. The detection of these evidence relies on a well-trained operator.

TABLE 2.5 Main experimental set-up characteristics of commonly used friction test machines.

Test method	Materials	Moving part	Fixed part	Roughness/hardness	Displacement type	Variables	Fixed characteristics	Range of load	Measurements	Quantity of EM per trial
BAM	All except liquids and pastes	porcelain plate	porcelain peg	Roughness 9–32 μm	sliding	weigh mass and position on the arm	/	5–10 - 20 - 40 - 60 -80 - 120 - 160 - 240 - 360 N	None	10 mm^3
Small BAM	Primary explosives and pyrotechnics	porcelain plate	porcelain peg	Roughness 9–32 μm	sliding	weigh mass and position on the arm	/	0.1–10 N	None	6 mm^3
Rotary	All except liquids and pastes	mild steel wheel	mild steel block	Surface preparation by a grit-blasting to a finish of 3.2 ± s0.4 μm	sliding	flywheel velocity and block-to-wheel loading pressure	/	40 - ~300–400 rpm under a loading pressure of 276 kPa 20–40 rpm under 34 kPa	Air pressure for the load and multiturn potentiometer for the flywheel velocity	15 mm^3
ABL	solid, semisolid, and powder substances	Steel anvil plate	Steel wheel	Both the wheel and anvil have a surface roughness of 1.3–1. 8 μm and a Rockwell C hardness of 55–62	sliding	Force applied on the wheel	Anvil velocity	250 N at anvil velocity of 2.4 m/s or 445 N at 1.2 m/s	None	15 mg
Steel/ Fiber Shoe	All (No restrictions mentioned)	Steel shoe or hard-fiber-faced shoe	Steel anvil plate	Not mentioned	sliding	Height of drop	/	Height of drop 50–200 cm	None	7.0 ± 0.1 g
Roto-Friction	All (No restrictions mentioned)	steel friction rod	1-in. tool steel cube (or similar shape with equivalent mass)	Not mentioned	rotational (during 60 s max)	Rotational velocity (torque)	Normal force weight of 25.0 ± 0.5 pounds	Rotational velocity up to 4800 ± 100 RPM	Torque measurement device and electronic timer	30 ± 5 mg
Russian friction test	All (No restrictions mentioned)	pendulum	pendulum holder	Not mentioned	sliding (1.5 mm)	Retaining pressure	/	Retaining pressure 30–1200 MPa	None	20 mg
Emery paper friction Test	Primary explosives	mild steel surface covered with Emery paper	Steel roller covered with the same Emery paper	Grade 0 Emery paper	sliding	Velocity strike	15 kg normal load	Velocity strike 2.5–12 ft./s (0.305–3.66 m/s)	None	5.5 mm^3

Credit: Author.

An additional measurement device may be added to the experimental set-up to help in the reaction assessment and make the test less dependent on the operator, e.g., a gas analyzer as in the ABL method or a high-speed camera as suggested by the participants of the round-robin test program organized in the frame of the European Defense Agency (EDA) subgroup European Network of National Authorities on Ammunition (ENNSA) [30].

In most cases, the Bruceton staircase technique in Skinner et al. [29] is used to determine the level of test stimulus which will cause a 50% probability that there is a positive event, be it ignition, burning, explosion, etc. Another method consists to perform a number of trials at a given level of load. In this case, the result is "pass" if there are no positive trials and "fail" if at least one positive trial is obtained at this level. In some other cases, the Probit method is used [31].

After the required number of trials has been conducted, the sensitivity to friction is expressed by different means, depending on the way the friction test apparatus operates. For the BAM apparatus, the result is expressed as the applied force, in Newton. Occasionally, it is expressed as the applied load, in kg. For the Rotary friction test, the parameter of interest is the median peripheral velocity V_{50}, but the final result is a dimensionless figure, the Figure of Friction (F of F), which formula includes the V_{50} for a reference quality of RDX:

$$F\,of\,F = 3.0\,\frac{V_{50}(sample)}{V_{50}(RDX)}$$

Hence, by definition, the reference quality of RDX has a Figure of Friction which equals 3.0.

Table 2.6 gathers the different ways of assessing the friction sensitivity in the eight friction test methods considered in this study.

The criteria for assessing the sensitivity of the energetic material under test vary according to the test method, the nation, the purpose, etc. The UN criteria given in Table 2.6 are those that determine if an article can be safely transported on public roads, according to the UN MoTaC [21].

Table 2.7 provides typical friction sensitivity results for common explosive substances, obtained at the BAM friction test and at the rotary friction test.

Given the variety of test machines and methods of assessment, it is hazardous to compare the results obtained with a test apparatus/method to results obtained with a different test apparatus/method.

It is however interesting to note a similar ranking of the tested substances for the BAM and the rotary friction tests (Table 2.7). A similar observation is reported by Wharton and Chapman [32] on another set of substances. This led to a linear relationship between the limiting load (in Newton) and the Figure of Friction:

$$Limiting\ load = K \times Figure\ of\ Friction$$

With K a constant falling within the envelope of 32–85. According to the authors, a value of 55 should prove adequate for most practical purposes, except pyrotechnic materials which were not included in the study. It is possible that they will respond differently, hence the relationship may not hold.

TABLE 2.6 Methods of assessment in commonly used friction tests.

Test Method	Criteria for a positive trial	Number of trials required	Result expressed in	Sensitivity Assessment	Reference material	Type of result	Method of assessment
BAM	Decomposition OR ignition OR crackling OR ignition	25–30	Applied force (N)	UN criterion: fails if <80 N	Type I or II, Class 5 RDX standard conforming to MIL-DTL-398	50% probability of event	Bruceton staircase method
Small BAM	Decomposition OR ignition OR crackling OR ignition	10–50, depending on the method	Applied force (N)	Not mentioned	lead styphnate conforming to MILL-757 and dextrinated lead azide conforming to Type I of MIL-L-3055	a functioning or nonfunctioning threshold, or a 50% probability of event	Trials performed at a given force or Bruceton staircase method
Rotary	(obvious audible evidence AND the sample is consumed) OR (sparks are observed), OR (the sample is partly consumed AND there is evidence of combustion)	50	V50 (m/s) or Figure of Friction (F of F) relative to RDX	Comparatively insensitive for a F of F>=6 Sensitive for 3=< F of F<6 Very sensitive for F of F<3 UN criterion: fails if <3	RDX	50% probability of event	Bruceton staircase method
ABL	any of the following events: visible sparks, visible flame, audible explosion, loud crackling, or the detection of reaction products by a gas analyzer	20	Threshold Initation Level or TIL (pounds)	Sensitive if TIL<250 pounds Very sensitive if TIL<50 pounds UN Criterion: fails if the lowest friction load at which at least one reaction occurs in six trials is 250 N at 2.4 m/s or 445 N at 1.2 m/s or less	Type I or II, Class 5 RDX standard conforming to MIL-DTL-398	50% probability of event	Bruceton staircase method
Steel/ Fiber Shoe	Any type of explosion, burning, crackling, or scorching	20 per height drop level	Pass/Fail	Pass if no reaction at the 20 trials with the steel shoe	Type I or II, Class 5 RDX standard conforming to MIL-DTL-398	Pass/Fail at a given load	Trials performed at a given load
Roto-Friction	Fire, spark or burn	Not mentioned	Energy value (= time to reaction multiplied by the torque)	Comparison with two reference materials	Class 7 PETN or Type I or II, Class 5 RDX standard conforming to MIL-DTL-398	Pass/Fail at a given energy	Trials performed at a given energy
Russian friction test	Sound effect, flash, or burn traces on the rollers	25 at each retaining pressure level	Pressure (Mpa)	UN criterion: fails if <200 MPa	/	The maximum retaining pressure at which there is no explosion in any of 25 trials	Trials performed at a given pressure level
Emery paper friction Test	Total or partial consumption of the material with a very audible report	max 20 trials at each velocity or a 50 shot up-and-down test	Velocity strike (m/s)	Very sensitive if velocity strike <10 ft./s or <3.05 m/s	/	Pass/Fail at a given energy or 50% probability of event	Trials performed at a given energy or Bruceton staircase method

Credit: Author.

TABLE 2.7 Typical friction sensitivity results for common explosive substances obtained at the BAM friction test and at the rotary friction test [18,21].

Substances	Result at the BAM friction test[a] (N)	Result at the rotary friction test[a] F of F[a]
Lead styphnate	2	0.005
Lead azide	10	0.07
PETN	60	1.3
HMX	80	1.5
RDX	120	3.0
TNT	360	6

[a] *Note that these values are only indicative. The substances listed in this table may not have been tested in the same aspect/morphology/particle size/environmental conditions.*
Credit: Author.

TABLE 2.8 Effect of temperature on the friction sensitivity of energetic materials Bailey et al. [37].

	Load (in kg) to give 50% event frequency			
	Test temperature			
Material	**20°C**	**50°C**	**80°C**	**100°C**
PETN	2.0	1.0	< 0.5	—
HMX	3.5	< 0.5	< 0.5	—
RDX	4.7	< 0.5	< 0.5	—
Pyrotechnic composition	1.0	< 0.5	-	—
Double base propellant	5.0	< 0.5	< 0.5	—
Blasting explosive	> 12	> 12	6.9	4.5

Credit: Author.

As a result, and despite the number of inter-laboratory round-robin tests ([30,33,34] and apple-to-apple comparisons available in the literature [35,36], no general correlation has been drawn so far.

2.3 Parameters affecting the friction sensitivity of energetic materials

The study conducted by Bailey et al. [37] shows that the environmental conditions, and especially the test temperature, have an effect on friction sensitivity. Friction tests performed on various explosives (β-HMX, RDX, PETN, a double base propellant, a pyrotechnic composition based on strontium nitrate, and an NG-based blasting explosive) over a range of temperature 20–100°C all show an increase in friction sensitivity with increasing temperatures, as can be seen in Table 2.8.

It is suspected that this trend would reverse in the cases where a binder or other additive softens or melts as the temperature increases. According to Chapman [11], such materials may well have a discontinuity in their sensitivity vs temperature dependency, and there could be a decrease in sensitivity at or about this temperature.

For Harris [38], the friction sensitivity of primary explosives depends on the crystals' hardness, their thermal conductivity, the shape and size of their rubbing surfaces, but also on whether the given material melts before decomposition or not. Hardness appears to be the most important physical characteristic for friction [38].

The presence of a coating may also strongly affect friction sensitivity. In general, the higher the amount of coating, the lower the sensitivity to friction is, as shown in Table 2.9. An exception is a lactose which seems to have no desensitization effect on PETN.

The role of particle size on friction sensitivity is less obvious. When tested alone, Song et al. [39] showed that the friction sensitivity increases gradually as particle size decreases for spherical β-HMX within the range 0.6–230 μm. This trend seems to reverse when the explosive crystals are included in an explosive formulation. Cohen et al. [40] showed that cast-cure formulations including RDX or HMX with a small monomodal particle size (5 μm) were characterized by better impact and friction sensitivities. Moreover, they demonstrated that the formulations with bimodal particle size distribution had improved safety characteristics when the greater part of filler's particles had a small size. The decrease in friction sensitivity with particle size is actually more in line with the assumption that the initiation under friction is attributed to the thermal ignition of internal defects in the material, also called hot spots.

For submicron-sized and nanoparticles, experimental studies tend to confirm the theory of hot spots as the friction sensitivity strongly decreases when the median particle size goes below 1 μm [17,41–43]. For Pessina and Spitzer [42], the noticeably lower sensitivity toward friction is also attributed to the self-lubricating effect, as small particles will tend to occupy small interstices instead of breaking. Table 2.10 gathers results that show the decreasing trend in friction sensitivity as a function of the particle size.

TABLE 2.9 Effect of coating on the friction sensitivity of PETN and RDX UN MoTaC [21].

Substances	Result at the BAM friction test (N)	Result at the Russian friction sensitivity test (MPa)
PETN (dry)	60	150
PETN/wax (95/5)	60	
PETN/wax (93/7)	80	
PETN/wax (90/10)	120	
PETN/water (75/25)	160	
PETN/lactose (85/15)	60	
PETN/Paraffin (95/5)		350
PETN/TNT (90/10)		350
RDX (dry)	120	200
RDX (water wet)	160	
RDX/water (75/25)		350

Credit: Author.

TABLE 2.10 Friction sensitivity for micrometer- and nanometer-sized [28] nergetics.

Substances and particle size	Friction sensitivity	Reference
ADN/HMX 100 μm	128 N	Berthe [17]
ADN/HMX few μm	> 360 N	Berthe [17]
Raw RDX M5 μm	160 N	Pessina and Spitzer [42]
Nanostructured RDX	> 360 N	Pessina and Spitzer [42]
CL-20 15 μm	6.4 kg	Bayat and Zeynali [41]
Nano-CL-20	No reaction	Bayat and Zeynali [41]

Credit: Author.

Finally, many authors have tried to predict the friction sensitivity by correlating it with other variables: molecular structure [44–46], detonation velocity [47–51], detonation pressure [48,49,52,53], molecular/surface electronic properties [54–56], particle size [57], thermal decomposition parameters [58–63] and impact sensitivity [59].

3 Shock sensitivity tests

The principle of the test is to initiate a block of high explosives with a given donor charge through a gap with varying lengths. This length is modified up to the value leading to the initiation of 50% of the tests. The thinner the gap, the more insensitive is explosive. This test can be used to estimate the shock pressure leading to the initiation of a high explosive.

There is an international standard NATO STANAG 4488 [64] for shock sensitivity. Several large-scale and small-scale experimental set-ups exist depending on the diameter, the gap, and the go/no go device [65].

This test alone gives information on the initiation pressure of a high explosive. It enables an evaluation of the detonic behavior at a small scale. It is performed on blocks of composition, contrary to the other tests, conducted on smaller samples. It implies the possibility to produce them.

A comparison of several tests can be found in Newgate's NATO MSIAC database [66].

3.1 Experimental set-up

Several tests can be found in the literature, proposed by several test centers, with different donor charges, gaps, confinements, dimensions, and methods to evaluate the go/no go results.

The geometry is always cylindrical. The diameter ranges between 5.1 and 200 mm to adapt to the large panel of a critical diameter of compositions. The donor charge was frequently Pentolite 50:50 in the past. This high explosive has been replaced by pressed RDX wax composition or Comp-B. The donor can be also inserted in metallic confinement.

Historically, the gap was made of sheets of cellulose acetate. This product has been replaced by PMMA cylinders, of a similar Hugoniot curve. The possibility to replace cellulose acetate with PMMA (also called Lucite) is justified in [67]. The past use of cards of cellulose acetate is conserved in the way of expressing the result, usually given in a number of cards of initiation in 50% of tests. The thickness of the gap is rarely given, however, it is possible to convert it. Other gap materials (water, aluminum, brass) are also specified in some cases.

The pipe containing the receiver can be in polyethylene (PE), brass, plastic, or steel, with various thicknesses.

Several possibilities are used to reveal the result of the test. The recovery of high explosives is not a valuable clue, because deflagration may occur, without transition to detonation. The more spread technic is to place a steel plate at the end of the receiver. A hole in the plate reveals a detonation, whereas the plate is found intact in case of no-detonation. Two tests use alternative solutions. For ISGT (Intermediate Scale Gap Test), a block of pressed RDX wax is intercalated between the receiver and the plate, to compare the effect of a constant explosive on the plate. For SSWGT (Small-Scale Water Gap Test), a detonating cord is attached at the end of the receiver. The detonating cord is in contact with a lead plate at the other end, in which a crater signs the detonation.

SSWGT is a German test [68], often used for very sensitive compositions [69] and helped to define the STANAG 4363 [70] to characterize detonating explosive components. ISGT is a French test (NF T 70-502, 2017 [71] for the last edition). The STANAG 4488 uses half-length for the donor, which may lead to different results for a given explosive. Two tests have been developed in LANL on a Small and Large Scale: SSGT and LSGT [72]. The NOL (Naval Ordnance Laboratory, today NSWC) has developed also an SSGT [73] and an LSGT [74]. The Insensitive High Explosives Gap Test (NOL IHEGT) is a small-scale heavily confined test constituting an alternative to the NOL Large-Scale Gap Test (LSGT) [75,76]. These scales are not sufficient for very insensitive compositions, leading Eglin to develop larger scales: ELSGT (Extended Large-Scale Gap Test) [73] and SLSGT (Super Large-Scale Gap Test) [73]. A modified ELSGT is proposed by Aubert et al. [77].

The detailed characteristics of each test are given in Table 2.11.

The configurations are so different from one test to another that the input pressure in the receiver between the two tests could be different, as illustrated in Fig. 2.9. Fig. 2.9 gives a plot of the pressure in the gap vs its length.

3.2 Experimental results

This section is focused on the gap test results and comparison for several high explosives.

The direct comparison of the gap thicknesses is pointless considering the variety of the configuration. We will focus on the initiation pressures.

At first, we evaluate the effect of the diameter (Table 2.12). ISGT and NOL-LSGT have similar diameters and confinement. The results on PBXN-109 give similar results.

If we compare ISGT and ELSGT-1 on B2238, ELSGT-2, and NOL-LSGT on PBXN-109 or mod-ELSGT and NOL-LSGT on Comp-B and Tritonal, similar results are obtained. These tests are comparable, due to their similar shapes, with a receiver confined in a steel pipe. The influence of the diameter increase is low.

SLSGT can be compared with NOL-LSGT and mod-ELSGT. The initiation pressure is quite half for SLSGT due to the larger diameter and thicker confinement.

We will now consider the effect of the confinement. For LSGT; the NOL one is confined whereas the LANL one is not for a similar diameter. For the second one, the initiation pressure on cast Comp-B, Pentolite, and RDX is four-time greater (two times for DATB, seven times for pressed Comp-B). NOL-LSGT is in proportion five times less confined than NOL-SSGT, in thickness ratio. The initiation pressure on CH-6, Comp-A5, Pentolite, RDX, and TNETB is four times greater for the NOL-LSGT. It is interesting to note that the initiation pressure factor between these tests is similar for different compositions, for example, Pentolite and RDX.

TABLE 2.11 Characteristics of the different gap tests.

	Donor		Gap		Receiver	
Test	Explosive % wt. Density (g/cc)	Dimensions (mm)	Nature	Dimension (mm)	Dimension (mm)	Witness % wt. (mm)
SSWGT	RDX wax 95:5 $s=1.60$	Phi 21 H 20	Water	Phi 21 H [1;62]	PE Phi 21/25 H40	Det cord Lead plate
NOL-SSGT	RDX $d=1.56$	Phi 5.1 H38.1 in brass pipe Phi 5.1/12.7	PMMA	Phi 12.7	Brass pipe Phi 5.1/12.7 H38.1	Steel plate [28]=12.7
LANL-SSGT	PBX-9407	Phi 7.62 H 5.26	Brass [28]=0.25	Phi 25.4	Plastic	Steel plate
ISGT	RDX wax 95:5 $d=1.60$	Phi 40 H 160	Cellulose Acetate [28]=0.19	Phi 46 H[1;76]	Steel pipe Phi40/48 H200	RDX wax 95:5 Phi 40 H 40 Steel plate [28]=10
NOL-LSGT	Pentolite 50:50 $d=1.56$	Phi 50.8 H 50.8	Cellulose Acetate [28]=0.25	Phi 50.8 H[0;100]	Steel pipe Phi 36.5/47.6 H140	Steel plate [28]=10
LANL LSGT	PBX 9205	Phi 41.3 H 102	Al dural	Phi 41.3	No casing Phi 41.3 H 102	Steel plate
NOL IHEGT	Pentolite 50:50 $d=1.56$	Phi 50.8 H 50.8	Cellulose Acetate [28]=0.25	Phi 50.8 H [0;100]	Steel pipe Phi 12.7/19.05 H50.8	Steel plate
ELSGT-1	RDX wax 95:5	Phi 95 H 95	PMMA	Phi 95	Steel pipe Phi 73/95 H 279	Steel pipe [28]=20
ELSGT-2	Pentolite 50:50	Phi 95 H 95	PMMA	Phi 95	Steel pipe Phi 73/95 H 279	Steel pipe [28]=20
Modified ELSGT	Comp-B	Phi 95 H 95	PMMA	Phi 95	Steel pipe Phi 73/95 H 279	Steel pipe [28]=20
SLSGT	Comp-B	Phi182 H 203 in steel pipe Phi 182/203	PMMA	Phi 203 H [0–400]	Steel pipe Phi 185/203 H 406	Steel plate $e=12.7$

Credit: Author.

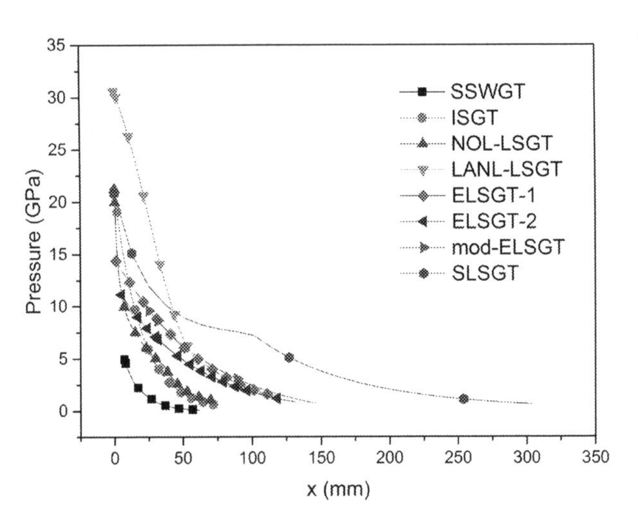

FIG. 2.9 Evolution of pressure in the gap. *Credit: NATO/MSIAC.*

TABLE 2.12 Comparison of gap thickness (L) and threshold initiation pressure (P) tests results for several high explosives.

Substance	rho0 (g/cm3)	SSWGT L (mm)	SSWGT P (GPa)	NOL SSGT L (mm)	NOL SSGT P (GPa)	ISGT L (mm)	ISGT P (GPa)	NOL-LSGT L (mm)	NOL-LSGT P (GPa)	LANL-LSGT L (mm)	LANL-LSGT P (GPa)	ELSGT-1 L (mm)	ELSGT-1 P (GPa)	ELSGT-2 L (mm)	ELSGT-2 P (GPa)	Mod. ELSGT L (mm)	Mod. ELSGT P (GPa)	SLSGT L (mm)	SLSGT P (GPa)
B-2238	1.572					41.8	2.50					110.0	1.65						
CH-6	1.739			6.3	0.39			63.0	1.30										
Comp A-5	1.700			8.8	0.32			65.2	1.21										
Comp-B (pressed)	1.660							60.5	1.42	43.2	9.50					132.0	1.14		
Comp-B (cast)	1.700							51.0	2.04	43.2	9.50					132.0	1.14		
DATB	1.700							33.5	4.61	45.4	8.64								
PBXN-109 (HAAP. Type II lot D)	1.660							44.5	2.79					105.7	1.60				
PBXN-109 (US Type II RDX)	1.660					42.8	2.38	49.4	2.20										
PBXN-109 (SME I-RDX)	1.660	9.0	4.18			28.0	4.99												
PBXN-109 (Dyno Type II RDX)	1.660					39.5	2.80	49.8	2.16										
PBXN-109 (Dyno "Type I" RDX)	1.660	11.0	3.62			44.0	2.24												
Pentolite	1.681			9.8	0.30			67.0	1.14	69.0	5.10								
PETN	1.744			9.8	0.30					60.0	5.28								
RDX	1.766			7.4	0.36			67.0	1.14	61.0	5.26								
TNETB	1.744			7.8	0.35			62.0	1.35										
Tritonal (80/20) pressed	1.720							46.0	2.59							103.0	2.30	233.0	1.40

Credit: NATO/MSIAC.

SSWGT and ISGT can be compared on PBXN-109. The order of magnitude of initiation pressure is similar. There might be a compensation between the presence or not of the confinement and the different natures of the gap.

We will now compare the results obtained for several high explosives on the same test. On NOL-LSGT, DATB requires a pressure three times greater than RDX, which seems logical. Classical compositions are listed from the more to the less sensitive Comp-B, PBXN-109, and Tritonal. It is coherent to the results obtained on pop-plot experiments. On LANL LSGT, we can compare RDX to Comp-B. We may note that the difference between these two compositions is larger than the one obtained on NOL-LSGT. On mod-ELSGT, one can find that Comp-B is easier to initiate than Tritonal. On NOL-LSGT, one can note the difference between pressed and cast Comp-B. The pressed formulation is here slightly less dense than the cast. The presence of porosities enables the apparition of hot spots, leading to initiate more easily than in the cast one.

The nanocrystalline Types A and B RDX show less shock sensitivity than the reference 4.8 μm RDX Table 2.2 for uncoated and even coated samples [14]. Type B is even less sensitive than type A, despite the mean size distribution.

The question of the influence of confinement has been studied at the Naval Ordnance Laboratory [78]. NOL-LSGT experiments were carried out with and without the confinement of the receiver for four high explosives. The relation between the thicknesses of the gap with

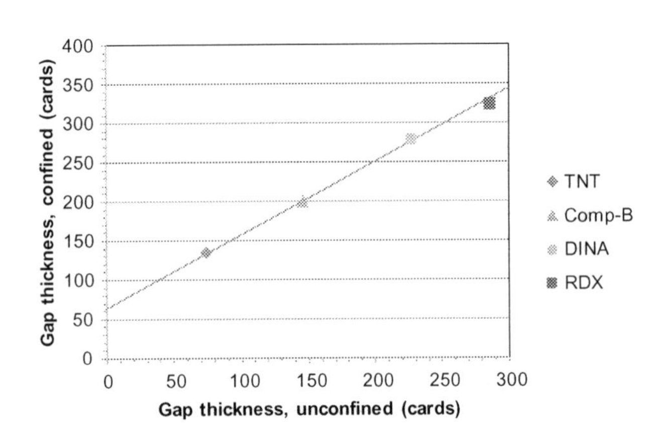

FIG. 2.10 Comparison of the gaps depending on the confinement. *Credit: Author.*

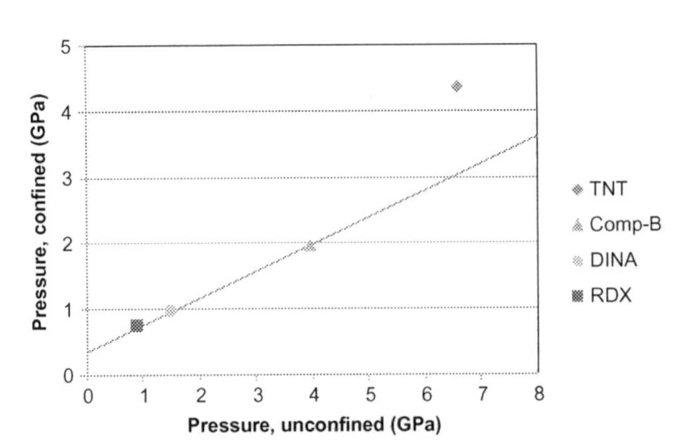

FIG. 2.11 Comparison of the initiation pressure depending on the confinement. *Credit: NATO.*

and without the confinement is linear for TNT, DINA, and Comp-B (Fig. 2.10). Only RDX is not on the line but is slightly more sensitive when unconfined.

If the same graph is plotted for shock pressure, the results are significantly different. TNT is far from the linear evolution (Fig. 2.11). This is explained by the author by the thickness of the gap, inferior to 35 mm. At this depth, the lateral rarefaction reaches the cylindrical axis of the unconfined attenuator, producing its breakup and modifying the way the shock is transmitted.

To conclude, the gap test is an interesting test to evaluate the ease of a composition initiation and is used also to generate damaged samples and evaluate their shock sensitivity [79]. It is less expensive than pop-plot experiments. It is not a safety test to analyze the possibility of composition manufacture. Differences exist between several tests described in the literature, making difficult comparisons between the different tests databases.

4 Electrostatic discharge sensitivity tests

The spark tests are small-scale electrostatic discharge tests depending on the apparatus and the environment with a small amount of granular or solid energetic materials about 100 mg or less in the range few mJoule to some Joule, addressing the electrostatic sensitivity possibly due to human handling in controlled temperature conditions.

The large-scale electrostatic discharge test is an international standard with the following Ref. [15] and addresses the sensitivity to repeated discharges with the same energy 15.6 J at ambient temperature or in specified temperature conditions, especially low temperature, for larger solid or powder samples about 1 kg or less (636 cm^3) possibly simulating larger discharge due to dielectric surface contacts with energetic materials during production handling.

4.1 Experimental set-up

Several small-scale set-ups exist depending on the apparatus:

- Three methods 1031, 1032, and 1033 for the MIL-STD-1751A [22]
- The UK EMTAP method 6
- The approaching needle test (ABL ESD Test; [80])
- Two French Norms [28,81]
- The EN 13938–2 European Norm (2004)
- Three Matasuzawa et al. [82] configurations for pyrotechnic compositions

The MIL-STD-1751A has already three methods: 1031, 1032, and 1033. The 1031 method has the same sample size as the 1033 method but a different apparatus. The single pulse current power supply is designed to charge the capacitors. The influence of the pulse duration especially and other parameters have been studied on different apparatuses [29,83,84]. In the capacitors bank, only one of the capacitors will discharge at a time. A point-to-plane electrode geometry is proposed between the needle and the grounded sample. The needle is set at a fixed distance from the cathode. The Threshold Initiation Limit (TIL) is defined as the maximum energy in Joules that can be applied without causing a reaction in 20 consecutive trials. The "up-and-down" Bruceton procedure is also applied to determine the 50% initiation level.

The TIL for RDX Class 1 is 0.095 J and the 50% initiation level is 0.162 J. The temperature and relative humidity are controlled. A positive result is given by a flash, spark, burn, odor, or noise other than instrument noise.

The 1032 method is divided into two parts: a screening test and a ranking test. The screening test is performed with a fixed energy of 0.25 J, obtained by the fixed distance between the needle and the cathode. A flash, spark, burn, or noise (other than instrument noise) defines a positive result. If there are no reactions in the 20 consecutive trials, the sample has passed the electrostatic test. If a reaction is obtained at any point, the test is discontinued and the ranking test is applied. The ranking test is operated by lowering the energy value after a positive value until no reaction is obtained in 20 trials. The Bruceton method can be applied. The sample cell volume is about 0.030 cm^3, sealed with Scotch tape, and is placed on a grounded cell holder. The pin is lowered onto the sample cell at a predefined position [29,85].

The 1033 method is designed to simulate a discharge through a thin layer of the sample. The discharge apparatus consists of a needle electrode and a grounded sample post which form a point-to-plane electrode geometry. Hygroscopic or granular materials should be dried before the test. Solid sample is 15.875 mm (0.625 in.) square or 15.875 mm diameter and 0.8382 mm thick (0.033 in.). The ESD sensitivity is addressed in terms of the maximum energy level in Joules that can be applied without causing a reaction in 20 consecutive trials. The Bruceton method can be applied [29].

The UK EMTAP method 6 is one standard with predetermined energy (4.5 J, 0.45 J, or 0.045 J) using Teledyne Reynolds apparatus, which mean deposited energy has been controlled recently around 4.54 J, 0.44 J, and 0.038 J [86].

The ABL ESD Test is based on an "approaching needle" apparatus. The standard delivered energy varies in the following range 0.001–6.25 J and could go up to 9.38 J and even 37.5 J with the instrument capabilities for the Australian set-up. The electric discharge is generated at a finite separation distance [87]. The MIL-STD 650 method 512 is also an approaching electrode sensitivity test. The energy is limited to 0.25 J. No reaction is assessed after 20 negative trials [33,34]. The needle electrode is charged before moving down to the fixed electrode gap (0.18 mm). The needle penetrates the sample, and discharge through it before rising again to its initial position.

Two French tests are also identified: the electrostatic discharges sensitivity test with the GEMO apparatus and the one with the SNPE apparatus, respectively [28,81]. There is a hemispherical tip for the positive electrode of the n°70-539 reference, the distance between anode and cathode is fixed to 2 mm, the sample is in between and the Bruceton method is applied for the experimental procedure with 30 different samples and the analysis of the results. The positive electrode is a needle for the n°70-540 reference. The distance between the two electrodes is 1.5 mm. The objective is to find the tension for 20 consecutive negative values. In both cases, it is possible to test a liquid or a gel.

Majzlík [88] analyzes the EN 13938-2 [89] norm with the ESZ KTTV apparatus, using the dedicated discharge energy formula taking into account internal and external capacitors. The oscillation discharge current is measured vs time and the influence of the grain size on the reaction delay is also studied [90]. The EN norm has a sample cell with two copper disks separated by a 3 mm thick plastic disk with a 6.3 mm diameter hole. Three given energies 5, 0.5, and 0.05 J are proposed to find 20 consecutive negative results. An interesting procedure is given to calibrate and control the capacitor, the tension with a high voltage coupling antenna, and the current discharge with B dot measurements.

Two electrostatic test devices (the Hosoyo apparatus and the Chegoku Kayaku apparatus) are used to measure minimum ignition energy [11,82]. With the first device, two closed electrode configurations are possible within a Teflon tube: the first one with two lower diameter flat nose stainless steel electrodes and a 1 mm gap, the second one with a 2 mm diameter spherical brass electrode and a flat brass electrode and a 2 mm gap. For the second device, there is an open brass electrode configuration with a 3 mm diameter spherical electrode and a flat electrode.

For Stepanov [16], the MIL-STD-1751A method 1032 [15] is used to perform the ESD tests on nanocrystalline RDX compared to micron size reference. The ESD threshold is given after 20 negative tests.

For Berthe [17], Impact, Friction, and ESD sensitivity thresholds are given after only six negative results. The ESD apparatus is OZM Research ESD 2008A with an energy range between 0.14 and 10 mJ and 7 mm^3 sample volume in a plastic tube between the anode and cathode.

The large-scale electrostatic discharge set-up is presented Fig. 2.12. Three samples are needed for a test. The compact sample is a cylinder with a diameter of 90 mm and a length of 100 mm. The granular sample is in a PMMA cylindrical container with an inner diameter of 90 mm and length 125 mm, filling the container without being compacted. The negative electrode is a brass disk, connected to the earthing cable and glued in the bottom of a PMMA box with a silicone resin. The sample is put in the PMMA box, in contact with the negative electrode. The painted face with a sprayed conductive silver lacquer is in contact with the negative electrode for the compact sample. The positive electrode is a brass rod with a 10 mm diameter and one 60° angle conical end in contact with the sample. This positive electrode is connected to the high voltage supply. The types of reaction are the following: N no reaction, S rising of the cover without flash, L rising of the cover with light flash, F block fragmentation, C combustion, E explosion. The material is labeled insensitive to electrostatic discharge if there is no reaction after 30 repeated discharges on the three samples. The material is labeled sensitive for all other types of reactions.

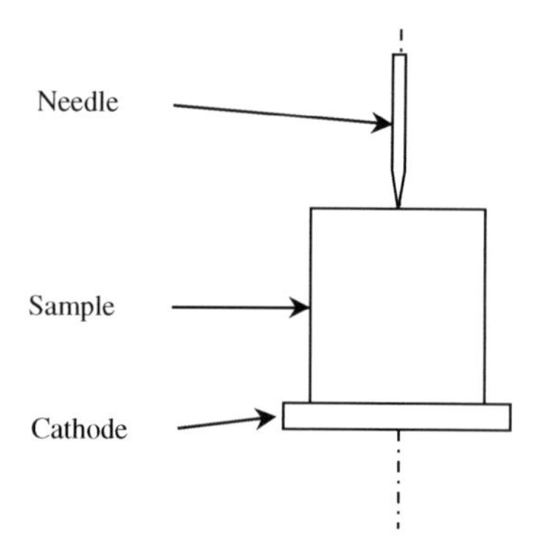

FIG. 2.12 The large-scale electrostatic discharge set-up for a compact sample. *Credit: Author.*

4.2 Experimental results

ESD results in Stepanov [16] and Berthe [17] are presented to highlight the possible correlation with nano and micron sizes and also the crystal phases.

Stepanov [16] showed that the reference 4.8 μm RDX (pure and coated) and the recrystallized Rapid Expansion of Supercritical Solutions method (RESS) type A around 150 nm nano RDX and the type B around 500 nm nano RDX have no ignition in 20 trials at 0.25 J with Picatinny method 1032, MIL-STD-1751A [15], see Table 2.2.

Berthe [17] presented an SFE nanocrystallization method for ADN, RDX, HMX, CL20, AND/RDX, AND/HMX, and AND/CL20 obtained in acetone at 110°C and 20 bar compared to μm size reference or slow recrystallization process, see Table 2.3. SFE HMX is evolving from γ phase to α and γ phase after 5 months of storage, the mean size is slightly evolving also. The sensitivity results are given after crystallization. The SFE nanocrystals with or without ADN are less sensitive than the μm size reference to the ESD test. The slow crystallizations with DNA are also less sensitive except for ADN/HMX. The water content of ADN in the slow process could explain some higher values.

The electrostatic sensitivity is improved by 18.3% with the nanosize CL20 compared to micrometer size (60 J vs 49 J; [41]). This desensitization trend has been observed also for hexolite compositions up to 25% (72 J vs 54 J; [91]).

5 Thermal self-initiation and decomposition sensitivity tests

Important information for laboratory pyrotechnic operators is melting and decomposition temperatures. These two values allow explosives substances safety handling. To study mechanical behavior, glass transition temperature must be evaluated too. For energetic materials manipulation or storage, it's necessary to estimate thermal conditions including thermal cycles, which can lead to a pyrotechnic reaction, with an indication of reaction delay.

Thermal properties may be different according to particles size and crystalline phase. So, these data need to be registered for each test. Moreover, sample preparation can be very influential, notably for small-scale analyses. Indeed, it's very important to be sure that the specimen used for the test is taken in a representative clean, and unpolluted area. As the sample amount is very small for certain experiments (just a few milligrams), it's preferable to repeat the test in order to check the sample representativeness. Humidity presence can modify results too, so sample conditioning needs to be the same for every part, and it can be useful to store it for around 15 h or more in an oven stabilized at low temperature (65°C for example) in order to dry it. Reactivity of materials can be modified too with the particle size reduction procedure. So the less aggressive method must be chosen, and every step of sample preparation must be noted.

The self-initiation and decomposition temperature are characterized by different national or international standardized small-scale tests, and some of them are presented, respectively, in Refs. [92,93]. Each of them gives information about a specific thermal property and deserves to be studied.

5.1 Decomposition temperature experimental set-up

The NATO STANAG 4515 [93] procedure is intended to determine thermal properties of energetic materials, especially the decomposition temperature, by small-scale laboratory thermal analyses: Differential Thermal Analysis (DTA), Differential Scanning Calorimetry (DSC), Thermo Gravimetric Analysis (TGA), and Heat Flow Calorimetry (HFC). Moreover, in France, the French norm NF T70-526 gives global recommendations about thermal analysis for explosives properties study by DSC.

All these technologies are based on the same principle: the sample is subjected to thermal conditions which involve property modification causing endothermic or exothermic reactions, recorded by thermal and/or weight loss detectors (temperature, heat flow, weight) within the sample cell and possibly also the reference cell for differential measurement. DTA, DSC, and TGA are generally used with dynamic methods and HFC with isothermal ones.

DTA, DSC, and TGA use a very small quantity of samples. Indeed, as the reaction violence is uncertain and can be dangerous for the operator or can damage the device, the energetic material sample amount must be adjusted with the heating speed and is generally less than 50 mg. It is important to avoid self-heating reactions, first for security but also for controlling the reaction and having enough data for kinetics simulation. The heating rate is usually then fixed between 1 and 10°C/min. Tests must be repeated in order to check the results. With HFC, sample weight is more important (around 200 mg) and the device is more sensitive than DSC. It can thus be operated at considerably lower temperatures with better sample representativeness. On the example of the DSC curve (Fig. 2.13), we can see an exothermic or endothermic reaction like glass transition, melting, or decomposition.

On this curve, there are different temperatures to study:

- Tg: the glass transition temperature is defined as the intersection between the DSC curve and the bisector of the angle made between the extrapolated baseline recorded before the transition and the extrapolated baseline recorded after the transition.

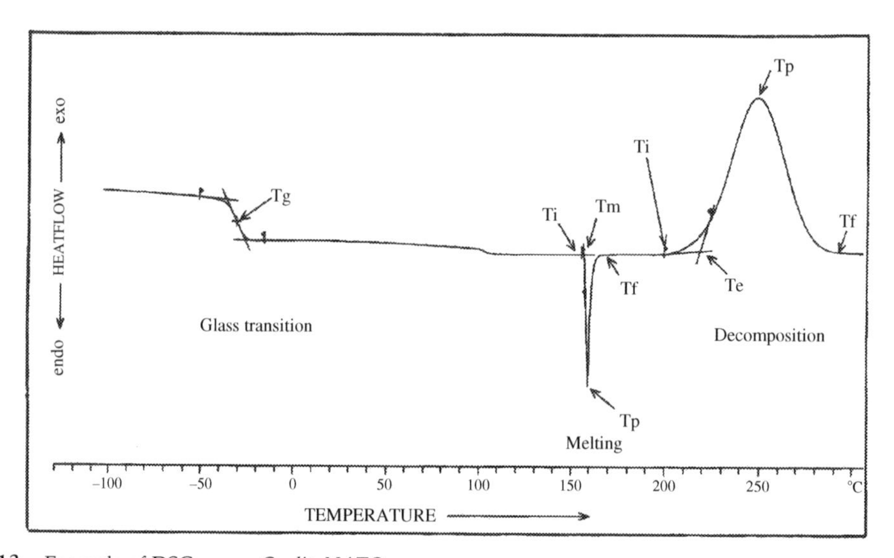

FIG. 2.13 Example of DSC curve. *Credit: NATO.*

- Ti: the initial temperature is the temperature at which the first deflection from the base line is observed for a chemical reaction or a phase transition.
- Tp: the peak maximum temperature is the temperature at which the peak maximum or minimum is observed.
- Tf: the final temperature is the temperature, after a peak, at which no further deflection from the base line is observed.
- Te: the extrapolated onset temperature is measured as the point of intersection of two tangents, one drawn to the curve before commencement of the event, the other to a point on the steepest part of the transition.
- Tm: the temperature of melting is measured as the point of intersection of two tangents, one drawn to the curve before the commencement of the event, the other to a point on the steepest part of the transition.

All of these values are important, but Te (or Tm) and Tp are the more representative ones. Ti can vary according to the operator and is so uncertain. Tf comes after reaction and so isn't very useful for pyrotechnic study, particularly for decomposition reaction. Comparing Te (or Tm) and Tp gives an indication of peak appearance: if these two values are closed, there is a dynamic reaction, but if they are detached, the reaction is slow. Decomposition of energetic material can present one or more steps. The first one is often the one that causes pyrotechnic reaction, but it can be useful to study the others.

It's important to note that these values depend on the type of crucible used for experimentation. In sealed ones, the pressure, which is rising in crucible, increases the violence of reaction. In an open crucible, it's possible to work with a stable flow of inert gas, which allows for example to study explosive properties without supplementary oxygen.

Kinetics study can be performed by repeating thermal tests like DSC, ATD, or ATG under different heating rates or isotherms. It is then possible, with complementary information about the material properties like thermal conductivity, to estimate activation energy and reaction model according to the reaction progress measurement. This allows to simulate reaction in given thermal conditions.

5.2 Temperature of ignition experimental set-up

NATO STANAG 4491 [92] proposes two processes for temperature of ignition evaluation, whose main difference is the heating rate.

The first test, also described in the French norm NF-70-504 [14] is a small-scale test with low confinement, in which an explosive sample is subjected to a heating rate of 5°C/min, up to 400°C. The lowest temperature at which reaction is observed is noted as the temperature of initiation.

To perform this test, a sample of explosive (around 200 mg) is loaded into a glass tube and is placed in a steel block or other suitable heat device. For unknown samples or those thought to be primary explosives, a smaller quantity (usually 50 mg) is loaded. Solid powders should all pass through a 3 mm standard sieve. Rubbery or gel-like samples are either cut into a cube shape of the correct mass or chopped to pass through the sieve. At least two tests for each sample must be performed.

Reaction can take the form of burning with the development of a flame, rapid decomposition (puff of smoke), or explosion, and the kind of reaction must be noted on the test sheet.

The difference between the two tests must be smaller than 3°C, and if it is not the case, another test must be done.

If no reaction is detected at 400°C, and if it's authorized in your company, it's possible to allow the sample to cool and then analyze it. It can reveal a beginning of decomposition, and then the test result can be noted as "slow decomposition." If it's impossible to analyze the sample, the result must be "no reaction up to 400°C."

The second test is performed in a Wood's metal bath. A 0.5 g sample is placed in glass tube, immersed in the bath, stabilized at 100°C, at a depth of 20 mm. Sample is then heated at 20°C/min to 360°C. Temperature at which a change is observed is noted. The change can be presence of smoke, gas, or flame. If there is no reaction, it is noted too.

5.3 Cook-off temperature experiments

NATO STANAG 4491 [92] or national norms propose several cook-off temperature tests, intended to determie the explosive sensitivity subjected to an important and continue thermal aggression, with a diverse rate of heating (Slow Cook-Off (SCO) and Fast Cook-Off (FCO) tests).

NATO STANAG 4491 [92] describes tube and variable confinement configurations, in SCO or FCO tests. The Koenon first configuration is dedicated to measure the thermal sensibility at small scale of solid energetic materials and is performed in a steel tube, closed by an orifice plate, in which the sample is hated at 3.3°C/s. The orifice size, varying between 1 and 20 mm, has an influence on the reaction type and can lead to combustion, deflagration, or detonation. Solid explosive substances are tested in the dry state. Powdered substances must be composed with particles size of 0.5–1.0 mm and are sieved. Pressed, cast, or otherwise compacted substances are crushed and then sieved. Most explosive substances will demonstrate a change from a burning response to an explosion response at a limiting orifice plate diameter and explode at all smaller orifice diameters. Thermal sensitivity increases with increasing limiting orifice diameter and decreasing time to event and time to explosion. Reaction can cause tube damage or not, and the result is classified according to nine categories: Tube undamaged, A: Tube bottom bulged, B: Tube bottom and wall bulged, C: Tube bottom severed, D: Tube torn open, E: Tube split into two fragments, F: Tube split into three or more mainly large fragments, G: Tube split into many mainly small fragments, H: as G; threaded collar, box nut and/or orifice plate damaged or fragmented. For the purposes of this test only F, G and H are positive reactions.

The variable confinement second configuration is intended to the measurement of the explosiveness of solid energetic materials and to determine the median tube thickness just resulting in a deflagration reaction rather than burning. For this test, the sample (about 50 g) is confined in an aluminum liner within a steel tube and is heated by means of electrical windings with the rate of SCO (3.3°C/h) or FCO (without fixed rate in the STANAG) tests. Steel tube thickness varies from 0.375 to 3 mm and the assembly is located between steel witness plates. The set-up is heated until reaction. After the test, reactions are classified according to five categories: burning, deflagration, explosion, partial detonation, and detonation.

The last test of NATO STANAG 4491 [92], is the tube test and is also destined to the measurement of the explosiveness of solid energetic materials, more in the objective to evaluate the level of explosiveness. The sample, confined in a steel tube (internal diameter from 30 to 50 mm, length from 200 to 254 mm, and thickness between 4 and 6 mm), is heated by means of an external fire for FCO (petrol or wood fire) and by an electrical winding for SCO. Tubes are sealed by threaded end caps so that in low explosiveness events the tube wall fails before the end caps fail. Again, the set-up is heated until reaction and reaction type is determined by the degree of tube fragmentation, from undamaged to fragmented parts.

In France, the French norm proposes a test with a rate of 5°C/min [94], and a test with a heating rate of 3.3°C/h [95]. In these tests, explosives are introduced in steel tubes, whose dimensions may vary and are given in another norm. Reaction violence is estimated by fragmentation analyses after the reaction. A soft steel control plate can be placed under the set-up to complement the expertise.

Another interesting thermal test is the determination of the critical temperature for thermal initiation, described in the French norm NF T70-526 [96]. This test isn't proposed in NATO STANAG but it allows collecting information about explosives with low confinement behavior kept in a stable environment, like during storage. It is identified as a cook-off test too. It can be performed on solid or powdered explosives.

The sample of explosive is placed in an aluminum carrying tube and is placed in a heating chamber conditioned at a specified isothermal temperature. During 360 h, the temperature is monitored in the chamber and in the sample in order to detect a self-heating reaction. Many tests are performed in order to find the maximal temperature without reaction if there isn't always a pyrotechnic reaction. If there is always a pyrotechnic reaction, the result corresponds to temperatures giving reaction after 20 and 100 h. For unknown samples, the first isothermal temperature tested is 160°C. Depending on whether there is a reaction or not, other isotherms are chosen by dichotomy, with ±10°C in a first time and ±5°C for more precision.

A pyrotechnic reaction is identified by the presence of deflagration, smoke, perturbation of the temperature profile curves (abrupt slope change, off-scale path resulting from the destruction of the thermocouple). It's necessary to perform at least eight tests at different temperatures.

5.4 Medium-scale experiments

Different international laboratories design complementary medium-scale tests in order to check the representativity of small-scale laboratory tests. For example, the LLNL proposed two tests. Hsu et al. [97] measure times to explosion and minimum ignition temperatures of energetic materials at elevated temperatures in the One Dimensional Time to eXplosion (ODTX) system. The high explosive spherical sample (solid, liquid, or powder) is placed in a 1.27 cm diameter spherical cavity between two aluminum anvils, heated at a predetermined isothermal temperature. The Scaled Thermal Explosion eXperiment (STEX) [98,99], is a small-scale thermal cook-off experiment, in which a vessel is heated and/or conditioned in isothermal till explosion. Le Gallo et al. [100] estimate reaction delay of a few kg high explosive block subjected to isothermal conditioning during a few hours with Controlled Thermal Aggression (CTA) tests. There are lots of other propositions around cook-off tests, which are interesting for pyrotechnics study.

5.5 Experimental results

The melting and decomposition temperatures for nano and μm size energetic materials are presented in Table 2.13. The decomposition temperature difference is less than 4°C for nanocrystalline and micron size RDX obtained by DSC with 5°C/min open crucible [14]. The decomposition temperature peak is also similar for RDX nano (obtained with TGA-DSC with an open alumina crucible and argon flux) and RDX μm with DSC closed gold-covered crucible [17]. The argon flux is given equally to 50 mL/min for the DSC. The second cell reference is empty for the two differential measurements. The closed crucible results show three peaks associated with liquid RDX decomposition, RDX evaporation, and evaporated RDX decomposition. The open crucible results show only one exotherm peak. Only slight differences in the decomposition temperatures are observed between Nano and μm HMX, CL20, and ADN. The ADN/Nitramine curves show, respectively, the double peaks associated with ADN and the Nitramine decompositions.

The ignition temperature results are presented in Table 2.14, which are different for each test procedure and heating rate.

Experimental results of the Koenon cook-off test are presented in Table 2.15 as a function of the orifice diameter of the end tube.

TABLE 2.13 Melting and decomposition temperatures for nano and μm size energetic materials.

Energetic materials	Melting temperature Tp (°C)	Decomposition temperature Tp (°C)
RDX 4.8 μm	204.6[b]	231.5[b]
RESS nano RDX 150 nm	206.6[b]	235.3[b]
α RDX ref. 3.69 μm	203[a]	230[a]
SFE α RDX 0.32 μm	203[a]	229 +
β HMX ref. 0.7 μm	273[a]	275[a]
SFE γ to α, γ HMX 0.15 μm	Not observed	282[a]
ε CL20 ref. 98 μm	—	233[a]
SFE β CL20 0.32 μm	—	228 +
α ADN ref. μm size	94.3 +	183 +
α Nano ADN 30 nm	93.6+(methyl acetate solvent)	176+(same solvent)
SFE α DNA/RDX 5 μm	89 +	179; 224 +
α ADN/RDX slow tens μm	93.7 +	185; 232 +
ADN/HMX slow 100 μm	95 +	188; 270 +
SFE α ADN/HMX few μm	93.3 +	188; 276 +
SFE α DNA/CL20 few μm	94 +	189; 240 +
α ADN/CL20 slow 40 μm	95 +	182; 244

[a] *5 °C/min closed crucible DSC; + open crucible TGA-DSC [17]*
[b] *5°C/min DSC [16].*
Credit: Author.

TABLE 2.14 Ignition temperature values for the two tests [92].

Energetic material	Temperature of ignition (°C) 5°C/min first test	Temperature of ignition (°C) 20°C/min Wood's	Reaction type
HMX	219	278	Inflammation
RDX	273	227	Inflammation
TNT	288	327	Inflammation
PETN	186	207	

Credit: Author.

TABLE 2.15 Cook-off Koenon test values [92].

Energetic material	Orifice diameter (mm)	Reaction categories
HMX	8	F
RDX	6	G
TNT	4	F
PETN	5	G

Credit: Author.

6 Conclusion

Numerous national and international sensitivity standardization methods have been identified to have a large overview of the detailed processes. The first difficulty of the sensitivity modeling challenge is to gather and understand the different experimental set-ups for every single test. Compared experimental configurations and results are listed in this chapter for this purpose, and will certainly help to improve the standardization.

References

[1] EN 13631-4, Explosives for Civil Uses—Determination of Sensitiveness to Impact of Explosives, 2003.
[2] H. Koenen, K.H. Ide, Exp. Dermatol. 9 (1961) 4–30.
[3] J. Köhler, R. Meyer, Explosives, VCH, Weinheim, 1993.
[4] NATO STANAG 4489, Explosives, Impact Sensitivity Tests, ed. 1, NATO Standard, 1999.
[5] NF T 70-500, Impact Sensitivity—BAM Impactor Test, AFNOR, 2007.
[6] GEMO FMD-410-B-1 (Groupe d'Etudes des Modes Opératoires), Sensibilité à l'Impact au Mouton de Bourges. French procedure, 1978.
[7] GEMO FMD-410-C, 1980. Sensibilité à l'Impact au Mouton Pendulaire de Sorgues. French procedure.
[8] GEMO FMD-410-D, Sensibilité à l'Impact au Mouton de 11 kg. French procedure, 1980.
[9] NF T 70-501, Impact Sensitivity—30 kg Impactor Test, AFNOR, 2019.
[10] GEMO SEN-217, Sensibilité à l'Impact des Matières Explosives très Sensibles. French procedure, 2010.
[11] D. Chapman, Sensitiveness of pyrotechnic compositions, sensitiveness testing, in: Pyrotechnic Chemistry, Vol. 17, 2004, pp. 1–22.
[12] M. Vaullerin, A. Espagnac, L. Morin-Allory, Prediction of explosives impact sensitivity, Propellants Explos. 23 (5) (1998).

[13] P. Morand, M. Vaullerin, A. Espagnac, Optimisation de la Méthode d'Evaluation de la Sensibilité à l'Impact (ISI) des Substances Energétiques peu Sensibles, in: EUROPYRO 99, 7ème Congrès International de Pyrotechnie, Brest, 1999.
[14] NF T 70-504, Matériaux Energétiques de Défense, Sécurité Vulnérabilité, Chauffage Progressif (Température d'Auto-Inflammation), AFNOR, 2008.
[15] NATO STANAG 4490, 2001. Explosives, Electrostatic Discharge Sensitivity Test(S). NATO Standard.
[16] V. Stepanov, Production of nanocrystalline RDX by RESS: development and material characterization, New Jersey Institute of Technology, 2008, p. 868. Dissertation.
[17] J.-E. Berthe, Amélioration des Explosifs par Ajustement de leur Balance en Oxygène lors de la Cristallisation par Evaporation Flash de Spray, Université de Strasbourg, 2018. Thèse.
[18] NATO STANAG 4487, Explosive, Friction Sensitivity Tests, ed. 2, 2009.
[19] UK EMTAP, 2016a. Test No. 2 Mallet Friction Test.
[20] G.R. Walker, The Technical Cooperation Program (TTCP)—Manual of Sensitiveness Tests, 1966. Published by the Canadian Armament Research and Development Establishment. On Behalf of TTCP Panel 0–2 (Explosives), Working Group on Sensitiveness.
[21] UN MoTaC, UN Recommendations on the Transport of Dangerous Goods—Manual of Tests and Criteria, 2019. 7th revision, ST/SG/AC.10/11/Rev.7.
[22] MIL-STD-1751A, Safety and Performance Tests for the Qualification of Explosives (High Explosives, Propellants, and Pyrotechnics), 2005. US DoD Department of Defense Test Method Standard.
[23] UK EMTAP, 2016b. Test No. 33 Rotary Friction Test.
[24] NATO AOP-7, Manual of Data Requirements and Tests for the Qualification of Explosive Materials for Military Use, ed. 2, 2004. Revision 1.
[25] NF T 70-509, Matériaux Energétiques de Défense—Sécurité—Vulnérabilité—Sensibilité à la Friction Linéaire des Matériaux très Sensibles, AFNOR, 2007.
[26] UK EMTAP, 2016c. Test No. 13 Emery friction test.
[27] UK EMTAP, 2016d. Test No. 6 ESD test.
[28] NF T70-540, Sensibilité à l'Etincelle Electrique, 2009. Appareil SNPE.
[29] D. Skinner, D. Olson, A. Block-Bolten, Electrostatic discharge ignition of energetic materials, Propellants Explos. Pyrotech. 23 (1997) 34–42.
[30] E. Krabbendam, E. van Arkel, W. de Klerk, Characterization of RDX, a propellant and a pyrotechnic mixture: round-robin between nine European test houses, in: Presented at the 13th AC/326 SG/A EMT Meeting, 2018.
[31] D.G. Finney, Probit Analysis, third ed., Cambridge University Press, 1971. ISBN 0-521-080421-X.
[32] R.K. Wharton, D. Chapman, The relationship between BAM friction and rotary friction sensitiveness data for high explosives, Propellants Explos. Pyrotech. 22 (1997) 71–73.
[33] S. Nicolich, M. Mezger, D.A. Geiss, E. Heider, CL-20 Round-Robin Sensitivity Testing Interim Report, 1998. IEM symposium.
[34] S.M. Nicolich, M. Mezger, D.A. Geiss, E. Heider, CL-20 round-robin sensitivity testing interim report, in: Proceedings of the IMEMTS Conference, session 3 B2, 1998.
[35] K.F. Warner, M.M. Sandstrom, G.W. Brown, D.L. Remmers, J.J. Phillips, T.J. Shelley, J.A. Reyes, P.C. Hsu, J.G. Reynolds, ABL and BAM friction analysis comparison, Propellants Explos. Pyrotech. 40 (2015) 583–589.
[36] R.K. Wharton, J.A. Harding, An experimental comparison of three documented methods for the evaluation of friction sensitiveness, J. Energ. Mater. II (1993) 51–65.
[37] A. Bailey, G. Miles, A.W. Train, R.K. Wharton, M.R. Williams, The effect of temperature on friction sensitiveness of explosives—A preliminary study, in: Proceedings of the 24th International Pyrotechnic Seminar, Monterey, CA, USA, 1998, pp. 707–709.
[38] J. Harris, Friction Sensitivity of Primary Explosives. Technical Report ARLCD-TR 82-012, 1983. Published by the US Army Development Research and Armament Command, Large Caliber Weapon Systems Laboratory, Dover, NJ, USA.
[39] X. Song, Y. Wang, C. An, X. Guo, F. Li, Dependence of particle morphology and size on the mechanical sensitivity and thermal stability of octahydro-1,3,5,7-tetranitro-1,3,5,7-tetrazocine, J. Hazard. Mater. 159 (2–3) (2008) 222–229.
[40] D. Cohen, S. Mandelbaum, E. Dreerman, The influence of size and shape of the explosive particles on a cure-cast explosive properties, in: Proceedings of the 31st International Conference of ICT, Karlsruhe, Germany, 2000.

[41] Y. Bayat, V. Zeynali, Preparation and characterization of Nano-CL-20 explosive, J. Energ. Mater. 29 (2011) 281–291.

[42] F. Pessina, D. Spitzer, The longstanding challenge of the nanocrystallization of 1,3,5-trinitroperhydro-1,3,5-triazine (RDX), Belstein J. Nanotechnol. 8 (2017) 452–466. https://doi.org/10.3762/bjnano.8.49.

[43] B. Siegert, M. Comet, D. Spitzer, Safer energetic materials by a nanotechnological approach, Nanoscale 3 (9) (2011) 3534–3544.

[44] S. Bénazet, G. Jacob, G. Pèpe, GenMolTM supramolecular descriptors predicting reliable sensitivity of energetic compounds, Propellants Explos. Pyrotech. 34 (2009) 120–135.

[45] M.H. Keshavarz, M.H. Moghadas, M.K. Tehrani, Relationship between the electrostatic sensitivity of nitramines and their molecular structure, Propellants Explos. Pyrotech. 34 (2009) 136–141.

[46] M.H. Keshavarz, M. Hayati, S. Ghariban-Lavasani, N. Zohari, A new method for predicting the friction sensitivity of nitramines, Cent. Eur. J. Energetic Mater. 12 (2) (2015) 215–227.

[47] M.H. Keshavarz, H.R. Pouretedal, A. Semnani, A simple way to predict electric spark sensitivity of nitramines, Indian J. Eng. Mater. Sci. 15 (2008) 505–509.

[48] G. Wang, H. Xiao, X. Ju, X. Gong, Detonation velocities and pressures, and their relationships with electric spark sensitivities of nitramines, Propellants Explos. Pyrotech. 31 (2006a) 102–109.

[49] G. Wang, H. Xiao, X. Ju, X. Gong, Calculation of detonation velocity, pressure, and electric sensitivity of nitro arenes based on quantum chemistry, Propellants Explos. Pyrotech. 31 (2006b) 361–368.

[50] V. Zeman, J. Koči, S. Zeman, Electric spark sensitivity of polynitro compounds. Part II. A correlation with detonation velocity of some polynitro arenes, HanNeng CaiLiao 7 (1999a) 127–132.

[51] V. Zeman, J. Koči, S. Zeman, Electric spark sensitivity of polynitro compounds: part III. A correlation with detonation velocity of some nitramines, HanNeng CaiLiao 7 (1999b) 172–175.

[52] M.H. Keshavarz, The relationship between the electric spark sensitivity and detonation pressure, Indian J. Eng. Mater. Sci. 15 (2008) 281–286.

[53] M.H. Keshavarz, H.R. Pouretedal, A. Semnani, Reliable prediction of electric spark sensitivity of nitramines: a general correlation with detonation pressure, J. Hazard. Mater. 167 (2009) 461–466.

[54] Z. Friedl, M. Jungova, S. Zeman, A. Husarova, Friction sensitivity of nitramines. Part IV: links to surface electrostatic potentials, HanNeng CaiLiao 19 (6) (2011) 613–615.

[55] Z. Jian-Ling, Z. Chun-Yan, Z. Feng, F. Shi-Quan, C. Xin-Lu, Relationship between electric spark sensitivity of cyclic Nitramines and their molecular electronic properties, Chinese J. Struct. Chem. 31 (2012) 1263–1270.

[56] C. Zhi, X. Cheng, The correlation between electric spark sensitivity of polynitroaromatic compounds and their molecular electronic properties, Propellants Explos. Pyrotech. 35 (2010) 555–560.

[57] N. Zohari, M.H. Keshavarz, S.A. Seyedsadjadi, The advantages and shortcomings of using Nano-sized energetic materials, Cent. Eur. J. Energ. Mater. 10 (1) (2013) 135–147.

[58] M. Jungova, S. Zeman, A. Husarova, Fiction sensitivity of nitramines. Part II: comparison with thermal reactivity, HanNeng CaiLiao 6 (2011a) 607–609.

[59] M. Jungova, S. Zeman, A. Husarova, Friction sensitivity of nitramines. Part I: comparison with impact sensitivity and heat of fusion, HanNeng CaiLiao 6 (2011b) 603–606.

[60] M.H. Keshavarz, N. Zohari, S.A. Seyedsadjadi, Relationship between electric spark sensitivity and activation energy of the thermal decomposition of nitramines for safety measures in industrial processes, J. Loss Prev. Process Ind. 26 (2013) 1452–1456.

[61] C. Zhao-Xu, X. Heming, Impact sensitivity and activation energy of pyrolysis for tetrazole compounds, Int. J. Quantum Chem. 79 (2000) 350–357.

[62] N. Zohari, M.H. Keshavarz, S.A. Seyedsadjadi, A novel method for risk assessment of electrostatic sensitivity of nitroaromatics through their activation energies of thermal decomposition, J. Therm. Anal. Calorim. 115 (1) (2014) 93–100.

[63] N. Zohari, M.H. Keshavarz, S.A. Seyedsadjadi, A link between impact sensitivity of energetic compounds and their activation energies of thermal decomposition, J. Therm. Anal. Calorim. 117 (2014) 423–432. https://doi.org/10.1007/s10973-014-3643-4.

[64] NATO STANAG 4488, Shock Sensitivity, ed. 2, NATO Standard, 2009.

[65] E.L. Baker, V. Pouliquen, M. Voisin, M. Andrews, in: Gap Test and Critical Diameter Calculations and Correlations, 16th International Detonation Symposium, Cambridge, MD, USA, 2018.

[66] F. Peugeot, Newgates Database v1.5. NATO/MSIAC, 2005.

[67] I. Jaffe, R.L. Beauregard, A.B. Amster, The attenuation of shock in Lucite, ARS J. 32 (1962) 22–25.
[68] German norm AZ 2.2-9/5901/82, Sensitivity to Shock Waves: BICT Gap Test—Test Description and Procedure, 1982.
[69] A. Lefrancois, R.S. Lee, C.M. Tarver, Shock desensitization effect in the confined explosive component water gap test defined by the Nato standardization agreement (STANAG) 4363, Propellants Explos. Pyrotech. 32 (3) (2007) 244–250.
[70] NATO STANAG 4363, Initiation Systems: Testing for the Assessment of Detonating Explosive Components, 2016.
[71] NF T 70-502, Matériaux Energétiques de Défense, Sécurité, Vulnérabilité, Amorçage de la Détonation à travers une Barrière, AFNOR, 2017.
[72] P.C. Souers, P. Vitello, Initiation pressure thresholds from three sources, Propellants Explos. Pyrotech. 32 (4) (2007) 288–295.
[73] T.P. Liddiard, D. Price, The Expanded Large-scale Gap Test, NAVAL SURFACE WEAPONS CENTER (NSWC) NSWC TR 86–32, United States Silver Spring, MD, 1987, pp. 443–452.
[74] D. Price, The NOL Large-Scale Gap Test. Part III. Compilation of Unclassified Data and Supplementary Information for Interpretation of Results, 1974. NOL Report AD-780 429.
[75] S.M. Caulder, P.J. Miller, K.D. Gibson, J.M. Kelley, Effect of particle size, particle morphology, and crystal quality on the critical shock initiation pressures of cast RDX/HTPB compositions, in: 37th International Annual Conference of ICT, 2006.
[76] J. Forbes, J. Watt, H. Adolph, The Insensitive High Explosives Gap Test. NWSCTR 86–58, 1986.
[77] S.A. Aubert, G.H. Parsons, J.G. Glenn, J.L. Thoreen, Gap tests as method of discriminating shock sensitivity, in: 9th International Symposium on Detonation, 1989.
[78] D. Price, F. Zerilli, Critical energy and pressure for initiation—wedge and gap test, in: Notes From Lectures on Detonation Physics, 1981, pp. 141–156.
[79] W. Arnold, T.R. Krawietz, J.L. Jordan, G. Sunny, M. Koch, Damage investigations and sensitivity tests of shock wave loaded plastic bonded explosives, in: 44th International Annual Fraunhofer Conference of Institut Chemische Technology (ICT), Karlsruhe, Germany, 2013.
[80] MIL-STD 650 method 512, n.d. Approaching Electrode Sensitivity Test.
[81] NF T70-539, Sensibilité à l'Etincelle Electrique, 2009. Appareil GEMO.
[82] T. Matasuzawa, M. Hoh, S. Hatanaka, A. Miyahara, T. Masamitu, H. Osada, Electric spark sensitivity for materials and compositions of fireworks, Kayaku Gakkaishi 55 (1) (1994) 37–45.
[83] R.J. Fisher, A Severe Human ESD Model for Safety and High Reliability System Qualification Testing, EOS/ESD, New Orleans, 1989.
[84] M. Roux, M. Auzanneau, C. Brassy, Electric spark and ESD sensitivity of reactive solids (primary or secondary explosive, propellant, pyrotechnics) part one: experimental results and reflection factors for sensitivity test optimization, Propellants Explos. Pyrotech. 18 (1993) 317–324.
[85] D.E.G. Jones, P.D. Lightfoot, R.C. Fouchard, Q. Kwok, A.-M. Turcotte, W. Ridley, Hazard characterization of KDNBF using a variety of different techniques, Thermochim. Acta 384 (2002) 57–69.
[86] R. Millar, Appropriateness of small-scale Hazard testing for pyrotechnics, J. Hazardous Mater. (2020). in preparation.
[87] L. Van Ieperen, A. Bates, B. Harris, Impact, friction, electrostatic and heat sensitiveness testing of energetic materials, in: Parari Conference, 2013.
[88] J. Majzlík, Comparison of two methods of determination of electrostatic discharge sensitivity of energetic materials, in: Fraunhofer International Conference of Institut Chemische Technology, 2011.
[89] EN 13938-2, Explosives for Civil Uses–Propellants and Rocket Propellants–Part 2: Determination of Resistance to Electrostatic Energy, European Committee for Standardization, Brussels, 2004. 16 pp.
[90] V. Pelikán, J. Majzlík, R. Matyáš, Study of the interaction between electrostatic discharge and the granular bed of the ultra-sensitive energetic materials, in: International Pyrotechnics Seminar, vol. 44, 2014.
[91] B. Risse, Continuous Crystallization of Ultra-Fine Energetic Particles by the Flash Evaporation Process, Ph.D. Thesis, Université de Lorraine, France, 2012.
[92] NATO STANAG 4491, Explosives, Thermal Sensitiveness and Explosiveness Tests, NATO Standardization Office, 2015.
[93] NATO STANAG 4515, Explosives, Thermal Analysis Using Differential Thermal Analysis (DTA), Differential Scanning Calorimetry (DSC), Heat Flow Calorimetry (HFC), and ThermoGravimetric Analysis (TGA), NATO Standardization Office, 2015.

[94] NF T 70-514, Matériaux Energétiques de Défense, Sécurité Vulnérabilité, Echauffement Lent 5 °C/min, AFNOR, 2004.
[95] NF T 70-515, Matériaux Energétiques de Défense, Sécurité Vulnérabilité, Echauffement très Lent 3,3 °C/min, AFNOR, 2004.
[96] NF T70-526, Matériaux Energétiques de Défense, Sécurité Vulnérabilité, Température Critique de Thermo-Initiation (Cook-Off), AFNOR, 2009.
[97] P.C. Hsu, G. Hust, M.X. Zhang, T.K. Lorenz, J.G. Reynolds, L. Fried, H.K. Springer, J.L. Maienschein, Study of thermal sensitivity and thermal explosion violence of energetic materials in the LLNL ODTX system, Lawrence Livermore National Laboratory, in: 18th APS-SCCM and 24th AIRAPT, Journal of Physics: Conference Series 500, 2014.
[98] J.M. Densmore, E.M. Kahl, E.A. Glascoe, M.A. McClelland, M.R. De Haven, N. Tan, M.A. Suda, LX-17 cook-off experiments, in: 16th International Detonation Symposium, Cambridge, MD, United States, Lawrence Livermore National Laboratory, 2018.
[99] J.F. Wardell, J.L. Maienschein, in: The Scaled Thermal Explosion Experiment, Proceedings of the Twelfth International Symposium on Detonation, Office of Naval Research ONR333-05-2, San Diego, CA, 2002, pp. 384–393.
[100] V. Le Gallo, M. Vaullerin, A. Osmont, Controlled thermal test on explosive cylinder, in: 50th International Annual Conference of the Fraunhofer Institut Chemische Technology, 2019.

Further reading

W. Andersen, Measurement of Critical Diameter, Shock and Impact Sensitivity of a Special Propellant, 1981. AD-A100 726.
J.C. Foster Jr., K.R. Forbes, M.R. Gunger, B.G. Craig, in: An Eight-Inch Diameter, Heavily Confined Card Gap Test, 8th International Symposium on Detonation, Albuquerque, 1985.
I. Jaffe, R.L. Beauregard, A.B. Amster, The attenuation of shock in Lucite, Navord 6876 (1960).
NF T70-368, Matériaux Energétiques de Défense, Sécurité Vulnérabilité, Détermination des Caractéristiques Physiques par DSC, 2012. French norm published by the Association Française de NORmalisation (AFNOR).
NF T 70-505, Matériaux Energétiques de Défense, Sécurité Vulnérabilité, Épreuve de Stabilité Thermique à 75°C, AFNOR, 2007.
D. Price, T.P. Liddiard, The Small-Scale Gap Test: Calibration and Comparison with the Large-Scale Gap Test, 1966. NOLTR 66–87 (Andersen, W., 1981).

PART II

Relationships with molecular structure

CHAPTER

3

Relationships with oxygen balance and bond dissociation energies

Betsy M. Rice and Edward F.C. Byrd*

U.S. Army Combat Capabilities Development Command (DEVCOM) Army Research Laboratory, Aberdeen Proving Ground, MD, United States

*Corresponding author: E-mail: betsy.rice.civ@army.mil

Research into explosive sensitivity has been conducted for decades, with fundamental questions remaining to this day. For a newcomer in any field, and for researchers looking for historical perspectives, one is drawn to overview treatises to aid in answering these questions. For example, in an overview provided by Sundberg [1], he addresses basic questions regarding chemical explosives and propellants such as "What makes an explosive explosive?" For questions such as this, there are accepted rules of thumb to use as guidelines, but often such overviews report hypotheses lacking sufficient validation. A novice to the field could potentially be inspired to conduct further studies that utilize the historical concepts as if complete validation had been achieved. Within the area of attempting to explain explosive sensitivity, one of the earliest examples of such an "accepted rule of thumb" is the trigger bond (or trigger linkage) concept. The origin of this concept came from MJ Kamlet, a prominent US Department of Defense chemist specializing in energetic material synthesis and characterization.

To understand the route Kamlet took to develop this concept, we start with his seminal work with Jacobs, in which a remarkably simple and useful method was given to predict the detonation properties of CHNO explosives [2]. Kamlet and Jacobs stated "To a chemist concerned with the synthesis of new high-explosive compounds the ability to compute detonation properties (detonation pressure, energy, and velocity as well as product composition) from a given molecular structure and the known or estimated crystal density is a problem of the utmost importance. The calculated properties could be meaningful in the decision as to whether it is worth the effort to attempt a new and complex synthesis." [2]. One can infer Kamlet used this same reasoning as the basis for his efforts to develop simple computational methods to assess the sensitivity of high explosives since a high-performing material would not be used if it cannot be safely handled.

Theoretical and Computational Chemistry, Volume 22
https://doi.org/10.1016/B978-0-12-822971-2.00005-X

In one of his first reports detailing the identification of factors that influence explosive sensitivity, Kamlet described his approach as one that explores "the reactions governing sensitivity in much the same manner as the physical organic chemist usually studies any reaction, i.e., by determining the effect of structure on reactivity." [3]. This paper describes the pros and cons of one of the most common tests used to assess explosive sensitivity, the drop hammer test. This test "involves subjecting samples of the explosives to the force of a series of hammer blows caused by a standard weight falling from varying heights. The sensitivity is reported as that height which has a 50% probability of causing an explosion", and results are denoted as $h_{50\%}$ values. As his organization had a large database of impact test data, Kamlet focused on identifying correlations of impact sensitivity with molecular structural parameters. These initial attempts were to simply develop a tool that reflected sensitivity-structure *trends*, as individual drop hammer results are known to be unreliable. To emphasize the variability in impact data, a separate group [4] stated that "while too much should not be made of the precise impact sensitivity of an individual compound, if a sufficiently large number of structurally related compounds are tested, these individual oddities and vagaries might tend to offset each other, and meaningful trends might evolve." This key point should be kept in mind whenever trying to develop a correlation based on noisy data, e.g., trying to predict specific numbers for impact tests when the experimental values themselves are suspect.

The first molecular structural parameter Kamlet explored was "oxidant" (oxygen) balance (OB_{100}), which he defined as "the number of equivalents of oxidant per hundred grams of explosive about[a] the amount required to burn all hydrogen to water and all carbon to carbon monoxide." The equation for OB_{100} is:

$$OB_{100} = \frac{100(2\,n_O - n_H - 2\,n_C - 2\,n_{COO})}{\text{Molecular Weight}}$$

where n_O, n_H,and n_C are the numbers of oxygen, hydrogen, carbon atoms in the explosive molecule, n_{COO} are the numbers of carboxyl groups, and the molecular weight of the compound is given in g/mol. In this study, Kamlet first plotted logarithms of $h_{50\%}$ values versus the corresponding OB_{100} for 78 polynitroaliphatic and alicyclic explosives, which showed a very rough trend of decreasing $h_{50\%}$ values with increasing OB_{100}. These 78 compounds were analyzed according to common structural properties, with 44 molecules having N-nitro linkages, and 34 having C-nitro linkages. Plots of logarithms of $h_{50\%}$ values versus the corresponding OB_{100} for the two groups that differed in nitro linkages showed distinct linear correlations (correlation coefficients are 0.95 or greater), with the groups having the N-nitro linkages being more impact sensitive.

It is at this point Kamlet posited hypotheses that assign a physical basis and mechanistic detail to the correlations in attempting to explain the differences in sensitivities between the two groups of molecules that would result from phenomena occurring during the impact event. He first raised an important point that is often misunderstood by newcomers to the field: that the "impact explosion is not a stable detonation" but instead, "a phenomenon resembling more closely a relatively low-temperature thermal decomposition than a detonation."

[a]The word "about" is a typographical error corrected in [5], and should be "above".

With this concept as a basis, he postulated that the difference in sensitivities between the two classes of compounds (C-nitro versus N-nitro groups) could be due to a variety of chemical and physical properties and behaviors, including reaction kinetics, heat release, heat capacity, thermal conductivity, latent heats of fusion and evaporation, crystal hardness, crystal shape and, in the case of liquids, surface tension, vapor pressure, and dissolved gases. Of these myriad possible explanations, he focused upon reaction kinetics and the concept of the "trigger linkage," which will be discussed in more detail shortly.

Soon thereafter, Kamlet and Adolph published a similar study in which structure-impact sensitivity relationships were explored, but which focused on polynitroaromatic explosives [5]. As in the earlier study for polynitroaliphatic explosives [3], a plot of logarithms of $h_{50\%}$ values versus corresponding OB_{100} for 38 compounds showed an overall trend of increased $h_{50\%}$ values with decreasing oxidant balance; however, there was wide scatter in the data. Upon apportioning the compounds into two classes that differed only in the presence or absence of an alpha C—H linkage, clear trends emerged (with linear-least squares fits having correlation coefficients of 0.96 or higher) with compounds containing alpha C—H linkages being more sensitive. A notable caution is given by the authors regarding three compounds that show anomalous behavior (and which were not included in the linear fits) which they attributed to "large secondary effects." While not including these compounds in the fits, Kamlet and Adolph saw these anomalies as an opportunity to gain a better understanding of factors controlling reactivity and underlying physics and chemistry through the isolation and study of "such exceptional behavior." In this work, Kamlet and Adolph also addressed how to treat compounds that contain structural features common to both sensitivity categories (i.e., molecules that contain both polynitroaromatic and polynitroaliphatic moieties) and found that in such cases, such compounds follow sensitivity trends of the more sensitive class [5]; a similar result was observed in the earlier study on polynitroaliphatic and alicyclic explosives [3].

Encouraged by Kamlet's success in establishing reasonable correlations of impact sensitivity and OB_{100}, other groups reported studies focused on structure/property relationships. Adolph et al. [6] explored whether relationships beyond oxygen balance existed between sensitivity and/or performance of explosive compounds. In this work, the authors attempted to find correlations between impact sensitivity (specifically logarithms of $h_{50\%}$ values) and molecular properties related to explosive performance. These properties include such descriptors as molecular weight (M), number of moles (N) of detonation product gases per 100 g of explosive as calculated by the CO_2 arbitrary [2], the quantity $NM^{1/2}$ [taken from Eq. (8) of Ref. [2]], the crystal density (ρ_0) of the compound, and $NO_2^{\#}$, defined as the molar number of nitro groups per 100 g of explosive (used here as an approximate measure of oxygen balance). In this study, the assessment involved 76 nitramine, 59 nitroaromatic, and 64 nitroaliphatic compounds. Single-linear fits relating individual properties to logarithms of $h_{50\%}$ values did not yield strong correlations, nor did 2-parameter fits using $NM^{1/2}$ and $NO_2^{\#}$. Instead, broad trends for relationships involving $NO_2^{\#}$ were noted, and a suggestion was given for future explorations of possible relationships between detonation energy and impact sensitivity.

Nearly a decade later, Storm et al. [7] reported correlations of impact sensitivity results with an empirical Sensitivity Index (SI) for multiple classes of explosives. The SI is based on ideas presented by Stine [8] in which any CHNO molecule can be topologically represented as a unique point in a tetrahedron whose corners represent carbon, hydrogen, nitrogen, and oxygen, and can be taken as an alternate definition of oxygen balance. This report compiled impact sensitivity results and calculated SI values for 258 explosives, conveniently categorized and published into the following structural types: nitroaromatics, nitroaromatics with an alpha C—H linkage, nitropyridines, nitroimidazoles, nitropyrazoles, nitrofurazans and nitrooxadiazoles, nitro-1,2,4-triazoles, nitro-1,2,3-triazoles, nitropyrimidines, nitroaliphatics, nitroaliphatics containing other functional groups, nitramines, nitrate esters, and miscellaneous nitroheterocyclic compounds. Storm et al. found that a correlation between SI and impact sensitivity (logarithms of $h_{50\%}$ values) showed a similar quality of fit as that of Kamlet and Adolph [5] for the 25 nitroaromatic compounds with no alpha C—H linkage listed in Table 1 of Kamlet and Adolph [5]. However, when plotting the SI against the impact sensitivity results for the full set of nitroaromatic compounds reported in Table 2 of Storm et al. (40 molecules), the regression was much worse. Upon partitioning this group into families (polynitroanilines, polynitrobenzenes, polyaminopolynitrobenzenes and polynitrophenols), outstanding correlations were obtained. This led Storm et al. to conclude that the scatter in the overall result was due to the series of families "crossing straight lines with rather different slopes and intercepts" [7]. They suggested that when estimating the impact sensitivity of a compound, one should choose a series and associated correlation for which the compound is as closely related as possible.

Following Kamlet and Adolph's approach [5], Wilson et al. [4] found correlations between drop hammer test results and oxygen balance for 39 polynitroaromatic explosives. The main distinction between the Wilson et al. [4] and Kamlet and Adolph [5] studies is the degree of nitration of the materials. In the Kamlet and Adolph study [5], all but two nitroaromatics contained no more than three nitro groups on one ring, whereas, in the Wilson et al. study [4], approximately half contained four or more nitro groups on the same ring. The compounds in the set used in the Wilson et al. study [4] allowed for comparison of sensitivities of isomeric species, such as tetranitrotoluidines, tetranitrotoluenes, and trinitrotoluenes, as well as for demonstrating the dramatic influence of number of amino groups in the polynitroaniline series on sensitivity. Wilson et al. also showed that the presence of an alpha C—H bond leads to increased sensitivity in these materials in agreement with Kamlet and Adolph [5], but only if compounds with the same oxygen balance are considered. Wilson et al. [4] noted that strong correlations exist for systems containing only nitro, amino, and furoxan functional groups, but if other functional groups, including azido, diazonium, methyl, and methlynitramino, are present, the correlation is not as robust. They concluded that the relation between drop hammer results and oxygen balance is "coincidental rather than causal" and that a more likely indicator of sensitivity is the energy of the weakest bond (i.e., "trigger linkage").

Kamlet's "trigger linkage" hypotheses presented in his study on sensitivity relationships in nitroaliphatic explosives [3] evolved from his initial thoughts regarding the relationship between structural "linkages" and sensitivities [9] toward a more refined version based on subsequent kinetics measurements challenging one of his main assumptions, chiefly that classes of compounds follow similar rate laws. Kamlet initially assumed that chemical reaction kinetics "play an important role under the impact hammer" and that aliphatic explosives having either N-nitro or C-nitro linkages followed similar rate laws for thermal decomposition. Following this, he initially hypothesized that the difference in sensitivities between the two families is simply due to differences in activation energies for thermolytic cleavage of the weakest bond, i.e., the "trigger" linkage. Because the N-nitro bonds are typically intrinsically weaker than C-nitro bonds, N-nitro bonds, therefore, have lower activation energies for thermolytic cleavage, resulting in faster reaction rates than for C-nitro compounds. After Kamlet's first study [9] was published, however, kinetics measurements for the two classes of compounds were reported which demonstrate that the two classes of compounds do not follow the same rate laws, contradicting one of Kamlet's key assumptions. Instead, the results suggested that the N-nitro compounds show strong autocatalytic effects, leading Kamlet to present a modified version of the "trigger linkage" hypothesis [3]. This version purports that the N-nitro bonds are "trigger linkages" because they provide alternative autocatalytic routes leading to "manifold rate accelerations" rather than because they are intrinsically weaker than C-nitro bonds. Kamlet and Adolph [5] also utilized the "trigger bond" hypothesis to explain differences in sensitivities between polynitroaromatic explosives based on the presence or absence of an alpha C—H linkage. They argued that for those compounds containing an alpha C—H bond, the preferred "site of inter- and intramolecular oxidative attack" is said bond. They also stated that the compounds in their study lacking the alpha C—H linkage are of "far more variegated structure, however, and from considerations of classical chemical reactivity, a number of widely differing decomposition mechanisms would be expected to apply." Kamlet and Adolph [10,11], in analyzing structure/sensitivity relations of a set of explosives containing trinitromethyl, fluorodinitromethyl, and gem-dinitroethyl groups, described factors that control the rate of the trigger linkage homolysis. These included the distribution of energy between rotations around and vibrations within the trigger linkages. They argue that hindered rotations of nitro groups about C-NO_2 bonds force energy to be deposited into the trigger linkages resulting in bond breaking, whereas freely rotating nitro groups "serve as an enthalpy sink" [11], by redirecting energy away from the vibrational mode that leads to breaking.

Storm et al. [7] examined the "trigger linkage" hypothesis in terms of a "single early dominant step". They argued that while such a step might be readily identified for sensitive explosives, identification might be difficult in insensitive explosives due to complex chemistry that precedes the rate-limiting step, using the insensitive explosive triaminotrinitrobenzene (TATB) as an example to argue their point. Wilson et al. [4], on the other hand, concluded that "impact sensitivity is more likely to be dependent on the energy of the weakest bond (the "trigger linkage"), including for insensitive polynitroaromatic explosives like TATB. Wilson et al. argued their point using the sensitivity results for the trinitrobenzene, 2,4,6-trinitroaniline, diaminotrinitrobenzene, TATB series, which shows that increasing the number of amine groups into a family of explosives results in decreased sensitivity. They argued that the C-NO_2 "trigger linkage" bond is strengthened and stabilized by the addition of the electron-donating amino groups. While proposing probable trigger linkages in the variety of polynitroaromatic species in the report, they stated that "identification of the weakest bond and estimation of its strength is not yet amenable to simple calculation."

In the modern era, the bond dissociation energy (BDE) is a molecular-based bond-strength metric that can be readily calculated using accurate quantum mechanical approaches. The direct calculation of the BDE is straightforward, which requires simply calculating the energy difference between a parent molecule and the products of its homolysis, as represented in the following expression:

$$\mathrm{BDE}(A-B) = [E(A) + E(B)] - E(A-B)$$

where E(A–B), E(A), and E(B) are the total energies (zero-point-energy corrected) of the A–B molecule and molecular fragments A and B. If one accepts Wilson et al.'s conclusion [4] regarding the trigger linkage (i.e., homolysis of the weakest bond), identifying it requires simply calculating the bond dissociation energy [12] of each bond in a molecule, and identifying which has the smallest value.

While the equation used to calculate the BDE is simple, its accuracy is dependent on the level of theory used (see Ref. [13] for a description of various quantum mechanical (QM) approaches used in EM research). The state of computational platforms and algorithms available at the time of the Wilson et al. study [4] severely limited the molecular sizes, number of systems, and level of theory that could be used to explore structure/sensitivity relationships. Many of the early molecular electronic structure studies [14–27] relied on low-levels of theory, such as the Hartree-Fock method [28,29] (which neglects electronic correlation) and minimal basis sets [30] or even more approximate semi-empirical methods. While these studies pushed the limits of computation at that time, the levels of theory were inadequate to accurately predict bond dissociation energies and electronic structure. That changed in the early 1990s with the confluence of powerful, inexpensive personal computers, the demonstrated accuracy of the computationally efficient density function theory (DFT) [31], and the implementation of DFT into easy-to-use commercial quantum chemistry software (Jaguar [32], DMol [33], Gaussian92/DFT [34]). This resulted in exponential growth of DFT-related publications during the 1990s, and a current doubling of DFT publications every 5–6 years [35]. The EM community readily embraced these computational advances, making DFT "the most often applied QM method in EM research" [13]. Its modest computational costs have led to numerous studies exploring molecular structure-sensitivity relationships [see [36–39] and references therein], and include those that predict the BDE as a measure of thermal stability [40–83] or for identifying trends [84–97]. Calculations of BDEs as key characteristics of explosive sensitivity have continued to proliferate in EM research, despite the early conclusions by Kamlet and others that factors other than bond strengths alone influence sensitivity.

We note that the computationally-inexpensive B3LYP functional [98] continues to enjoy widespread use (including in many of the cited studies above) despite it being one of the older functionals. It was one of the first exchange-correlation functionals that showed significant improvement over Hartree Fock, and was found to be "a good compromise between computational cost, coverage, and accuracy of results. It has become a standard method used to study organic chemistry in the gas phase." [99]. Unfortunately, it has since been shown to be inaccurate in its prediction of BDE, with more suitable functionals available for this purpose [100–102]. We are aware of a few studies that report C-NO_2 BDEs calculated using the B3LYP functional that were in worse agreement with experimental values [103–105] in comparison with other functionals, a finding consistent with a larger study involving a more diverse set of chemical bonds [106]. A more recent study [107] compares BDEs for a few energetic materials (EMs) calculated using B3LYP and the newer M06-2X functional [108], which was shown to be the best performing functional among many for predicting BDEs for reactions that produce radicals [101]. Moxnes et al. [107] found that BDEs for the N-NO_2 group calculated using the B3LYP functional were "between 40 kJ mol^{-1} and 135 kJ mol^{-1} lower" than that calculated using the more accurate M06-2X functional or the non-DFT CBS-4M method [109]. An example of how such inaccuracy in a functional manifests itself in applications is illustrated in a theoretical characterization of decomposition pathways for hexahydro-1,3,5-trinitro-s-triazine (RDX) [110]. Pathways for N—N homolysis and HONO elimination were calculated [110] using the "gold standard" of quantum mechanical methods [13], coupled-cluster singles, doubles, and perturbative triples [CCSD(T)] [111,112] with complete basis set (CBS) extrapolation. The CCSD(T) calculations were compared with results calculated using the B3LYP density functional [113,114]. The B3LYP results show that the two mechanisms had comparable reaction barriers, with the N—N homolysis being slightly lower in energy. The CCSD(T)/CBS predictions for HONO elimination were in reasonable agreement with the B3LYP predictions. However, the CCSD(T)/CBS prediction of N—N homolysis was larger by at least 10 kcal/mol, leading the authors to conclude that N—N homolysis is not the initiating reaction in a thermal decomposition event in RDX [110]. The reader is cautioned that "no practical present-day density functional is universally accurate" [115], which is the reason for the vigorous activity in the development of new functionals. This has resulted in a "zoo" [116] of highly-specialized functionals used for prediction of specific properties. Determining the most suitable functional from such a large and diverse set of density functionals is daunting and care must be taken to select the most accurate functional for predicting a property of interest. The aforementioned examples serve as a warning to the user that inadequate choices of theoretical treatment could produce misleading predictions, and due consideration of method should be given for any theoretical study.

As new molecular property-sensitivity relations are pursued, we point to cautionary words regarding correlations from Brill [117], who committed much of his experimental career toward understanding the chemical details involved in the thermal decomposition of EMs, a subject for which he wrote prolifically. He extensively studied the thermal decomposition of a variety of different EMs using T-Jump/FT-IR spectroscopy [118], a method he developed to simulate the conditions an explosive would experience under combustion and drop hammer tests. In expositions on relating molecular properties of explosives with macroscale energy release events (decomposition, combustion, or explosion) [117,119], Brill provides a newcomer to the field with concise synopses of the complexities of the energy release events of explosives, which involve numerous chemical and physical factors occurring under extreme conditions. Within these reviews, he also urged care in establishing these relationships, using examples to highlight pitfalls to avoid and factors to consider in relating molecular properties to macroscale events. One notable example is illustrated from his study of the series composed of the series composed of trinitrobenzene, 2,4,6-trinitroaniline, diaminotrinitrobenzene, and TATB [120], for which he stated that there are 153 positive correlations among molecular, electronic, crystal, and explosive data. He stated "With 153 positive correlations among the data, single correlations are unlikely to provide the true controlling factors. Perhaps all of these factors (or none of them) contribute to the sensitivity to some degree" [119]. He also emphasized that the sensitivity of a material only partially depends on its molecular and electronic structure, noting that sensitivity is also dependent on properties specific to the condensed phase, both in terms of reaction mechanisms and physical properties such as void volume and density. He stated "while gas-phase studies of individual molecules can be a useful guide to the behavior in other phases, the most important reactions arise when the molecules are in the condensed phase in the presence of like molecules." [119]. To expand upon this thought, we emphasize that single-molecule calculations cannot encompass environmental effects and that both reaction mechanisms and molecular properties might differ from those in the condensed phase. Brill tempered his remarks by concluding that his "rather pessimistic assessment of molecular correlations with macro should be taken as a cautionary note as opposed to a call to avoid these types of studies altogether" [117], and acknowledged the importance of developing capabilities for sensitivity predictions for the design of new EMs. We are less pessimistic than Brill, believing there is value in continuing to pursue molecular structure-sensitivity relationships, including using data-based methods for extensive interrogation of electronic structure of a large series of molecules that might yield new molecular structure-sensitivity correlations. Our optimism is based on Tufte's assessment of the popular phrase "Correlation is not causation" as being incomplete. Tufte argued that the more appropriate statement is that "Correlation is not causation but it sure is a hint." [121]. Molecular structure-sensitivity correlations help frame appropriate questions for further explorations, evidenced by numerous physically-based studies designed to understand the factors affecting explosive sensitivity, many of which are described in later chapters in this book. Taken together, we believe these will lead to understanding factors that control sensitivity in EMs, and ultimately result in predictive capabilities for use in the design of new EMs.

References

[1] R.J. Sundberg, The Chemical Century, Apple Academic Press, New York, 2017.
[2] M.J. Kamlet, S.J. Jacobs, Chemistry of detonations I. a simple method for calculating detonation properties of C-H-N-O explosives, J. Chem. Phys. 48 (1968) 23–35.
[3] M.J. Kamlet, The relationship of impact sensitivity with structure of organic high explosives. I. Polynitroaliphatic explosives, in: Proceedings 6th Symposium (International) on Detonation, San Diego, California, Aug. 24–27, 1976, ONR Report ACR 221, 1976. p. 312.
[4] W.S. Wilson, D.E. Bliss, D.J. Knight, Explosive Properties of Polynitroaromatics, NWC TP 7073, China Lake, CA, 1990.
[5] M.J. Kamlet, H.G. Adolph, The relationship of impact sensitivity with structure of organic high explosives. II. Polynitroaromatic explosives, Propellants, Explos., Pyrotech. 4 (1979) 30–34.
[6] H.G. Adolph, J.R. Holden, D.A. Chicra, Relationships between the impact sensitivity of high energy compounds and some molecular properties which determine their performance, in: NSWC TR 80-495, White Oak, MD, 1981.
[7] C.B. Storm, J.R. Stine, J.F. Kramer, Sensitivity relationships in energetic materials, in: S.N. Bulusu (Ed.), Chemistry and Physics of Energetic Materials, Springer Netherlands, Dordrecht, 1990.
[8] J.R. Stine, On predicting properties of explosives – detonation velocity, J. Energ. Mater. 8 (1990) 41–73.
[9] M.J. Kamlet, Sensitivity relationships, in: Proceedings 3rd Symposium (International) on Detonation, Princeton, New Jersey, ONR Report ARC-52, Vol. III, 1960, p. 671.
[10] M.J. Kamlet, H.G. Adolph, Some comments regarding the sensitivities, thermal stabilities, and explosive performance characteristics of Fluorodinitromethyl compounds, in: Proceedings 7th Symposium (International) on Detonation, Annapolis, MD, June 16–19, 1981, NSWC MP 82–334, 1981. p. 84.
[11] K.F. Mueller, R.H. Renner, W.H. Gilligan, H.G. Adolph, M.J. Kamlet, Thermal stability/structure relations of some polynitroaliphatic explosives, Combust. Flame 50 (1983) 341–349.
[12] IUPAC, Compendium of Chemical Terminology, 2nd Ed. (the "Gold Book"), 1997th ed., Blackwell Scientific Publications, Oxford, 1997.
[13] D.E. Taylor, B.M. Rice, Chapter five - Quantum-Informed Multiscale M&S for energetic materials, in: J.R. Sabin (Ed.), Advances in Quantum Chemistry, Vol. 69, Academic Press, 2014.
[14] A. Delpuech, J. Cherville, Relation between shock sensitiveness of secondary explosives and their molecular electronic-structure. 1. Nitroaromatics and Nitramines, Propellants Explos. 3 (1978) 169–175.
[15] A. Delpuech, J. Cherville, Relation between shock sensitiveness of secondary explosives and their molecular electronic-structure. 2. Nitrate esters, Propellants Explos. 4 (1979) 121–128.

[16] P.C. Hariharan, J.J. Kaufman, A.H. Lowrey, R.S. Miller, Ab initio MODPOT/VRDDO/MERGE calculations on energetic compounds. IV. Nitrocubanes: mononitro to octanitro quantum chemical calculations and electrostatic molecular potential contour maps, Int. J. Quantum Chem. 28 (1985) 39–59.
[17] P.C. Hariharan, W.S. Koski, J.J. Kaufman, R.S. Miller, Ab initio MODPOT/VRDDO/MERGE calculations on energetic compounds. III. Nitroexplosives: polyaminopolynitrobenzenes (including DATB, TATB, and Tetryl), Int. J. Quantum Chem 23 (1983) 1493–1504.
[18] P.C. Hariharan, W.S. Koski, J.J. Kaufman, R.S. Miller, A.H. Lowrey, Ab initio MODPOT/VRDDO/MERGE calculations on energetic compounds. II. Nitroexplosives: RDX and α-, β- and δ-HMX, Int. J. Quantum Chem., Quantum Chem. Symp. 16 (1982) 363–375.
[19] J.S. Murray, P. Politzer, Computational studies of energetic nitramines, in: S.N. Bulusu (Ed.), Chemistry and Physics of Energetic Materials, Springer Netherlands, Dordrecht, 1990.
[20] J.S. Murray, P. Politzer, Structure-sensitivity relationships in energetic compounds, in: S.N. Bulusu (Ed.), Chemistry and Physics of Energetic Materials, Springer Netherlands, Dordrecht, 1990.
[21] F.J. Owens, Relationship between impact induced reactivity of trinitroaromatic molecules and their molecular-structure, J. Mol. Struct. (THEOCHEM) 22 (1985) 213–220.
[22] F.J. Owens, K. Jayasuriya, L. Abrahmsen, P. Politzer, Computational analysis of some properties associated with the nitro-groups in polynitroaromatic molecules, Chem. Phys. Lett. 116 (1985) 434–438.
[23] P. Politzer, L. Abrahmsen, P. Sjoberg, Effects of amino and nitro substituents upon the electrostatic potential of an aromatic ring, J. Am. Chem. Soc. 106 (1984) 855–860.
[24] P. Politzer, R. Bar-Adon, Electrostatic potentials and relative bond strengths of some nitro- and nitrosoacetylene derivatives, J. Am. Chem. Soc. 109 (1987) 3529–3534.
[25] D. Sheng, H. Huang, The quantum chemical study of tetrazene, in: Proc. Int. Pyrotech. Semin., 11th, 1986, pp. 523–536.
[26] H. Xiao, B. Feng, Y. Sun, Quantum chemistry of the sensitivity and stability of aromatic nitro explosives. III. Two-atom interaction energy and delocalization energy, Baozha Yu Chongji 6 (1986) 253–257.
[27] H. Xiao, Z. Wang, J. Yao, Quantum chemical study on the sensitivity and stability of aromatic nitro explosives. I. Nitro derivatives of aminobenzenes, Huaxue Xuebao 43 (1985) 14–18.
[28] V. Fock, Näherungsmethode zur Lösung des quantenmechanischen Mehrkörperproblems, Zeitschrift für Physik 61 (1930) 126–148.
[29] D.R. Hartree, The wave mechanics of an atom with a non-coulomb central field. Part I. theory and methods, Math. Proc. Camb. Philos. Soc. 24 (1928) 89–110.
[30] W.J. Hehre, R.F. Stewart, J.A. Pople, Self-consistent molecular-orbital methods. I. Use of Gaussian expansions of slater-type atomic orbitals, J. Chem. Phys. 51 (1969) 2657–2664.
[31] R.G. Parr, W. Yang, Density-Functional Theory of Atoms and Molecules, Oxford University Press, New York, 1989.
[32] A.D. Bochevarov, E. Harder, T.F. Hughes, J.R. Greenwood, D.A. Braden, D.M. Philipp, D. Rinaldo, M.D. Halls, J. Zhang, R.A. Friesner, Jaguar: a high-performance quantum chemistry software program with strengths in life and materials sciences, Int. J. Quantum Chem. 113 (2013) 2110–2142.
[33] B. Delley, An all-electron numerical method for solving the local density functional for polyatomic molecules, J. Chem. Phys. 92 (1990) 508–517.
[34] M.J. Frisch, G.W. Trucks, H.B. Schlegel, P.M.W. Gill, B.G. Johnson, M.W. Wong, J.B. Foresman, M.A. Robb, M. Head-Gordon, E.S. Replogle, R. Gomperts, J.L. Andres, K. Raghavachari, J.S. Binkley, C. Gonzalez, R.L. Martin, D.J. Fox, D.J. Defrees, J. Baker, J.J.P. Stewart, J.A. Pople, Gaussian 92/DFT, Gaussian, Inc., Pittsburgh, PA, 1993.
[35] R. Haunschild, A. Barth, W. Marx, Evolution of DFT studies in view of a scientometric perspective, J. Cheminformatics 8 (2016) 52.
[36] G. Li, C. Zhang, Review of the molecular and crystal correlations on sensitivities of energetic materials, J. Hazard. Mater. 398 (2020) 122910.
[37] P. Politzer, H.E. Alper, Detonation initiation and sensitivity in energetic compounds: some computational treatments, in: Computational Chemistry: Reviews of Current Trends, Vol. 4, World Scientific, 1999.
[38] Q.-L. Yan, S. Zeman, Theoretical evaluation of sensitivity and thermal stability for high explosives based on quantum chemistry methods: a brief review, Int. J. Quantum Chem. 113 (2013) 1049–1061.
[39] S. Zeman, M. Jungová, Sensitivity and performance of energetic materials, Propellants, Explos., Pyrotech. 41 (2016) 426–451.

[40] J. Bai, W.-J. Chi, L.-L. Li, T. Yan, X.-E. Wen, B.-T. Li, H.-S. Wu, F.-L. Ma, Quantum chemical study of Aminonitrocyclopentanes as possible high energy density materials (HEDMs), Cent. Eur. J. Energetic Mater. 10 (2013) 467–480.
[41] F. Bao, S. Jin, Y. Li, Y. Zhang, K. Chen, L. Li, Design and properties of N,N′-linked bis-1,2,4-triazoles compounds as promising energetic materials, J. Mol. Model. 26 (2020) 130.
[42] Q. Cao, Dinitroamino benzene derivatives: a class new potential high energy density compounds, J. Mol. Model. 19 (2013) 2205–2210.
[43] W.-J. Chi, L.-L. Li, B.-T. Li, H.-S. Wu, Density functional calculations for a high energy density compound of formula $C_6H_{6-n}(NO_2)_n$, J. Mol. Model. 18 (2012) 3695–3704.
[44] W.-J. Chi, L.-L. Li, B.-T. Li, H.-S. Wu, Looking for high energy density compounds among polynitraminecubanes, J. Mol. Model. 19 (2013) 571–580.
[45] W.-J. Chi, Z.-S. Li, Molecular design of prismane-based potential energetic materials with high detonation performance and low impact sensitivity, C. R. Chim. 18 (2015) 1270–1276.
[46] A. Devi, S. Deswal, S. Dharavath, V.D. Ghule, Molecular design and screening of energetic nitramine derivatives, J. Mol. Model. 21 (2015) 298.
[47] M. Du, Computational study of the structure and properties of bicyclo[3.1.1]heptane derivatives for new high-energy density compounds with low impact sensitivity, Journal of Molecular Modeling 24 (2018) 17.
[48] M. Du, T. Han, F. Liu, H. Wu, Theoretical investigation of the structure, detonation properties, and stability of bicyclo[3.2.1]octane derivatives, Journal of Molecular Modeling 25 (2019) 253.
[49] M. Du, X. Wang, Z. Guo, Theoretical design of bicyclo[2.2.1]heptane derivatives for high-energy density compounds with low impact sensitivity, Comput. Theor. Chem. 1095 (2016) 54–64.
[50] T. Fei, Y. Du, C. He, S. Pang, Theoretical investigations on azole-fused tricyclic 1,2,3,4-tetrazine-2-oxides, RSC Adv. 8 (2018) 27235–27245.
[51] V.D. Ghule, Computational studies on the triazole-based high energy materials, Comput. Theor. Chem. 992 (2012) 92–96.
[52] V.D. Ghule, D. Srinivas, S. Radhakrishnan, P.M. Jadhav, S.P. Tewari, Computational study on energetic properties of nitro derivatives of furan substituted azoles, Struct. Chem. 23 (2012) 749–754.
[53] V.D. Ghule, D. Srinivas, R. Sarangapani, P.M. Jadhav, S.P. Tewari, Molecular design of aminopolynitroazole-based high-energy materials, J. Mol. Model. 18 (2012) 3013–3020.
[54] Y.-Y. Guo, W.-J. Chi, Z.-S. Li, Q.-S. Li, Molecular design of N–NO_2 substituted cycloalkanes derivatives $C_m(NO_2)_m$ for energetic materials with high detonation performance and low impact sensitivity, RSC Adv. 5 (2015) 38048–38055.
[55] Y. Jiao, Z. Liu, W. Zhu, Searching for a new family of modified CL-20 cage derivatives with high energy and low sensitivity, Struct. Chem. 29 (2018) 837–845.
[56] M. Jing, H. Li, J. Wang, Y. Shu, X. Zhang, Q. Ma, Y. Huang, Theoretical investigation on the structure and performance of N, N′-azobis-polynitrodiazoles, J. Mol. Model. 20 (2014) 2155.
[57] X.-H. Li, Z.-M. Fu, X.-Z. Zhang, Computational DFT studies on a series of toluene derivatives as potential high energy density compounds, Struct. Chem. 23 (2012) 515–524.
[58] X.-H. Li, R.-Z. Zhang, X.-Z. Zhang, Theoretical studies on a series of 1,2,3-triazoles derivatives as potential high energy density compounds, Struct. Chem. 22 (2011) 577–587.
[59] H. Liu, H. Du, G. Wang, X. Gong, Theoretical studies on the structures, densities, detonation properties and pyrolysis mechanism of energetic compounds containing pyridine ring, Struct. Chem. 23 (2012) 479–486.
[60] H. Liu, F. Wang, G.-X. Wang, X.-D. Gong, Theoretical studies on 2-(5-amino-3-nitro-1,2,4-triazolyl)-3,5-dinitropyridine (PRAN) and its derivatives, J. Phys. Org. Chem. 26 (2013) 30–36.
[61] Y. Liu, X. Gong, L. Wang, G. Wang, H. Xiao, Substituent effects on the properties related to detonation performance and sensitivity for 2,2′,4,4′,6,6′-Hexanitroazobenzene derivatives, J. Phys. Chem. A. 115 (2011) 1754–1762.
[62] Y. Liu, L. Wang, G. Wang, H. Du, X. Gong, Theoretical studies on 2-diazo-4,6-dinitrophenol derivatives aimed at finding superior propellants, J. Mol. Model. 18 (2012) 1561–1572.
[63] Y. Pan, W. Zhu, Theoretical design on a series of novel bicyclic and cage Nitramines as high energy density compounds, J. Phys. Chem. A. 121 (2017) 9163–9171.
[64] Y. Pan, W. Zhu, Designing and looking for novel cage compounds based on bicyclo-HMX as high energy density compounds, RSC Adv. 8 (2018) 44–52.
[65] Y. Pan, W. Zhu, H. Xiao, Comparative theoretical studies of dinitromethyl- or trinitromethyl-modified derivatives of CL-20, Can. J. Chem. 91 (2013) 1243–1251.

[66] Y. Pan, W. Zhu, H. Xiao, DFT studies on trinitromethyl- or dinitromethyl-modified derivatives of RDX and β-HMX, Comput. Theor. Chem. 1019 (2013) 116–124.
[67] Y. Pan, W. Zhu, H. Xiao, Theoretical studies of a series of azaoxaisowurtzitane cage compounds with high explosive performance and low sensitivity, Comput. Theor. Chem. 1114 (2017) 77–86.
[68] C. Qi, R.-B. Zhang, X.-J. Zhang, Y.-C. Li, Y. Wang, S.-P. Pang, Theoretical investigation of 4,4′,6,6′-tetra(azido)azo-1,3,5-triazine-N-oxides and the effects of N→O bonding on organic azides, Chem. Asian J. 6 (2011) 1456–1462.
[69] L. Qiu, X. Gong, J. Zheng, H. Xiao, Theoretical studies on polynitro-1,3-bishomopentaprismanes as potential high energy density compounds, J. Hazard. Mater. 166 (2009) 931–938.
[70] Z. Rui-Zhou, L. Xiao-Hong, Z. Xian-Zhou, Theoretical studies on a series of 1,2,4-triazoles derivatives as potential high energy density compounds, J. Chem. Sci. 124 (2012) 995–1006.
[71] H.J. Singh, M.K. Upadhyay, S.K. Sengupta, Theoretical studies on benzo[1,2,4]triazine-based high-energy materials, J. Mol. Model. 20 (2014) 2205.
[72] X. Su, X. Cheng, S. Ge, Theoretical investigation on structure and properties of 2,4,5-trinitroimidazole and its three derivatives, J. Mol. Struct. (THEOCHEM) 895 (2009) 44–51.
[73] X. Su, X. Cheng, C. Meng, X. Yuan, Quantum chemical study on nitroimidazole, polynitroimidazole and their methyl derivatives, J. Hazard. Mater. 161 (2009) 551–558.
[74] F. Wang, H.-C. Du, J.-Y. Zhang, X.-D. Gong, Comparative theoretical studies of energetic azo s-triazines, J. Phys. Chem. A. 115 (2011) 11852–11860.
[75] T. Wei, J. Wu, W. Zhu, C. Zhang, H. Xiao, Characterization of nitrogen-bridged 1,2,4,5-tetrazine-, furazan-, and 1H-tetrazole-based polyheterocyclic compounds: heats of formation, thermal stability, and detonation properties, J. Mol. Model. 18 (2012) 3467–3479.
[76] T. Wei, W. Zhu, X. Zhang, Y.-F. Li, H. Xiao, Molecular design of 1,2,4,5-Tetrazine-based high-energy density materials, J. Phys. Chem. A. 113 (2009) 9404–9412.
[77] L. Xiaohong, C. Qingdong, Z. Xianzhou, Density functional theory study of several nitrotriazole derivatives, J. Energ. Mater. 28 (2010) 251–272.
[78] L. Xiaohong, Z. Ruizhou, Z. Xianzhou, Computational study of imidazole derivative as high energetic materials, J. Hazard. Mater. 183 (2010) 622–631.
[79] T. Yan, G. Sun, W. Chi, B. Li, H. Wu, Looking for high energy density compounds among polynitra-minepurines, J. Mol. Model. 19 (2013) 3491–3499.
[80] C. Zhang, W. Zhu, H. Xiao, Density functional theory studies of energetic nitrogen-rich derivatives of substituted carbon-bridged diiminotetrazoles, Comput. Theor. Chem. 967 (2011) 257–264.
[81] X. Zhang, X. Gong, A simple, fast and convenient new method for predicting the stability of nitro compounds, J. Comput. Aided Mol. Des. 29 (2015) 471–483.
[82] W. Zhu, C. Zhang, T. Wei, H. Xiao, Computational study of energetic nitrogen-rich derivatives of 1,1′- and 5,5′-bridged ditetrazoles, J. Comput. Chem. 32 (2011) 2298–2312.
[83] W. Zhu, C. Zhang, T. Wei, H. Xiao, Theoretical studies of furoxan-based energetic nitrogen-rich compounds, Struct. Chem. 22 (2011) 149–159.
[84] T. Atalar, M. Jungová, S. Zeman, A new view of relationships of the N–N bond dissociation energies of cyclic Nitramines. Part II. Relationships with Impact Sensitivity, J. Energ. Mater. 27 (2009) 200–216.
[85] V.K. Golubev, Influence of Structural and Energetic Factors on Impact Sensitivity of Aromatic Nitro Compounds, in: Proceedings of the 14th Seminar on New Trends in Research of Energetic Materials. Part I. Pardubice, Czech Republic, 2011, pp. 195–204.
[86] T.L. Jensen, J.F. Moxnes, E. Unneberg, D. Christensen, Models for predicting impact sensitivity of energetic materials based on the trigger linkage hypothesis and Arrhenius kinetics, J. Mol. Model. 26 (2020) 65.
[87] G.M. Khrapkovskii, D.D. Sharipov, A.G. Shamov, D.L. Egorov, D.V. Chachkov, B. Nguyen Van, R.V. Tsyshevsky, Theoretical study of substituents effect on C-NO_2 bond strength in mono substituted nitrobenzenes, Comput. Theor. Chem. 1017 (2013) 7–13.
[88] J. Li, A multivariate relationship for the impact sensitivities of energetic N-nitrocompounds based on bond dissociation energy, J. Hazard. Mater. 174 (2010) 728–733.
[89] J. Li, Relationships for the impact sensitivities of energetic C-nitro compounds based on bond dissociation energy, J. Phys. Chem. B 114 (2010) 2198–2202.
[90] X.H. Li, D.F. Han, X.Z. Zhang, Investigation of correlation between impact sensitivities and bond dissociation energies in benzenoid nitro compounds, J. Struct. Chem. 54 (2013) 499–504.
[91] H. Liu, F. Wang, G.-X. Wang, X.-D. Gong, Theoretical studies of -NH_2 and -NO_2 substituted dipyridines, J. Mol. Model. 18 (2012) 4639–4647.

[92] Z. Mei, F. Zhao, S. Xu, X. Ju, A simple relationship of bond dissociation energy and average charge separation to impact sensitivity for nitro explosives, J. Serb. Chem. Soc. 84 (2019) 27–40.
[93] P. Politzer, J.S. Murray, Relationships between dissociation energies and electrostatic potentials of C-NO_2 bonds: applications to impact sensitivities, J. Mol. Struct. 376 (1996) 419–424.
[94] B.M. Rice, S. Sahu, F.J. Owens, Density functional calculations of bond dissociation energies for NO_2 scission in some nitroaromatic molecules, J. Mol. Struct. (THEOCHEM) 583 (2002) 69–72.
[95] X.-S. Song, X.-L. Cheng, X.-D. Yang, B. He, Relationship between the bond dissociation energies and impact sensitivities of some nitro-explosives, Propellants, Explos., Pyrotech. 31 (2006) 306–310.
[96] X. Song, X. Cheng, X. Yang, D. Li, R. Linghu, Correlation between the bond dissociation energies and impact sensitivities in nitramine and polynitro benzoate molecules with polynitro alkyl groupings, J. Hazard. Mater. 150 (2008) 317–321.
[97] B. Tan, X. Long, R. Peng, H. Li, B. Jin, S. Chu, H. Dong, Two important factors influencing shock sensitivity of nitro compounds: bond dissociation energy of X-NO_2 (X=C, N, O) and Mulliken charges of nitro group, J. Hazard. Mater. 183 (2010) 908–912.
[98] P.J. Stephens, F.J. Devlin, C.F. Chabalowski, M.J. Frisch, Ab initio calculation of vibrational absorption and circular dichroism spectra using density functional force fields, J. Phys. Chem. 98 (1994) 11623–11627.
[99] J. Tirado-Rives, W.L. Jorgensen, Performance of B3LYP density functional methods for a large set of organic molecules, J. Chem. Theory Comput. 4 (2008) 297–306.
[100] Y. Zhao, D.G. Truhlar, Density Functionals with broad applicability in chemistry, Acc. Chem. Res. 41 (2008) 157–167.
[101] Y. Zhao, D.G. Truhlar, How well can new-generation density functionals describe the energetics of bond-dissociation reactions producing radicals? J. Phys. Chem. A 112 (2008) 1095–1099.
[102] Y. Zhao, D.G. Truhlar, Applications and validations of the Minnesota density functionals, Chem. Phys. Lett. 502 (2011) 1–13.
[103] H. Chen, X. Cheng, Z. Ma, X. Su, Theoretical studies of C-NO_2 bond dissociation energies for chain nitro compounds, J. Mol. Struct. (THEOCHEM) 807 (2007) 43–47.
[104] J. Shao, X. Cheng, X. Yang, Density functional calculations of bond dissociation energies for removal of the nitrogen dioxide moiety in some nitroaromatic molecules, J. Mol. Struct. (THEOCHEM) 755 (2005) 127–130.
[105] J. Shao, X. Cheng, X. Yang, The C-NO_2 bond dissociation energies of some nitroaromatic compounds: DFT study, Struct. Chem. 17 (2006) 547–550.
[106] X.-Q. Yao, X.-J. Hou, H. Jiao, H.-W. Xiang, Y.-W. Li, Accurate calculations of bond dissociation enthalpies with density functional methods, J. Phys. Chem. A 107 (2003) 9991–9996.
[107] J.F. Moxnes, Ø. Frøyland, T. Risdal, A computational study of ANTA and NTO derivatives, J. Mol. Model. 23 (2017) 240.
[108] Y. Zhao, D.G. Truhlar, The M06 suite of density functionals for main group thermochemistry, thermochemical kinetics, noncovalent interactions, excited states, and transition elements: two new functionals and systematic testing of four M06-class functionals and 12 other functionals, Theor. Chem. Acc. 120 (2008) 215–241.
[109] J.A. Montgomery, M.J. Frisch, J.W. Ochterski, G.A. Petersson, A complete basis set model chemistry. VII. Use of the minimum population localization method, J. Chem. Phys. 112 (2000) 6532–6542.
[110] R.W. Molt, T. Watson, A.P. Bazanté, R.J. Bartlett, N.G.J. Richards, Gas phase RDX decomposition pathways using coupled cluster theory, Phys. Chem. Chem. Phys. 18 (2016) 26069–26077.
[111] I. Shavitt, R.J. Bartlett, Many-Body Methods in Chemistry and Physics: MBPT and Coupled-Cluster Theory, Cambridge University Press, Cambridge, 2009.
[112] J.D. Watts, J. Gauss, R.J. Bartlett, Coupled-cluster methods with noniterative triple excitations for restricted open-shell Hartree–Fock and other general single determinant reference functions. Energies and analytical gradients, J. Chem. Phys. 98 (1993) 8718–8733.
[113] D. Chakraborty, R.P. Muller, S. Dasgupta, W.A. Goddard, The mechanism for unimolecular decomposition of RDX (1,3,5-Trinitro-1,3,5-triazine), an ab initio study, J. Phys. Chem. A 104 (2000) 2261–2272.
[114] M. Miao, Z.A. Dreger, J.E. Patterson, Y.M. Gupta, Shock wave induced decomposition of RDX: quantum chemistry calculations, J. Phys. Chem. A 112 (2008) 7383–7390.
[115] N. Mardirossian, M. Head-Gordon, Thirty years of density functional theory in computational chemistry: an overview and extensive assessment of 200 density functionals, Mol. Phys. 115 (2017) 2315–2372.
[116] L. Goerigk, A. Hansen, C. Bauer, S. Ehrlich, A. Najibi, S. Grimme, A look at the density functional theory zoo with the advanced GMTKN55 database for general main group thermochemistry, kinetics and noncovalent interactions, Phys. Chem. Chem. Phys. 19 (2017) 32184–32215.

[117] T.B. Brill, Connecting molecular properties to decomposition, combustion and explosion trends, in: R.W. Shaw, T.B. Brill, D.L. Thompson (Eds.), Overviews of Recent Research on Energetic Materials, Advanced Series in Physical Chemistry, Vol. 16, World Scientific, 2005.
[118] T.B. Brill, P.J. Brush, K.J. James, J.E. Shepherd, K.J. Pfeiffer, T-jump/FT-IR spectroscopy: a new entry into the rapid, isothermal pyrolysis chemistry of solids and liquids, Appl. Spectrosc. 46 (1992) 900–911.
[119] T.B. Brill, K.J. James, Kinetics and mechanisms of thermal decomposition of nitroaromatic explosives, Chem. Rev. 93 (1993) 2667–2692.
[120] T.B. Brill, K.J. James, Thermal decomposition of energetic materials. 61. Perfidy in the amino-2,4,6-trinitrobenzene series of explosives, J. Phys. Chem. 97 (1993) 8752–8758.
[121] E.R. Tufte, The Cognitive Style of Power Point: Pitching Out Corrupts Within, Connecticut, Graphics Press LLC, Cheshire, 2006.

Dr. Betsy M. Rice serves as Leader of the Multiscale Reactive Modeling Team at the DEVCOM Army Research Laboratory. Dr. Rice earned her B.S. in Chemistry from Cameron University (1984) and a Ph.D. in Chemistry from Oklahoma State University (1987) before joining the Army research community first as a National Research Council postdoctoral fellow, then as a civilian hire (1990). Her expertise is in the areas of ab initio and Density Functional Theory (DFT) for materials characterization used in Quantitative Structure Property Relations or machine-learned models, and atomistic molecular dynamics simulations (both DFT and classical), most directed toward advanced modeling of energetic materials. Dr. Rice has received four U.S. Army Research and Development Achievement Awards (1997, 2003, 2008, 2012) and two ARL Achievement Awards (1999, 2007). Dr. Rice has published over 100 peer-reviewed journal articles and eleven book chapters relating to molecular simulations of materials of interest to the Army. Dr. Rice was named as an ARL Fellow in 2000.

Dr. Edward F.C. Byrd serves as Research Area Leader of the Disruptive Energetics and Propulsion Technologies program for DEVCOM Army Research Laboratory. Dr. Byrd received his B.S. in chemistry from Duke University in 1996 before continuing his education and receiving his Ph.D. in 2001 from the University of California, Berkeley. He subsequently joined ARL in January 2002 as a National Research Council postdoctoral fellow and was hired to a government position in 2004. Dr. Byrd has received multiple awards for his research, such as an ARL Award for Science (2016), ARL Award for Performance (2009), and Army Research and Development Achievement Award for Technical Excellence (2008). Dr. Byrd has over 90 publications cited in excess of 6600 times. Dr. Byrd has been active in developing computational screening tools that will allow for prediction of properties associated with performance and sensitivity for notional energetic materials as well as atomistic modeling of energetic and protection materials. He is responsible for initiating, planning, and personally performing research investigations to determine the controlling physical and chemical processes of energetic materials, formulations, and component technologies of interest to the Army.

CHAPTER

4

Properties of molecular charge distributions affecting the sensitivity of energetic materials

*Itamar Borges, Jr**, *Roberta Siqueira Soldaini Oliveira*, *and Marco Aurélio Souza Oliveira*

Departamento de Química, Instituto Militar de Engenharia (IME), Rio de Janeiro, RJ, Brazil

***Corresponding author: E-mail: itamar@ime.eb.br**

1 Introduction

Explosives, propellants, and pyrotechnics are energetic materials that have important roles in technology and, in a sense, are controllable storage systems of chemical energy [1]. They are primarily molecular solids composed of polyatomic organic molecules arranged in complicated crystal structures [2–4].

The most useful energetic materials have low sensitivity to prevent an accidental explosion by different stimuli and high energy content [5–7]. On the other hand, there is evidence that a good performance of an energetic material goes along with an enhanced sensitivity [8,9]. In other words, an insensitive explosive usually does not exhibit top performance. These seemingly incompatible requirements additionally imply the necessity of comprehensive knowledge of the sensitivity of energetic materials.

Therefore, the safety and great energy content of explosives, pyrotechnics, and propellants are major concerns in the area of energetic materials. In particular, the ability of these materials to release large amounts of energy in a short time interval due to different types of stimuli is related to their sensitivity [5,10,11], which can be classified into different categories: thermal sensitivity, impact sensitivity, friction sensitivity, shock sensitivity and electrostatic sensitivity [12]. Given the variety of chemical, structural and physical factors that

Theoretical and Computational Chemistry, Volume 22
https://doi.org/10.1016/B978-0-12-822971-2.00013-9

determine sensitivity in these diverse situations and the complexity of the problem, it is crucial to elaborate "meaningful evaluation of proposed target molecules" [13].

To measure the stability of an explosive compound quantitatively, frequently the impact sensitivity parameter h_{50} is used [14–16]. This parameter is determined by the height (h) from where a given standard weight dropped over a small amount of explosive material will start an explosion with a likelihood of 50%. Therefore, insensitive materials have large values of h_{50} and vice-versa. This parameter indicates the ignitability of the explosive.

The measurements of impact sensitivity have several sources of uncertainty affecting the reproducibility and limiting the reliability of the experimental h_{50} values. The diversity of possible errors includes atmospheric conditions, particle size distribution, humidity, and even the operator technique [14,17]. However, it was shown before in different molecular models that the correlation of different types of sensitivity, especially impact sensitivity, with chemical structure depends on a large body of data of chemically related explosives, and "individual out-of-line points tended to offset one another" [15]. Therefore, evaluations of quantitative errors are not available, and the diversity of sensitivity-structure models in the literature assumes this fact. Hence, experimental error bars for h_{50} values are not considered [13,16,18–21].

The stability of the molecule of an energetic material is connected to its sensitivity, thus, this property "is primarily due to the chemical character of the materials" [8]. Therefore, accurate computational quantum chemistry methods have been used extensively to investigate the molecular properties of explosive compounds [22]. Furthermore, in the design of a new explosive before the synthesis, it is desirable to investigate those properties to eliminate poor candidates.

The quantum-chemical approaches, in most cases, have been used to rationalize the properties and mechanisms of known substances. The application of these methods can also contribute to the design of new materials that combine high performance and low sensitivity without the drawbacks and the hazards associated with the synthetic work in the laboratory [23]. In particular, the impact sensitivity is a macroscopic property that is not known to be directly dependent on intermolecular interactions on the bulk [18].

The fundamental theorems of density functional theory (DFT) imply that the electron distribution (or density) of a molecule determines the external potential hence the Hamiltonian operator H. Knowledge of H allows one by applying approximations of the Schrödinger equation to compute the molecular wave function or to obtain the electron density in the framework of DFT. Therefore, ultimately *all* the properties of a molecule, including the molecular charge distribution, can be in principle calculated [24,25].

In the literature, the terms "charge density" and "electron density" are often used interchangeably [26] even when considering the electron distribution alone. Hence, the reader should bear this in mind when we discuss both concepts in this chapter [27], and the words "distribution" and "density" will be used interchangeably.

The electron distribution is then a central property directly accessed experimentally [28]. Both the measured and the computed electron distribution, and consequently, the molecular charge density—distinct from the electron distribution only by the inclusion of the positive point charge values of the nuclei, can be analyzed in different ways. Frequently, to accomplish this topological analysis, a global (e.g., electrostatic potential) molecular charge density property is decomposed into a local one (e.g., charges) that, if is precise, should reflect the global property accurately. This approach has been used extensively to develop quantitative models to correlate molecular properties to the sensitivity of energetic materials.

Among the possibilities to describe the molecular charge distribution, methods that divide it into atomic contributions belong to a large family of charge density analysis approaches that provides detailed knowledge of topological bonding properties. These methods have undergone "a major renaissance in the last two decades" [29]. A textbook on the subject with authoritative reviews was edited [30].

The **trigger linkage** is an important concept in the research of energetic materials included frequently in sensitivity models based on the molecular charge distribution. It is defined as the chemical bond, which broken by an external stimulus, is the key step in the detonation initiation [19,23]. In nitro explosives, the C–NO_2, N–NO_2, and O–NO_2 bonds are usually the trigger linkages [31,32]. These bond ruptures activate further exothermic self-sustaining decomposition processes characterizing the detonation or explosion [33].

In this chapter, we present a diversity of attempts carried out over more than 40 years to correlate sensitivity (mostly due to impact) of an explosive, a bulk property, and molecular properties of components of energetic materials [34]. The focus is on molecular properties derived from the molecular charge (electron) densities found to correlate with the sensitivity of different families of explosives. We do not present the specific mathematical expressions developed to reach these conclusions—they can be found in the original papers. Among these approaches, we discuss our contributions to the subject based on the distributed multipole analysis of Stone [35].

Although those attempts may be subject to criticism [36] given that many solid-state properties can influence the sensitivity of energetic materials, other factors may play a role, and correlations of this type can mask the dominant chemical mechanisms of initiation reactions [37]. However, we follow the advice of Rice and Hare: this approach should not be used to interpret them but "rather identify molecular properties that indicate sensitivity to impact" [18].

2 Theoretical foundations and methods

In this section, we discuss a plethora of theoretical methods to analyze the molecular charge density employed to develop quantitative models of sensitivity. They may be used for developing other models.

We considered the following properties to be related or derived from the molecular charge distribution: electrostatic potential, electronegativity, chemical hardness, multipole moments, atomic or group charges, bond orders, and bond dissociation energies, as depicted in Fig. 4.1. We also classify these properties as either global (the first three) or local (the remaining ones), despite the nuances involved in this separation. Moreover, we should bear in mind that from the molecular charge (electron) distribution, other properties can be derived.

Atomic units (a.u.) will be used throughout the chapter unless stated otherwise.

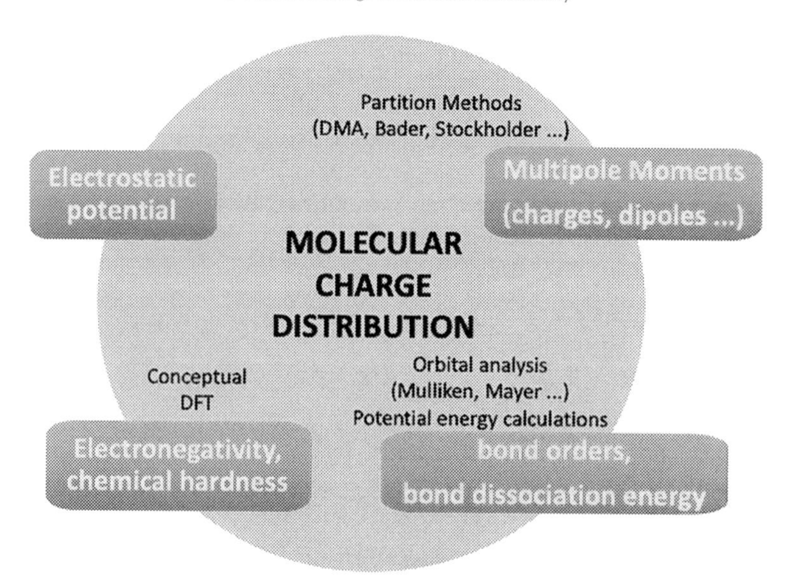

FIG. 4.1 Global (*light brown* (*dark gray* in the print version)) and local (*yellow* (*light gray* in the print version)) properties derived from the molecular charge distribution.

2.1 The molecular electron and charge distribution

A time-dependent normalized wave function ψ is a function of the position in space $\boldsymbol{r}$ and the spin projection (up or down). The physical interpretation of ψ given by Max Born is embodied in the expression:

$$|\psi(\boldsymbol{r})|^2 dxdydz = |\psi(\boldsymbol{r})|^2 dV \tag{4.1}$$

which gives the probability of finding an electron, or any other quantum system, at position $\boldsymbol{r}$ in an infinitesimal volume dV around $\boldsymbol{r}$. The probability density function of the electron is $|\psi(\boldsymbol{r})|^2$.

A molecule is a many-electron system described by the wave function $\Psi(\boldsymbol{r}_1, \boldsymbol{r}_2, \ldots, \boldsymbol{r}_N)$, which square modulus gives the probability density of finding the electron 1 at $\boldsymbol{r}_1$, electron 2 at $\boldsymbol{r}_2$, …, electron N at $\boldsymbol{r}_N$ simultaneously. For instance, the probability of finding electron 1 at $\boldsymbol{r}_1$ without specifying the position of the remaining $N-1$ electrons (i.e., irrespective of their positions), is determined by integrating $\Psi(\boldsymbol{r}_1, \boldsymbol{r}_2, \ldots, \boldsymbol{r}_N)$ over the coordinates of all electrons excepting electron 1:

$$\int |\Psi(\boldsymbol{r}_1, \boldsymbol{r}_2, \ldots, \boldsymbol{r}_N)|^2 d\boldsymbol{r}_2 \ldots d\boldsymbol{r}_N \tag{4.2}$$

where the summation over the spin is implied.

If we multiply the above integral by the number N of electrons in the molecule, we obtain a one-electron density function known as the **electron density**:

$$\rho(\boldsymbol{r}) = N \int |\Psi(\boldsymbol{r}_1, \boldsymbol{r}_2, \ldots, \boldsymbol{r}_N)|^2 d\boldsymbol{r}_2 \ldots d\boldsymbol{r}_N \tag{4.3}$$

which gives the probability of finding any single electron at $\boldsymbol{r}$ with spin s weighted by the total number of electrons N, where we have dropped the subscript "1." The function $\rho(\boldsymbol{r})$ is non-negative, depends only on the three spatial variables, and vanishes at infinity: $\rho(\boldsymbol{r} \longrightarrow \infty)=0$.

Strictly speaking, $\rho(\boldsymbol{r})$ is a probability density, but it is usual to call it electron density. The reader should have in mind that the multiple integrals in Eq. (4.3) represents the probability of finding a particular electron in the interior of the volume element $d\boldsymbol{r}$. However, considering that electrons are indistinguishable, the probability of finding any electron at position $\boldsymbol{r}$ is just equal N times the probability for one particular electron.

By integrating the electron density $\rho(\boldsymbol{r})$ of Eq. (4.3) over all space with respect to the $\boldsymbol{r}$ coordinate, we obtain the total number of electrons in the molecule:

$$\int \rho(\boldsymbol{r}) d\boldsymbol{r} = N \tag{4.4}$$

because the wave function $\Psi(\boldsymbol{r}_1, \boldsymbol{r}_2, \ldots, \boldsymbol{r}_N)$ is normalized.

The probability density $|\Psi(\boldsymbol{r}_1, \boldsymbol{r}_2, \ldots, \boldsymbol{r}_N)|^2$ then defines a scalar field. In a multielectron atom or molecule, only the *total* electron density $\rho(\boldsymbol{r})$ can be observed experimentally [38–40].

The total charge of a molecule is an integral number given by

$$Q_{\text{total}} = \sum_{A\ (\text{nuclei})} Z_A - \int \rho(\boldsymbol{r}) d\boldsymbol{r} \tag{4.5}$$

where Z_A is the electric charge on nucleus A and $\rho(\boldsymbol{r})$ is the electron density function of the molecule. If the molecule is neutral, as usually is the case in the models of sensitivity, $Q_{total}=0$. The charges of the nuclei are described by point charges.

The molecular charge *density* P is the sum of the point charges of the nuclei with the electron density:

$$\mathrm{P} = \frac{1}{V} \sum_{A\ (\text{nuclei})} Z_A - \rho(\boldsymbol{r}) \tag{4.6}$$

where V is the molecular volume, which can be defined in several ways [41]. Note that in contrast with the molecular electrostatic potential (see next subsection), the molecular charge density P is not an integrated property, hence it can provide finer information on the molecular charge density.

2.2 Molecular electrostatic potential

The molecular electrostatic potential (MEP) $V(\boldsymbol{r})$ at a point, $\boldsymbol{r}$ is given by

$$V(\boldsymbol{r}) = \sum_{A}^{nuclei} \frac{Z_A}{|\boldsymbol{R}_A - \boldsymbol{r}|} - \int \frac{\rho(\boldsymbol{r}') d\boldsymbol{r}'}{|\boldsymbol{r} - \boldsymbol{r}'|} \tag{4.7}$$

where Z_A is the electric charge on nucleus A located at $\boldsymbol{R}_A$ and $\rho(\boldsymbol{r})$ is the electron density function of the molecule. There are two important features of this expression [42]: (i) it assumes no polarization of the molecule in response to the test charge [25]; (ii) the value of $V(\boldsymbol{r})$ depends on the vicinity of $\boldsymbol{r}$, and it includes explicit contributions from each nucleus and each element of the electronic charge of the entire molecule [43]. The MEP can indicate the reactivity behavior of molecules: an approaching electrophile will be attracted to negative regions of $V(\mathbf{r})$ (electron-rich regions), whereas positive regions (electron-deficient) are prone to nucleophiles [44].

When visualized on surfaces or regions of space, the MEP is especially useful because it informs on the local polarity of a molecule. The electrostatic "surface" is conveniently defined according to Bader as a 0.001 a.u. contour value of the molecule's electronic density [45]. In Fig. 4.2, we depict three illustrative examples of MEPs of energetic molecules highlighting different degrees of positive values of the C–NO_2 and N–NO_2 bonds, which according to Politzer, Murray, Rice, and coworkers are related to the sensitivity of explosives. These researchers also found that energetic molecules have dominant positive central regions (Fig. 4.2), in contrast with a typical organic molecule—these works are discussed below.

2.3 Electric multipole moments

The expectation values of multipole moment operators in Cartesian coordinates for a molecule described by the wave function $\Psi(\mathbf{r})$ are computed from Eq. (4.8) [25]

$$\langle x^k y^l z^m \rangle = \sum_{i}^{atoms} Z_i x_i^k y_i^l z_i^m - \int \Psi(\mathbf{r}) \left(\sum_{j}^{s} x_i^k y_i^l z_i^m \right) \Psi(\mathbf{r}) d\mathbf{r} \tag{4.8}$$

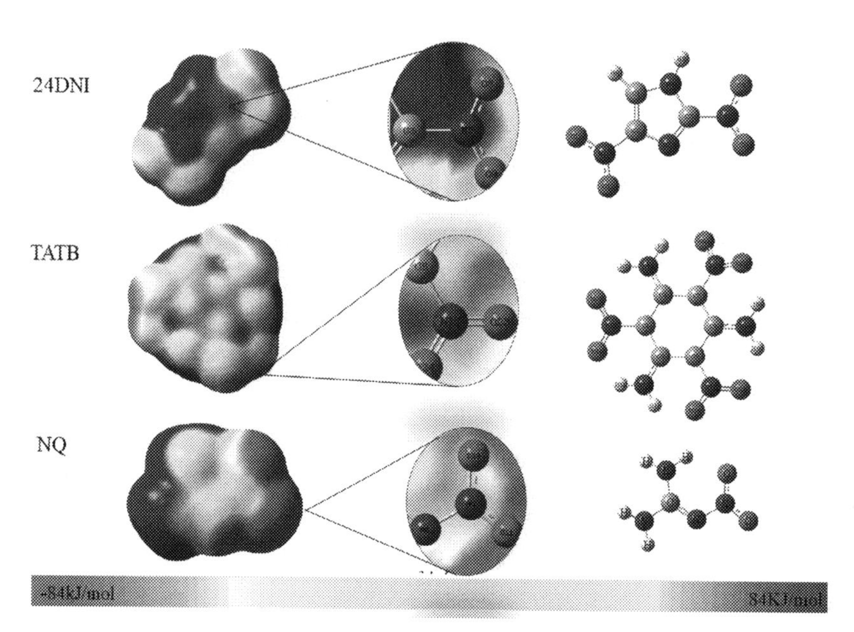

FIG. 4.2 B3LYP/6–311+G(d) molecular electrostatic potential (MEP) of the of 2,4-dinitro-1H-imidazole (24DNI), 1,3,5-trinitro-2,4,6-triamino-benzene (TATB) and 1-nitroguanidine (NQ) explosive molecules. Insets highlight the C–NO_2 and N–NO_2 bonds. Contour value = 0.001 a. u.

where the sum $k+l+m$ determines the type of moment, namely, monopole (0), dipole (1), quadrupole (2), etc. Z_i is the charge on atom i, and the integration coordinate $\boldsymbol{r}$ represents the Cartesian coordinates of all j electrons.

We show below that in the distributed multipole analysis (DMA) framework of the molecular charge density, the interpretation of atom-centered electric multipoles is chemically intuitive.

2.4 Partial (or net) electric atomic charges

The idea of associating polarity with charge build-up or depletion on the atoms of a molecule is a very convenient way to rationalize bonding and other chemical properties. The reason is that the same atoms or groups of atoms in different molecules display different amounts of charges given the different chemical and bonding environments.

However, although it is possible to define unambiguously quantum mechanical operators for computing (partial) atomic charges, they cannot be measured experimentally. Moreover, it has been identified over 30 different computational methods [42], including approaches to calculate charges from the molecular electrostatic potential. Hence, there are too many possibilities to represent a molecular property bearing different degrees of accuracy [46].

Of the most popular methods, we discuss the Mulliken charges because Zhang largely used them to develop sensitivity models for nitro explosives [12,20]. However, the Mulliken charge analysis method has five significant deficiencies, namely: (a) orbital populations with nonphysical values smaller than zero or greater than two, which violates the Pauli Exclusion Principle; (b) electrons shared in a bond are evenly distributed among the atoms, thus ignoring electronegativity differences; (c) a considerable dependence on the basis set, a complicating feature for comparisons between different calculation levels; (d) the electron density may be counted as belonging to an atom even if the basic functions effectively describe the wave functions far from the atom and; (e) a set of atomic charges does not reproduce the original multipole moment [25]. The charges resulting from the DMA decomposition of the molecular charge distribution that we have been using extensively do not suffer from these deficiencies [46].

2.5 Bond strengths, bond order, and electronegativity

The strength of a bond is a useful concept in chemistry, especially in the investigation of sensitivity via the trigger bond concept. It depends on the degree of overlap between interacting atomic orbitals and bond polarity; the latter is related to the difference in the energies of the involved atomic orbitals. In other words, bond strengths depend on covalency (overlap) and bond polarity [47].

In general, the bond dissociation energy (BDE) is used to measure bond energy and bond strength [48–52]. However, some authors argue that relying only on BDE parameters for computing bond strengths can lead to erroneous results [47,53].

In the framework of the Atoms-in-Molecules (AIM) theory of Bader, the strength of a bond is assumed to be directly proportional to the electron density value at the bond critical point [54]. It would be interesting to explore this widely used concept in developing sensitivity models, but most works concentrated on discussing specific molecules, not sensitivity models (see below).

Bond orders are one of those concepts in chemistry that does not have a unique definition, a sound physical basis, and, moreover, it is not a physical observable. However, like other chemical constructs (e.g., atomic charges), it is often very useful.

The Bayse group and others explored the Wiberg bond index (WBI) concept for assigning trigger bonds [55,56] claiming it to be a better predictor of trigger bond strengths compared to bond dissociation energies. WBIs are determined as the sum of the squares of the off-diagonal elements of the density matrix that have magnitudes similar to bond orders as expected from valence bond theory [57].

To compute the strength of trigger bonds using WBIs, Bayse defined a relative scale for bond activation determined by the difference of the WBI for specific bond types of energetic materials and reference molecules, labeling it %ΔWBI. The reference molecule should have the same bond type, hybridization, and explosophore of the investigated energetic materials. The trigger bond would display the most negative %ΔWBI relative to the reference molecule. They argue in favor of defining %ΔWBIs by showing that the use of absolute WBIs would lead to erroneous results concerning the determination of the correct trigger linkage in explosives [56].

The problem with the concept of %ΔWBI is the choice of the reference molecule considering that there is always more than one possible choice. In spite of that, this issue could be overcome if the authors tested the consistency of their results by employing different types of reference molecules for the same family of explosives to compute %ΔWBIs.

A conceptual basis for defining chemical bonding, a central concept in chemistry, was given by Pauling: chemical bond arises "from the tendency of an atom to attract electrons in a bond" [58]. In other words, from electronegativity.

Scales of electronegativity have been employed to describe chemical mechanisms, bond polarity, band gap, chemical hardness, among other properties [59]. We mention it in the context of sensitivity models (Section 3) for two reasons: due to the interesting work of Mullay on sensitivity and to inspire future work.

2.6 Methods to partition the molecular charge density including the distributed multipole method (DMA)

The methods to partition the molecular charge distribution can be classified in two main categories: (i) space-partition schemes that divide the whole density into disjoint regions, each one corresponding to an atomic domain, and; (ii) atom-centered schemes in which each atomic contribution extends over the whole space [60].

The widely used Atoms-in-Molecules (AIM) theory of [40,61] belongs to category (i). According to AIM, the electron density distribution of a molecule, given by Eq. (4.3), is characterized by the stationary points of the electron density function together with the gradient paths that connect these points. The critical point in the middle of a bond is a minimum of the electron density along the line connecting the atoms, reflecting the atomic sizes and being closer to the smaller atom. The theory recovers central operational concepts of the molecular structure, such as a functional group of molecules with additive properties and chemical bonds. Rice and collaborators employed the AIM method for establishing correlations with the sensitivity of representative energetic materials [16].

In the deformed atoms-in-molecules (DAM) method [62], the partition of molecular density assigns to each atomic fragment electron density distribution centered on its nucleus plus the part of the closest two-center distributions. The atomic fragments are determined by the least deformation criterion [60], which retains as much as possible the sphericity of each atom fragment, characteristic of isolated atoms. The two-center distribution is partitioned according to the fast convergence of the multipolar expansion of the molecular fragments' long-range potential toward the potential of the full (i.e., molecular) distribution [63,64]. The method furnishes detailed pictures of the molecular electron density as regions of electron accumulation and depletion thereby reproducing the expected chemical concepts. This approach is similar to the electron density difference pictures in solid-state, e.g., [65,66], being a powerful method to interpret molecular phenomena. We employed the DAM partition to investigate the decomposition processes and the electronic density of four conformers of RDX, hexahydro-1,3,5-trinitro-1,3,5-triazine [67], and FOX-7, 1,1-diamino-2,2-dinitroethylene [68]. Fig. 4.3B depicts the DAM partition of the electron distribution of the FOX-7 molecule. In both works, nuclear Fukui response functions in the framework of conceptual DFT [69] were used.

The DMA was originally developed for investigating molecular interactions [35,70,71]. It is a method of the category (ii) that has been explored by our group in different works, hence it is discussed in more detail.

According to the DMA, the molecular charge density is divided into a sum of products of atom-centered Gaussian basis functions, with coefficients determined from the one-electron density matrix given by Eq. (4.3). Any individual product of the atom-centered basis functions corresponds to a sum of multipole moments of ranks up to its polynomial degree. Therefore, the overlap of two *s* functions represents a pure charge (monopole), the product of an *s* with a *p* function describes a charge plus a dipole, the overlap of two *p* functions generates a

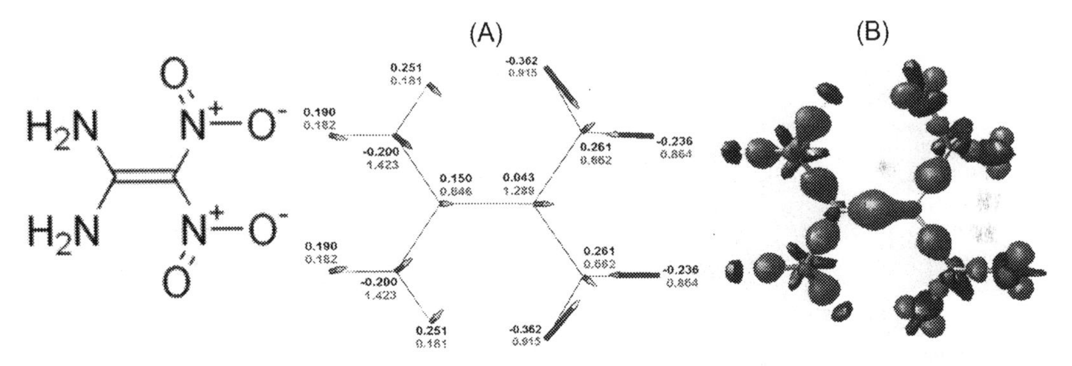

FIG. 4.3 B3LYP/aug-cc-pVTZ partition of the molecular charge density of FOX-7. (A) DMA multipole values. Numbers written in the upper position of each site are monopole values (*black*) in units of *e* (1.602×10^{-19} C). Those in the lower position (*red* (*gray* in the print version)) are the quadrupole values in ea_0^2 (4.486×10^{-40} C m^2) units. Site dipoles are represented by vectors drawn at the corresponding atomic nuclei. (B) FOX-7 DAM regions of electron accumulation (*red* (*dark gray* in the print version)) and depletion (*blue* (*black* in the print version)).

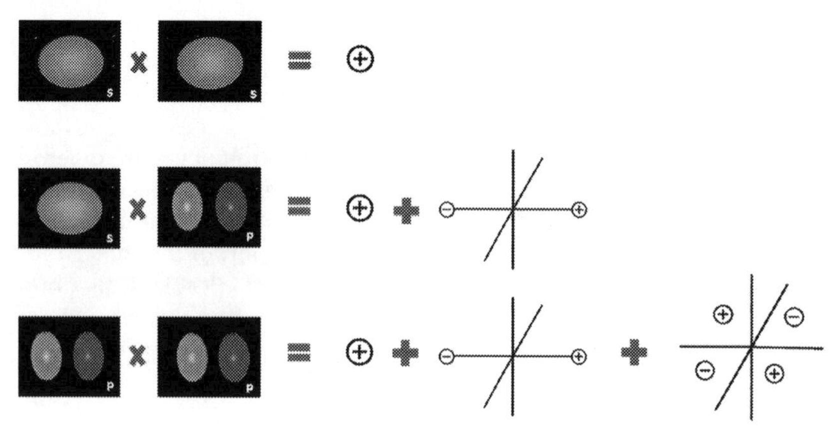

FIG. 4.4 Schematic representation of the electric multipoles computed from the distributed multipole analysis (DMA).

charge, dipole, and quadrupole moments, and so on—see Fig. 4.4. If the orbitals are on different atoms, then each pair of Gaussian functions produces a finite multipole series at a point between the two atoms determined by the exponents of the involved Gaussian orbitals. These multipoles are described by a series on the nearest atom or another expansion site. DMA evaluates these exact representations and approximates each of them by an electric multipole expansion, usually centered on the atomic nuclei. This series rapidly converges due to the expansion on different points of the electron charge distribution. The molecular charge density is obtained by combining the electron charge densities with the positive charge density of the nuclei.

The expansion terms of the DMA method have a clear-cut chemical interpretation. The monopole term represents charges localized on the atomic sites, with bonds between adjacent atoms usually exhibiting some degree of charge separation. Dipole moment vectors pictorially represent charge displacement in the form "−→+," i.e., by a vector pointing from a negative charge "−" to a positive one "+" of the same magnitude—see examples in Figs. 4.3 and 4.5. Bond densities can produce significant site dipole moments depending on the different electronegativity values of adjacent atoms and the remaining molecular environment. Dipole moments express atomic polarization, in general, followed by charge separation in the opposite direction [38,72]. An isolated atom would have a perfectly spherical electron cloud thus no dipole vector.

The quadrupole moment is the first electrostatic moment to include contributions from the "out-of-plane density." Therefore, it is associated with delocalized π electrons and lone pairs of electrons [71].

Given that DMA provides a detailed description of the molecular charge density, the method can be used to study intermolecular interactions and rationalize chemical bonding in different problems [38,72–75]. Fig. 4.4 illustrates these concepts, whereas Fig. 4.5 shows examples from three different chemical families of explosive molecules.

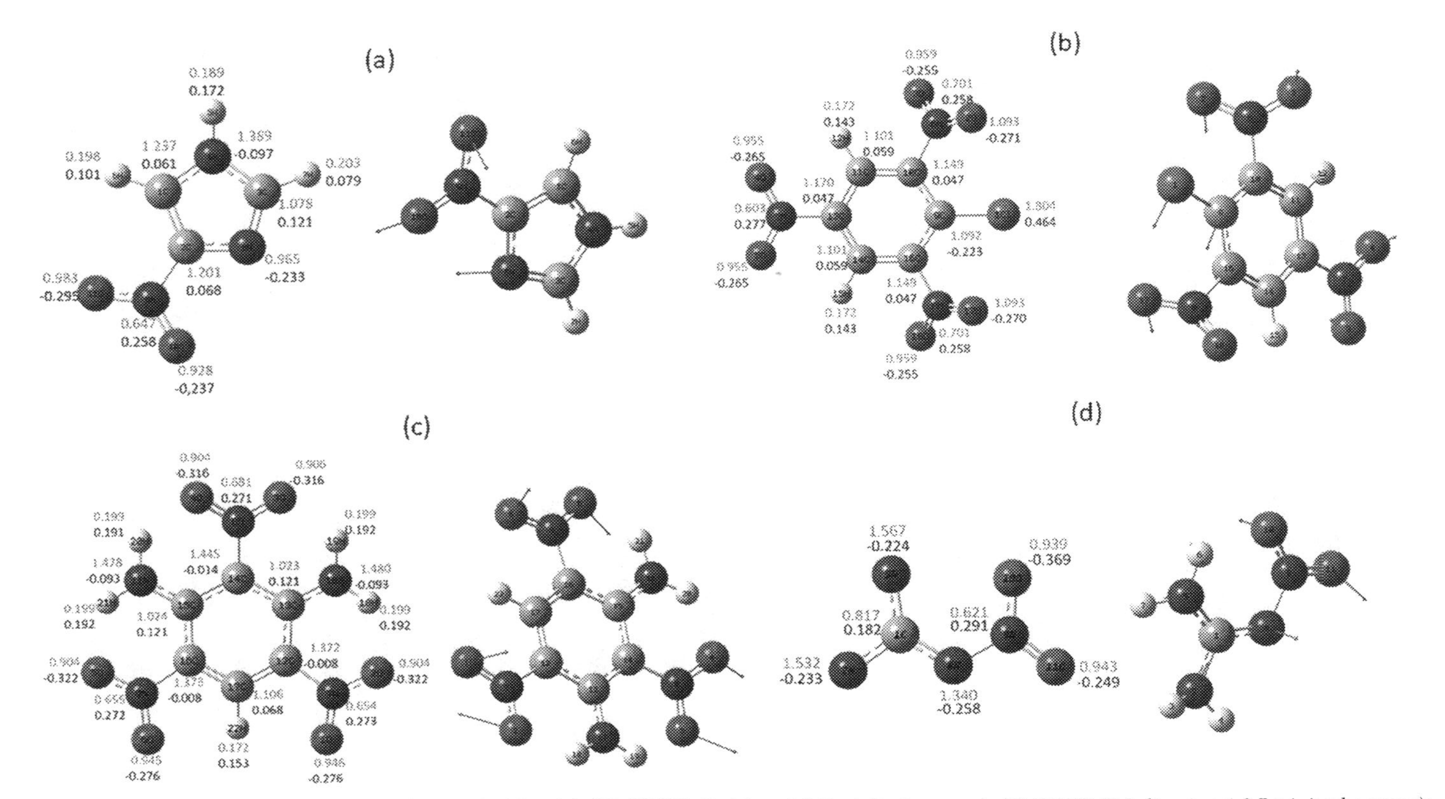

FIG. 4.5 DMA pictures of the (A) 4NI (4-nitro-imidazole), (B) ClTNB (2-chloro-1,3,5-trinitrobenzene); (C) DATB (2,4-diamino-1,3,5-trinitrobenzene); NQ (1-nitroguanidine) molecules. Atomic dipoles are represented by vectors drawn at the corresponding atomic nuclei. Relative values only. Monopole values (numbers in *black*) are shown in units of elementary charge e (1.602×10^{-19} C) and quadrupole values (numbers in *red* (*gray* in the print version)), in units of 4.486×10^{-40} C m^2 (this unit correspond to ea_0^2).

For dipole moments, a vector property, their magnitudes, and vectors can be drawn, whereas for quadrupole moments, a tensor, a number corresponding to the square root of the sum of all tensor components squared, is used [76].

The DMA approach then provides a precise description of the molecular charge density. The method can be considered in a certain sense a generalized population analysis without the drawbacks of the most popular ones [25]. DMA is complete considering that it "offers an exact representation of the potential outside the molecular charge distribution" [70]. Moreover, we showed that electric multipoles from the DMA partition of the molecular charge density are quite independent of the size of the Gaussian basis set [76]. In contrast, AIM displays a considerable variation of the electron distribution topology and cannot currently be used to correlate the AIM results with the impact sensitivity of energetic materials [16].

3 Modeling of sensitivity based on the molecular charge density

In the following subsections, the main results concerning the sensitivity of energetic materials obtained from properties of the molecular charge (electron) distribution (Fig. 4.1) are discussed.

3.1 Molecular electrostatic potentials

Relevant contributions employing molecular electrostatic potentials (MEPs) to investigate impact sensitivity were produced by the groups of Politzer and Murray, and Rice.

The first contribution from the group of Politzer employing MEPs used SCF/STO-3G wave functions combined with crystallography molecular geometries. They investigated impact and shock sensitivities (the latter interpreted as directly proportional to the maximum gap width through which a given shock wave will still detonate the compound) of seven polinitroaromatic molecules [77]. Both sensitivity properties correlated quite well with the V_{mid} value of the MEP at the midpoint of the largest C–NO_2 bond (hence the weakest) in each molecule, with the largest $V_{mid,\ max}$ value and even with the average V_{mid} value, which is a consequence of the electron-withdrawing effects of nitro groups. The values of V were estimated from Mulliken charges of the C and N atoms employing the general expression $V = Q_C Q_N / R$.

The positive value of V_{mid} increases with increasing sensitivity toward shock or impact. For nitroaromatic molecules, V_{mid} is always positive for C–NO_2 bonds because of the positive charges of the C and N atoms (see Fig. 4.2 for examples)—hence these bonds display certain instability. This pattern is "an additional indication that breaking of this bonds is a key part of the decomposition pathways" and rupture of the C–NO_2 bond may be rate-controlling. This work was extended and modified by including 20 more molecules considering $V_{mid,\ max}$, the largest V_{mid}, and also a good correlation with impact sensitivity was found [78].

The same group computed, not estimated, as before, the HF/STO/(6-31G* or STO-5G*) MEPs using Eq. (4.7) for 14 nitroaromatics (using crystallographic geometries) and nitroheterocycles (using HF/3-21G* geometries) which are stabilized by electronic charge delocalization [79]. They employed two properties derived from the MEP in their models: the average deviation of the potential on the surface Π (a measure of the internal charge separation, or local polarity), and their most positive values $V_{S,\,max}$(local maxima). It was suggested that a critical factor in determining the impact sensitivity of a compound would be the degree to which the stabilization effect of the charge delocalization is balanced with respect to the characteristic electron-withdrawing effect of nitro groups and that the impact sensitivity increases as the strength of the C–NO_2 bond decreases. The impact sensitivity of both families of molecules was related to the degree of charge separation and the presence of strongly positive electrostatic potential maxima on the molecular surface.

In another work, they related the impact sensitivity of nitroaromatics, nitramine, and nitroheterocyclic explosives to the Π and $V_{S,\,max}$ properties of the MEPs. Their main finding, similarly to previous works, was that "impact sensitivities of the three classes of energetic compounds can be related to the degree of imbalance between their characteristically stronger positive surface electrostatic potentials and weaker negative ones" [13].

Rice and Hare [18] analyzed the B3LYP/6-31G* MEPs of 34 polynitroaromatic and benzofuroxan molecules, probing different aspects of previous ideas from Politzer and Murray and confirming overall their findings also by developing five models to correlate impact sensitivity h_{50} with parameters related to the electrostatic potential [13]. They suggested that impact sensitivity would be connected to the *degree* of electron deficiency over covalent bonds within the molecular structure's inner skeleton. For nitroaromatics and benzofuroxan derivatives, the sensitivity would apparently be described by the degree of the positive potential distribution located in the aromatic ring or on the C–NO_2 bond. They found that the build-up of positive charges located over covalent bonding regions of the compounds containing nitro groups is characteristic of highly sensitive explosives. In contrast, insensitive explosives do not display this behavior.

The aforementioned works of Politzer, Murray, Rice, and collaborators employed properties of the MEPs quantified by several descriptors, known as general interaction properties functions (GIPFs) [80]. The GIPFs have been employed successfully to describe a variety of other bulk properties, including sensitivities—see Chapter 8 of this book for more details of their applications.

3.2 Atomic electric charges

Atomic electric charges of selected atoms or groups of atoms belonging to trigger linkages of energetic molecules have been employed as indicators of molecular stability and sensitivity [12]. We discuss below works mostly associated with impact sensitivity.

The use of quantum-chemical atomic charges to rationalize the sensitivity of explosives has a long history. In the 1970s, Schroeder and Haskins's attempts using atomic charges from semiempirical methods were published in the restricted literature or in inaccessible conference proceedings, as reported by [32].

Based on the fact that C–NO_2, N–NO_2, and O–NO_2 are trigger bonds in nitro explosives, Delpuech and Cherville employed the INDO and CNDO/CI semiempirical methods to develop relationships between thermal sensitivities and the Mulliken charges of the atoms of those bonds [81–83]. From these charges, they defined for trigger bonds a dissymmetry of charge per unit length for the ground and the first excited states, and from these values a relative variation of polarity. Considering these properties, they showed that it is the dissymmetry of the charge per unit length of the excited state which is correlated with the impact sensitivity of nitric esters, nitramines, and nitroaromatics. They suggested further investigations of the other examined properties, a recommendation not strictly followed by other studies, though correlations with HUMO-LUMO gap were attempted [84], including for describing spark sensitivity [85].

Owens investigated trinitro aromatic molecules with the semiempirical CNDO method using crystal geometries to find that a larger positive charge on the carbon atom bonded to a NO_2 group corresponds to a stronger C–N bond, hence to a less sensitive compound [86]. He also used the charges to calculate the MEP using $V = Q_C Q_N / R$ and found it also correlating with impact sensitivity, apparently being the first to do it.

Mullay examined several polynitroaromatic, polynitroaliphatic, nitramines and, aliphatic ester molecules components of organic high explosives. He established that the amount of charge transferred to nitro groups via the C–NO_2, N–NO_2, and O–NO_2 trigger linkages allows one to classify and correlate impact sensitivity trends of the compounds [87,88]. The charge values were computed directly from the electronegativity χ of atoms, groups of atoms or, for the whole molecule, calculated from the ionization potential and the electron affinity.

Mullay concluded that a *global quantity*—related to the whole molecule, in his case, the total electronegativity of the molecule, and a *local quantity*, namely, atomic charges related to the trigger linkage, are necessary to rationalize the behavior of impact sensitivity. In particular, he found that local quantities are only good indicators when global quantities are similar, i.e., the investigated compounds are from the same family. This far-reaching conclusion explains why most correlation studies of sensitivity-molecular properties provide good results only for the same family of explosives. The use of molecular charge (electron) density analysis, such as the distributed multipole analysis (DMA), by providing global or semiglobal properties (e.g., for a whole aromatic ring) overcomes the deficiencies of employing only atomic charges [32].

Chen and Xiao, upon investigating tetrazole compounds with semiempirical and DFT methods, found that for some specific groups (e.g., NO_2), correlation of charge values with impact sensitivity only works if: (i) the compounds have similar structures; (ii) their detonation are triggered by the same initiation mechanisms (e.g., by the rupture of a similar trigger linkage) and: (iii) molecular packing, crystal structure, and heat transfer—what he named "aggregate" effects resemble [89,90]. The latter is another general result that confirms why most molecular structure–sensitivity correlation models only work within the same families. In another work, the same group obtained for polynitroadamantane explosives that activation energies for the rupture of C–NO_2 bonds are related to the respective bond order and the charge of NO_2 groups, hence to the impact sensitivity [91].

Stephen and collaborators employed AIM and Chelp-G methods from DFT calculations to compute charges and the electrostatic potential V_{mid} of the TNB and TATB molecules. They confirmed that C–NO_2 bonds are the weakest due to an extensive charge depletion, thus are trigger bonds and correlated with impact sensitivity [92,93].

Zhang and collaborators developed the nitro group charge method (NGCM) to assess impact sensitivity, molecular stability, and nitration of nitro explosives applying it to over 40 molecules [19,20,94,95]. The method is based on the defining role of the electron-acceptor NO_2 groups in different nitro explosives. The explosophore nitro groups common to this family of explosives are the root for their detonation properties, and their atomic charges reflect different chemical environments. The fact that different amounts and signs of the total charge Q_{nitro} of a weakest bonded nitro group in an explosive molecule is correlated with the aforementioned properties forms the method's physical basis. Their approach also worked for correlating the electron-acceptor azide group charge values with impact sensitivities of metal-azide compounds [20].

The conclusions drawn by these authors from the total Mulliken charges of the nitro groups are in complete agreement with one of our sensitivity models based on DMA charges [76,96]. Therefore, despite the aforementioned drawbacks of Mulliken charges, Zhang et al. have captured the essence of the molecular basis of the sensitivity of the energetic materials composed of molecules bearing nitro groups, which our work using the more precise DMA representation of the molecular charge density has confirmed and extended, as it will be discussed.

The two main findings of Zhang and collaborators concerning impact sensitivity are: (a) The more negative Q_{nitro} values correspond to more insensitive explosives (i.e., higher h_{50}); (b) for all nitro explosives for which R–NO_2 is the weakest bond, Q_{nitro} can be used to predict impact sensitivity. On the other hand, the method has the following limitations [20]: (1) it applies to nitro compounds with stabilities determined by the chemical bonds connecting nitro groups, consequently, it does not work when the bond of the nitro group is not the weakest in the molecule; (2) when applied to rationalize impact sensitivity of nitro explosives, it considers only molecular properties overlooking other factors affecting impact sensitivity such as crystal properties and; (3) given that Q_{nitro} values greatly depends on the calculation methods, comparison and development of models should employ the same charge method.

Cao and Gao found that the oxygen balance OB_{100} (property closely related to the decomposition products) and the nitro charge Q_{nitro} are two molecular properties affecting impact sensitivity of nitrobenzenes, saturated nitro compounds as well as of proposed new dinitroamino benzene derivative molecules [97,98]. For C–H–N–O molecules, OB_{100} is given by a mathematical expression that subtracts the number of oxygen atoms from the number of the other atoms in the energetic molecule including the carboxyl groups [15]. To obtain Q_{nitro}, they used Mulliken charges computed with the AM1 semiempirical method [97] in the first work and DFT/B3LYP in the second [98]; the smallest absolute value of Q_{nitro} was considered. The model developed for describing the impact sensitivity was quadratic in Q_{nitro} and linear in OB_{100}.

The group of Cheng in the first work on 28 polynitroaromatics found a correlation between spark sensitivity and three properties: (a) DFT Mulliken charge values of the nitro groups Q_{nitro}; (b) number of aromatic rings and number of nonnitro (substituted) groups attached to the aromatic ring and; (c) the LUMO orbital energy in eV [99]. Overall, they found that for compounds with the same substituted group, the electric spark sensitivity increases the larger is Q_{nitro}. Furthermore, groups attached to the aromatic ring such as –CH_3, –OH, and –NH_2 groups, especially the first, affect the spark sensitivity. In another work, the DFT Mulliken Q_{nitro} values of the nitro groups of 14 cyclic nitramines were combined with two structural parameters, the ratio of hydrogen to oxygen, and the ratio of carbon to oxygen, parameters used before in other sensitivity models [100,101]. The group obtained a good correlation with spark sensitivity by separating the molecules into two classes, one containing methylenenitramin (–$CH_2N(NO_2)$–) groups with symmetric structures and the other bearing other groups without symmetric structures [85].

Zhang and coworkers proposed in 2019 an interesting approach based on a self-ignition propagation coefficient (SIPC) that represents the capability for the molecular decomposition of energetic materials. The SIPC is based on heat release and energy barrier, shown to be highly correlated with impact sensitivities for 150 diverse energetic molecules [84]. The SIPC is the number of molecules ignited by the decomposition of a given molecule, being defined as Q/E_a, where Q is the heat released by the decomposition of the molecule and E_a is the activation energy required for molecular decay. The SIPC model's success was attributed to the combination of thermodynamic and kinetic parameters as a better representation of impact sensitivity. SIPC has similarities with the model of impact sensitivity developed before by Mathieu [102–104] and applied successfully to a huge number of molecules from different families. The SIPC model is presented because it is compared with the Q_{nitro} model of the same group and have similar accuracy, though the latter is only feasible for NO_2 compounds. The method can be used for fast screening and design of new energetic compounds.

We have explored in our own work sensitivity models based on charge values but computed with the atom-centered DMA monopoles which represent the atomic charges. These works will be discussed in the appropriate subsection below.

3.3 Bond strengths, bond order, and electronegativity

Politzer and coworkers used the reciprocal of bond lengths R as a measure of relative bond strengths of N–NO_2 and/or C–NO_2 bonds, combined with the overall size of the molecule (given by the molecular weight) and the number of these bonds, to investigate separately shock sensitivity of nitramines and nitroaliphatics of diverse structural types [105]. They found good correlations from the combination of these properties with sensitivity. In contrast, using only the longest value of R (corresponding presumably to the weakest bond) or the average R_{aveg} values of all N–NO_2, instead of the other combination of parameters, did not show any correlation and did not improve the results.

Mathieu has combined in a ratio the weakest bond dissociation energy (BDE)—a measure of bond strength, in this case of the trigger linkage, and the decomposition enthalpy per covalent bond to establish a shock sensitivity index [106]. He gave well-grounded physical arguments for considering the latter property as a measure of energy content. Moreover, it was substantially correlated with the thermal conductivity of the material κ that controls the diffusion rate which affects the shock response of the explosives. In that work, he provided new insights on the sensitivity of explosive mixtures and the surprising stability of TATB and FOX-7.

Xiao and coworkers proposed the principle of the smallest bond order (PSBO) that states that among structurally similar compounds having the same detonation mechanism, the smaller the bond order of the trigger linkage (i.e., the weaker the bond), the larger the impact sensitivity [32]. Conversely, larger bond orders correspond to stronger chemical bonds thus to an explosive less sensitive to an external stimulus. They computed π-bond orders employing the Huckel method and Mulliken and Wiberg bond orders using the CNDO and (UHF-)AM1 semiempirical methods and DFT in latter works. Their findings were based on the prediction of the impact sensitivity of different classes of explosives, including nitro compounds, nitramine, and nitrate esters [107–109]. Pyrolysis initiation reactions of nitro derivatives of aminobenzenes via C–NO_2 homolysis were also investigated through their reaction paths using the Wiberg bond order index to identify the trigger linkage in the reactant molecule and to establish a correlation with the activation energy E_a [110]. Given that impact sensitivity depends on the thermal decomposition rate, they found a relationship of this property with E_a (kinetic variable) and the bond order (thermodynamic index) of the weakest C–NO_2 bond. A similar investigation was carried out for six nitro derivatives of phenol compounds, which in this case are more easily initiated via isomerization reactions through the transfer of the phenolic hydrogen to the O atom of the *ortho*-NO_2 group [111].

In a work already discussed in the previous subsection on atomic charges, Wiberg indices were employed to compute bond orders to identify the weakest bond, which was considered the breaking (trigger) bond according to the principle of the smallest bond order [98]. It was found that for dinitroamino benzene derivative molecules the bond order of the C–N is larger as compared with N–N and the corresponding bond dissociation energies follow the same trends.

Bayse and coauthors evaluated the bond strengths of various explosive molecules using the %ΔWBI index computed from DFT, a concept that was discussed in the previous section. Focusing on the X–NO_2 (X=N, C, O) trigger bonds, they found that %ΔWBI correlate well with impact sensitivity, especially for energetic materials which have intramolecular hydrogen bonds. In the case of 63 nitroaromatic molecules, %ΔWBI would be a better predictor of trigger bond strength compared to bond dissociation energies [56]. In another work on the common explosives RDX, HMX, TNT, PETN, including different conformations, and newly synthetized tetrazole molecules, the same group assigned trigger bonds based on the %ΔWBI index [55]. They found that axial and/or crowded nitro groups are more activated in cyclic nitramines (RDX and HMX), and for the novel tetrazole molecules, a relative scale of sensitivity could be established according to electronic effects of the functionalization of the molecule on the weakening of different bonds. Applying the same approach, the group also investigated the sensitivity of aromatic azide-based and organic-cage energetic materials [112,113].

3.4 Partition of the molecular charge density, especially the distributed multipole analysis (DMA) method

Klapötke et al. synthetized the 1,5-diamino-4-methyl-1H-tetrazolium dinitramide energetic material and reported employing AIM that the O···O and O···N interactions could be trigger bonds [114]. Pinkerton and coworkers also used AIM for crystals to investigate the electronic structure of pentaerythritol tetranitrate (PETN) and the β-phase of 1,3,5,7-tetranitro-1,3,5,7-tetraazacyclooctane (HMX). They found that several intermolecular bonding interactions, including O···H, O···O, and O···C, as determined and characterized by AIM analyses and indices, could be related to the sensitivity of the energetic material [54,115].

Rice and collaborators [16] looked for the origin of inconsistencies in the AIM description of the electronic density of isolated molecules of representative energetic materials assessing the feasibility of using AIM to model sensitivity. The authors employed DFT to nine molecules of various chemical families using the PW91, B3LYP, PBE exchange-correlation functionals and the 6-31G*, 6-311++G(2d,2p) and 6-311++G(3df,3pd) Gaussian basis sets. They found a significant variation in the electronic density AIM topology of molecules that form energetic materials crystals, concluding that these calculations could not currently be used to correlate the AIM results with the impact sensitivity of energetic materials.

We have been employing two methods to partition a molecular charge density computed from DFT, the deformed atoms-in-molecules (DAM) and mostly the distributed multipole analysis (DMA), to investigate the sensitivity and the electron (charge) distribution of energetic molecules. Both methods to partition the molecular charge density (global property) overcome the deficiency of atomic charges (local property) [32,46] by providing a much more accurate and detailed representation of the density that reflects more faithfully via atom-centered electric multipoles the properties of the molecular charge distribution. The DAM partition also has a similar accuracy. These works have been propelled and inspired by previous attempts, several of them described in this chapter.

We used the DAM partition and conceptual DFT to examine the electronic structure and the onset of decomposition processes of four conformers (with equatorial (E) or axial (A) nitro groups) of RDX [67]. We showed that equatorial NO_2 groups increase the delocalization of electrons in the ring, thus the larger the number of equatorial (E) nitro groups in RDX, the lower is the stability of the conformer, hence the lower the impact sensitivity. Additionally, we showed that the decomposition process of three conformers occurs through the cleavage of an N–NO_2 trigger bond. In contrast, for the AAE conformer, HONO elimination is favored, in agreement with previous works.

As described in the last section, DMA generates atom-centered monopole, dipole, and quadrupole electric multipoles that provide a detailed picture of the molecular charge distribution and have a clear chemical interpretation. In our studies, the molecular charge density was computed using DFT.

In a first work, we decomposed using DMA the molecular charge density of three diazocyclopropane molecules which bear a three-atom strained ring and are potential energetic materials [116]. We found that electron-withdrawing from the C atoms of the ring and charge build-up on the N atoms bonded to the ring increase with the number of diazo groups (which varies from one to three). These two effects were related to an easiness of cleavage of C–N bonds and increased sensitivity to impact.

In another investigation employing the DAM, and also DMA, we investigated the electronic structure and the decomposition process of the 1,1-diamino-2,2-dinitroethylene (FOX-7) molecule, a component of a recently developed Swedish explosive with low impact sensitivity—see [68]. In FOX-7 the central C=C bond plays the role of the ring in nitroaromatics. The carbon atoms of this bond have large quadrupole values, a measure of electron delocalization, and small positive charge values. Accordingly, DAM indicates an accumulation of electron density over the C=C bond—see Fig. 4.3.

According to a Coulombic model of bond strengths using DMA charges, in FOX-7 the strength of the C–NO_2 bonds are about three times smaller as compared with the strengths of the C–NH_2 bonds, which agrees with other results of dissociation barrier heights and indicates that the decomposition should start with the rupture of a C–NO_2 bond. Concerning sensitivity, the electron-donor amino groups, by donating electron density to the central C=C bond, tend to decrease the impact sensitivity. The molecular origin of the low sensitivity of FOX-7, taking into account also the existing intra- and intermolecular hydrogen bonding, could be rationalized.

We also investigated the electronic structure and developed different impact sensitivity models of 17 nitroaromatic molecules employing DMA multipoles [76]. This work was extended to 50 molecules [96], and the original conclusions were further confirmed for the larger set and by new sensitivity models. The main conclusions for nitroaromatic molecules are: (1) the charges on the carbon atoms of the aromatic ring become more positive with the increase of the number of nitro groups corresponding to carbon electrons being transferred to the electron-withdrawing nitro groups; (2) the dipole vectors (indicators of atomic polarization) of a ring carbon atom bonded to a nitro group is affected by the adjacent groups, with the largest dipole values being due to neighbor C–H bonds; (3) the total quadrupole moment of the aromatic ring decreases with the increase of the number of nitro groups, thus π-electron density of the rings decreases. From the proposed sensitivity models based on DMA multipoles, we found that the simplest model, which includes only the DMA Q_{nitro} nitro charges, predicted good h_{50} values. For this reason, we consider Q_{nitro} as the main molecular property derived from the molecular charge density related to the impact sensitivity of nitroaromatic explosives, which is in complete agreement with the proposal of Zhang and coworkers employing the not so precise Mulliken charges [20]. All the developed mathematical models of sensitivity included the number of nitro groups of the molecules because they confer the explosive character to materials and a different number of NO_2 groups per molecule modify the properties of explosives.

In those works, we also developed more complex models of impact sensitivity that showed that larger delocalized electron densities in the aromatic ring, quantified by larger values of the total quadrupole moment (equal to sum of the values of the ring C atoms), correspond to less sensitive explosives. In contrast, less electron injecting groups (e.g., an amine group) and more electron-withdrawing groups bonded to a ring result in less delocalized electrons in the ring hence a smaller h_{50} value (i.e., the material is more sensitive). Our findings agree with the discoveries of Politzer, Murray, and coworkers employing MEPs (the integrated nuclear and electron distributions over the volume) that indicated that the imbalance between charge delocalization and the electron-withdrawing effects of NO_2 groups are determinant for the impact sensitivity of an energetic compound. It also agrees with Rice and Hare that correlated the magnitude of the impact sensitivity to the degree of charge build-up over covalent bonds within the inner framework. The idea of positive electrostatic potential over nitroaromatic C–NO_2 was first established by the Politzer and Murray group in several works since 1984, as discussed above.

In more recent work, the impact sensitivity of 14 cyclic nitramine molecules was investigated using the DMA [117]. The set included primary nitramines, which have reactive acidic hydrogen, and secondary nitramines which do not have it. These molecules have N–NO_2 bonds and bear in the same structure C–NO_2 bonds, with cyclic, cage-like, and strained-ring nitramines being particularly a focus of interest. RDX and HMX are among the most useful and powerful explosives built from nitramine molecules. Several nitramine explosives are more sensitive as compared to nitroaromatic ones; a possible molecular reason for that is the fact that the former does not have aromatic rings characterized by large electronic delocalization—we found that quadrupole values of the carbon ring atoms in a nitramine molecule are typically about 40% smaller as compared with the ring carbons in the nitroaromatic molecules. In contrast, for secondary nitramines, the nitrogen atom of the nitramine group has quadrupole values comparable to nitroaromatic ring carbon atoms (in contrast with the C atoms in the heterocyclic ring of the nitramines) due to the resonance structure observed throughout the nitramine group, which confers a strong unsaturation character to the N–N bond. Furthermore, the positive charge values on the ring carbon atoms of the studied nitramines are appreciably larger (in some cases twice) as compared with the nitroaromatic values, thereby indicating lower electron concentration in the ring. The overall result is the typical smaller sensitivity of nitramine explosives as compared with the nitroaromatics. We showed that for the complete set of cyclic nitramine molecules, the charge of the nitramine group, and for two five-atom strained-ring molecules (TNAZ and TNTriCB), the quadrupole values of the heterocyclic ring atoms are the most important properties of the molecular charge distribution affecting impact sensitivity.

In another recent work, we employed DMA to study 33 nitrogen-rich heterocyclic nitroazole derivative molecules which contain a five-atom ring with one or three nitrogen atoms that results in three different chemical families: nitroimidazoles, nitrotriazoles, and nitropyrazoles [118]. These molecules bear a nitrogen-rich heterocyclic ring and have potential applications as energetic materials. Similarly to the aforementioned works on other families of energetic molecules, we found that a decrease of charge (electron) delocalization of ring atoms, and also the charge of the nitro groups, lead to more sensitive explosives. This property strongly depends on the ring position of the N atoms (because of the repulsion of their electron pairs in adjacent positions) and the bonding site of the substituent groups. The presence of electron injector groups (e.g., NH_2 and CH_3) bonded to the carbon atoms, similarly to nitroaromatic molecules, increases the charge of the nitro groups. In this case, the nitro group is less able to attract electrons from the ring; hence, electron delocalization increases. This effect contributes to the lower sensitivity of the molecules bearing these groups. The N/C ratio for the ring atom and the repulsion of the nonbonding electron pairs of the vicinal nitrogen atoms of the ring also plays an important role in the stability of nitroazoles, hence to their sensitivity because the larger this ratio, the smaller the electron delocalization, thus the larger the sensitivity.

In the aforementioned works, we have shown that DMA provides an accurate and detailed picture of the molecular density distribution for studying sensitivity. In contrast with other partition methods such as AIM, it is quite insensitive to the size of the basis set. Therefore, we could rationalize the electron structure of energetic molecules and develop models of impact sensitivity, which allowed us to establish the main molecular features contributing to this property, in agreement with other works and most importantly extending them.

4 Final remarks

In this chapter, we presented and discussed different findings obtained from sensitivity models of energetic materials to rationalize this macroscopic property. We have concentrated on those works which employed theoretical properties derived from molecular charge densities. These properties include atomic charges, electrostatic potentials, partition methods of the molecular charge (electron) distribution, bond orders, bond dissociation energies, and electronegativity.

We showed the development of the area from the beginning in the 1970s when crude, semiempirical, quantum-chemical methods using experimental crystal geometries were employed for determining those properties to the most recent works using sophisticated DFT methods. These results have brought information on the properties of the *molecular* charge distribution that affect the sensitivity of energetic materials, especially impact sensitivity. In particular, we discussed our group's results employing the accurate DMA partition of the molecular charge distribution, a nonintegrated property of the electron density that provides atom-centered electric multipoles that have straightforward chemical interpretation. The impact sensitivity models developed from the DMA multipoles confirmed and extended previous works thereby providing valuable information on the molecular charge density features that affect this important property of energetic materials.

Acknowledgments

One of the beauties of Science is being a collective enterprise. Our warmest thanks to researchers and anonymous referees that, throughout time, have inspired and improved so much of our work in the area of energetic materials. We have tried to recognize in this chapter their efforts concerning sensitivity models based on properties of the molecular charge densities. Past and present (CNPq research grants 304148/2018-0 and 409447/2018-8 and FAPERJ grant E26/201.197/2021) funding of this work I. B. J. from different Brazilian Agencies is greatly acknowledged as well as the support of the Brazilian Army through our Institute (IME).

References

[1] L.E. Fried, M.R. Manaa, P.F. Pagoria, R.L. Simpson, Design and synthesis of energetic materials, Ann. Rev. Mater. Res. 31 (2001) 291–321.

[2] M.M. Kuklja, E.V. Stefanovich, A.B. Kunz, An excitonic mechanism of detonation initiation in explosives, J. Chem. Phys. 112 (7) (2000) 3417–3423.

[3] A.A. Dippold, T.M. Klapotke, A study of dinitro-Bis-1,2,4-triazole-1,1 '-diol and derivatives: design of high-performance insensitive energetic materials by the introduction of N-oxides, J. Am. Chem. Soc. 135 (26) (2013) 9931–9938.

[4] T.M. Klapotke, B. Krumm, A. Widera, Synthesis and properties of tetranitro-substituted adamantane derivatives, ChemPlusChem 83 (1) (2018) 61–69.

[5] P. Politzer, J.S. Murray, Energetic Materials. Part 1. Decomposition, Crystal and Molecular Properties, Elsevier, Amsterdan, 2003.

[6] P. Politzer, J.S. Murray, Energetic Materials. Part 2. Detonation, Combustion, Elsevier, Amsterdan, 2003.

[7] R.W.B. Shaw, B. Thomas, D.L. Thompson, Overviews of Recent Research on Energetic Materials, World Scientific, New Jersey, 2005.

[8] S. Zeman, M. Jungova, Sensitivity and performance of energetic materials, Propellants Explos. Pyrotech. 41 (3) (2016) 426–451.

[9] P. Politzer, J.S. Murray, High performance, low sensitivity: conflicting or compatible? Propellants Explos. Pyrotech. 41 (3) (2016) 414–425.

[10] E.R. Bernstein, R.W. Shaw, T.B. Brill, D.L. Thompson, Overviews of Recent Research on Energetic Materials, Vol. 16, Advanced Series in Physical Chemistry, https://doi.org/10.1142/5759 2005.

[11] N. Kubota, Propellants and Explosives: Thermochemical Aspects of Combustions, second ed., WILEY-VCH Verlag GmbH & Co. KGaA, Weinheim, 2007.

[12] G. Li, C. Zhang, Review of the molecular and crystal correlations on sensitivities of energetic materials, J. Hazard. Mater. 398 (2020) 122910.

[13] J.S. Murray, P. Lane, P. Politzer, Effects of strongly electron-attracting components on molecular surface electrostatic potentials: application to predicting impact sensitivities of energetic molecules, Mol. Phys. 93 (2) (1998) 187–194.

[14] M.J. Kamlet, The relationship of impact sensitivity with structure of organic high explosives. 1. Polynitroaliphatic explosives, in: 6th International Symposium on Detonation, ONR Report ACR 221, San Diego, California, 1976, p. 312.

[15] M.J. Kamlet, H.G. Adolph, Relationship of impact sensitivity with structure of organic high explosives: polynitroaromatic explosives, Propellants Explos. 4 (2) (1979) 30–34.
[16] A.D. Yau, E.F.C. Byrd, B.M. Rice, An investigation of KS-DFT Electron densities used in atoms-in-molecules studies of energetic molecules, J. Phys. Chem. A 113 (21) (2009) 6166–6171.
[17] P.J. Rae, P.M. Dickson, Some observations about the drop-weight explosive sensitivity test, J. Dyn. Behav. Mater. 7 (2020) 414–424.
[18] B.M. Rice, J.J. Hare, A quantum mechanical investigation of the relation between impact sensitivity and the charge distribution in energetic molecules, J. Phys. Chem. A 106 (9) (2002) 1770–1783.
[19] C.Y. Zhang, Y.J. Shu, Y.G. Huang, X.D. Zhao, H.S. Dong, Investigation of correlation between impact sensitivities and nitro group charges in nitro compounds, J. Phys. Chem. B 109 (18) (2005) 8978–8982.
[20] C.Y. Zhang, Review of the establishment of nitro group charge method and its applications, J. Hazard. Mater. 161 (1) (2009) 21–28.
[21] J.S. Murray, M.C. Concha, P. Politzer, Links between surface electrostatic potentials of energetic molecules, impact sensitivities and C-NO_2/N-NO_2 bond dissociation energies, Mol. Phys. 107 (1) (2009) 89–97.
[22] P. Politzer, J.S. Murray, J.M. Serninario, P. Lane, M.E. Grice, M.C. Concha, Computational characterization of energetic materials, Theochem. J. Mol. Struct. 573 (2001) 1–10.
[23] B.M. Rice, E.F.C. Byrd, Theoretical chemical characterization of energetic materials, J. Mater. Res. 21 (10) (2006) 2444–2452.
[24] W.H. Koch, M.C. Holthausen, A Chemist's Guide to Density Functional Theory, second ed., Wiley-VCH, Weinheim, 2002.
[25] C.J. Cramer, Essentials of Computational Chemistry: Theories and Models, John Wiley & Sons, Chichester, 2004.
[26] M. A. Spackman. (Chapter 5). Charge densities from X-ray diffraction data, Annu. Rep. Sect. C Phys. Chem., 94(0), 177–207, 1998.
[27] C. Gatti, P. Macchi, A guided tour through modern charge density analysis, in: C. Gatti, P. Macchi (Eds.), Modern Charge-Density Analysis, Springer, Dordrecht, 2011.
[28] P. Coppens, A. Volkov, The interplay between experiment and theory in charge-density analysis, Acta Crystallogr. Sect. A 60 (2004) 357–364.
[29] D. Chopra, Advances in understanding of chemical bonding: inputs from experimental and theoretical charge density analysis, J. Phys. Chem. A 116 (40) (2012) 9791–9801.
[30] C. Gatti, P. Macchi, Modern Charge-Density Analysis, Springer, New York, 2012.
[31] D. Mathieu, Toward a physically based quantitative modeling of impact sensitivities, J. Phys. Chem. A 117 (10) (2013) 2253–2259.
[32] Z.X. Chen, H.M. Xiao, Quantum chemistry derived criteria for impact sensitivity, Propellants Explos. Pyrotech. 39 (4) (2014) 487–495.
[33] P. Politzer, J.S. Murray, Detonation performance and sensitivity: a quest for balance, in: R.S. John (Ed.), Advances in Quantum Chemistry, Academic Press, 2014, pp. 1–30 (Chapter one).
[34] S. Zeman, Sensitivities of high energy compounds, in: High Energy Density Materials, Springer-Verlag Berlin, Berlin, 2007, pp. 195–271.
[35] A.J. Stone, The Theory of Intermolecular Forces, Oxford University Press, Oxford, 2000.
[36] D.D. Dlott, Fast molecular processes in energetic materials, in: P. Politzer, J.S. Murray (Eds.), Energetic Materials. Part 2. Detonation, Combustion, Elsevier, Amsterdam, 2003, pp. 125–191.
[37] T.B. Brill, K.J. James, Thermal decomposition of energetic material. 61. Perfidy in the Amini-2,4,6-Trinitrobenzene series of explosives, J. Phys. Chem. 97 (34) (1993) 8752–8758.
[38] R.J. Gillespie, P.L.A. Popelier, Chemical Bonding and Molecular Geometry: From Lewis to Electron Densities, Oxford University Press, Oxford, 2001.
[39] C.F. Matta, R.J. Gillespie, Understanding and interpreting molecular electron density distributions, J. Chem. Educ. 79 (9) (2002) 1141–1152.
[40] R.F.W. Bader, Atoms in Molecules: A Quantum Theory, Clarendon Press, Oxford, 1994.
[41] C.J. Eckhardt, A. Gavezzotti, Computer simulations and analysis of structural and energetic features of some crystalline energetic Materials, J. Phys. Chem. B 111 (13) (2007) 3430–3437.
[42] J.S. Murray, P. Politzer, The electrostatic potential: an overview, Wiley Interdiscip. Rev. Comput. Mol. Sci. 1 (2) (2011) 153–163.

[43] J.S. Murray, P. Politzer, Molecular electrostatic potentials and noncovalent interactions, Wiley Interdiscip. Rev. Comput. Mol. Sci. 7 (6) (2017) 10.
[44] P. Politzer, L.N. Domelsmith, L. Abrahmsen, Electrostatic potentials of strained systems—cubane, homocubane, and bishomocubane, J. Phys. Chem. 88 (9) (1984) 1752–1758.
[45] P. Politzer, P. Lane, J.S. Murray, Sensitivities of ionic explosives, Mol. Phys. 115 (5) (2017) 497–509.
[46] F.D. Botelho, R.S.S. Oliveira, J.S.F.D. Almeida, T.C.C. Franca, I. Borges Jr., Comparação Entre Métodos Para Determinação De Cargas Atômicas Em Sistemas Moleculares: A Molécula N-{N-(Pterina-7-Il)Carbonilglicil}-L-Tirosina (NNPT), Química Nova 44 (2) (2021) 161–171.
[47] R. Kalescky, E. Kraka, D. Cremer, Identification of the strongest bonds in chemistry, J. Phys. Chem. A 117 (36) (2013) 8981–8995.
[48] X.F. Su, X.L. Cheng, C.M. Meng, X.L. Yuan, Quantum chemical study on nitroimidazole, polynitroimidazole and their methyl derivatives, J. Hazard. Mater. 161 (1) (2009) 551–558.
[49] L. Turker, T. Atalar, Quantum chemical study on 5-Nitro-2,4-Dihydro-3h-1,2,4-Triazol-3-one (NTO) and some of its constitutional isomers, J. Hazard. Mater. 137 (3) (2006) 1333–1344.
[50] Z. Jun, C. Xin-lu, H. Bi, Y. Xiang-dong, Neural networks study on the correlation between impact sensitivity and molecular structures for nitramine explosives, Struct. Chem. 17 (5) (2006) 501–507.
[51] X.S. Song, X.L. Cheng, X.D. Yang, D.H. Li, R.F. Linghu, Correlation between the bond dissociation energies and impact sensitivities in nitramine and polynitro benzoate molecules with polynitro alkyl groupings, J. Hazard. Mater. 150 (2) (2008) 317–321.
[52] T. Atalar, M. Jungova, S. Zeman, A new view of relationships of the N-N bond dissociation energies of cyclic nitramines. Part ii. Relationships with impact sensitivity, J. Energ. Mater. 27 (3) (2009) 200–216.
[53] E. Kraka, D. Cremer, Characterization of CF bonds with multiple-bond character: bond lengths, stretching force constants, and bond dissociation energies, ChemPhysChem 10 (4) (2009) 686–698.
[54] E.A. Zhurova, A.I. Stash, V.G. Tsirelson, V.V. Zhurov, E.V. Bartashevich, V.A. Potemkin, A.A. Pinkerton, Atoms-in-molecules study of intra- and intermolecular bonding in the pentaerythritol tetranitrate crystal, J. Am. Chem. Soc. 128 (45) (2006) 14728–14734.
[55] L.K. Harper, A.L. Shoaf, C.A. Bayse, Predicting trigger bonds in explosive materials through Wiberg bond index analysis, ChemPhysChem 16 (18) (2015) 3886–3892.
[56] A.L. Shoaf, C.A. Bayse, Trigger bond analysis of nitroaromatic energetic materials using Wiberg bond indices, J. Comput. Chem. 39 (19) (2018) 1236–1248.
[57] O.V. Sizova, L.V. Skripnikov, A.Y. Sokolov, Symmetry decomposition of quantum chemical bond orders, Theochem. J. Mol. Struct. 870 (1–3) (2008) 1–9.
[58] L. Pauling, The Nature of the Chemical Bond, Cornell University Press, Ithaca, NY, 1960.
[59] G.D. Sproul, Evaluation of electronegativity scales, ACS Omega 5 (20) (2020) 11585–11594.
[60] J.F. Rico, R. Lopez, I. Ema, G. Ramirez, Deformed atoms in molecules: analytical representation of atomic densities for Gaussian type orbitals, Theochem. J. Mol. Struct. 727 (1–3) (2005) 115–121.
[61] R.F.W. Bader, A quantum theory of molecular structure and its applications, Chem. Rev. 91 (5) (1991) 893–928.
[62] R. Lopez, J.F. Rico, G. Ramirez, I. Ema, D. Zorrilla, DAMQT: a package for the analysis of electron density in molecules, Comput. Phys. Commun. 180 (9) (2009) 1654–1660.
[63] J.F. Rico, R. Lopez, G. Ramirez, Analysis of the molecular density, J. Chem. Phys. 110 (9) (1999) 4213–4220.
[64] J.F. Rico, R. Lopez, I. Ema, G. Ramirez, Analysis of the molecular density: STO densities, J. Chem. Phys. 117 (2) (2002) 533–540.
[65] R.S. Alvim, I. Borges, D.G. Costa, A.A. Leitao, Density-functional theory simulation of the dissociative chemisorption of water molecules on the MgO(001) surface, J. Phys. Chem. C 116 (1) (2012) 738–744.
[66] T. Guerra, I. Borges, Adsorption of trinitrotoluene on a MgO(001) surface including surface relaxation effects, J. Chem. 359202 (2013) 1–8.
[67] T. Moraes, I. Borges, Nuclear Fukui functions and the deformed atoms in molecules representation of the electron density: application to gas-phase RDX (Hexahydro-1,3,5-Trinitro-1,3,5-Triazine) electronic structure and decomposition, Int. J. Quantum Chem. 111 (7–8) (2011) 1444–1452.
[68] T. Giannerini, I. Borges, Molecular electronic topology and fragmentation onset via charge partition methods and nuclear Fukui functions: 1,1-diamino-2,2-dinitroethylene, J. Braz. Chem. Soc. 26 (5) (2015) 851–859.
[69] P. Geerlings, E. Chamorro, P.K. Chattaraj, F. De Proft, J.L. Gázquez, S. Liu, C. Morell, A. Toro-Labbé, A. Vela, P. Ayers, Conceptual density functional theory: status, prospects, issues, Theor. Chem. Acc. 139 (2) (2020) 36.

[70] A.J. Stone, Distributed multipole analysis, or how to describe a molecular charge-distribution, Chem. Phys. Lett. 83 (2) (1981) 233–239.
[71] A.J. Stone, M. Alderton, Distributed multipole analysis—methods and applications, Mol. Phys. 56 (5) (1985) 1047–1064.
[72] S.L. Price, A.J. Stone, A distributed multipole analysis of charge-densities of the azabenzene molecules, Chem. Phys. Lett. 98 (5) (1983) 419–423.
[73] I. Borges, A.M. Silva, A.P. Aguiar, L.E.P. Borges, J.C.A. Santos, M.H.C. Dias, Density functional theory molecular simulation of Thiophene adsorption on MoS_2 including microwave effects, Theochem. J. Mol. Struct. 822 (1–3) (2007) 80–88.
[74] I. Borges, A.M. Silva, Probing topological electronic effects in catalysis: thiophene adsorption on NiMoS and CoMoS clusters, J. Braz. Chem. Soc. 23 (10) (2012) 1789–1799.
[75] A.J. Stone, The Theory of Intermolecular Forces, Clarendon Press, Oxford, 1997.
[76] G. Anders, I. Borges, Topological analysis of the molecular charge density and impact sensitivity models of energetic molecules, J. Phys. Chem. A 115 (32) (2011) 9055–9068.
[77] F.J. Owens, K. Jayasuriya, L. Abrahmsen, P. Politzer, Computational analysis of some properties associated with the nitro groups in polynitroaromatic molecules, Chem. Phys. Lett. 116 (5) (1985) 434–438.
[78] J.S. Murray, P. Lane, P. Politzer, P.R. Bolduc, A relationship between impact sensitivity and the electrostatic potentials at the midpoints of C-NO2 bonds in nitroaromatics, Chem. Phys. Lett. 168 (2) (1990) 135–139.
[79] J.S. Murray, P. Lane, P. Politzer, Relationships between impact sensitivities and molecular-surface electrostatic potentials of nitroaromatic and nitroheterocyclic molecules, Mol. Phys. 85 (1) (1995) 1–8.
[80] J.S. Murray, T. Brinck, P. Lane, K. Paulsen, P. Politzer, Statistically-based interaction indexes derived from molecular-surface electrostatic potentials—a general interaction properties function (GIPF), Theochem. J. Mol. Struct. 113 (1994) 55–64.
[81] A. Delpuech, J. Cherville, Relation between shock sensitiveness of secondary explosives and their molecular electronic-structure. 1. Nitroaromatics and nitramines, Propellants Explos. 3 (6) (1978) 169–175.
[82] A. Delpuech, J. Cherville, Relation between shock sensitiveness of secondary explosives and their molecular electronic-structure. 2. Nitrate esters, Propellants Explos. 4 (6) (1979) 121–128.
[83] A. Delpuech, J. Cherville, Relation between shock sensitiveness of secondary explosives and their molecular electronic-structure. 3. Influence of crystal environment, Propellants Explos. 4 (3) (1979) 61–65.
[84] X. Xiong, X. He, Y. Xiong, X. Xue, H. Yang, C. Zhang, Correlation between the self-sustaining ignition ability and the impact sensitivity of energetic materials, Energ. Mater. Front. 1 (2020) 40–49.
[85] J.L. Zhao, C.Y. Zhi, F. Zhao, S.Q. Feng, X.L. Cheng, Relationship between electric spark sensitivity of cyclic nitramines and their molecular electronic properties, Chin. J. Struct. Chem. 31 (9) (2012) 1263–1270.
[86] F.J. Owens, Relationship between impact induced reactivity of Trinitroaromatic molecules and their molecular-structure, Theochem. J. Mol. Struct. 22 (1985) 213–220.
[87] J. Mullay, A relationship between impact sensitivity and molecular electronegativity, Propellants Explos. Pyrotech. 12 (2) (1987) 60–63.
[88] J. Mullay, Relationships between impact sensitivity and molecular electronic structure, Propellants Explos. Pyrotech. 12 (4) (1987) 121–124.
[89] Z.X. Chen, H.I. Xiao, S.L. Yang, Theoretical investigation on the impact sensitivity of tetrazole derivatives and their metal salts, Chem. Phys. 250 (3) (1999) 243–248.
[90] Z.X. Chen, H.M. Xiao, Impact sensitivity and activation energy of pyrolysis for tetrazole compounds, Int. J. Quantum Chem. 79 (6) (2000) 350–357.
[91] X.J. Xu, H.M. Xiao, X.D. Gong, X.H. Ju, Z.X. Chen, Theoretical studies on the vibrational spectra, thermodynamic properties, detonation properties, and pyrolysis mechanisms for polynitroadamantanes, J. Phys. Chem. A 109 (49) (2005) 11268–11274.
[92] A.D. Stephen, P. Kumaradhas, R.B. Pawar, Charge density distribution, electrostatic properties, and impact sensitivity of the high energetic molecule TNB: a theoretical charge density study, Propellants Explos. Pyrotech. 36 (2) (2011) 168–174.
[93] A.D. Stephen, P. Srinivasan, P. Kumaradhas, Bond charge depletion, bond strength and the impact sensitivity of high energetic 1,3,5-Triamino 2,4,6-Trinitrobenzene (TATB) molecule: a theoretical charge density analysis, Comput. Theor. Chem. 967 (2–3) (2011) 250–256.
[94] C.Y. Zhang, Y.J. Shu, Y.G. Huang, X.F. Wang, Theoretical investigation of the relationship between impact sensitivity and the charges of the nitro Group in Nitro Compounds, J. Energ. Mater. 23 (2) (2005) 107–119.

[95] C.Y. Zhang, Investigation of the correlations between nitro group charges and some properties of nitro organic compounds, Propellants Explos. Pyrotech. 33 (2) (2008) 139–145.
[96] R.S.S. Oliveira, I. Borges Jr., Correlation between molecular charge properties and impact sensitivity of explosives: nitrobenzene derivatives, Propellants Explos. Pyrotech. 46 (2) (2021) 308–321
[97] C.Z. Cao, S. Gao, Two dominant factors influencing the impact sensitivities of nitrobenzenes and saturated nitro compounds, J. Phys. Chem. B 111 (43) (2007) 12399–12402.
[98] Q. Cao, Dinitroamino benzene derivatives: a class new potential high energy density compounds, J. Mol. Model. 19 (6) (2013) 2205–2210.
[99] C.Y. Zhi, X.L. Cheng, The correlation between electric spark sensitivity of polynitroaromatic compounds and their molecular electronic properties, Propellants Explos. Pyrotech. 35 (6) (2010) 555–560.
[100] M.H. Keshavarz, Theoretical prediction of electric spark sensitivity of nitroaromatic energetic compounds based on molecular structure, J. Hazard. Mater. 153 (1–2) (2008) 201–206.
[101] M.H. Keshavarz, M.H. Moghadas, M.K. Tehrani, Relationship between the electrostatic sensitivity of nitramines and their molecular structure, Propellants Explos. Pyrotech. 34 (2) (2009) 136–141.
[102] D. Mathieu, T. Alaime, Predicting impact sensitivities of nitro compounds on the basis of a semi-empirical rate constant, J. Phys. Chem. A 118 (41) (2014) 9720–9726.
[103] D. Mathieu, Physics-based modeling of chemical hazards in a regulatory framework: comparison with quantitative structure-property relationship (QSPR) methods for impact sensitivities, Ind. Eng. Chem. Res. 55 (27) (2016) 7569–7577.
[104] D. Mathieu, Sensitivity of energetic materials: theoretical relationships to detonation performance and molecular structure, Ind. Eng. Chem. Res. 56 (29) (2017) 8191–8201.
[105] P. Politzer, J.S. Murray, P. Lane, P. Sjoberg, H.G. Adolph, Shock-sensitivity relationships for nitramines and nitroaliphatics, Chem. Phys. Lett. 181 (1) (1991) 78–82.
[106] D. Mathieu, Theoretical shock sensitivity index for explosives, J. Phys. Chem. A 116 (7) (2012) 1794–1800.
[107] H.M. Xiao, Z.Y. Wang, J.M. Yao, Quantum chemical study on sensitivity and stability of aromatic nitro explosives. 1. Nitro-derivatives of aminobenzenes, Acta Chim. Sin. 43 (1) (1985) 14–18.
[108] H. Xiao, Y. Li, Banding and electronic structures of metal azides—sensitivity and conductivity, Sci. Sin. B 38 (5 Series B) (1995) 538–545.
[109] H.J. Song, B.H. Yu, H.M. Xiao, The ab initio confirmation of "the principle of the smallest bond order"—nitro derivatives of benzene and aminobenzene, Chin. J. Chem. Phys. 16 (5) (2003) 337–338.
[110] H.-M. Xiao, J.-F. Fan, Z.-M. Gu, H.-S. Dong, Theoretical study on pyrolysis and sensitivity of energetic compounds: (3) nitro derivatives of aminobenzenes, Chem. Phys. 226 (1) (1998) 15–24.
[111] J. Fan, Z. Gu, H. Xiao, H. Dong, Theoretical study on pyrolysis and sensitivity of energetic compounds. Part 4. Nitro derivatives of phenols, J. Phys. Org. Chem. 11 (3) (1998) 177–184.
[112] A.L. Shoaf, C.A. Bayse, The effect of nitro groups on N-2 extrusion from aromatic azide-based energetic materials, New J. Chem. 43 (38) (2019) 15326–15334.
[113] C.A. Bayse, M. Jaffar, Bonding analysis of the effect of strain on trigger bonds in organic-cage energetic materials, Theor. Chem. Acc. 139 (6) (2020) 11.
[114] T.M. Klapötke, P. Mayer, A. Schulz, J.J. Weigand, 1,5-Diamino-4-methyltetrazolium dinitramide, J. Am. Chem. Soc. 127 (7) (2005) 2032–2033.
[115] E.A. Zhurova, V.V. Zhurov, A.A. Pinkerton, Structure and bonding in B-HMX-characterization of a trans-annular N···N interaction, J. Am. Chem. Soc. 129 (45) (2007) 13887–13893.
[116] I. Borges, Conformations and charge distributions of diazocyclopropanes, Int. J. Quantum Chem. 108 (13) (2008) 2615–2622.
[117] M.A.S. Oliveira, I. Borges, On the molecular origin of the sensitivity to impact of cyclic nitramines, Int. J. Quantum Chem. 119 (8) (2019) 14.
[118] R.S.S. de Oliveira, I. Borges, Correlation between molecular charge densities and sensitivity of nitrogen-rich heterocyclic nitroazole derivative explosives, J. Mol. Model. 25 (10) (2019) 314.

CHAPTER

5

Estimation methods for sensitivities to various stimuli

Aurélien Demenay[a], *Laurent Catoire*[a,*], *and Antoine Osmont*[b,*]

[a]ENSTA Paris, Institut Polytechnique de Paris, Palaiseau, France [b]CEA, DAM, Gramat, Gramat, France

*Corresponding authors: E-mail: laurent.catoire@ensta-paris.fr; antoine.osmont@cea.fr

1 Introduction

Because of the continuous need for improved energetic materials for various applications, scientists try to design molecules having greater energetic potential. This performance request is coupled with an insensitivity request that is difficult to obtain at the same time. Sensitivity refers to the degree of response of energetic materials caused by external stimuli. The most important stimuli include impact, shock, electrostatic discharges, heat, and friction [1]. The ease of energetic materials manipulation is limited because of potential hazards due to the high energy of molecules.

Synthesizing a new molecule in large amounts is very demanding work and the sensitivity evaluation requires several dozens of grams of the new compound. Furthermore, once the first crystals are synthesized, the question of the safety of handling after drying is difficult to answer. Therefore, it is interesting to use numerical tools to evaluate *in silico* the performance of a new molecular structure. Empirical and semi-empirical methods have been developed to reduce costs generated by experimental devices, reduce hazards related to energetic materials handling and allow investigations of a larger range of compounds. Large efforts are made on performance, sensitivity, and physical and thermodynamic properties determination. Computer power development contributes to the development of these methods. However theoretical investigation is difficult to undertake because of numerous parameters including molecular structure, crystal structure, electronic structure, energy rate transfer, and chemical bonds [2].

https://doi.org/10.1016/B978-0-12-822971-2.00003-6

Safe handling in real conditions of usage, transport, and storage for scientists and engineers requires the development of insensitive energetic materials. The selection of usable compounds needs good accuracy of prediction in order to classify molecules by level of danger for a large range of molecules, even specific ones. Therefore the purpose of this paper is to review some predicting models established for the sensitivity of energetic materials depending on a wide range of stimuli. For completeness and to avoid redundancies, our review of methods to estimate sensitivities to impact, shock, and heat focus on methods not mentioned in detail in the other chapters. More advanced approaches to such properties are reported in Chapters 7, 10, and 14. Some advantages and drawbacks of each method will be highlighted.

2 Estimation methods

Nowadays, energetic materials are qualified thanks to normalized safety tests which reproduce real conditions, as described in Chapter 2. This chapter mainly focuses on empirical models derived from data obtained for different kinds of stimuli.

The drop hammer test is used to study an impact. The drop height at which an energetic material decomposes with a probability equal to 50% ($h_{50\%}$) is recorded.

Concerning shock sensitivity, a shock wave of tunable amplitude is applied on an energetic material using a variable thickness attenuator to determine the threshold pressure leading to the initiation of 50% of the samples studied.

For the spark sensitivity test, an electrostatic discharge is applied to the energetic compound. Energy at which energetic materials decomposed is recorded (E_{ES}).

Thermal stability and heat sensitivity are analyzed by differential scanning calorimeter and differential thermal analysis devices which allow measurement of decomposition temperature (T_D) and activation energy (E_A).

The last test is the evaluation of friction sensitivity, which is important when extracting an explosive block from a mold, or when an explosive is present in a screwed assembling for example.

Because the microscopic decomposition processes of energetic materials are not well understood, numerous more or less empirical approaches are employed to estimate the sensitivity of energetic materials. The simplest involves elementary composition and structural parameters. Another one uses the assumption that decomposition is initiated from the breaking of the weakest bond of molecules. Some neural networks and QSPR methods use a large set of compounds to identify the main properties involve. Some methods involve resonance energy or electronic properties. Physicochemical models are also developed. The following sections summarize these methods to predict energetic materials sensitivity.

3 Impact sensitivity

Impact sensitivity is the most studied property and it is virtually impossible to refer to all the papers published yearly. Therefore this article is not an exhaustive review of existing methods as new ones are regularly proposed [3].

3.1 Methods based on elementary composition and structural parameters

Simple empirical correlations are used to predict impact sensitivity. Their major advantage is the simplicity of calculations without the need for quantum chemistry and a high-speed computer.

Initial works done on impact sensitivity are simple relations between $h_{50\%}$ and elementary composition and specific groups of CHNO energetic materials. In that way, Kamlet et al. developed linear correlations between $\log(h_{50\%})$ and the oxygen balance (OB_{100}) for nitramine, nitroaliphatic, and nitroaromatic with a distinction in subclasses for molecules having α—CH linkage [4]. As seen in Eq. (5.1), OB_{100} is calculated from elementary composition, a number of COO groups, and molecular weight (MW). Eqs. (5.2)–(5.5) are reported in [4].

$$OB_{100} = \frac{100(2n_o - n_H - 2n_c - n_{coo})}{MW} \tag{5.1}$$

Nitroaromatic:

$$\log h_{50} = 1.73 - 0.32\ OB_{100} \tag{5.2}$$

Nitroaromatic with α—CH linkage:

$$\log h_{50} = 1.33 - 0.26\ OB_{100} \tag{5.3}$$

Nitroaliphatic:

$$\log h_{50} = 1.74 - 0.23\ OB_{100} \tag{5.4}$$

Nitramine:

$$\log h_{50} = 1.37 - 0.17\ OB_{100} \tag{5.5}$$

Storm et al. [5] established a large experimental database for several classes of energetic materials. They include the sensitivity index (SI) which takes into account elementary composition and the number of CO groups. Eq. (5.6) defines the SI parameter and corresponding relations (5.6)–(5.10) developed by Storm et al. are given below. Results from OB_{100} and SI parameters are reasonable linear correlations for families of molecules having similar structures but they have no external validation and deviation in some cases is large.

$$SI = 100\frac{\left(n_o - n_c - \left(\frac{n_H}{2}\right) - n_{co}\right)}{n_H(n_c + n_H + n_N + n_o)} \tag{5.6}$$

- Polynitroanilines:

$$\log h_{50} = 1.75 - 0.24SI;\ R^2 = 0.997 \tag{5.7}$$

– Polynitrobenzenes:

$$\log h_{50} = 1.70 - 0.13\ \mathrm{SI};\ R^2 = 0.938 \tag{5.8}$$

– Polyaminopolynitrobenzenes:

$$\log h_{50} = 0.58 - 0.84\ \mathrm{SI};\ R^2 = 0.998 \tag{5.9}$$

– Polynitrophenols:

$$\log h_{50} = 1.72 - 0.28\ \mathrm{SI};\ R^2 = 0.895 \tag{5.10}$$

Keshavarz et al. [6] kept the idea that sensitivity has to be predicted in a simple way and developed empirical correlations for large classes of explosives. Relations (5.11)–(5.13) appear using elementary composition and molecular weight only for $C_aH_bN_cO_d$ polynitroaromatics, benzofuroxanes, nitramines, and polynitroaliphatics.

– Polynitroaromatics (and benzofuranes):

$$\log h_{50} = \frac{11.76a + 61.72b + 26.89c + 11.48d}{\mathrm{MW}} \tag{5.11}$$

– Polynitroaromatics with α—CH and α—N—CH bonds and nitramines:

$$\log h_{50} = \frac{47.33a + 23.50b + 2.357c - 1.105d}{\mathrm{MW}} \tag{5.12}$$

– Polynitroaliphatics:

$$\log h_{50} = \frac{81.40a + 16.11b - 19.08c + 1.089d}{\mathrm{MW}} \tag{5.13}$$

As seen in Eq. (5.14) in including a specific parameter such as number of R—C(NO_2)$_2$—CH_2- group ($n_{R-C(NO_2)_2-CH_2-}$) where R is an alkyl group, they predict impact sensitivity of nitroaliphatic and nitrate explosives [7,8].

– nitroaliphatics, nitroaliphatics having other functional groups and explosives nitrates:

$$\log h_{50} = 2.47 + 0.371\left[\frac{100\left(a + \frac{b}{2} - d\right)}{\mathrm{MW}}\right] - 0.485\left[\frac{100c}{\mathrm{MW}}\right] + 0.185 n_{R-C(NO_2)_2-CH_2-} \tag{5.14}$$

As regards nitroheterocycles including nitropyridines, nitroimidazoles, nitropyrazoles, nitrofurazanes, nitrotriazoles, and nitropyrimidines, a good correlation is found using two structural parameters. There are the numbers of—CNC- (n_{-CNC-}) and —CNNC- (n_{-CNNC-}) moieties in aromatic rings divided by molecular weight [7,8]. Model (5.15) is valid for $C_aH_bN_cO_d$ nitroheterocycles compounds.

$$\log h_{50} = \frac{\begin{array}{c}46.2923a + 35.6305b - 7.7005c + 7.9425d \\ +44.4167n_{-CNC-} + 102.2749n_{-CNNC-}\end{array}}{MW} \quad (5.15)$$

Then a model for insensitive molecules as polynitroheterocyles takes into account the sum of specific structural parameters [9]. In order to have a general model including nitroaromatics, benzofuroxanes, nitroaliphatic, and nitrate esters, Keshavarz used elementary composition, molecular weight, DSSP, and ISSP properties. Values of the two last parameters depend on the family of compound, presence of specific groups, and presence of specific linkage [10]. The general model is presented in Eq. (5.16)

$$\log h_{50} = \frac{\begin{array}{c}48.81a + 25.94b + 13.73c - 4.786d \\ +111.6\mathrm{DSSP} + 132.3\mathrm{ISSP}\end{array}}{MW} \quad (5.16)$$

To stay on this topic, Lai et al. [11] added a factor indicating the connective position of groups in the global Eq. (5.17). They found a better agreement compared to Keshavarz global model.

$$\log h_{50} = Aa' + Bb' + Cc' + Dd' + E + F \quad (5.17)$$

$$F = \sum Nn'$$

where

- A, B, C, D, E, and N: coefficients allowing having the lowest error in comparison with experimental data
- a', b', c', d' and n': ratio of the atomic numbers of C, H, N, and O and groups with different connections to the molecular weight
- F: factor indicating the effect of connective position of groups

This approach is commonly used for qualitative research and ranks energetic molecules in the function of their impact sensitivity. In order to have accurate results, it is important to determine how much energy is needed to initiate such compounds. The following section concerns the quantification of energy to break the weakest bond of molecules.

3.2 Methods based on bond dissociation energy

Decomposition of energetic materials occurs through many mechanisms sometimes in a condensed phase which makes associated activation energy difficult to estimate. In some cases, assumptions are exposed in order to link the sensitivity of compounds and molecular characteristics. One of them admits that decomposition occurs through the rupture of weak bonds. Generally, these linkages are X—NO_2 bonds for energetic materials (X=C, N, O). This phenomenon is extremely important concerning decomposition kinetic. Studying nitramines and polynitroarenes, Zeman and Krupka exposed that sensitivity is linked to the first reaction during impact [12,13].

Song et al. established correlations between bond dissociation energy (BDE) and impact sensitivity using DFT calculation with B3LYP/6-31G* level of theory for some explosives [14]. For such purposes, BDE is calculated for removing NO_2 group from the compound. As an example, the BDE of 1-dinitromethyl-3 nitrobenzene is determined as followed in Eq. (5.18).

$$\mathrm{BDE} = [E(C_7H_5N_2O_4) + E(NO_2) - E(C_7H_5N_3O_6)] \tag{5.18}$$

The first conclusion explains that C—NO_2 bond is weaker than N—NO_2 bond. When an energetic material incurs a mechanical deformation, thanks to Arrhenius law, the kinetic of reaction depends on the ratio of activation barrier and temperature [15]. Thereby Fried et al. explained that decomposition energy is related to the molecular total energy (E) for compounds having similar structures. In that case, relations between $h_{50\%}$ and the ratio BDE/E by Song et al. give reasonable predictive results for nitramines, nitroaromatics, nitrobenzoates, and nitroesters [16]. The good predictability of this ratio is confirmed by Atalar et al. in a nitramines study thanks to DFT calculations [17]. Zhang et al. investigated nitro groups charge on the basis of DFT and general gradient approximation (GGA) [18]. The associated nitro group charge method (NGCM) involves Q_{nitro} calculation through the following equation for X—NO_2 bond:

$$Q_{\mathrm{nitro}} = Q_N + Q_{O1} + Q_{O2} \tag{5.19}$$

This factor represents the capacity of nitro group to attract electrons. The more negative the charge, the less the NO_2 group attracts electrons, and the more the compound is stable. Correlation between Q_{nitro} and the opposite of $h_{50\%}$ are not in good agreement) because more properties have to be involved. However, they indicated that an energetic material is sensitive for $h_{50\%}$ inferior to 40 cm which corresponds to Q_{nitro} equal to 0.23e. In a further study, Zhang et al. compute by DFT theory method SIESTA the bandgap of several polynitroaromatic explosives for which the weakest bond is the one between the aromatic ring and the NO_2 group [18]. They explained BDE alone cannot predict impact sensitivity accurately. Whereas considering BDE and the band gap value of the crystal state together can allow good results.

Another approach based on acid–base reaction formalism is used by Koch to predict sensitivity of organic as well as inorganic explosives [19,20]. The hard and soft acid–base principle (HSAB) is based on a number of electrons transferred in an acid–base reaction (ΔN) during the scission of the weakest bond. Results for 1,3,5-trinitrobenzene derivatives organic compounds give a good linear correlation coefficient between $h_{50\%}$ and ΔN. Metal azides inorganic compounds are studied *via* electrons transfer for the reaction of M^{n+}/N_3^- to establish correlation. Sensitivity seems to well correlate with covalent bond degree. Azides having a high covalent bond degree are really sensitive whereas ionic azides are relatively insensitive.

3.3 Methods based on the electrostatic potential

Electronic properties are studied by many researchers to correlate with impact sensitivity. In that way, Owens and his team investigate electrostatic potential using quantum chemistry for CHNO explosives [21]. This parameter is due to electronic charges distribution in molecules and is usually used in identifying sites within molecules that might be subject to nucleophilic or electrophilic attack. Some further works by Murray, Politzer et al. [22–24] develop generalized interaction property function (GIPF). Parameters involved in this method are used by Rice and Hare to develop five models. The first model used the electrostatic potential at the midpoint of each bond ($\overline{\overline{V}}_{\mathrm{mid}}$). The difference between the magnitudes of the averages of the positive and the negative values of the electrostatic potential on the isosurface $\left(\left|\overline{V}_S^+ - \left|\overline{V}_S^-\right|\right|\right)$ is the basis of the second model. The balance parameter (v) in model number three is a statistical quantity associated with the electrostatic potential of the molecule.

The fourth model uses heats of detonation (Q_{det}). A hybrid model combining the two previous models (v and Q_{det}) corresponds to model 5. These 5 models are obtained using 34 energetic materials and are presented below:

Model 1:

$$h_{50\%} = y_0 + a \exp\left(-b\overline{\overline{V}}_{mid}\right) + c\overline{\overline{V}}_{mid} \tag{5.20}$$

Model 2:

$$h_{50\%} = a_1 + a_2 \exp\left(-a_3\left|\overline{V}_S^+ - \left|\overline{V}_S^-\right|\right|\right) \tag{5.21}$$

Model 3:

$$h_{50\%} = a_1 + a_2 \exp(a_3 v) \tag{5.22}$$

Model 4:

$$h_{50\%} = a_1 + a_2 \exp[-a_3[Q - a_4]] \tag{5.23}$$

Model 5:

$$h_{50\%} = a_1 \exp(a_2 v - a_3[Q - a_4]) \tag{5.24}$$

Thanks to previous works, Murray et al. [24] showed that NO_2 and aza group are responsible for the unusual value of surface potential and for the weakness of C—NO_2 and/or N—NO_2 bonds. Previous researches explained the initiation of decomposition for nitroaromatics and nitroaliphatics from C—NO_2 bond scission. Concerning nitramines, nitrate esters, and organic azides, the first split is done, respectively from N—NO_2, O—NO_2, and N—N_2 breaking bonds. They finally connect v and positive variance on the molecular surface (σ_+^2).

An important point is updated by Pospisil et al. in a recent study [25]. It concerns the existence of a relation between sensitivity and property that does not necessarily implicate the latter as a causative factor. It can simply be symptomatic of something else. As an example bond strengths and lengths are affected by polarization of electronic charge density which evolves in function of electrostatic potential, atomic charges, electronegativities, NMR shifts, and substituent constants. Therefore it is important to consider correlation causative parameters. Besides according to the complexity of sensitivity, it should not reduce its estimation to a single parameter. In that way, Pospisil used DFT-B3PW91/6-31G calculation on 20 energetic materials from different families to determine the available free space per molecule unit cell (ΔV). It is expressed as a difference between the effective volume per molecule that would be required to completely fill the unit cell (V_{eff}) and the intrinsic gas-phase molecular volume (V_{int}) [25]. Finally, a good correlation between $h_{50\%}$ and both crystal and electronic properties (ΔV and σ_{tot}^2) is obtained for this set of compounds. σ_{tot}^2 being the total variance on the molecular surface.

3.4 QSPR and neural networks methods

The quantitative structure–property (QSPR) and artificial neural network both select geometric, molecular, electronic, crystalline, or topological properties called descriptors to predict sensitivity. A large set of compounds is necessary for both methods and permit to cover various kinds of energetic molecules from several families.

The artificial neural network is developed copying the nervous system where multiple neurons are interconnected. In science, neurons are considered as a processor operating in parallel. They are programmed to build mathematic models including descriptors. Some scientists used this approach to predict sensitivity [26]. It provides advantages as computing speed, ability to obtain information, and fault tolerance. Using a database of 204 molecules, Nefati et al. [27] observed a reduced number of descriptors and non-linear terms compared to linear regression provide better results [27]. Cho et al. pointed to the importance of electronic descriptors including LUMO_MOPAC (lowest unoccupied molecular orbital from MOPAC), HOMO_MOPAC (highest occupied molecular orbital from MOPAC), Dipole_MOPAC (dipole moment from MOPAC), and HF_MOPAC (heat of formation from MOPAC) [26].

The QSPR principle is to establish a quantitative relationship between descriptors, and a macroscopic observable (physicochemical property, *etc.*), for a set of similar compounds thanks to analysis treatment methods. Afanas'ev et al. obtained non-accurate results at the molecular level because of the lack of a clear physical model to describe the decomposition mechanism [28]. Electronic properties such as the energy of the highest occupied molecular orbital (ε_{HOMO}), dipole moment (DP), and ionization potential (IP) are involved by Badders et al. in correlations with $\log(h_{50\%})$ [29]. Fayet et al. used the QSPR method to build models for several families (nitramines, nitroaliphatics, and nitroaromatics) and a global one [30]. Good modeling results are found after multi-linear regressions and are validated with the OECD principle. Apart from nitroaromatics, the global model is reliable. Fayet et al. explain chosen descriptors are judicious in the sense that they are directly related to the properties of the nitro group.

3.5 Methods based on crystal framework

Energetic materials are in most cases in a solid-state. Therefore crystal structure must be taken into account in sensitivity prediction. The presence of defects might provoke hot spot formation [31]. The hot spot theory explains energy can accumulate in some zones in the crystalline structure. Under impact, the energy transferred is stored in these areas with temperature rises in these specific points. Under high temperatures, hot spots decomposition corresponds to ignition of energetic material. Hot spot formation is governed by material morphology, structural defects concentration, and rheological properties. In a microscopic view, these heterogeneities are found either punctually, either linearly as vacancy, dislocation, inclusion, or pore. Then in order to understand structural defects' influence, two approaches are explained. The presence of defects influences the crystal density. The higher the density of the material is, the more insensitive it is because of the reduction of cavities. In other words, it is a reduction of active specific surface of energetic material. Nevertheless, it might be possible that a material having a lot of defects is insensitive. A part of the impact energy transferred will be divided throughout all cavities. The increase of temperature will be less important and might be insufficient to initiate decomposition of hot spots (in the case where the energy is spread homogeneously on all defects). The existence of a threshold value of the number of defects is possible.

In 1993 Odiot et al. highlighted the importance to take into account the crystalline environment to have an accurate predictive model [32]. They estimated that the C—N bond of nitromethane is strengthened by 4.7 kcal.mol^{-1} due to crystalline environment compared to linkage in a gaseous phase. Therefore they refer that models only based on molecular parameters have to be moderate, especially the BDE approach as the binding energies depend on the physical state of the substance.

Kohno et al. studied HMX polymorphs by SCF *ab initio* calculations [33]. They explained that compressibility of molecules affects total energy which is linked to impact sensitivity. Besides compressibility is not only governed by intermolecular interactions and intramolecular interactions have a non-negligible role. Thus they establish a new parameter called crystal effect parameter (δD) to link compressibility and impact sensitivity. δD is defined according to Eq. (5.25):

$$\delta D = (\text{length } N-N \text{ bond optimised}) - (\text{length of } N-N \text{ bond in crystal}) + \text{crystal} \quad (5.25)$$

This parameter is considered as a measure of compressibility of N—N bond in the crystalline environment and compares to the length of this link in a gaseous phase. Nevertheless, no quantitative relationship is presented.

These last years, an idea of the hybrid model is proposed by Haskins et al. including three main parameters: the energy of the weakest bond, the energy of crystalline structure, and the energy of decomposition reaction [34]. They justify their statement by suggesting energy provided from impact is first communicated to a crystalline structure. In a second step, the energy excess (if there is) will cause the rupture of the weakest bond. Third, the decomposition of the molecule will release energy allowing the chemical reaction to be self-sustaining. The chemical reaction rate is clearly linked to a percentage of decomposition. They finally proposed to calculate activation energy (E_a) thanks to DFT–UB3LYP/6-31G* by the following sum $E_a = E_{bond} + E_{lattice}$. The ratio E_a/Q takes into account all three parameters and a good correlation is found with impact sensitivity.

Inorganic azides are specific molecules that do not contain C, H, and O atoms. Correlations developed for CHNO compounds or involving OB_{100} are unusable for inorganic energetic molecules. However inorganic azides have a sensitivity range from lithium azide (low sensitive) to copper (I) azide (extremely sensitive). In this case, Cartwright and Wilkinson [35] use the molecular modeling program Cerius 3.0 to determine interatomic distances in the ideal unit cell structure, especially the distance between non-bonded nitrogen atoms on neighboring azides ions. Cerius 3.0 and Universal Force Field (UFF) approach allows taking into account structural defects. A good correlation is found between impact sensitivity and the minimum non-bonded nitrogen to nitrogen distance across a large variety of azides. Comparison of calculated results for the ion positions with the literature measures valid the modeling process. Moreover, this relation seems to be consistent with the theory exposing minimum atomic movement is required to produce the reaction products. For example considering sodium and copper azides, two monovalent cations, the first one will induce a controlled decomposition with gaseous loosening, the second one will produce a great detonation. Then they develop a mechanism for interpreting differences within properties of azides. Reasons will be the reduction of the potential of the metal cations (electronic affinities) and rearrangement of the ions in the crystal.

Dienes uses a statistical crack mechanics program to model friction effect on energetic materials under impact [36]. They examined the assumption that the frictional sliding between the faces of a closed crack can form a hot spot that causes molecule ignition. Dienes et al. test the effect of crack orientation and the temperature dependence of viscosity of the melt obtained after decomposition. Results show that crack orientation has a significant effect on defects behavior, especially under compressive loading where interfacial friction plays an important role. Besides particle velocities calculated compare well with those recorded using embedded gauges by using reasonable crack orientation and temperature dependence of viscosity. In the same way, Wu et al. study shear bands formation monitoring by hot spot formation under impact [37]. The ballistic impact chamber (BIC) device and the arbitrary Lagrangian–Eulerian (ALE3D) code are employed. Results explained that without friction, deformation caused by impact induces a homogeneous plastic deformation. Whereas friction generated at the interface plays an important role in shear bands formation which is responsible for hot spot formation. Friction provokes a local increase of temperature proportional to the yield stretch. Numbers of zones have a temperature exceeding this threshold inducing ignition of decomposition. This is consistent with experimental observation showing binders addition reduce the yield stretch and also impact sensitivity.

Finally, the impact can provoke hot spot formation at structural defects and induce decomposition. Another way to predict impact sensitivity is to quantify the difference between energy provided by impact and energy really received by energetic materials.

3.6 Energy transfer rate methods

Most secondary explosives are solid and stable organic molecules having a large energetic barrier to a chemical reaction. When a molecule is shocked under an impact or a shock, molecular transfer of energy and chemical reactivity phenomenon appears. Phonons perform this energetic transfer from the shock front to the intern vibrational states of the molecule. Vibrational energy is a fundamental effect of chemical reactivity. Few of this energy are required to pass through this energetic barrier and trigged reaction. Shock and impact initiation process is quite a complex process. Mechanical excitation in a crystalline structure, the excess mechanical energy is dissipated into a bath consisting of the low-frequency mode of lattice vibrations. Thus the initial energy transferred to phonons must be deposited into molecular vibrations to induce bonds breaking. Phonons correspond to crystalline structure vibrations and generally have frequencies around 200 cm^{-1} whereas molecular vibrations allowing a bond breaking have a greater frequency of about 1000 cm^{-1}. It is clear that phonon energy must be converted to higher vibrations by a process called multiphonon up-pumping. The main mechanism for up-pumping is the anharmonic coupling of an excited phonon with low-frequency molecular vibrations. These low-frequency vibrational modes are expressed as the "doorway mode". In other words, they match molecular vibrations located just above the frequency limit. The mechanical energy provided by a perturbation is deposited in the external modes and is distributed among all phonons. This rich phonons area is attained within roughly 1ps after perturbations. Vibrons are high frequencies vibrations of molecular structure and are not excited during this step. In the up-pumping zone, the energy in phonon modes is transferred to the vibrations by multiphonon up-pumping processes. The two baths (phonons and "doorway vibrations") equilibrate within roughly 100 ps. After the excitation of vibrational modes in the doorway states, the vibration to vibration (V—V) up-pumping processes excite higher vibrational states that provoke bond breaking. This phenomenon happens roughly 1 ns behind the front shock and leads to a series of exothermic reactions to reach a Chapman-Jouguet state.

Tokmakoff et al. are responsible for determining the time to induce the decomposition [38]. This time is found abnormally long attributed to the large gap between the highest phonon frequency and the lowest frequency of vibration of the molecule. Then they undertake DFT-B3P86/6-31(d,p) calculations to link energy transfer rate with the numbers of normal vibrational modes of molecules. No relationship has been developed. However, they supposed that a linear relation between atomization energy and the number of doorways modes for compounds having similar structures could be possible. On the same theme, Fried and Ruggiero [39] derive a simple formula for the total energy transfer rate into a given vibron band in terms of the density of vibrational states and the vibron-phonon coupling. In examining existing inelastic neutron scattering data, they succeed in estimating the phonon upconversion rate for a variety of energetic materials. Then MacNesby and Coffey suggest determining energy transfer rate using Fermi's Golden Rule and result from simple theories of near-resonant energy transfer [40]. Raman spectra are required to construct vibrational energy level diagrams and serve as input for a model to calculate the rate of energy transfer from phonon and near-phonon vibrational energy levels to higher energy vibrational levels. Seven explosives are analyzed and results are in agreement with drop weight impact test data. Ye et al. found that the number of doorway modes shows a strong correlation with impact sensitivity assuming the rate is proportional to a product of the number of states and the rate of population relaxation [41]. The good modeling results confirmed the rate of population relaxation is almost the same for all explosives investigated in their work (PETN, B-HMX, RDX, tetryl, TNT, FOX-7, m-DNB, ANTA, PN, NQ, NTO, and DMN). Ge et al. exposed some correlations involving the number of doorway modes [42]. The approach is based on the simple theory in which the energy transfer rate is proportional to the number of normal mode vibrations. Normal vibrational frequencies are determined by DFT-B3P86/6-31G(d,p) for a set containing several explosives. Three doorway regions are considered for which the number of vibrational states in low-frequency vibrational modes j (doorway mode) is a function of $h_{50\%}$. The following Eqs. (5.26)–(5.28) are given by Ge et al. [42].

- Region (200–700 cm^{-1})

$$j = 0.127\, h_{50\%} + 16.426;\ R^2 = 0.955 \tag{5.26}$$

- Region (200–600 cm^{-1})

$$j = 0.075\, h_{50\%} + 19.372;\ R^2 = 0.908 \tag{5.27}$$

- Region (200–500 cm^{-1})

$$j = 0.063\, h_{50\%} + 16.766;\ R^2 = 0.846 \tag{5.28}$$

These relations found for explosives are consistent with experimental data. This same approach is employed by Ge et al. in another study for nitramines [43].

3.7 Methods based on topological approach

In pure explosives, intramolecular oxydoreduction reactions have a prominent role in the detonation process. In that way, chemical topology of an energetic material brings the main contribution to detonation and quantum mechanical properties.

Some amino and methyl-substituted 1,3,5-TNT are studied by Türker with a topological approach [44]. The molecular orbital method of Hückel provides the total energy of π electrons (E_π) which is an important point for conjugated molecules. E_π is defined as the sum of product between molecular orbital occupied and the occupancy number. U is the upper limit of the deepest lying molecular orbital (x_1) energy and has the highest contribution to E_π. Therefore the following Eq. (5.30) allows U calculation where e is the number of edges in the molecular graph associated with the p-skeleton of the conjugated system and a4 is the respective coefficient of the secular polynomial $P(x)$ (from the topological approach). Then Türker [44] predicts the sensitivity of compounds through linear Eq. (5.31).

$$x_1 \leq \sqrt[4]{e^2 - 2a_4} \equiv U \tag{5.29}$$

$$S = 15.82.227U - 4246.844 \tag{5.30}$$

This approach is time-saving and practical to understand how skeletal changes affect the impact sensitivity for TNT derivatives.

In order to extend impact sensitivity determination to a large set of compounds, Türker involves in his relation a term for the gross topology of the system (m) and structural parameters as the number of NO_2 groups (k_{NO2}), the number of NH_2 groups k_{NH2} and on the number of heteroatom non-considered in k_{NO2} and k_{NH2} (k_X) [45]. The following Eq. (5.31) exposed correlation established where A, B, C, and D are regression coefficients for a series of compounds.

$$h = A\sqrt{\sqrt{4ma_4} + e} + Bk_{NO_2} + Ck_{NH_2} + Dk_X + S \tag{5.31}$$

Then Türker [45] uses electronic properties such as HOMO and LUMO energies (ε_{HOMO} and ε_{LUMO}) added to the difference between these two values ($\Delta\varepsilon$) to predict impact sensitivity. DFT-B3LYP/6-31G(d) calculations have allowed the establishment of Eqs. (5.32) and (5.33).

$$h = 1261.7770 + 306.0026\varepsilon_{LUMO} \tag{5.32}$$

$$h = 1584.0180 + 53.4677\varepsilon_{HOMO} + 274.2812\varepsilon_{LUMO} \tag{5.33}$$

LUMO energy has a higher contribution compared to ε_{HOMO} in the second equation and is only available for nitrobenzenoid. Further studies are needed to extend models to other classes of compounds.

3.8 Physicochemical model

Dubovik develop a complex physicochemical model based on the general concepts of mechanical deformation and destruction of a thin layer of viscoplastic explosive material under impact and the resulting dissipative and chemical heat release [46,47]. Thereby he defines two critical states and estimates related critical properties (impact energy and created pressure). H_{min} corresponding to the minimum height at which the load is dropped is evaluated in function of viscoplastic layer parameters. $h_{50\%}$ value is then calculated for Hmin equal to 100% of ignition probability. Critical properties result from this value. Other critical parameters considered as sensitivity indexes such as critical initiation pressure and critical charge thickness are then estimated through this physicochemical model for HM, TNT, mixtures of ammonium perchlorate, and mixtures of polymethylmethacrylate.

Intermolecular reactional energies are used by DeCarlos to compare the relative stability of explosives taking HNB as reference [48]. Perturbation theory and his "adapted symmetry" approach allow estimation for a set composed of Fox-7, HNB, TNT, TNB, TNA, DATB, and TATB. The expected value is the total interaction energy which is determined by DFT calculation with a SAPT approach. HNB is the most stable compound toward an impact is taken as reference. The following Eq. (5.34) allows estimation of relative stability:

$$\text{relative stability} = \frac{E_{SAPT}}{E_{SAPT,HNB}} \tag{5.34}$$

The experimental trend is found excluding for Fox-7 compound, especially for nitroaromatic and nitroaniline. This simple model is able to have a global view of the stability of energetic materials. Further studies are needed to fix the limits of this method.

A paper by Vaullerin et al. discusses a new sensitivity criterion (CS) based on maximum enthalpy of reaction (ΔH_{max}) to screen all families of energetic substances [49]. CHETAH code deduces the CS value from thermodynamic data and compounds products of the reaction. The relation (5.35) involves ΔH_{max} the molecular weight (M) and the number of atoms per gram (n).

$$CS = \frac{10\Delta H_{max}^2 . M}{n} \tag{5.35}$$

In comparison with experimental data, three sensitivity zones emerge (sensitive, fairly sensitive, and insensitive). Nevertheless, it is difficult to distinguish the sensitivity of molecules in the same zone. This model is not only usable for energetic molecules but for all CHNO compounds.

Lately, the relationship found between thermodynamics and sensitivity by Smirnov et al. [50] allow classification of studied explosive in four hazards levels using references compounds well known: (1) less sensitive than TNT, (2) less sensitive than RDX, but more than TNT, (3) less sensitive than PETN, but more than RDX and (4) more sensitive than PETN [50]. The "Soviet Union Impact Machine" device is employed to measure the sensitivity of compounds. This tool is developed for solids, liquids, and pasty explosive materials. Experiments are carried out at three configurations called devices 1,2 and 3 which give different impact sensitivity parameters:

- LL: the low limit of the impact sensitivity of solids explosives (defined by the maximum drop height of the load) (kg.m)
- A (dev.1), A (dev.2): the probability of explosions for solid explosives in devices 1 and 2 (only for explosives that have the low limit of the impact sensitivity of 50 mm and more for the load of 10 kg) (%)

- h_{50} (dev.1), h_{50} (dev.2): height at which the explosions in devices 1 and 2 have 50% probability (for solid explosives) (kg.m)

Then, these are properties involved in correlations by Smirnov et al.:

- ρ: monocrystal density (g.cm^{-3})
- B: "gross-sum of gram-atoms of all chemical elements containing in 1 kg of the explosive under consideration" (g-atom.kg^{-1})
- α: factor of surplus oxidizer
- T_{mel}: melting point (°C)
- Q_{max}: maximum heat of the explosion

Below are presented equations for calculation of the impact sensitivity in common with parameters previously described.

$$A_{(dev.1)} = (\rho.B)^{-1.270} Q_{max}{}^{2.801} a^{1.380} (500 - T_{mei})^{-1.644}; R^2 = 0.971 \tag{5.36}$$

$$A_{(dev.2)} = (\rho.B_{max})^{2.786} a^{0.9801} (500 - T_{mei})^{-3.134}; R^2 = 0.941 \tag{5.37}$$

$$LL = 0.1(\rho.B/Q_{max})^{1.699} Q_{max}{}^{2.801} a^{-1.013} (500 - T_{mei})^{1.351}; R^2 = 0.989 \tag{5.38}$$

$$h_{50\,(dev.1)} = 0.01(\rho.B)^{1.617} Q_{max}{}^{-0.9022} (500 - T_{mei})^{0.5565}; R^2 = 0.994 \tag{5.39}$$

$$h_{50\,(dev.2)} = 0.01(\rho.B)^{0.8902} Q_{max}{}^{-0.9383} (500 - T_{mei})^{1.218}; R^2 = 0.990 \tag{5.40}$$

Smirnov et al. [50] evoke that thermodynamic and kinetic factors should be enough to predict excitation of explosion and, hence, the sensitivity of energetic materials. In their study presented here, they only use the thermodynamics of explosions. As an example, the rise of the melting temperature is caused by the heightened durability of intermolecular bonds and therefore by thermodynamic stability. Added to that role of kinetic factors cannot be neglected. It is known that organic azides have roughly the same power of consumption as nitrocompounds whereas they are for the most part more sensitive. This might be connected with the features of decomposition kinetics in hot spots.

4 Shock sensitivity

Small and large-scale gap tests are used to predict experimental shock sensitivity which is less studied compared to impact sensitivity. Furthermore, in their paper presented in the previous chapter, Storm et al. [5] tried to link experimental shock and impact data. Taking 21 compounds into account, the following relation (5.41) is found.

$$h_{50} = 7.23 P_{90} - 50.00; R^2 = 0.87 \tag{5.41}$$

This correlation is reliable because of differences between molecules in the basis set. The main conclusion is that shock sensitivity is a complex phenomenon independent of impact sensitivity.

4.1 Energy transfer rate methods

Multiphonons up-pumping mechanism presented earlier is investigated by McGrane et al. [51] since it might govern shock initiation of detonation by vibrational excitation. As for an impact, a shock wave is assumed to excite only delocalized phonons (<200 cm^{-1}). This hot bath comes into equilibrium with the higher frequency intramolecular vibrational states (200–3500 cm^{-1}) in a picosecond and leads to chemical reactions that can release free energy to support the shock wave until a steady reactive shock is formed and steady-state detonation proceeds.

Anharmonic interactions between vibrations are the source of energy transfer by multiphonons up-pumping. No normal vibration modes are in an anharmonic system, thus McGrane et al. [51] consider vibrations as a normal harmonic mode with anharmonic correction terms handled by the perturbation theory. Thereby they use temperature-dependent Raman spectroscopy to link shock sensitivity to anharmonic couplings between thermally populated phonons and higher frequency vibrations relevant to shock up-pumping. Obtained data are compared with those calculated by collision theory for PETN, HMX, TATB, and inert naphthalene. As in perturbation theory, no fundamental differences are observed between the three explosives and naphthalene. Calculations estimating the multiphonon densities of states also failed to correlate clearly with shock sensitivity. Nevertheless, the data presented suggest the further assumption that hindered vibrational energy transfer in the molecular crystals is a significant factor in shock sensitivity.

4.2 Methods based on bond dissociation energy

Previous studies show that BDE of C—NO_2 bonds is not only responsible for bond strength; they are a key factor for the sensitivity of energetic compounds [52]. N—NO_2 and O—NO_2 linkages are generally weaker than C—NO_2 bond but their influence is not negligible for sensitivity prediction. DFT-UB3P86 calculations are carried out by Li et al. at the zero-point energy for 11 common energetic compounds. BDE_{ZPE} values obtained are involved with experimental $\ln(P_{98\%})$ values in a polynomial correlation (5.42).

$$\ln P_{98} = a_1 + a_2 \mathrm{BDE}_{\mathrm{ZPE}} + a_3 \mathrm{BDE}^2_{\mathrm{ZPE}} + \ldots + a_m \mathrm{BDE}^n_{\mathrm{ZPE}} \quad (5.42)$$

BDE_{ZPE} and $P_{98\%}$ are respectively in kJ.mol^{-1} and kbar. At $n = 4$ the regression is very successful. The few differences compared to experimental data are due to other factors involved. But a good quantitative relationship is found between molecular structure and shock sensitivity.

Tan's works register in shock mechanism decomposition understanding. Fourteen nitro compounds are investigated with DFT/BLYP/DNP calculations. The reactive center of such molecules is found around nitro groups. Thus BDE of X-NO2 (X = C, N, O) and Mulliken charges if nitro group is estimated to linearly fit with shock sensitivity values. The following equation is established and is reliable for nitro compounds where pressures are in kbar and BDE in kJ.mol^{-1}.

$$\ln p_{\mathrm{cal}} = 6.84.10^{-4}\mathrm{BDE} + 0.532\, q_{\mathrm{NO_2,max}} - 1.28.10^{-2}\mathrm{BDE}.q_{\mathrm{NO_2,max}} + 2.09 \ln p_{\mathrm{exp}};\ R^2 = 0.948 \quad (5.43)$$

Further studies should help to understand the shock mechanism.

4.3 Methods based on bond resonance energy and isodesmic and homodesmic reactions

A new starting point is taken by Tan et al. in shock sensitivity prediction. It consists in using a global property, the resonance energy (RE), compared to local methods (study of nitro groups) [53,54]. Eight nitro molecules that contain benzene rings including three azaheterocycles and one nitrogen-rich energetic compound are investigated. The resonance energy is stabilization energy due to electron's delocalization. Homodesmotic and isodesmic reactions are used to calculate the RE. Energy difference in a homodesmotic reaction equation denotes the resonance energy. Homodesmotic reactions involve reactants and products for which the numbers of carbon atom having the same hybridization states is conserved and linked to the same number of hydrogen atoms. Bonds and valence are also conserved. Systematic errors are significantly diminished during the use of these reactions. In the case of isodesmic reactions, the number and type of linkages to the reactants and products are maintained. After DFT BLYP/DNP calculations, RE and experimental data follow the same trend. In general, the higher the RE, the lower the shock sensitivity. Explosives molecules having a resonance structure can delocalize electrons and disperse energy from shock. RE has a significant role in determining shock sensitivities of sensitive and insensitive molecules. In addition strain energy is an important factor to investigate shock-induced chemical resonance energy. Tan et al. have worked on this line of research [54]. Four-membered and six-membered heterocyclic compounds are studied. Ring strain energy and bond rotational energy are determined. The ring strain energy of four-membered heterocycles is higher than six-membered heterocycles. Ring-opening reactions occur preferentially for molecules containing higher strain energy under shock. Besides rotational energy threshold of $N—NO_2$ is higher than $C—NO_2$ bond. Thus the last one easily can disperse energy from shock by rotation. On the contrary, shock energy will cause rupture of $N-NO_2$ bond. Then the introduction of groups such as —C=O, —C=NH and $-NH_2$ provoke a change in the charge distribution. Especially the addition of —C=O and —C=NH diminishes nitro group charge (in absolute value) and rises strain energy of heterocycles. Thereby compounds are more sensitive against a shock.

4.4 Methods based on elementary composition and structural parameters

As for impact, Keshavarz and his team developed simple relations based on elementary composition and structural parameters thanks to gap-test experimental values [55,56]. They use results from small and large-scale devices and obtain the following equations for $C_aH_bN_cO_d$ (5.44)–(5.46).

$$P_{x\%TMD} = x_1 + x_2\left(a + \frac{b}{2} - d\right) + x_3 E^0_{aCH/NNO_2} + x_4 (An_{NH_2} - n_{NO_2})_{pure} \tag{5.44}$$

$$G_{50} = x_1 + x_2\rho_0 + x_3\left(a + \frac{b}{2} - d\right) + x_4 \text{Void}_{theo} + x_5 (C - N(NO_2) - C)_{pure} \tag{5.45}$$

$$\text{Void}_{theo} = \frac{1/\rho_0 - 1/\rho_{TM}}{1/\rho_0} \times 100 \tag{5.46}$$

with

$-x_1-x_4$: Adjustable parameters to obtain the best fit with experimental data
$-(a+\frac{b}{2}-d)$: parameter that shows the distribution of oxygen between carbon and hydrogen to form carbon monoxide and water
$-E^0_{aCH/NNO_2}$: parameter that shows the existence of α—C—H or N—NO_2 in nitroaromatic. Equal to 1 for nitramines or nitroaromatics having α—C—H linkage. Equal to 0 for EMs which do not have N—NO_2 bond
$-(An_{NH_2}-n_{NO_2})_{pure}$: difference of number of amino and nitro groups for amino compounds where A is a constant
$-\rho_0$: initial density of explosives
$-(C-N(NO_2)-C)_{pure}$: presence of C—N(NO_2)—C group (for pure explosives)
$-Void_{theo}$: theoretical percent of void
$-\rho_{TM}$: theoretical maximum density

Multiple linear regression methods were used to find adjustable parameters. These reliable correlations do not require knowledge of any measured physical, chemical, or thermochemical properties. The second relation (5.45) is only used for pure and composite mixtures that are prepared under vacuum cast, cast, hot-pressed, and pressed conditions. Deviations can be large for creamed, granular, and flake situations.

4.5 Methods based on crystal framework

Complex energetic processes occur during a shock in on EMs from initiation to a detonation. The features of these processes highly depend on the mechanical (orientation, gaps, defects, microstructure interface structure, *etc.*) and chemical (reaction rate, species, composition, energy, *etc.*). The temperature is the most important thermodynamic variable influencing physical and chemical properties under shock compression. Thus Yoo et al. recorded shock temperature, detonation velocity, and detonation distance of PETN single crystals using a nanosecond time-resolved spectropyrometric system between 350 and 700 nm [57]. Results show a strong dependence on crystal orientation. PETN is sensitive along with the shock propagation to the (110) plane and insensitive along the (100) plane where (hkl) are Miller indices. Moreover, detonation temperature seems independent of crystal orientation.

Bellitto et al. [58,59] study 7 varieties of RDX from five manufacturers. RDX is taken because many characterizations are made on this molecule including shock sensitivity and HMX impurity levels. Differential scanning calorimeter and atomic force microscopy are used to reveal the difference between RDX compounds investigated. Results show a disparity of surface defects. Thanks to statistical analyses, a relation between surface roughness and shock sensitivity are found. But it is observed that HMX impurity level affected the thermal property of RDX but was uncorrelated with the shock sensitivity. A crystalline factor such as surface roughness should be involved in sensitivity-determined correlation in order to take into account crystalline effects.

4.6 Physicochemical model

Price tried proposed empirical correlations of sensitivity data of EMs [60]. The goal was to involve four variables potentially in relation to shock sensitivity:

- a: radius of exter primer in Minimum Priming Test. $a^3 = $(masse Extex, mg)/3.20445
- x^*(8.3): run distance to detonation in the wedge test when the initial pressure is 8.3 GPa
- d_C: critical diameter for the propagation of steady-state detonation
- E: critical flying plate kinetic energy fluence

In order to have a linear equation, three parameters are finally kept and following equation are exposed. L matches to $G_{50\%}$ value and all dimensions are in mm.

$$x^*(8.3) = a \quad (5.47)$$

$$d_c = 0.685a \quad (5.48)$$

$$\log d_c = 2.0316 - 0.03437\ L \quad (5.49)$$

The relationships cited above are not applicable to all EMs. Most explosives that do not conform are those that are relatively insensitive and that have long reaction zones. Detonation properties are then taken up by Smirnov in a recent study involving thermodynamic and physicochemical parameters [61]. Shock sensitivity of explosives is most of the time unacceptably high (1.5–5 kbar) that matches to the maximum of sensitivity of EMs (for PETN) and no way to reduce this hazardous characteristic is known. The critical pressure of detonation initiation is estimated (P_{cr}) is estimated by Smirnov with characteristics describing the stability of the material to shock, to penetrative actions, and with parameters of combustion to explosion transfer. Eq. (5.50) is developed thanks to gap-test experimental data. Properties involve are the same used by these authors for impact sensitivity

$$P_{cr} = (\rho.B)^{2.732} Q_{max}{}^{-1.534} a^{-1.105} - 5 \quad (5.50)$$

Reliability of this relation is significant but as said earlier, results are unacceptable in terms of safety.

5 Electrostatic discharge sensitivity

Electrostatic performance is an important point since electrostatic discharge or electric current is used as a trigger for most EMs initiators. In another point of view, in a propitious environment to electrostatic discharge, prediction of sensitivity of this kind of stimuli is fundamental in terms of safety.

5.1 Methods based on decomposition properties

Sensitivity of EMs is experimentally determined thanks to the estimation of electrostatic energy needed to ignite decomposition (E_{ES}). Some studies show a possible conversion of spark energy into a thermal effect. This leads to a thermolytic mechanism of the initiation by electric spark. However numerous authors consider this initiation to be a multidisciplinary problem and involve molecular structure, thermal reactivity, and sensitivity to mechanical stimuli.

In that way, Zeman et al. use a form of Evans-Polanyi-Semenov equation linking the square of detonation velocities (D^2) in $km.s^{-1}$ [62]. 32 Polynitro arenes are investigated and the linear Eq. (5.51) is exposed.

$$D^2 = A.E_{ES} + B \tag{5.51}$$

A and B factors are estimated for different classes of compounds. Results suggest that intermolecular reactions affect spark sensitivity. The knowing mechanism of transfer of discharge into the center of a molecule could help in this way.

Zeman et al. then work on nitramine family development relations for two subclasses [63]. Eq. (5.52) concerns explosives with methylenenitramine units ($—CH_2—C(NO_2)—$) in their molecules (including TNAD and DNDC). Relation (5.53) exposes molecules that cannot be considered as multiples of the DIGEN.

$$D^2 = -2.649.E_{ES} + 0.9842;\ R^2 = 0.9842 \tag{5.52}$$

$$D^2 = -3.4601.E_{ES} + 73.821;\ R^2 = 0.867 \tag{5.53}$$

The lack of experimental data and structure diversity explain the inaccuracy of the second equation. More parameters should be involved.

Thermally stable explosives, especially aromatics, are investigated in an electric-spark test by Hosoya et al. [64]. They observed that E_{50} values (50% of ignition energy probability) decrease with adiabatic flame temperature (Tf). They exposed the relation (5.54) but no validation was performed.

$$E_{ES} = \exp\left(-1.1076.10^{-3}\mathrm{Tf} + 3.1354\right) \tag{5.54}$$

Skinner et al. [65] then show that secondary explosives in powdered form are mainly dependent upon their chemical characteristics. Authors recorded ESD values corresponding to the energy needed to ignite explosives with a probability equal to 50%. Four correlations are tested involving different properties. The ESD values are poorly correlated with the sensitivity index and the decomposition activation energy. However, the critical temperature for thermal runaway of 20μm particles is calculated by the Franck-Kamenetskii equation and a well model is found between ESD and the ratio 1000 K/Tc. Tc is expressed by relation (5.55).

$$T_c = \frac{E_a/R}{\ln\left[\frac{r^2 \rho Q A E_a}{Tc^2 \lambda \mathrm{Sh} R}\right]} \tag{5.55}$$

with

- T_c: critical temperature (K)
- E_a: activation energy ($J.mol^{-1}$)
- A: Pre-exponential factor (s^{-1})
- r: radius or half-thickness (cm)
- ρ: density ($g.cm^{-3}$)
- Q: Heat of decomposition reaction ($J.g^{-1}$)
- λ: thermal conductivity (W/(cm K))
- Sh: shape factor, 3.32 for a sphere

Then, assuming that decomposition is described using first-order kinetic, ESD is equated with the temperature at which the decomposition rate coefficient of each explosive reaches a given value ($k = 10^3 s^{-1}$ matching to 5% needed to ignite EMs). This method has been tested by Zeman's team on 53 polynitroaromatics [66]. Results show that the relation is generally not validated for this set of molecules (non-powder). This equation has different forms for different groups of EMs. In addition to the decomposition rate kinetics, intermolecular interactions in crystals are decisive factors.

In the case of primary explosives, low static energy (2–3 mJ) is able to cause safety troubles. Talawar et al. [67] expose electrostatic sensitivity values in terms of zero ignition probability data (E_{SE0}) of initiatory explosives. On the one hand, as Skinner et al. [65] observed, a possible link between spark energy and the ratio $\ln(E_A/T_E)$ where T_E is the explosion energy is found. On the other hand, the square of the detonation velocity or heat of explosion well correlates with this ratio. So thermal, detonation, and mechanical properties can be connected with spark sensitivity.

Attempts have been tried to link the mechanism of impact energy transition to the center of their molecule and transition of spark energy. Forty-one polynitroaromatics are investigated to correlate EES and impact sensitivity expressed as drop energy E_d of the first reaction. The conclusion expresses that intermolecular interactions differ from impact and spark perturbations and decomposition mechanisms are different. No direct connection is possible between these two stimuli sensitivity.

Kamlet-Jacobs relations are used by Wang et al. to calculate velocity, pressure, and heat of detonation on 18 nitramines and 39 polynitroarenes [68,69]. Here are the corresponding equations:

$$D = \Phi^{0.5}(1.011 + 1.312\rho) \tag{5.56}$$

$$\Phi = N\bar{M}^{0.5}Q^{0.5} \tag{5.57}$$

$$P = 1.558\Phi\rho^2 \tag{5.58}$$

with

- P: detonation pressure (GPa)
- D: detonation velocity (km.s^{-1})
- ρ: density (g.cm^{-3})
- Φ: characteristic value of explosives
- N: number of moles of gas produced per gram of explosives (mol)
- $\bar{M}$: average molar weight of detonation products (g.mol^{-1})
- Q: heat of detonation (kJ.g^{-1})

Linear relations similar to those exposed by Zeman et al. are found between EES and D^2 or $\log(P)$. But equations presented by Wang et al. [68,69] do not involve all compounds because in some case D^2 diminish with spark sensitivity. Thereby they investigated other structural parameters such as Mulliken bond order (M_{NN}), and the heat of formation ($\Delta_f H$). Using multiple linear regressions for 10 methylenenitramines are done and Eq. (5.59) is presented.

$$E_{ES} = -29.743 + 216.547\, M_{NN} - 1.02.10^{-2}\, \Delta_f H \tag{5.59}$$

Modeling results are remarkably improved. Nevertheless, for the eight other nitramines, the M_{NN} parameters do not have the same influence. Although the front line orbit energy (E_{HOMO} and E_{LUMO}) seems to be important. It is possible that the pyrogenation and detonating of compounds differ from the molecules involve in the relation.

Then a new aspect of relationship between E_{ES} and thermal stability of polynitroarenes is exposed by Zeman. The thermostability threshold (T_{max}), corresponding to the maximum temperature that is applied to the given explosive for 6 h without destroying its functionality, is calculated by Eq. (5.60) thanks to Arrhenius parameters.

$$T_{max} = \frac{E_a}{2.303.R.(\log A - \log k)} \tag{5.60}$$

Relationship found is approximate and fall for several sub-group.

5.2 Methods based on decomposition properties and structural parameters

This part focuses on Keshavarz et al. works who try to have an accurate model easy to use. First, he finds a new method to determine the velocity of detonation [70]. The corresponding relation (5.61) is usable for $C_aH_bN_cO_d$ EMs where n_{NR} is the number of –N=N- or NH4+ in molecules and n_{MN} is the number of nitro groups attached to a carbon atom.

$$D_{max} = 7.68 - 0.198a - 0.111b + 0.294c + 0.0742d - 0.635n_{NR} - 0.735n_{mN} \tag{5.61}$$

Thereby D_{max} is correlated with EES according to 17 nitroarenes as can be seen in Eq. (5.62).

$$E_{ES} = 28.18 - 3.052D_{max} + 11.87C_{SG} + 4.395n_{-NH_2,-HNCO-}; R^2 = 0.82 \tag{5.62}$$

In this relation CSG contributions are:

- $C_{SG} = 0.5$ for a —OH group attached to ring
- $C_{SG} = 1.5$ for a alkoxy group attached to ring
- $C_{SG} = 0.25$ for —N=N- and —S(O2)- groups attached to ring

Predictability of this model is roughly the same compared to Zeman or Wang models but it is applicable to a larger set of molecules and easier to establish.

Afterward, Keshavarz et al. studied nitramines spark sensitivity with the maximum detonation pressure in kbar (P') [71]. Thanks to Kamlet-Jacobs equations for detonation pressure and structural properties the following relation is found for a $C_aH_bN_cO_d$ compound.

$$P' = 221.53 - 20.437a - 2.2538b + 17.216c + 16.140d - 79.067C_{SSP} - 66.335n_N \tag{5.63}$$

C_{SSP} is equal to 1 for explosives which contain at least —N=N-, —O—NO_2, NH_4^+, N_3 in the molecular structure and equal to 0 otherwise. n_N is equal to $n_{NO2}/2 + 1.5$ where n_{NO2} is the number of NO_2 groups. P' is then involved in a correlation with experimental sensitivity data established for 20 nitramines.

$$E_{ES} = 21.19 - 0.0422P' - 3.257SSP^- + 6.498SSP^+; R^2 = 0.94 \tag{5.64}$$

SSP^+ and SSP^- are positive and negative contributions of specific structural parameters. Added to that Keshavarz explains that the effect of shape and size of nitramine crystals influence energy transfer in the center of the molecule and consequently spark sensitivity. However, there is no sufficient data to incorporate particles sizes on relations.

A similar correlation is exposed by Keshavarz this time it is for polynitroaromatics. The following reliable relation (5.65) is established for 36 compounds where SG is respectively equal to 0.21, 0.5, and 1.5 for —N=N-, —OR and —OH groups and n' is the number of amines or amid groups.

$$E_{ES}(J) = 13.465 - 0.0305P_{max} + 11.81SG + 4.167n'; R^2 = 0.90 \quad (5.65)$$

Keshavarz simplified spark sensitivity determination by involving only structural parameters. Eqs. (5.66) and (5.67) are respectively developed for 11 nitramines and 17 polynitroaromatics.

$$E_{ES} = 3.460 + 6.504r_{c/o} - 4.059C_{CH_2NNO_2 \geq 3, C(=O)(Oou\,NH)}; \; R^2 = 0.86 \quad (5.66)$$

$$E_{ES} = 4.60 - 0.733n_C + 0.724n_O + 9.16R_{nH/nO} - 5.14C_{R,OR} \quad (5.67)$$

with:

- $r_{C/O}$: ratio of carbon to oxygen
- $C_{CH_2NNO_2 \geq 3, C(=O)(Oou\,NH)}$: indicates presence of methylenenitramine groups greater or equal to 3 in cyclic nitramines or the presence of COO or CONH groups.
- n_C and n_O: number of carbon and oxygen atoms
- $R_{nH/nO}$: ratio of number of hydrogen and n_O
- $C_{R,OR}$: presence of certain group such as alkyl (−R) ($C_{R,OR} = 1$) or alkoxy (–OR) ($C_{R,OR} = -2$) attached to the aromatic ring

5.3 Methods based on electronic properties

The electronic structure of a molecule is put to use when subjected to an electrical or electrostatic discharge. Electronic properties can then be related to sensitivity to such stimulus. Based on the lowest unoccupied molecular orbital energy, the Mulliken charges of the nitro group and structural parameters used by Keshavarz previously presented, Zhi et al. [72] studied nitroaromatics and nitramines compounds. In both cases, only a geometry optimization is done by DFT-B3LYP calculations. The first equation concerns nitroaromatics molecules (5.68):

$$E_{ES}(J) = (-1)^{n_1} 10.16Q_{nitro} - 1.05n_1n_2E_{LUMO} - 0.20; R^2 = 0.97 \quad (5.68)$$

where n_1 is the number of aromatic rings and n_2 is the number of substituted groups attached to the aromatic ring (excluding NO_2). Sensitivity rises with Q_{nitro} and decreases with E_{LUMO}. In this correlation, some kinds of groups are not implicated and only their number is involved. Thereby this model is reliable for molecules having less than two different substituted groups.

The next equations concerns nitramines (5.69) (5.70):

$$E_{ES}(J) = 9.82n_1Q_{nitro} - 2.35n_2 + 22.95n_3 - 0.96; R^2 = 0.929 \quad (5.69)$$

$$E_{ES}(J) = 4.69E_{LUMO} + 27.07n_1 - 6.46n_2 + 7.73; R^2 = 0.929 \quad (5.70)$$

here n_1 is the number of nitro groups, n_2 is the ratio of hydrogen to oxygen and n_3 is the ratio of carbon to oxygen.

The last two relations are related to cyclic nitramines (5.71) (5.72):

$$E_{ES}(J) = 20.96n_1Q_{nitro} - 6.98n_2 + 39.95n_3 - 3.3; R^2 = 0.959 \quad (5.71)$$

$$E_{ES}(J) = 7.1E_{LUMO} + 0.36n_1 - 14.99n_2 + 43.07; R^2 = 0.646 \quad (5.72)$$

where n_1 is the ratio of carbon to oxygen and n_2 is the ratio of hydrogen to oxygen. It can be seen that E_{LUMO} property failed in prediction of spark sensitivity of cyclic nitramines. This shows the limitations for the study of sensitivity to electrostatic discharge from this factor.

A new approach for nitramine spark sensitivity determination is exposed by Türker [73]. Molecular orbital characteristics are implicated with experimental spark data *via* ionic forms of compounds. In a few words when an electric field is applied to one molecule, the molecular geometry and electron distribution are distorted. The measure of this distortion is called polarizability. It changes the dipole moment of bonds and consequently the dipole moment of the molecule. With an increase of the electric field, the ionization of molecules occurs. Depending on the structure and other factors, various anionic and cationic species are generated which cause initiation of decomposition. In order to understand this decomposition mechanism, cyclic and acyclic nitramines are investigated. An electric discharge applied on organic molecules provokes the structure gains or loses electron(s) prior to the disintegration of the molecule into fragments. In other words, decomposition occurs thanks to anionic and cationic forms. Thereby as seen below, initiation of decomposition of neutral but polarized or singly or double charged structures can easily be expressed as single electron transfer process involving HOMO and or LUMO orbital.

5.4 Electron transfer from the HOMO

Cations

$$A^+ - e^- \rightarrow A^{2+} \quad (5.73)$$

Neutre

$$A - e^- \rightarrow A^+ \quad (5.74)$$

Anions

$$A^- - e^- \rightarrow A \quad (5.75)$$

5.5 Electron transfer to the LUMO

Cations

$$A^+ + e^- \rightarrow A \quad (5.76)$$

Neutre

$$A + e^- \rightarrow A^- \quad (5.77)$$

Anions

$$A^- + e^- \rightarrow A^{2-} \quad (5.78)$$

UHF/PM3 calculations give better results for correlations between E_{ES} and HOMO or LUMO energies.

In a second time, Türker considers the ionic attraction energy. The weakest bond between NO_2 and an organic atom is generally heteropolar because of the inductively and mesomerically electron attracting nature of the NO_2 group. As seen below, this energy is calculated thanks to the opposite charges accumulated on atoms (Q_1 and Q_2) of a particular bond having a length of r.

$$E = \frac{Q_1 Q_2}{r} \tag{5.79}$$

Once again UHF/PM3 calculations allow the most reliable equations between E_{ES} and ionic attraction energies. The findings of this study are that nitramines compounds preferentially decompose *via* anionic states in an electric field. Besides their behavior is complex due to the specific properties of each molecule.

5.6 QSPR method

A QSPR study is used for electrostatic discharge sensitivity determination by Fayet et al. [74]. Twenty-six polynitroaromatics are investigated by DFT/PBE0/6-31G(d,p). From 300 descriptors, 4 are selected as the most reliable to be correlated with E_{ES} whose 3 are linked to the loss of the NO_2 group from a C—NO_2 bond.

6 Heat sensitivity

Thermal stability of EMs is always evaluated by the degree of response of the cook-off events which is range from a benign rupture of the confinement to a violent detonation [75]. Thus properties such as shelf life, explosion delay, critical temperature, thermostability threshold, 500-day cook-off temperature, and approximate time to explosion characterize the thermal stability of compounds. Concerning sensitivity, a different approach similar to other kinds of sensitivity is developed.

6.1 Methods based on bond dissociation energy

Zeman et al. studied thermal sensitivity by the DTA method and data were analyzed according to the Kissinger method. The thermal reactivity is expressed as the E_aR^{-1} of the Kissinger relationship [76]. The homolytic character and the identity of the primary fission in ignition of EMs are formulated by Zeman et al. using Evans-Polanyi-Semenov (EPS) equation. They finally modified the EPS relation in order to involve the thermal reactivity E_aR^{-1} in Kelvin. As seen below, the square of the detonation velocity D^2 (km.s^{-1}) is involved for sensitivity determination.

$$E_aR^{-1} = b \pm aD^2 \tag{5.80}$$

In the present study, Zeman study 11 commercial explosives with different nitric ester contents, oxidizing systems of 10 classical emulsion explosives (water in oil type). Results show that the thermal reactivity of the oxidizing system replaces the primary fission of explosophore groups in an EM.

Another study of Zeman and Friedl [77] on 11 nitramines involves the electronic charges at nitrogen atoms of nitro groups with thermal reactivity. DFT B3LYP/6-31G** method permits electronic charge (q^N) by Mulliken population analysis of electron densities. Finally, the primary fission on nitro groups describes thanks to q^N is linearly linked to E_aR^{-1} as seen in Eq. (5.81).

$$E_aR^{-1} = aq^Nb \tag{5.81}$$

The relation is valid for nitramines having a closely molecular structure. The previous relationship is also used by Zeman et al. to study the initiation of highly thermostable polynitroarenes [1].

During triacetone triperoxyide investigation by Zeman and Bartei, the EPS equation is modified in order to treat non-isothermal DTA results [78,79]. The following (5.82) relation expressed the thermal reactivity in the function of the rate of temperature increase ϕ, the onset temperature of the exothermic decomposition T, the universal gas constant R, and the pre-exponential factor.

$$\ln\left(\frac{\phi}{T^2}\right) = -\left(E_aR^{-1}\right)\frac{1}{T} = \ln\left(A\frac{R}{E_a}\right) \tag{5.82}$$

In terms of performance the heat of explosion Q is expressed in function of the detonation velocity D and the polytropic coefficient. This property is then used to calculate the pre-exponential factor according to relations (5.83) (5.84) and link thermal reactivity and performance.

$$Q = \frac{D^2}{2(\gamma^2 - 1)} \tag{5.83}$$

$$A = Q\left(1 - (V_1/V_2)^{\gamma-1}\right) \tag{5.84}$$

V_1 and V_2 are respectively the initial volumes of gaseous products of detonation and their final volume.

6.2 Methods based on elementary composition and structural parameters

As seen previously, Keshavarz et al. make the effort to link all types of sensitivity with elementary composition and structural parameters. Studies are made on polynitroarenes and nitramines to evaluate thermal stability, activation energy, and decomposition temperature. They make the effort to build simple accurate relations easy to use without quantum chemistry calculations. The first relation (5.85) is obtained after multi-linear regression on 23 polynitroarenes.

$$\log(E_a) = 2.25 + 0.0337\mathrm{OEC} + 0.146n_{\mathrm{NHCOCONH,NN_2}>1} + 0.124E_{a-\mathrm{CH_{ou}CH_3O-[C(NO_2)-CH-C(NO_2)]}};\quad R^2 = 0.87 \tag{5.85}$$

where

–OEC: optimized elementary composition for polynitroarenes having $C_aH_bN_cO_d$ type.

$-n_{NHCOCONH,NN_2>1}$: number of —NH(C=O)—C(=O)NH— groups or amino groups so that more than one –NH2 group should be attached to aromatic rings
$-E_{a-CH_{ou}CH_3O-[C(NO_2)-CH-C(NO_2)]}$: existence of either one a—CH or methoxy group attached to one aromatic ring in form α—CH or CH_3O—[C(NO_2)—CH—C(NO_2)]

For nitramines, the Eqs. (5.86) and (5.87) are established.

$$\ln(E_a)_{core} = x_1 + x_2a + x_3b + x_4c + x_5d \quad (5.86)$$

$$\ln(E_a) = 0.5385 + 0.8951 \ln(E_a)_{core} + 0.1698P_{>5}; R^2 = 0.94 \quad (5.87)$$

with

$-(E_a)_{core}$: core activation energy expressed by equation () in the function of elementary composition
$-(E_a)$: corrected activation energy
$-P_{>5}$: existence of cyclic nitramines that contain more than five member ring

Then decomposition temperature is given by relation (5.88) for 12 polynitroarenes.

$$T_D(K) = 571.17 + 30.63n_C - 21.29n_H + 32.57n_N - 43.11n_O + 15.98(n_{NH_2} - n_{NO_2}) + 50.69(|n_{TNB} - 2| - E_{TNB\text{-}CH_2\text{-}TNB}); R^2 = 0.95 \quad (5.88)$$

$-n_C$, n_H, n_N, and n_O: number of carbon, hydrogen, nitrogen, and oxygen atoms
$-(n_{NH_2} - n_{NO_2})$: difference between the number of amino and nitro groups
$-(|n_{TNE} - 2| - E_{TNB\text{-}CH_2\text{-}TNB})$: n_{TNB} is the number of 1,3,5-trinitrobenzene et $E_{TNB\text{-}CH_2\text{-}TNB}$ is equal to 1 or 0 for the existence or absence of CH_2-*bridge* between two 1,3,5-trinitrobenzene units

The previous relationships show that no heavy computation is needed thanks to the Keshavarz methods.

6.3 QSPR method

These last years, the QSPR method was often used to build models allowing the prediction of thermal decomposition parameters. It is the case of Saraf et al. who worked on a database composed of 19 nitrocompounds [80]. Onset temperature and energy of reaction are experimentally recorded. After the selection of the descriptors involved in a relationship, Semi-empirical calculations (DFT/B3P86 and AM1) are used to calculate these parameters. Finally, onset temperature is expressed in function of the highest positive charge, the delocalizability index, and the dipole moment. Added to that, Saraf et al. estimate that the energy of reaction strongly correlates with the number of nitro groups for this type of molecule ($-\Delta H = 75n_{NO2} \pm 5$ for mononitro and $-\Delta H = 150n_{NO2} \pm 14$ for dinitro compounds).

Then Fayet et al. investigated thermal stabilities of energetic compounds. Nitroaromatics are studied and relationships have been presented. In all cases QSPR method allows the selection of parameters to involve topological, geometrical, geometrical charge related, electronic, constitutional, thermodynamic, and quantum chemistry descriptors. Their initial studies are done on 22 nitroaromatics and underlined the importance of chemical and concerning the NO_2 group descriptors [81]. Then a large set of a molecule containing 77 nitroaromatics is used [82]. Descriptors showing the importance of chemical reactivity are involved. The relationship is reliable but cannot predict the decomposition enthalpy of molecules taken out from the database. Finally, the database is divided into two subgroups to improve the accuracy of predictions. Compounds substituted in ortho position are separated from non-ortho substituted molecules.

Sang et al. employed QSPR and descriptors derived from the electrostatic potential [83]. Twenty-two nitroaromatics are analyzed by DFT-HF/6-31G*. Results reflect to a great extent the hydrogen bond-donating tendency or hydrogen bond acidity of a molecule. Thereby thermal stability decreases with an increase of bond-donating ability of hydrogen.

A quantitative structure–activity relationship (QSAR) represents a potential tool for predicting thermal stability properties. Li et al. [84] (make the effort to investigate the relationship between the molecular structures and decomposition enthalpy of 77 nitrobenzene derivatives with various functional groups. A new genetic algorithm (QSARINS) allows variable selection and validation. A first model containing seven descriptors and a reduced one having four descriptors are built. These parameters are related to the reactivity and thermal stability of nitrocompounds. In comparison with other QSPR models, models of Li are more robust and externally more predictive. As seen previously, some models have only developed thanks to 22 molecules. So that the basis set is entirely used for equation construction. Saraf et al. used a validation set. Although the equation found is reliable, the model failed in its validation. This is also the case for Fayet's model developed for 77 compounds. Finally, Li et al. used the same basic set that Fayet et al. but selection and validation through the genetic algorithm provides obtainment of reliable equations which can be extended to external compounds.

7 Friction sensitivity

Friction is a complex phenomenon and is often generated by the action of other perturbations as the impact of a shock wave which makes it a complicated study. Nowadays, some recent experimental devices are built to investigate the friction effect but models establishment is a new topic.

In their recent study of impact sensitivity, Smirnov and his team realized experimental tests on explosives with the Soviet Union Impact Test machine wherein friction parameters are recorded [50]. There are:

$P_{fr.ll}$: the low limit of the friction sensitivity is the maximum pressure on the explosive placed between two steel planes at which there is no explosion (MPa)
$P_{fr.50\%}$: the pressure of operation with 50% probability is calculated on the curve necting the pressure with the probability of explosions (MPa)

Probabilities are defined during 25 tests at the shock shift of pressing between two steel planes. Thermodynamic parameters previously presented in Smirnov's studies [50,61] allow the development of these two following models.

$$P_{fr,ll} = (\rho,B)^{1.616}\left(Q_{max}{}^{0.5}\right)^{-0.8618}(500 - T_{mei})^{0.4016}/10;\ R^2 = 0.990 \quad (5.89)$$

$$P_{fr,50\%} = (\rho,B)^{1.647}\left(Q_{max}{}^{0.5}\right)^{-0.5646}(500 - T_{mei})^{0.3138}/10;\ R^2 = 0.999 \quad (5.90)$$

Regressions coefficients are satisfying and parameters involved combined with chemical kinetic seem enough to well predict friction sensitivity as impact and shock sensitivity. Thereby when the friction is caused by an impact, a possible link between impact sensitivity and friction sensitivity is to consider. Also, it may be possible to link the sensitivity according to the disturbance that engenders.

8 Conclusion

We have presented numerous empirical or semi-empirical methods proposed for evaluating the sensitivity of energetic materials to impact, shock, electrostatic discharge, heat, and friction. This review is however not exhaustive because existing methods are improved continuously and also because new methods are proposed almost every year. It is beyond the scope of this review to recommend any of these methods. Various results are obtained, depending on the molecules and the methods, with small validation basis sets due to the small number of explosive molecules. The major drawbacks of these methods are their empirical or semi-empirical forms, which hinder the treatment of new molecules. That is why approaches based on *ab initio* simulations are of particular interest.

These methods are interesting for a chemist, designing new molecules. However, explosive crystals are never used alone but added into formulations. The form, defects, the granulometry of the crystals, the type and proportion of the binder, the fabrication process are important parameters having a great influence on sensitivity. The challenge of the future is to take them into account.

References

[1] S. Zeman, Z. Friedl, M. Roháč, Molecular structure aspects of initiation of some highly thermostable polynitro arenes, Thermochim. Acta 451 (1–2) (2006) 105–114.
[2] A.K. Sikder, G. Maddala, J.P. Agrawal, H. Singh, Important aspects of behaviour of organic energetic compounds: a review, J. Hazard. Mater. 84 (1) (2001) 1–26.
[3] T.L. Jensen, J.F. Moxnes, E. Unneberg, Christensen, D. models for predicting impact sensitivity of energetic materials based on the trigger linkage hypothesis and Arrhenius kinetics, J. Mol. Model. 26 (4) (2020). Article number 65.
[4] M.J. Kamlet, H.G. Adolph, The relationship of impact sensitivity with structure of organic high explosives. II. Polynitroaromatic explosives, Propellants Explos. Pyrotech. 4 (2) (1979) 30–34.
[5] C.B. Storm, J.R. Stine, J.F. Kramer, Sensitivity relationships in energetic materials, in: S.N. Bulusu (Ed.), Chemistry and Physics of Energetic Materials, NATO ASI Series (Series C: Mathematical and Physical Sciences), vol 309, Springer, Dordrecht, 1990.
[6] M.H. Keshavarz, H.R. Pouretedal, Simple empirical method for prediction of impact sensitivity of selected class of explosives, J. Hazard. Mater. 124 (1–3) (2005) 27–33.
[7] M.H. Keshavarz, H.R. Pouretedal, A. Semnani, Novel correlation for predicting impact sensitivity of nitroheterocyclic energetic molecules, J. Hazard. Mater. 141 (3) (2007) 803–807.
[8] M.H. Keshavarz, Prediction of impact sensitivity of nitroaliphatic, nitroaliphatic containing other functional groups and nitrate explosives, J. Hazard. Mater. 148 (3) (2007) 648–652.

[9] M.H. Keshavarz, A. Zali, A. Shokrolahi, A simple approach for predicting impact sensitivity of polynitroheteroarenes, J. Hazard. Mater. 166 (2–3) (2009) 1115–1119.
[10] M.H. Keshavarz, Simple relationship for predicting impact sensitivity of nitroaromatics, nitramines, and nitroaliphatics, Propellants Explos. Pyrotech. 35 (2) (2010) 175–181.
[11] W.-P. Lai, P. Lian, B.-Z. Wang, Z.-X. Ge, New correlations for predicting impact sensitivities of nitro energetic compounds, J. Energ. Mater. 28 (1) (2010) 45–76.
[12] S. Zeman, New aspects of the impact reactivity of nitramines, Propellants Explos. Pyrotech. 25 (2) (2000) 66–74.
[13] S. Zeman, M. Krupka, New aspects of impact reactivity of polynitro compounds, part II. Impact sensitivity as "the first reaction" of polynitro arenes, Propellants Explos. Pyrotech. 28 (5) (2003) 249–255.
[14] X.-S. Song, X.-L. Cheng, X.-D. Yang, B. He, Relationship between the bond dissociation energies and impact sensitivities of some nitro-explosives, Propellants Explos. Pyrotech. 31 (4) (2006) 306–310.
[15] L.E. Fried, M.R. Manaa, P.F. Pagoria, R.L. Simpson, Design and synthesis of energetic materials, Annu. Rev. Mater. Sci. 31 (2001) 291–321.
[16] X. Song, X. Cheng, X. Yang, D. Li, R. Linghu, Correlation between the bond dissociation energies and impact sensitivities in nitramine and polynitro benzoate molecules with polynitro alkyl groupings, J. Hazard. Mater. 150 (2) (2008) 317–321.
[17] T. Atalar, M. Jungova, S. Zeman, A new view of relationships of the N–N bond dissociation energies of cyclic nitramines. Part II. Relationships with impact sensitivity, J. Energ. Mater. 27 (3) (2009) 200–216.
[18] C. Zhang, Review of the establishment of nitro group charge method and its applications, J. Hazard. Mater. 161 (1) (2009) 21–28.
[19] E.-C. Koch, Acid-Base interactions in energetic materials: I. the hard and soft acids and bases (HSAB) principle–insights to reactivity and sensitivity of energetic materials, Propellants Explos. Pyrotech. 30 (1) (2005) 5–16.
[20] E.-C. Koch, Erratum: Acid-Base interactions in energetic materials: I. the hard and soft acids and bases (HSAB) principle–insights to reactivity and sensitivity of energetic materials, Propellants Explos. Pyrotech. 30 (2) (2005) 164.
[21] F.J. Owens, K. Jayasuriya, L. Abramhsen, P. Politzer, Computational analysis of some properties associated with the nitro groups in polynitroaromatic molecules, Chem. Phys. Lett. 116 (5) (1985) 434–438.
[22] J.S. Murray, P. Lane, P. Politzer, Relationships between impact sensitivities and molecular surface electrostatic potentials of nitroaromatic and nitroheterocyclic molecules, Mol. Phys. 85 (1) (1995) 1–8.
[23] J.S. Murray, P. Lane, P. Politzer, Effects of strongly electron-attracting components on molecular surface electrostatic potentials: application to predicting impact sensitivities of energetic molecules, Mol. Phys. 93 (2) (1998) 187–194.
[24] J.S. Murray, M.C. Concha, P. Politzer, Links between surface electrostatic potentials of energetic molecules, impact sensitivities and $C–NO_2/N–NO_2$ bond dissociation energies, Mol. Phys. 107 (1) (2009) 89–97.
[25] M. Pospisil, P. Vavra, M.C. Concha, J.S. Murray, P. Politzer, A possible crystal volume factor in the impact sensitivities of some energetic compounds, J. Mol. Model. 16 (5) (2010) 895–901.
[26] S.G. Cho, K.T. No, E.M. Goh, J.K. Kim, J.H. Shin, Y.D. Joo, S. Seong, Optimization of neural networks architecture for impact sensitivity of energetic molecules, Bull. Korean Chem. Soc. 26 (3) (2005) 399–408.
[27] H. Nefati, J.-M. Cense, J.-J. Legendre, Prediction of the impact sensitivity by neural networks, J. Chem. Inf. Comput. Sci. 36 (4) (1996) 804–810.
[28] G.T. Afanas'Ev, T.S. Pivina, D.V. Sukhachev, Comparative characteristics of some experimental and computational methods for estimating impact sensitivity parameters of explosives, Propellants Explos. Pyrotech. 18 (6) (1993) 309–316.
[29] N.R. Badders, C. Wei, A.A. Aldeeb, W.J. Rogers, M.S. Mannan, Predicting the impact sensitivities of polynitro compounds using quantum chemical descriptors, J. Energ. Mater. 24 (1) (2006) 17–33.
[30] G. Fayet, P. Rotureau, V. Prana, C. Adamo, Global and local quantitative structure–property relationship models to predict the impact sensitivity of nitro compounds, Process Saf. Prog. 31 (3) (2012) 291–303.
[31] A.K. Sikder, N. Sikder, A review of advanced high performance, insensitive and thermally stable energetic materials emerging for military and space applications, J. Hazard. Mater. 112 (1–2) (2004) 1–15.
[32] S. Odiot, M. Blain, E. Vauthier, S. Fliszar, Influence of the physical state of an explosive on its sensitivity. Is nitromethane sensitive or insensitive? J. Mol. Struct. (THEOCHEM) 279 (C) (1993) 233–238.
[33] Y. Kohno, K. Maekawa, T. Tsuchioka, T. Hashizume, A. Imamura, A relationship between the impact sensitivity and the electronic structures for the unique NN bond in the HMX polymorphs, Combust. Flame 96 (4) (1994) 343–350.

[34] P.J. Haskins, M.D. Cook, H. Flower, A.D. Wood, Ab-initio prediction of impact sensitivity, AIP Conf. Proc. 845 (2006) 527–530.

[35] M. Cartwright, J. Wilkinson, Correlation of structure and sensitivity in inorganic Azides I effect of non-bonded nitrogen nitrogen distances, Propellants Explos. Pyrotech. 35 (4) (2010) 326–332.

[36] J.K. Dienes, Q.H. Zuo, J.D. Kershner, Impact initiation of explosives and propellants via statistical crack mechanics, J. Mech. Phys. Solids 54 (6) (2006) 1237–1275.

[37] C.J. Wu, T. Piggott, J. Yoh, J. Reaugh, Numerical Modeling of Impact Initiation of High Explosives, LLNL report, UCRL-TR-221760, 2006.

[38] A. Tokmakoff, M.D. Fayer, D.D. Dlott, Chemical reaction initiation and hot-spot formation in shocked energetic molecular materials, J. Phys. Chem. 97 (9) (1993) 1901–1913.

[39] L.E. Fried, A.J. Ruggiero, Energy transfer rates in primary, secondary, and insensitive explosives, J. Phys. Chem. 98 (39) (1994) 9786–9791.

[40] K.L. McNesby, C.S. Coffey, Spectroscopic determination of impact sensitivities of explosives, J. Phys. Chem. B 101 (16) (1997) 3097–3104.

[41] S. Ye, K. Tonokura, M. Koshi, Energy transfer rates and impact sensitivities of crystalline explosives, Combust. Flame 132 (1–2) (2003) 240–246.

[42] S.-H. Ge, X.-L. Cheng, L.-S. Wu, X.-D. Yang, Correlation between normal mode vibrations and impact sensitivities of some secondary explosives, J. Mol. Struct. (THEOCHEM) 809 (1–3) (2007) 55–60.

[43] S.-H. Ge, X.-L. Cheng, X.-X. Wang, G.-X. Dong, G.-H. Sun, Energy transfer rates and impact sensitivities of two classes nitramine explosives molecules, J. Struct. Chem. 18 (6) (2007) 985–991.

[44] L. Türker, Structure-impact sensitivity relation of some substituted 1, 3, 5-trinitrobenzenes, J. Mol. Struct. (THEOCHEM) 725 (1–3) (2005) 85–87.

[45] L. Türker, Structure-impact sensitivity relation of certain explosive compounds, J. Energ. Mater. 27 (2) (2009) 94–109.

[46] A.V. Dubovik, Estimating impact breakup and initiation parameters of condensed explosives, Combust. Explos. Shock Waves 35 (2) (1999) 191–197.

[47] A.V. Dubovik, Explosion-like chemical reactions in solids initiated by mechanical impact, Russ. J. Phys. Chem. B 10 (6) (2016) 966–972.

[48] D.C.E. Taylor, Prediction of the Impact Sensitivity of Energetic Molecules Using Symmetry Adapted Perturbation Theory, Army Research Laboratory Report ARL-TR-5550, 2011.

[49] M. Vaullerin, A. Espagnacq, L. Morin-Allory, Prediction of explosives impact sensitivity, Propellants Explos. Pyrotech. 23 (5) (1998) 237–239.

[50] A. Smirnov, O. Voronko, B. Korsunsky, T. Pivina, Impact sensitivity investigations of individual explosives: some experimental and calculating approaches, New Trends Res. Energ. Mater. (2013) 340–352.

[51] S.D. McGrane, J. Barber, J. Quenneville, Anharmonic vibrational properties of explosives from temperature-dependent Raman, J. Phys. Chem. A 109 (44) (2005) 9919–9927.

[52] J. Li, A quantitative relationship for the shock sensitivities of energetic compounds based on X–NO2 (X = C, N, O) bond dissociation energy, J. Hazard. Mater. 180 (1–3) (2010) 768–772.

[53] B. Tan, X. Long, J. Li, F. Nie, J. Huang, Insight into shock-induced chemical reaction from the perspective of ring strain and rotation of chemical bonds, J. Mol. Model. 18 (12) (2012) 5127–5132.

[54] B. Tan, R. Peng, X. Long, H. Li, B. Jin, S. Chu, An important factor in relation to shock-induced chemistry: resonance energy, J. Mol. Model. 18 (2) (2012) 583–589.

[55] M.H. Keshavarz, H. Motamedoshariati, H.R. Pouretedal, M.K. Tehrani, A. Semnani, Prediction of shock sensitivity of explosives based on small-scale gap test, J. Hazard. Mater. 145 (1–2) (2007) 109–112.

[56] M.H. Keshavarz, H.R. Pouretedal, M.K. Tehrani, A. Semnani, Predicting shock sensitivity of energetic compounds, Asian J. Chem. 20 (2) (2008) 1025–1031.

[57] C.S. Yoo, N.C. Holmes, P.C. Souers, C.J. Wu, F.H. Ree, J.J. Dick, Anisotropic shock sensitivity and detonation temperature of pentaerythritol tetranitrate single crystal, J. Appl. Phys. 88 (1) (2000) 70–75.

[58] V.J. Bellitto, M.I. Melnik, D.N. Sorensen, J.C. Chang, Predicting the shock sensitivity of cyclotrimethylene-trinitramine, J. Therm. Anal. Calorim. 102 (2) (2010) 557–562.

[59] V.J. Bellitto, M.I. Melnik, Surface defects and their role in the shock sensitivity of cyclotrimethylene-trinitramine, Appl. Surf. Sci. 256 (11) (2010) 3478–3481.

[60] D. Price, Examination of some proposed relations among HE sensitivity data, J. Energ. Mater. 3 (3) (1985) 239–254.

[61] E.B. Smirnov, A.N. Averin, B.G. Loboiko, O.V. Kostitsyn, Y.A. Belenovskii, K.M. Prosvirnin, A.N. Kiselev, Dynamics of the detonation-wave front in solid explosives, Combust. Explos. Shock Waves 48 (3) (2012) 309–318.
[62] V. Zeman, J. Kočí, S. Zeman, Electric spark sensitiviy of polynitro compounds: part II. A correlation with detonation velocities of some polynitro arenes, Energ. Mater.-CHENGDU 7 (3) (1999) (131–132+136).
[63] V. Zeman, J. Kočí, S. Zeman, Electric spark sensitivity of polynitro compounds: part III. A correlation with detonation velocities of some nitramines, Energ. Mater.-CHENGDU 7 (4) (1999) 173–175.
[64] F. Hosoya, K. Shiino, K. Itabashi, Electric-spark sensitivity of heat-resistant polynitroaromatic compounds, Propellants Explos. Pyrotech. 16 (3) (1991) 119–122.
[65] D. Skinner, D. Olson, A. Block-Bolten, Electrostatic discharge ignition of energetic materials, Propellants Explos. Pyrotech. 23 (1) (1998) 34–42.
[66] S. Zeman, J. Kočí, Electric spark sensitivity of polynitro compounds: part IV. A relation to thermal decomposition parameters, Energ. Mater.-CHENGDU 8 (1) (2000) 23–26.
[67] M.B. Talawar, A.P. Agrawal, M. Anniyappan, D.S. Wani, M.K. Bansode, G.M. Gore, Primary explosives: electrostatic discharge initiation, additive effect and its relation to thermal and explosive characteristics, J. Hazard. Mater. 137 (2) (2006) 1074–1078.
[68] G. Wang, H. Xiao, X. Ju, X. Gong, Calculation of detonation velocity, pressure, and electric sensitivity of nitro arenes based on quantum chemistry, Propellants Explos. Pyrotech. 31 (5) (2006) 361–368.
[69] G.-X. Wang, H.-M. Xiao, X.-J. Xu, X.-H. Ju, Detonation velocities and pressures, and their relationships with electric spark sensitivities for nitramines, Propellants Explos. Pyrotech. 31 (2) (2006) 102–109.
[70] M.H. Keshavarz, Estimating heats of detonation and detonation velocities of aromatic energetic compounds, Propellants Explos. Pyrotech. 33 (6) (2008) 448–453.
[71] M.H. Keshavarz, H.R. Pouretedal, A. Semnani, Reliable prediction of electric spark sensitivity of nitramines: a general correlation with detonation pressure, J. Hazard. Mater. 167 (1–3) (2009) 461–466.
[72] C. Zhi, X. Cheng, F. Zhao, The correlation between electric spark sensitivity of polynitroaromatic compounds and their molecular electronic properties, Propellants Explos. Pyrotech. 35 (6) (2010) 555–560.
[73] L. Türker, Contemplation on spark sensitivity of certain nitramine type explosives, J. Hazard. Mater. 169 (1–3) (2009) 454–459.
[74] G. Fayet, P. Rotureau, Predicting explosibility properties of chemicals from quantitative structure-property relationships, Joubert, L., Adamo, C, Process Saf. Prog. 29 (4) (2010) 359–371.
[75] Q.-L. Yan, S. Zeman, Theoretical evaluation of sensitivity and thermal stability for high explosives based on quantum chemistry methods: a brief review, Int. J. Quantum Chem. 113 (8) (2013) 1049–1061.
[76] S. Zeman, P. Kohlíček, A. Maranda, A study of chemical micromechanism governing detonation initiation of condensed explosive mixtures by means of differential thermal analysis, Thermochim. Acta 398 (1–2) (2003) 185–194.
[77] S. Zeman, Z. Friedl, Relationship between electronic charges at nitrogen atoms of nitro groups and thermal reactivity of nitramines, J. Therm. Anal. Calorim. 77 (1) (2004) 217–224.
[78] S. Zeman, C. Bartei, Some properties of explosive mixtures containing peroxides: part II. Relationships between detonation parameters and thermal reactivity of the mixtures with triacetone triperoxide, J. Hazard. Mater. 154 (1–3) (2008) 199–203.
[79] M. Chovancová, S. Zeman, Study of initiation reactivity of some plastic explosives by vacuum stability test and non-isothermal differential thermal analysis, Thermochim. Acta 460 (1–2) (2007) 67–76.
[80] S.R. Saraf, W.J. Rogers, M.S. Mannan, Prediction of reactive hazards based on molecular structure, J. Hazard. Mater. 98 (1–3) (2003) 15–29.
[81] G. Fayet, L. Joubert, P. Rotureau, C. Adamo, A theoretical study of the decomposition mechanisms in substituted *o*-Nitrotoluenes, J. Phys. Chem. A 113 (48) (2009) 13621–13627.
[82] G. Fayet, P. Rotureau, L. Joubert, C. Adamo, Development of a QSPR model for predicting thermal stabilities of nitroaromatic compounds taking into account their decomposition mechanisms, J. Mol. Model. 17 (10) (2011) 2443–2453.
[83] P. Sang, J.-W. Zou, X. Lin, P. Zhou, Linear and nonlinear QSPR models for predicting thermal stabilities of nitroaromatic compounds, Chem. Res. Chin. Univ. 27 (5) (2011) 891–895.
[84] J. Li, H. Liu, X. Huo, P. Gramatica, Structure-activity relationship analysis of the thermal stabilities of Nitroaromatic compounds following different decomposition mechanisms, Mol. Inform. 32 (2) (2013) 193–202.

CHAPTER

6

General quantitative structure–property relationships and machine learning correlations to energetic material sensitivities

Jason A. Morrill[a], *Brian C. Barnes*[b], *Betsy M. Rice*[b], *and Edward F.C. Byrd*[b,*]

[a]Department of Chemistry and Biochemistry, William Jewell College, Liberty, MO, United States
[b]U.S. Army Combat Capabilities Development Command (DEVCOM) Army Research Laboratory, Aberdeen Proving Ground, MD, United States
*Corresponding author: E-mail: edward.f.byrd2.civ@army.mil

1 Introduction

Well before quantitative structure–property relationship (QSPR) approaches became tools routinely used in material discovery [1,2], their potential for rapid, computationally efficient screening of candidate materials appealed to the community of energetic materials (EMs) researchers, due to the extensive cost, time, and hazard [3] associated with the discovery, synthesis, scale up, formulation, and fielding of new EMs. Since the 1970s, EM researchers have used structure–property relationships with differing degrees of success to predict performance and sensitivity parameters of candidate materials [4,5], with sensitivity prediction [6,7] being arguably the least successful in quantitative prediction (not overly surprising as experimental small-scale sensitivities are typically noisy). The majority of QSPR studies have focused on impact sensitivity using results from drop hammer tests, which usually have a larger and more granular available dataset compared to thermal stability, electrostatic discharge, and friction, where data is either sparse or not easily compiled. It is notable that over the past few years we have seen a reduction in the use of traditional QSPR modeling techniques for EM sensitivity prediction, perhaps due to advances in high performance

Theoretical and Computational Chemistry, Volume 22
https://doi.org/10.1016/B978-0-12-822971-2.00001-2

computing which make physics-based modeling (often utilizing quantum-mechanical (QM) methods) computationally feasible. In the past, QM-based semiempirical theories were widely used to obtain molecular descriptors used in QSPR studies; however, they lacked the fidelity and accuracy of rigorous quantum mechanical theories. With the progress in computational platforms and algorithms, modeling the electronic structure for large numbers of molecules with a significant degree of fidelity is readily achievable using modest computational resources. Thus, currently, calculations are not limited to obtaining accurate molecular descriptors for traditional QSPR studies but are readily amenable to exploration of mechanism-based arguments to explain sensitivities, many of which require a depiction of the potential energy surface. Such physics-based modeling approaches to explore explosive sensitivity will be highlighted in other parts of this book, and will not be covered here. Additionally, there are a number of earlier articles reviewing QSPR applications to predict properties of explosives (including sensitivities) [8–15], including a recent review article [16] that nicely categorizes and summarizes the various molecular and crystal correlations with EM sensitivities covering the past two decades. Due to the availability of these, we will restrict our discussion to the more recent notable advances in traditional QSPR approaches applied to EM sensitivity. We will also discuss the expanding role of machine learning approaches in QSPR modeling, envisioning them as a successor technique to traditional multilinear regression techniques.

2 Traditional QSPR methods for energetic materials sensitivity

Quantitative structure–property relationships for energetic material sensitivity, which are built from large a priori pools of descriptors are generally derived using a common set of steps. First, a property of interest is identified and then a set of molecular structures exhibiting promise in idealizing the property of interest is identified. Then, in order to derive some relationship between the structures and their properties a means for quantitatively describing the molecular structures (descriptors) is determined. Finally, a multilinear equation that fits the descriptors to the property of interest and has predictive capability beyond the set of compounds whose descriptors were used in the fitting is derived and validated. EMs exhibit several modes of sensitivity including impact, shock, friction, and sensitivity to thermal degradation, with concomitant instability and can be dangerous to handle. Thus, much work has been done to predict and understand EM sensitivity by way of computationally derived QSPRs [15]. Multilinear QSPRs typically involve the use of some commercially available software for computing hundreds to thousands of descriptors followed with a selection algorithm for either narrowing the descriptor pool to some predictive and manageable set, or weighting noncorrelating descriptors so minimally that they essentially make no contribution to the predictive model. Some commercial packages for descriptor computing and QSPR derivation are DRAGON and CODESSA. More recently, there has been a surge in open-source packages based on the popular Python programming language and its libraries, many of which compute descriptors from molecules represented as Simplified Molecular-input Line Entry System (SMILES) [17] strings. The explosive growth in popularity of the Python Programming language with the development of libraries that automate many QSPR development tasks has only fueled the open-source approach.

Typically QSPR development for EM sensitivity involves developing predictive models for certain classes of compounds, reflecting the generally accepted notion that the QSPR approach works best when descriptor space is continuous. For example, toward developing a model of the heat of decomposition of 38 structurally similar organic peroxides Prana et al. derived a QSPR model solely based on constitutional and topological descriptors [18]. Their models were developed in order to be simple and easily accessible, employing easy to compute descriptors. The authors also developed their models to be in compliance with QSAR development best practices as defined by European Union regulations. Specifically, REACH (Registration, Evaluation, Authorization, and restriction of CHemicals) requires evaluation of various physicochemical, toxicological, and environmental properties of large quantities of chemicals transported within EU countries. Whereas experimentally determining such properties can be costly and dangerous, REACH regulations allow the use of QSPR/quantitative structure–activity relationship (QSAR) studies to determine property data, provided that it is based on sound methodology. In their work the authors made use of CODESSA for descriptor computing and QSPR derivation on the heat of decomposition in J/g divided by the concentration. The algorithm used for QSPR derivation was the Best Multilinear Regression algorithm, a forward-stepping algorithm, which builds models based on the multiple coefficient of determination, provided an acceptable minimal intercorrelation among descriptors in the model. They obtained a four-descriptor model consisting of constitutional and topological descriptors for the number of O—O bonds, O—OH bonds, relative number of oxygen atoms, and the bonding information content order 1. With a training set of 25 organic peroxides and a test set of 12, this model gave $R^2=0.95$ and root-mean-squared error (RMSE) = 117 J/g for the training set, and for the test set $R^2=0.68$ with RMSE = 345 J/g. The authors also derived a QSPR-based solely on constitutional descriptors of a similar nature and found $R^2=0.92$ with RMSE = 103 J/g for the training set and $R^2=0.80$ with RMSE = 303 J/g for the test set.

Due to the thermal sensitivity of organic peroxides Pan et al. investigated the thermal risk hazards associated with a series of a small set of 14 peroxides of similar molecular structure [19]. Because processing temperature can be an issue in handling peroxides, the authors derived a Probability Index (PI) that is predictive of the likelihood that a thermal decomposition can occur. They also derived a Severity Index (SI) that is predictive of the severity of a thermal decomposition. The measures of thermal hazard used by the authors were the heat of thermal decomposition reaction (ΔH), temperature of thermal decomposition onset (T_o), and time (in minutes) to maximum decomposition reaction rate under adiabatic conditions (TMR_{ad}). The descriptor values used to derive models for ΔH and T_o were taken from previously published QSPR studies, but the authors derived their own QSPR model for TMR_{ad}. In their model development the authors optimized the structures of the 14 peroxides using the AM1 semi-empirical quantum mechanical model [20] as implemented in the program Hyperchem. Computing of descriptors was achieved using DRAGON with a total of 4885 descriptors computed. The authors then made use of their own genetic algorithm (GA) for deriving multiple linear equations using a maximum of three descriptors to develop a model of TMR_{ad}. The fitness for the GA was determined as the root-mean-square error of leave-one-out cross-validation for all 14 of the compounds of interest. For this model they report $R^2=0.934$, $F=47.374$, $Q^2_{LOO}=0.865$, average absolute error (AAE) of 13.37 min, and root-mean-squared error of ±16.827. The authors went on to develop predictors of PI by combining descriptors for TMR_{ad} and T_o. The predictive model for SI was based on descriptors used to describe ΔH. The training set for PI and SI consisted of 67 organic peroxides taken from the literature. Again DRAGON was used to compute descriptors, but they were standardized and then weighted using the entropy weight method, which is derived from Shannon information theory [21]. The authors found that their indices predicted hazard levels reliably as compared to previously published methods [22].

In a related study Pan and coworkers developed a QSPR for the thermal decomposition temperature for a dataset of 168 imidazolium ionic liquids (ILs) [23]. Again, they used DRAGON for computing descriptors and their own genetic algorithm for descriptor selection. For the ionic liquids they also included a topological descriptor designed for the interaction between anions and cations, which had been published previously [24]. Using a training set of 80% of their dataset with a test set of remaining 20%, the authors again fitted descriptors primarily on the basis of R^2, RMSE, and AAE. The authors obtained a model having seven descriptors that were primarily constitutional and connectivity in nature. For the training set, the model gave $R^2=0.875$, RMSE $=\pm 23.1$ K, and AAE $=19.0$ K. For the test set, the model was assessed according to values for $Q^2_{F1}=0.862$ [25,26], $Q^2_{F2}=0.862$, $R^2=0.862$, RMSE $=\pm 25.7$ K, and AAE $=20.0$ K. The authors assessed the applicability domain (AD) by plotting the standardized residuals vs the model leverage (Williams plot). From the Williams plot the authors found that all ILs predicted within $\pm$ three standardized residuals of zero, and there were five ILs that exceeded the warning threshold for the AD. The authors concluded that the five ILs that exceeded warning threshold had structures that were somewhat different than the rest of the model structures and were outliers, or nearly outliers.

Another common QSPR predictive target, the enthalpy of formation of a compound in its standard state, is a measure of the relative stability of that compound with respect to its constituent elements in their standard states. As such, the enthalpy of formation provides a general measure of the sensitivity of a compound. Toward design of green nitrogen-rich energetic materials Sheibani derived QSPRs for the enthalpy of formation of azido compounds [27]. In the study, the author modeled structures using the semiempirical AM1 method [20] as implemented in the program, Hyperchem. Descriptors were calculated using DRAGON software and the QSPR equations were derived using a step-wise regression method and another best subsets method, though the authors gave no further details on the selection criteria used in these methods. For a training set of 33 compounds and a test set of 7, the author settled on a set of 8 descriptors. Two of the descriptors were the ratios of the oxygen to nitrogen as well as carbon to nitrogen. The remaining six descriptors were selected from the DRAGON descriptors. Not surprisingly, given the large number of descriptors relative to the training set size, the author obtained $R^2=0.9821$, RMSE $=55.88$ kJ/mol, and average absolute error of 45.05 kJ/mol. The author explored the possibility of overfitting by plotting leave-one-out cross-validated predictions vs fitted values and found similar results in both sets of plotted values. Although the standard R^2_{CV} was discussed (as Q^2_{LOO}), it was not reported. The predictive capability of the model was also explored by examining results for the test set of compounds and the author found $Q^2_{EXT}=0.9387$, RMSE $=72.10$ kJ/mol, and average absolute error of 63.51 kJ/mol. Finally, it was found that four compounds were outside the applicability domain in the training set and that there was one such compound in the test set.

In an attempt to develop a model for electric spark sensitivity, that might also clarify the mechanism of electricity-initiated explosive detonation, Tan et al. sought correlations between 32 descriptors and electric spark energy [28]. The authors investigated descriptors, such as extrema and average deviation in electrostatic surface potential, dipole moments, polarizabilities, and others. The quantum mechanically derived descriptors were derived from the results of optimizing molecular structures of 12 explosives using density functional theory (B3LYP [29]) and the aug-cc-pVDZ basis set [30–32] as implemented in the program, Gaussian 09 [33]. To derive their multiple correlation model the authors made use of a genetic function approximation and found a relationship between a 4-descriptor model and electrostatic spark energies for all 12 explosives in their study. Interestingly, the authors applied a ramp function to some of the descriptors in their model, but gave no rationale as to why. Not surprisingly, the authors found good agreement between the 4 descriptors and the 12 training set electric spark energy values, with $R^2=0.9921$. No validation or other statistical analysis was applied to this model.

In 2015, two authors of this chapter published a QSPR model, which at that time was the most comprehensive and systematic QSPR for drop height sensitivities published [34]. Our model was based on the drop height impact sensitivities ($h_{50}\%$) of EMs from the database of Storm, Stine, and Kramer [7]. In that study, our primary approach involved automatic descriptor selection from a wide variety of descriptor classes (constitutional, topological, geometric, quantum mechanical, and thermodynamic) for a structurally diverse set of 220 nitroorganic compounds. We modeled each EM as an isolated molecular structure using the computationally inexpensive semiempirical AM1 [20] method as implemented in the program, AMPAC. In our approach we made no assumptions about the mechanisms of mechanically induced EM decomposition. For an 8-descriptor model regressed on a training set of 212 EMs we obtained $R^2=0.8141$ and $R^2_{CV}=0.7951$. For an external test set of eight EMs distributed across the range of experimental $h_{50}\%$ values we obtained $R^2=0.6911$. In response to this prior-published work, Mathieu and Alaime [10] developed a semiempirical model that was based on the assumption that mechanically induced decomposition is mainly the result of bond scission of explosophores within the molecule. Thus, they developed semiempirical descriptors for bond dissociation energies for various types of explosophore bonds (i.e., X—NO_2 bonds). They tested their approach against our previously published data set and for the 212 EM training set found $R^2=0.8085$ and $R^2_{CV}=0.7934$. For the external validation set they found $R^2=0.8804$. Their model performs quite well, given its simplicity and the generalization of the bond dissociation approach. On expanding their work to a larger data set the authors found several compounds for which there were no reliable parameters or the compounds were found to be outside of the AD as determined by the methods of 3D and 4D convex hulls. For some of the compounds the authors found that exotic functional groups were not well-treated by their approach, suggesting that models derived from quantum mechanical calculations represent a more generally applicable approach. In testing this semiempirical approach against datasets from several other published models Mathieu found that the approach performed quite well [35]. It was noted in a later study [36] that a slight error had been found in this study, but was subsequently corrected and will be described hereafter.

We will touch upon other models for sensitivity prediction that are based on QSPR approaches, but do not utilize automated descriptor selection algorithms to select optimal sets for fitting. These include several studies that correlate experimental sensitivities models that use elemental composition as descriptors augmented by correction terms that are dependent on specific classes of energetic materials [37–41]. For example, Keshavarz et al. [40] developed two models to predict decomposition temperatures of imidazolium-based energetic ionic liquids based on molecular structural descriptors. The descriptors in the first model were solely elemental composition of the cations and anions in the imidazolium compounds, and the second included correction terms based on specific cations and anions. The training and test sets for this study were composed of 164 and 17 imidazolium-based ionic liquids, respectively. The correlations were established using the multiple linear regression method with the first model having $R^2 = 0.9953$ and a standard error of 38.6 K; the model that included the correction terms had an $R^2 = 0.9977$ and a standard error of 27.0 K. No cross-validation was reported. Further, we note that the mean absolute percent error (MAPE) is reported as a convenient measure of error. In this case, however, MAPE could lead to incorrect conclusions about the quality of the predictions. For example, a 5% error in decomposition temperature has vastly different meanings for systems having decomposition temperatures of, e.g., 300 K (where a MAPE of 5% implies a 15 K error) vs 1000 K (where a MAPE of 5% implies a 50 K error). A better indicator of error might be RMSE, as pointed out in the Morrill and Byrd [34] discussion of QSPR predictions of melting temperatures. Keshavarz and coworkers next applied this approach to predict impact sensitivities of quaternary ammonium-based energetic ionic liquids [37], again using elemental composition of the cations and anions as descriptors, and ionic-species-dependent correction terms. The training set used for fitting was composed of 72 ionic molecular systems, with 3 ionic systems used for testing the correlations. As would be expected, R^2 and RMSE values are better for those fits which include correction terms, with $R^2 = 0.908$ and RMSE = 6 J. Application of this model to the test set produces an RMSE of 6 J as well. Similar approaches were used to predict electrostatic sensitivities of polynitroarenes [42], nitramine compounds [38], and quaternary ammonium-based energetic ionic liquids or salts [41]. Zohari et al. [43] derived a QSPR model based on molecular structural descriptors and the multiple linear regression method to predict the decomposition temperature of azole-based energetic materials. These descriptors included elemental composition, nitroamino, imino and azido groups; an initial model using only the aforementioned descriptors was augmented with terms based on structural moieties on the azoles to correct for deviations of predictions from experiment. The training set was composed of 116 azole-based energetic materials for which experimental data were available; application of this model to the training set resulted in an $R^2 = 0.936$ with RMSE = 16.23 K. Cross-validation using the leave-many-out approach (25% in this work) produced an Q^2_{LMO} of 0.937. External validation of the model was accomplished by application of the model to a test set composed 30 azole molecules that were not in the original training set, and produced RMSE = 16.46 K with $Q^2_{EXT} = 0.925$.

The specialized QSPR models described in the aforementioned paragraph as well as more generally applicable QM-based QSPR models for which empirical parameters do not exist are examples of models that fall outside of standard and widely used QSPR approaches implemented in commercial software. Utilization of these alternate QSPR models requires an investment in software development, often a dissuading factor in pursuing their use. To counter this, Mathieu [36] argues that building QSPR models using the python programming language and its wide array of libraries is a straightforward way to implement arbitrary models, and demonstrates this for two alternate QSPR approaches described above: the Keshavarz model to predict spark sensitivities of nitroarenes [42] and his own earlier approach to predict impact sensitivities [10]. Mathieu first implemented and reproduced the Keshavarz model for spark sensitivity predictions with a ~75-line Python code and the open-source RDKit library. Mathieu then demonstrated a less-straightforward implementation of a previously developed general impact sensitivity model using two Python scripts. The first script performed a nonlinear least-squares fit of logarithms of $h_{50}\%$ values to provide revised model parameters, and the second was used to evaluate impact sensitivities for molecules represented as SMILES strings [17]. Despite the differences in results between the revised model and those from the earlier study [10], the results of the two studies are similar, with the new model giving an overall comparable accuracy of test system predictions of $\log(h_{50}\%)$ values, as well as quality of fit for the 212-molecule training set ($R^2 = 0.72$ and $\text{RMSE} = 0.237$).

Finally, we note that two QSPR-type models of sensitivity predictors using quantum-chemistry descriptors [44] relating to molecular charge distribution have been reported [45,46] that are similar in spirit to studies that showing correlations between features of electrostatic surface potentials (ESPs) [5,47–49] and impact sensitivities. In these, the distributed multipole analysis (DMA) method was used to partition DFT-molecular charge densities of cyclic nitramines [46] and nitroazoles [45], from which models using DMA multipole values were fitted to small training sets of molecules for which experimental drop-hammer results are available. The nitramine study [46] explored the transferability of similar DMA-based models fitted to nitroaromatic explosives [50] to predict impact sensitivities of 14 nitramine explosives, of which they use only 6 measured $h_{50}\%$ values for evaluation of the quality of the models. Deviations of the predictions from measured values were large for some of the molecules, leading the authors to reparameterize the models using results from DMA multipole values for the nitramines molecules. Only 4 of the nitramines for which experimental data were available were used as the training set; the test set was composed of the remaining 10 molecules, of which only 2 had experimental $h_{50}\%$ values. The models well predicted $h_{50}\%$ values for the four molecules in the test set, but the predictions for the two molecules in the test set for which experimental data were available was mixed: 1,3,3,-trinitroazetidine (TNAZ) was predicted poorly, 4,10-dinitro-2,6,8,12-tetraoxa-4, 10-diazatetracyclo[5.5.0.0^{5,9}.0^{3,11}]-dodecane (TEX) was fairly well predicted. In the nitroazole study [45], 7 molecules made up the training set, and 26 molecules comprised the test set, although only 4 of these had experimental $h_{50}\%$ values to assess the quality of the models. While the major conclusions of the nitroazole study were similar to other studies that explored possible correlations between molecular charge descriptors and impact sensitivities, the errors in the predictions (as well as the earlier study on cyclic nitramines [46]) indicated that DMA multipole values might not be suitable descriptors for predictive models of impact sensitivity due to the significant errors for some molecules in the test set. The large deviations could, however, be a result of a very small number of training data used for fitting. Similarly, one of the authors of this chapter was also unsuccessful in developing a predictive capability using a somewhat similar idea, to wit: Byrd [51] used Bader's Atoms-in-Molecules theory [52] to partition the molecular ESPs of 21 aromatic and nonaromatic EMs into atom- or group-specific areas. Statistical descriptors used in analyses of molecular surface ESPs [47–49] were then calculated for each atom- or group-specific ESP and a variety of correlations between the statistical descriptors and experimental impact sensitivities were attempted. Byrd found slight correlations, always accompanied by significant outliers, rendering the approach as insufficient to accurately predict impact sensitivity. Notably, the use of electrostatics as descriptors for models of impact sensitivity are called into question by the recent study by Aina et al. [53]. This study explores the changes in molecular charge distribution due to conformational changes of the NO_2 groups of four widely studied energetic molecules, 1,3,5-trinitro-1,3,5-triazacyclohexane (RDX), trinitrotoluene (TNT), 1-3-5-trinitrobenzene (TNB), and hexanitrobenzene (HNB). The study used the iterated stockholder atom (ISA) [54–56] distributed multipole analysis to calculate changes in electrostatic properties (including the ESP) of these molecules due to conformational changes involving NO_2 groups. The results clearly showed distinct changes in the ESPs due solely to the conformational changes of molecules when isolated vs within the crystalline environment. The authors argue that these significant changes have strong implications for any QSPR model that is based on conformational-dependent information, such as the ESP. However, a significantly larger study is warranted to prove or disprove this argument. The indirect sampling of the electronic wavefunction through these select features (e.g., DMA, ESPs, Mulliken charges, etc.) to date have not yielded any accurate correlation. However, it remains an open question as to whether direct probing of the electronic wavefunction and related density can yield a suitable correlation. Such a direct interrogation of the enormous data associated with the electron density of a variety of molecular conformations might be possible using "big data" approaches, i.e., machine learning.

3 QSPR machine learning models for energetic materials sensitivity

In the traditional QSPR approach described above, a multilinear regression is typically created to model the response of interest. What is popularly referred to as "Machine Learning" (ML) in computational chemistry QSPR efforts often eschews parameterization of either linear coefficients or nonlinear parameters within predetermined, physically inspired functional forms. Instead, algorithms first pioneered by the computer science and statistics communities [57–59], and now developed by a larger ML community [60–62], are applied to computational chemistry and materials problems, including EM [63–66]. First, let us frame sensitivity QSPR problems in the typical language of ML. Typically this will be for "supervised" learning: a dataset containing some ground truth is known for a relationship of inputs and desired outputs. That is in contrast to "unsupervised" learning, which attempts to detect patterns, correlations, or clusters within a dataset for which no particular relationship between the set and a ground truth is trying to be determined. The supervised learning problem may be framed as a "regression" problem for prediction of continuous values (such as an impact sensitivity $h_{50}\%$ value), or "classification" problem for prediction among a set of discrete classes (e.g., sensitive or insensitive EM). Such regression or classification models are categorized as "discriminative" models, as they are inference models reliant on the underlying data to make predictions given specific inputs. In contrast, there are also "generative" models which are capable of creating new examples of inputs (in this case, molecules) which may have a desired value for the target property; generative models are rarely used for solely predictive purposes in QSPR of EM sensitivity. To round out the introductory discussion, we mention there are also "reinforcement" learning models, which discover a policy optimal for a given state of a well-defined system (such as a game-playing or car-driving strategy). Reinforcement models are less relevant for the problem of creating EM QSPRs, but we mention them due to their popular connotation with artificial intelligence or ML, so readers can avoid conflating the different types of models.

If a primary difference between traditional QSPR and ML-driven QSPR is the choice of algorithm, then a brief survey of popular algorithms is warranted. The most popular methods for supervised learning can be broadly categorized as being: (a) neural networks [60], (b) decision-tree based [57], or kernel-based [62] and refer the reader to the review articles cited as starting points. There is a wide and ever-expanding collection of neural network types, including dense feed-forward networks (often equivalent to multilayer perceptrons), recurrent neural networks (often used with time-dependent signals), convolutional neural networks (CNN; popular when there is local, spatially correlated information such as in image recognition), and more recently message-passing neural networks (MPNN; which operate on mathematical graphs). Neural networks are popular for their ability to discover relevant representations (as in "deep learning") and empirically proven effectiveness when working with large datasets. Decision-tree based methods include random forests and gradient-boosted tree methods (e.g., XGBoost), and are particularly popular for their ease of interpretation and computational efficiency. Popular kernel-based methods include support vector machines, kernel ridge regression, and the closely related Gaussian process regression. The kernel in these methods serves to transform the data into a higher dimensional space where the classification or regression problem is more easily solved. This choice of kernel is an expert-level decision based upon the loss function for the problem and dataset at hand, but, by default, a radial basis function kernel that often works well for chemistry or materials problems. All of the above methods, when used as part of model training, may be refined through a hyperparameter or architecture optimization process for the problem and dataset at hand, subject to the same caveats regarding overfitting and validation as a traditional QSPR. The models above are "data-driven" in that they do not assume a particular relationship (e.g., a form needing a single linear coefficient) between an input variable and the target variable. Instead, the training process discovers a useful relationship.

While the same descriptors as used in traditional QSPR may be used in ML-based QSPR, that is not necessarily common practice. ML algorithms may work more effectively than multilinear models when using descriptor vectors such as extended-connectivity (aka "Morgan") fingerprints [67] as input representations. In chemistry software such as RDKit, there is a plethora of fingerprints available that may be immediately constructed from skeletal formulas. Further, there is extremely active research in creating new molecular representations that are particularly useful when used in conjunction with ML methods [68], and some ML methods (in particular, CNNs and MPNNs) are designed to effectively learn new intermediate representations from "raw" data [69]. Through either user-selected cutoff-based downselection or, more typically, regularization, ML methods may effectively reduce the original input space when faced with a large set of potential descriptors, as is done with regularized methods for multilinear regression.

Many ML models are nonparametric models, in the sense that the structure of the model is not fixed, and the complexity of the model depends on the size of the underlying dataset. Such is the case for Gaussian process regression. Likewise, the risk of overfitting a neural network is not solely dependent on its number of parameters compared to the size of the dataset, due to both their architecture and any regularization (or related methods, e.g., dropout) applied. Thus, like in traditional QSPR, test and validation of models must be performed on data not used in training. For neural networks, "early stopping," or ceasing training when loss on the validation set bottoms out, is a common practice to avoid overfitting training data. The model should still be evaluated against held-out test data, although in practice, some publications use simple cross-validation instead. When stochastically trained models are used with large datasets, and in particular with multiple splits (or cross-validation) on those sets, it is common practice to "ensemble" different training results (different models!) into a larger set of models and to use the average of the resulting predictions as the final prediction. In fact, ensembling lies at the heart of the strategy for random forests and gradient-boosted decision trees, which are collections of individual decision trees trained on different subsets of data. This stands in contrast to traditional multilinear QSPR models that typically only use one parameter set for the function at hand. All the practices described above, aside from particular chemical fingerprints, also widely apply to use of ML models to address problems in nonchemistry fields.

We will now briefly walk the reader through how a simple multilayer perceptron works (Fig. 6.1). This particular example is also a densely connected, feed-forward neural network. For each sample in a dataset, the neural network input is a vector **x** of real numbers, integers, or a mixed set thereof. Each element of the input vector is assigned to a single input node (*yellow* (*light gray* in the print version)) in the neural network; the size of **x** determines the size of the input layer in the neural network. The values for each input node are passed to each hidden node (*green* (*dark gray* in the print version)) in the arbitrarily sized hidden layer. As they are passed, they are multiplied by a weight from the weight matrix **W**; this weight depends on the identity of the input node and hidden node. At each hidden node, the incoming values are summed, a bias value is added (specific to each hidden node), and the result is

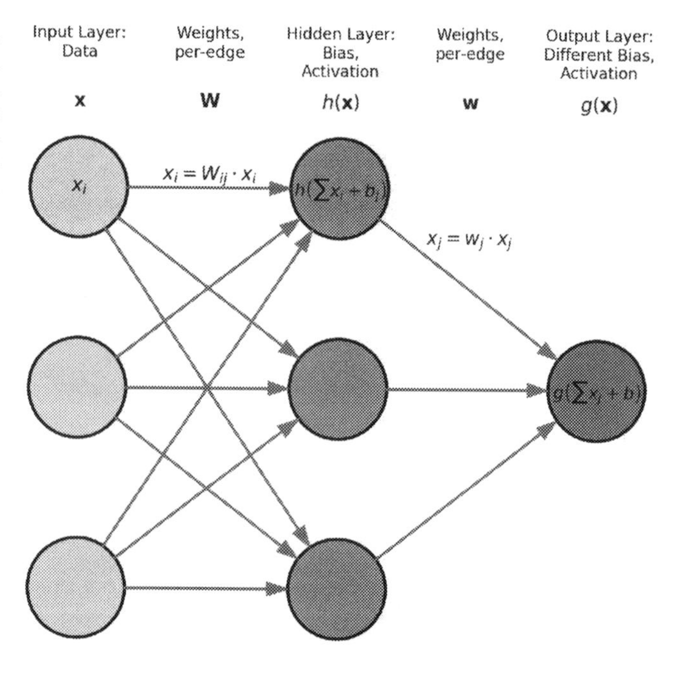

FIG. 6.1 Information flow of a small feed-forward neural network (also known as multilayer perceptron), drawn with nodes (*circles*) and edges (*arrows*). Input nodes are drawn in *yellow* (*light gray* in the print version), hidden nodes in green (*dark gray* in the print version), and the output node in orange (*dark gray* in the print version). Math operations are shown for a subset of the network. *Reprinted with permission from B.C. Barnes, A primer on machine learning for materials and its relevance to Army challenges, ARL-SR-0411, 2018.*

passed through a nonlinear activation function h(**x**) chosen by the user. This activation function is often a sigmoid function or rectified linear unit, but there are a variety of possible choices. Finally, the results are passed to the output node (*orange* (*dark gray* in the print version)), and once again a set of weights from vector **w** are multiplied on the values before they are summed at the output node, a bias is added, and another function g(**x**) is applied (which may be linear for regression problems). In this example, we start with a vector of numbers, and end with a single number. The range is not specified, but normalization is typically applied to the inputs (to scale around zero or between zero and one), and the output will have a target value between zero and one for sigmoid-activated classification problems or some arbitrary (potentially normalized) value for linear-activated regression problems. The process of training the neural network is the stochastic, global optimization of the weights and bias values used in transforming values as they are passed between different layers of the network. In supervised learning for QSPR of EM sensitivity, the *yellow* (*light gray* in the print version) input nodes would be the descriptor vector for a molecule, and the *orange* (*dark gray* in the print version) output node would be a sensitivity value for a molecule. In general, the size and number of hidden layers (*green* (*dark gray* in the print version)) and choice of activation function h(**x**) is a matter of neural network architecture design that may be subject to hyperparameter optimization.

Unfortunately, there has been little recent work in prediction of EM sensitivity using ML methods, most likely due in part to the difficulty of properly training and validating ML models when faced with small datasets. This particular problem can be addressed in part through multitarget training (training one model architecture to predict several target properties simultaneously), or through transfer learning. When used with neural networks, transfer learning trains a model to predict one set of quantities, then the model's weights are frozen except for in a small number of tail layers (sometimes just the final layer itself), and the model is retrained on the smaller dataset to reoptimize that subset of weights. The inspiration for transfer learning is that the earlier layers are learning generally useful representations, while the final layer(s) are learning the correlation for the specific problem at hand, and this technique has proven useful for other molecular applications. In the remainder of this section, we will discuss some specific EM ML-based QSPR papers studying sensitivity and stability.

Seeking to better predict thermal stability of EMs, Mage et al. [70] combined multiple types of models in the analysis of differential scanning calorimetry (DSC) data for 197 curves. Unsupervised hierarchical clustering of image data (DSC traces) was used to group DSC results into seven unique groups, thus creating a ground truth "label" associated with individual molecules (their unique group label). A decision tree was created using molecular group descriptors as input and the aforementioned cluster labels as target outputs, thus creating a model that may be used for molecules that do not yet have DSC data. If the decision tree assigns a model to a particular cluster, its thermal response is predicted using the reference data local to that cluster. The classification accuracy of the decision tree was evaluated and found to be above 66% for 4 of the 7 clusters (also referred to as families). This work highlights the need for larger, uniformly curated DSC datasets spanning both non-EM and EM molecules. The DSC simulations for some clusters were claimed to be relatively accurate, with R^2 above 0.8 on validation data. However, DSC validation data was not analyzed in detail for all clusters. The authors create a risk matrix for EM risk evaluation, and its usefulness entirely depends on classification accuracy of the decision tree, which is unreliable for nearly half its validation samples. Nevertheless, this work provided a new, stacked approach to EM model creation and showed a good pathway toward creating a practical model that may be useful to experimentalists. On a similar topic, the prediction of DSC thermal decomposition onset temperatures, Beste and Barnes [71] have performed work using a 124-molecule dataset for EM, curated to have consistent measurement of onset temperatures and experimental methods. Using fivefold cross-validation, the performance of methods such as kernel ridge regression (KRR), random forests, multilayer perceptrons, support vector machines are compared with a suite of linear methods, for two different input fingerprints (Morgan and MACCS keys). The authors were able to predict onset temperatures within a RMSE of 32 K on test data, similar to the range of experimental error in the collected data, with an R^2 above 0.8. For this dataset, the physically inspired MACCS key and kernel ridge regression with a radial basis function kernel had optimal performance; the success of KRR is consistent with the success of Gaussian process regression in other molecular applications.

Fathollahi and Sajady [72] created a multilayer perceptron neural network model for thermal decomposition of energetic cocrystals. The cocrystal dataset contained only 30 samples, many of which contained the same molecule as one of the coformers (e.g., 12 cocrystals containing TNT). Other molecules (e.g., A5 and A8 in Table 6.1 of Fathollahi and Sajady) contain extremely similar molecular scaffolds, and one is present in the training set, while the other is in the validation set. Using three descriptors and a four-node hidden layer with sigmoid activation, the authors report a MSE of under 0.3 K. The descriptors selected were produced by the DRAGON software, downselected out of over 1600 potential descriptors—while using the whole dataset. Only one example in the validation set, BTF/DNB does not have scaffolds manifestly similar to a molecule in the training set, which still contains nitro groups (present in training set and inescapable in EM data) and a ~6.5 K error, with a prediction similar to the mean of the TNT cocrystal data. The authors provide parameters for their network.

One of the earliest applications of a neural network to prediction of impact sensitivity was by Nefati and coworkers [73], who used a dataset of 204 CHNO molecules to train a small collection of multilayer perceptron (feed-forward) neural networks. Their architectures contained 11–13 input descriptors downselected from an initial pool of 39, and 1–4 hidden nodes and logistic activation functions, and were trained using leave-one-out cross-validation. The target metric was log of h_{50}%, and results contained R^2 values above 0.8, with 5%–10% of the results being identified as outliers. Nevertheless, the neural network approach was generally more robust than linear models created from the same descriptor pool. This general approach: using correlations, or parametric sensitivity to downselect among a large pool of parameters, and then using the new subset to feed into a simple dense feed-forward neural network (or a linear model!), is used by many other authors. It has the advantage of being highly likely to succeed in some respectable degree for small sets of training data and molecules extremely similar to the training data. But, as the dataset becomes more diverse, or newly tested molecules diverge from training scaffolds, trust in the approach is also tested.

A neural network was also created by Keshavarz and Jaafari [74] to predict impact sensitivity for pure materials. Using a dataset of 291 molecules for which experimental information is available, the authors randomly split the data into a training set and test set (with the test set containing 14 compounds). The neural network architecture is a multilayer perceptron with 1 hidden layer of 15 nodes, and sigmoid activation. The input representation is vector of

10 quantities that may be determined from molecular skeleton alone, including dependence on elemental formula and discrete binary variables for the presence of groups such as aromatics and nitramines. The hidden layer size had been optimized as a hyperparameter, and the network was trained until loss failed to improve. On the test set, the authors report a RMSE of 56 cm for $h_{50}\%$ values (slightly above the RMSE of 41 cm on the training set). This work was performed before operations such as scaffold splitting or cross-validation became commonplace in molecular machine learning studies, although cross-validation and nested cross-validation were known to the cheminformatics community at the time.

In 2012, Xu and coauthors [75] use a 156 compound dataset to create both linear models and a multilayer perceptron model. DRAGON is used to calculate 735 different 3D descriptors, and the dataset is split into train/test with 29 members in the test set. A multilayer perceptron architecture is trained with 10 of the original 735 parameters, and 7 hidden nodes. The model is trained on log of $h_{50}\%$, and no mean-squared error metric is reported. An R^2 value of above 0.84 is reported for both training and test sets. Realizing the difficulty of using the model outside of its domain of training, the authors devote a section to discussion of applicability domain and create Williams plots of residuals vs $h_{50}\%$ values, so that readers (and the authors) may investigate how failures of the model correlate to particular molecular structures, but no trend is commented upon. The neural network does outperform a similarly constructed multilinear model in the same work. The authors compare its result favorably to a work by Wang et al. [76], with a multilayer perceptron model constructed on the same dataset, but which used 16 electrotopological indices as input representation and a larger (12-node) hidden layer. The results are similar to another study by Wang et al. [77], that uses a multilayer perceptron on a dataset of 186 nonheterocyclic, nitro-containing EM compounds, in this case with 9 input descriptors and 9 nodes in the hidden layer, and 36 of the 186 molecules in the randomly selected test set. For these works, there may be a limited domain of applicability of each model due to the descriptor downselection process and random splitting among self-similar molecules. That problem is common for small dataset model creation, and often faced by EM researchers.

In order to better leverage these modern data-driven methods, the first step would be curation of larger, more robust datasets for the experimental properties of interest. Such curation would also benefit traditional QSPR methods. Uniformity in experimental procedures is also beneficial for data curation. But, that is challenging when using historical data. For example, there are a wide variety of DSC experimental set-ups and heating profiles, and it is well-known that impact sensitivity measurements may be difficult to compare across labs due to factors such as local humidity or the roughness of drop weights. Those, and other challenges, serve to make data curation challenging. It is perhaps a problem that may be solved in the future at scale through automated, continuous flow synthesis and new, in-line diagnostics—if the safety hazards of working with energetics are satisfied. Alternatively, large-scale computational datasets using trusted physics-based methods could be created and archived. As demonstrated in work recently published by DEVCOM ARL and Purdue University for EM [78] and other authors for non-EM [79–81], this may provide an alternative way to interrogating quantum mechanical data. It is also often used as a way to create surrogate models for physics-based techniques with simpler input representations previously discussed in this chapter.

4 Concluding remarks

In this tour of QSPR work on EM sensitivity, a recurring theme occurs where input representations (descriptors) are created from skeletal formula or gas-phase quantum chemistry information. The models reviewed, while able to be published with reasonable R^2 values [82], are often restricted to a small set of self-similar molecules and may contain significant outliers, while also typically not having the ability to provide per-prediction uncertainty estimates for novel molecules. Features derived from microstructure, particle size distribution, or crystal lattice symmetry is generally not included in these models, as they require much more detail than is quickly available for hypothetical molecules of interest. Yet, those quantities are believed to have a significant effect on EM sensitivity [83]. Coupling the typical limitations on descriptors with small, noisy datasets turns in QSPR prediction of EM sensitivity into a quite challenging task when compared to the success of QSPR methods for other applications [84]. Researchers rightfully refuse to let the perfect be the enemy of the good. While the models available today may be of limited utility, they are still useful for some applications. This utility may be increased in the future through many avenues, including: continued expansion of high-quality machine-readable experimental datasets, leveraging of high-fidelity physics-based predictions to expand datasets where experimental data is not available, continued innovation in descriptors that go beyond what is available in commercial pharmaceutical packages, and application of machine learning techniques to more effectively capture the highly nonlinear relationship between features and EM response. Theorists should also adhere to rigorous best-practice statistical validation methodology [85–87] as used in other fields that leverage machine learning. Finally, work in this area should be done in conjunction with experimentalists who will use and validate models for diverse classes of molecules.

References

[1] A.R. Katritzky, M. Kuanar, S. Slavov, C.D. Hall, M. Karelson, I. Kahn, D.A. Dobchev, Quantitative correlation of physical and chemical properties with chemical structure: utility for prediction, Chem. Rev. 110 (2010) 5714–5789.

[2] T. Le, V.C. Epa, F.R. Burden, D.A. Winkler, Quantitative structure–property relationship modeling of diverse materials properties, Chem. Rev. 112 (2012) 2889–2919.

[3] M.S. Miller, B.M. Rice, A.J. Kotlar, R.J. Cramer, A new approach to propellant formulation: minimizing life-cycle costs through science-based design, Clean Prod. Process. 2 (2000) 37–46.

[4] S.N. Bulusu (Ed.), Chemistry and Physics of Energetic Materials, Springer Netherlands, Dordrecht, 1990.

[5] P. Politzer, J.S. Murray, J.M. Seminario, P. Lane, M. Edward Grice, M.C. Concha, Computational characterization of energetic materials, J. Mol. Struct. (THEOCHEM) 573 (2001) 1–10.

[6] J.S. Murray, P. Politzer, Structure-sensitivity relationships in energetic compounds, in: S.N. Bulusu (Ed.), Chemistry and Physics of Energetic Materials, Springer Netherlands, Dordrecht, 1990.

[7] C.B. Storm, J.R. Stine, J.F. Kramer, Sensitivity relationships in energetic materials, in: S.N. Bulusu (Ed.), Chemistry and Physics of Energetic Materials, Springer Netherlands, Dordrecht, 1990.

[8] Z.-X. Chen, H.-M. Xiao, Quantum chemistry derived criteria for impact sensitivity, Propellants Explos. Pyrotech. 39 (2014) 487–495.

[9] G. Fayet, P. Rotureau, How to use QSPR models to help the design and the safety of energetic materials, in: M. Shukla, V.M. Boddu, J.A. Steevens, R. Damavarapu, J. Leszczynski (Eds.), Energetic Materials—From Cradle to Grave (Challenges and Advances in Computational Chemistry and Physics, Vol. 25), Springer International Publishing, 2017.

[10] D. Mathieu, T. Alaime, Impact sensitivities of energetic materials: exploring the limitations of a model based only on structural formulas, J. Mol. Graph. Model. 62 (2015) 81–86.
[11] K. Mohammad Hossein, M.K. Thomas, The Properties of Energetic Materials—Sensitivity, Physical and Thermodynamic Properties, De Gruyter, Berlin, Boston, 2018.
[12] C. Nieto-Draghi, G. Fayet, B. Creton, X. Rozanska, P. Rotureau, J.-C. de Hemptinne, P. Ungerer, B. Rousseau, C. Adamo, A general guidebook for the theoretical prediction of physicochemical properties of chemicals for regulatory purposes, Chem. Rev. 115 (2015) 13093–13164.
[13] P. Politzer, J.S. Murray, Detonation performance and sensitivity: a quest for balance, in: J.R. Sabin (Ed.), Advances in Quantum Chemistry, Vol. 69, Academic Press, 2014.
[14] P. Politzer, J.S. Murray, High performance, low sensitivity: the impossible (or possible) dream? in: M.K. Shukla, V.M. Boddu, J.A. Steevens, R. Damavarapu, J. Leszczynski (Eds.), Energetic Materials: From Cradle to Grave (Challenges and Advances in Computational Chemistry and Physics, Vol. 25), Springer International Publishing, 2017.
[15] S. Zeman, M. Jungova, Sensitivity and performance of energetic materials, Propellants Explos. Pyrotech. 41 (2016) 426–451.
[16] G. Li, C. Zhang, Review of the molecular and crystal correlations on sensitivities of energetic materials, J. Hazard. Mater. (2020), 122910.
[17] D. Weininger, SMILES, a chemical language and information system. 1. Introduction to methodology and encoding rules, J. Chem. Inf. Comput. Sci. 28 (1988) 31–36.
[18] V. Prana, P. Rotureau, D. Andre, G. Fayet, C. Adamo, Development of simple QSPR models for the prediction of the heat of decomposition of organic peroxides, Mol. Inform. 36 (2017) 1700024.
[19] Y. Pan, R. Qi, P. He, R. Shen, J. Jiang, L. Ni, J. Jiang, Q. Wang, Thermal hazard assessment and ranking for organic peroxides using quantitative structure–property relationship approaches, J. Therm. Anal. Calorim. 140 (2020) 2575–2583.
[20] M.J.S. Dewar, E.G. Zoebisch, E.F. Healy, J.J.P. Stewart, Development and use of quantum mechanical molecular models. 76. AM1: a new general purpose quantum mechanical molecular model, J. Am. Chem. Soc. 107 (1985) 3902–3909.
[21] C.E. Shannon, The Mathematical Theory of Communication, University of Illinois Press, Champaign, IL, 1949.
[22] D. Yang, H. Koseki, K. Hasegawa, Predicting the self-accelerating decomposition temperature (SADT) of organic peroxides based on non-isothermal decomposition behavior, J. Loss Prev. Process Ind. 16 (2003) 411–416.
[23] X. Zhao, Y. Pan, J. Jiang, S. Xu, J. Jiang, L. Ding, Thermal hazard of ionic liquids: modeling thermal decomposition temperatures of imidazolium ionic liquids via QSPR method, Ind. Eng. Chem. Res. 56 (2017) 4185–4195.
[24] F. Yan, S. Xia, Q. Wang, P. Ma, Predicting the decomposition temperature of ionic liquids by the quantitative structure–property relationship method using a new topological index, J. Chem. Eng. Data 57 (2012) 805–810.
[25] K. Roy, I. Mitra, S. Kar, P.K. Ojha, R.N. Das, H. Kabir, Comparative studies on some metrics for external validation of QSPR models, J. Chem. Inf. Model. 52 (2012) 396–408.
[26] G. Schüürmann, R.-U. Ebert, J. Chen, B. Wang, R. Kühne, External validation and prediction employing the predictive squared correlation coefficient—test set activity mean vs training set activity mean, J. Chem. Inf. Model. 48 (2008) 2140–2145.
[27] N. Sheibani, Heat of formation assessment of organic azido compounds used as green energetic plasticizers by QSPR approaches, Propellants Explos. Pyrotech. 44 (2019) 1254–1262.
[28] B. Tan, Z. Li, X. Guo, J. Li, Y. Han, X. Long, Insight into electrostatic initiation of nitramine explosives, J. Mol. Model. 23 (2017) 10.
[29] P.J. Stephens, F.J. Devlin, C.F. Chabalowski, M.J. Frisch, Ab initio calculation of vibrational absorption and circular dichroism spectra using density functional force fields, J. Phys. Chem. 98 (1994) 11623–11627.
[30] T.H. Dunning, Gaussian basis sets for use in correlated molecular calculations. I. The atoms boron through neon and hydrogen, J. Chem. Phys. 90 (1989) 1007–1023.
[31] R.A. Kendall, T.H. Dunning, R.J. Harrison, Electron affinities of the first-row atoms revisited. Systematic basis sets and wave functions, J. Chem. Phys. 96 (1992) 6796–6806.
[32] D.E. Woon, T.H. Dunning, Gaussian basis sets for use in correlated molecular calculations. IV. Calculation of static electrical response properties, J. Chem. Phys. 100 (1994) 2975–2988.
[33] M.J. Frisch, G.W. Trucks, H.B. Schlegel, G.E. Scuseria, M.A. Robb, J.R. Cheeseman, G. Scalmani, V. Barone, B. Mennucci, G.A. Petersson, H. Nakatsuji, M. Caricato, X. Li, H.P. Hratchian, A.F. Izmaylov, J. Bloino,

G. Zheng, J.L. Sonnenberg, M. Hada, M. Ehara, K. Toyota, R. Fukuda, J. Hasegawa, M. Ishida, T. Nakajima, Y. Honda, O. Kitao, H. Nakai, T. Vreven, J.A. Montgomery Jr., J.E. Peralta, F. Ogliaro, M. Bearpark, J.J. Heyd, E. Brothers, K.N. Kudin, V.N. Staroverov, T. Keith, R. Kobayashi, J. Normand, K. Raghavachari, A. Rendell, J.C. Burant, S.S. Iyengar, J. Tomasi, M. Cossi, N. Rega, J.M. Millam, M. Klene, J.E. Knox, J.B. Cross, V. Bakken, C. Adamo, J. Jaramillo, R. Gomperts, R.E. Stratmann, O. Yazyev, A.J. Austin, R. Cammi, C. Pomelli, J.W. Ochterski, R.L. Martin, K. Morokuma, V.G. Zakrzewski, G.A. Voth, P. Salvador, J.J. Dannenberg, S. Dapprich, A.D. Daniels, O. Farkas, J.B. Foresman, J.V. Ortiz, J. Cioslowski, D.J. Fox, Gaussian 09, Revision D.01. Wallingford, CT, 2009.

[34] J.A. Morrill, E.F.C. Byrd, Development of quantitative structure property relationships for predicting the melting point of energetic materials, J. Mol. Graph. Model. 62 (2015) 190–201.

[35] D. Mathieu, Physics-based modeling of chemical hazards in a regulatory framework: comparison with quantitative structure-property relationship (QSPR) methods for impact sensitivities, Ind. Eng. Chem. Res. 55 (2016) 7569–7577.

[36] D. Mathieu, Modeling sensitivities of energetic materials using the Python language and libraries, Propellants Explos. Pyrotech. 45 (2020) 966–973.

[37] M.H. Keshavarz, K. Esmaeilpour, H. Khoshandam, Z. Keshavarz, H. Hafizi Atabak, S. Damiri, A. Afzali, A novel method for prediction of impact sensitivity of quaternary ammonium-based energetic ionic liquids, Cent. Eur. J. Energ. Mater. 14 (2017) 520–533.

[38] M.H. Keshavarz, Two novel correlations for prediction of electric spark sensitivity of nitramines based on the experimental data of the new instrument, Z. Anorg. Allg. Chem. 644 (2018) 1607–1610.

[39] M.H. Keshavarz, S. Damiri, V. Bagheri, Recent advances for prediction of electric spark and shock sensitivities of organic compounds containing energetic functional groups to assess reliable models, Process Saf. Environ. Prot. 131 (2019) 9–15.

[40] M.H. Keshavarz, H.R. Pouretedal, E. Saberi, A new method for predicting decomposition temperature of imidazolium-based energetic ionic liquids, Z. Anorg. Allg. Chem. 643 (2017) 171–179.

[41] B. Nazari, M.H. Keshavarz, M. Jafari, F. Jafari, A novel approach for prediction of sensitivity toward the electrical discharge of quaternary ammonium-based energetic ionic liquids or salts, Z. Anorg. Allg. Chem. 644 (2018) 1153–1157.

[42] M.H. Keshavarz, A novel approach for the prediction of electric spark sensitivity of polynitroarenes based on the measured data from a new instrument, Cent. Eur. J. Energ. Mater. 16 (2019) 65–76.

[43] N. Zohari, F. Abrishami, V. Zeynali, Prediction of decomposition temperature of azole-based energetic compounds in order to assess of their thermal stability, J. Therm. Anal. Calorim. (2019), https://doi.org/10.1007/s10973-019-09127-2.

[44] M. Karelson, V.S. Lobanov, A.R. Katritzky, Quantum-chemical descriptors in QSAR/QSPR studies, Chem. Rev. 96 (1996) 1027–1044.

[45] R.S.S. de Oliveira, I. Borges Jr., Correlation between molecular charge densities and sensitivity of nitrogen-rich heterocyclic nitroazole derivative explosives, J. Mol. Model. 25 (2019) 314.

[46] M.A.S. Oliveira, I. Borges Jr., On the molecular origin of the sensitivity to impact of cyclic nitramines, Int. J. Quantum Chem. 119 (2019) e25868.

[47] J.S. Murray, P. Lane, P. Politzer, Effects of strongly electron-attracting components on molecular surface electrostatic potentials: application to predicting impact sensitivities of energetic molecules, Mol. Phys. 93 (1998) 187–194.

[48] J.S. Murray, P. Lane, P. Politzer, Relationships between impact sensitivities and molecular surface electrostatic potentials of nitroaromatic and nitroheterocyclic molecules, Mol. Phys. 85 (1995) 1–8.

[49] B.M. Rice, J.J. Hare, A quantum mechanical investigation of the relation between impact sensitivity and the charge distribution in energetic molecules, J. Phys. Chem. A 106 (2002) 1770–1783.

[50] G. Anders, I. Borges, Topological analysis of the molecular charge density and impact sensitivy models of energetic molecules, J. Phys. Chem. A 115 (2011) 9055–9068.

[51] E.F.C. Byrd, On the Failure of Correlating Partitioned Electrostatic Surface Potentials Using Bader's Atoms-in-Molecules Theory to Impact Sensitivities, US Army Research Laboratory, Aberdeen Proving Ground, MD, 2013.

[52] R.F.W. Bader, Atoms in molecules, Acc. Chem. Res. 18 (1985) 9–15.

[53] A.A. Aina, A.J. Misquitta, M.J.S. Phipps, S.L. Price, Charge distributions of nitro groups within organic explosive crystals: effects on sensitivity and modeling, ACS Omega 4 (2019) 8614–8625.
[54] T.C. Lillestolen, R.J. Wheatley, Redefining the atom: atomic charge densities produced by an iterative stockholder approach, Chem. Commun. (2008) 5909–5911.
[55] T.C. Lillestolen, R.J. Wheatley, Atomic charge densities generated using an iterative stockholder procedure, J. Chem. Phys. 131 (2009), 144101.
[56] A.J. Misquitta, A.J. Stone, ISA-Pol: distributed polarizabilities and dispersion models from a basis-space implementation of the iterated stockholder atoms procedure, Theor. Chem. Acc. 137 (2018) 153.
[57] L. Breiman, Random forests, Mach. Learn. 45 (2001) 5–32.
[58] L. Breiman, Statistical modeling: the two cultures, Stat. Sci. 16 (2001) 199–231.
[59] C.E. Rasmussen, C.K.I. Williams, Gaussian Processes for Machine Learning, MIT Press, Cambridge, MA, 2006.
[60] I. Goodfellow, Y. Bengio, A. Courville, Deep Learning, MIT Press, 2016.
[61] T. Hastie, R. Tibshirani, J. Friedman, The Elements of Statistical Learning, second ed., Springer, New York, 2009.
[62] T. Hofmann, B. Schölkopf, A.J. Smola, Kernel methods in machine learning, Ann. Stat. 36 (2008) 1171–1220.
[63] B.C. Barnes, A primer on machine learning for materials and its relevance to army challenges, in: ARL-SR-0411, 2018.
[64] B.C. Barnes, Deep learning for energetic material detonation performance, in: 21st Biennial Conference of the APS Topical Group on Shock Compression of Condensed Matter, AIP Conference Proceedings 2272, 070002 (2020).
[65] B.C. Barnes, D.C. Elton, Z. Boukouvalas, D.E. Taylor, W. Mattson, M.D. Fuge, P.W. Chung, Machine learning of energetic material properties, in: Proceedings of the 16th International Detonation Symposium, arXiv:1807.06156 [cond-mat.mtrl-sci], 2018.
[66] D.C. Elton, Z. Boukouvalas, M.S. Butrico, M.D. Fuge, P.W. Chung, Applying machine learning techniques to predict the properties of energetic materials, Sci. Rep. 8 (2018) 9059.
[67] D. Rogers, M. Hahn (Eds.), Extended-connectivity fingerprints, J. Chem. Inf. Model. 50 (2010) 742–754.
[68] M. Rupp, Machine learning for quantum mechanics in a nutshell, Int. J. Quantum Chem. 115 (2015) 1058–1073.
[69] K. Yang, K. Swanson, W. Jin, C. Coley, P. Eiden, H. Gao, A. Guzman-Perez, T. Hopper, B. Kelley, M. Mathea, A. Palmer, V. Settels, T. Jaakkola, K. Jensen, R. Barzilay, Analyzing learned molecular representations for property prediction, J. Chem. Inf. Model. 59 (2019) 3370–3388.
[70] L. Mage, N. Baati, A. Nanchen, F. Stoessel, T. Meyer, A systematic approach for thermal stability predictions of chemicals and their risk assessment: pattern recognition and compounds classification based on thermal decomposition curves, Process Saf. Environ. Prot. 110 (2017) 43–52.
[71] A. Beste, B.C. Barnes, Prediction of thermal decomposition temperatures using statistical methods, in: 21st Biennial Conference of the APS Topical Group on Shock Compression of Condensed Matter, AIP Conference Proceedings, Portland, OR, 2020.
[72] M. Fathollahi, H. Sajady, QSPR modeling of decomposition temperature of energetic cocrystals using artificial neural network, J. Therm. Anal. Calorim. 133 (2018) 1663–1672.
[73] H. Nefati, J.-M. Cense, J.-J. Legendre, Prediction of the impact sensitivity by neural networks, J. Chem. Inf. Comput. Sci. 36 (1996) 804–810.
[74] M.H. Keshavarz, M. Jaafari, Investigation of the various structure parameters for predicting impact sensitivity of energetic molecules via artificial neural network, Propellants Explos. Pyrotech. 31 (2006) 216–225.
[75] J. Xu, L. Zhu, D. Fang, L. Wang, S. Xiao, L. Liu, W. Xu, QSPR studies of impact sensitivity of nitro energetic compounds using three-dimensional descriptors, J. Mol. Graph. Model. 36 (2012) 10–19.
[76] R. Wang, J. Jiang, Y. Pan, H. Cao, Y. Cui, Prediction of impact sensitivity of nitro energetic compounds by neural network based on electrotopological-state indices, J. Hazard. Mater. 166 (2009) 155–186.
[77] R. Wang, J. Jiang, Y. Pan, Prediction of impact sensitivity of nonheterocyclic nitroenergetic compounds using genetic algorithm and artificial neural network, J. Energ. Mater. 30 (2012) 135–155.
[78] A.D. Casey, S. F. Son, I. Bilionis, B. C. Barnes, J. Chem. Inf. Model. 60 (2020) 4457–4473.
[79] S. Kajita, N. Ohba, R. Jinnouchi, R. Asahi, A universal 3D voxel descriptor for solid-state material informatics with deep convolutional neural networks, Sci. Rep. 7 (2017) 16991.
[80] K. Yao, J. Parkhill, Kinetic energy of hydrocarbons as a function of electron density and convolutional neural networks, J. Chem. Theory Comput. 12 (2016) 1139–1147.

[81] Y. Zhou, J. Wu, S. Chen, G. Chen, Toward the exact exchange–correlation potential: a three-dimensional convolutional neural network construct, J. Phys. Chem. Lett. 10 (2019) 7264–7269.
[82] D.L. Alexander, A. Tropsha, D.A. Winkler, Beware of R(2): simple, unambiguous assessment of the prediction accuracy of QSAR and QSPR models, J. Chem. Inf. Model. 55 (2015) 1316–1322.
[83] C.A. Handley, B.D. Lambourn, N.J. Whitworth, H.R. James, W.J. Belfield, Understanding the shock and detonation response of high explosives at the continuum and meso scales, Appl. Phys. Rev. 5 (2018), 011303.
[84] L.D. Hughes, D.S. Palmer, F. Nigsch, J.B.O. Mitchell, Why are some properties more difficult to predict than others? A study of QSPR models of solubility, melting point, and log P, J. Chem. Inf. Model. 48 (2008) 220–232.
[85] G.C. Cawley, N.L.C. Talbot, On over-fitting in model selection and subsequent selection bias in performance evaluation, J. Mach. Learn. Res. 11 (2010) 2079–2107.
[86] D. Krstajic, L.J. Buturovic, D.E. Leahy, S. Thomas, Cross-validation pitfalls when selecting and assessing regression and classification models, J. Cheminform. 6 (2014) 10.
[87] T. Scior, A. Bender, G. Tresadern, J.L. Medina-Franco, K. Martínez-Mayorga, T. Langer, K. Cuanalo-Contreras, D.K. Agrafiotis, Recognizing pitfalls in virtual screening: a critical review, J. Chem. Inf. Model. 52 (2012) 867–881.

Further reading

AMPAC 8 with Graphical User Interface, Semichem Inc., Box 1649, Shawnee, KS 66216 n.d.
CODESSA 3, Semichem Inc., Box 1649 Shawnee, KS, 66222 n.d...
DRAGON for Windows (Software for Molecular Descriptor Calculations), Version 6.0 n.d.
HyperChem(TM) Professional 7.51, Hypercube, Inc., 1115 NW 4th Street, Gainesville, Florida 32601, USA, n.d.
RDKit n.d.: Open-Source Cheminformatics, http://www.rdkit.org.

CHAPTER

7

Thermal initiation and propagation of the decomposition process

Romain Claveau[a,*], *Julien Glorian*[b], *and Didier Mathieu*[a]

[a]CEA, DAM, Le Ripault, Monts, France [b]French-German Research Institute of Saint-Louis, Saint-Louis, France

*Corresponding author: E-mail: romain.claveau@cea.fr

1 Introduction

As detailed in Chapter 2, the ability of an energetic material (EM) to decompose under representative loads is reflected by sensitivity criteria derived using standardized measurement protocols. For instance, differential scanning calorimetry (DSC) and one-dimensional time to explosion (ODTX) experiments reflect different aspects of thermal stability, namely a decomposition temperature T_d that depends on the DSC heating rate and a critical temperature T_c at which a thermal explosion occurs. Regarding mechanical sensitivities, the drop weight impact height h_{50} is widely used as an impact sensitivity indicator due to its relative ease of determination compared to shock sensitivity measured in gap test experiments. Predicting the values of such sensitivities from the structural formula of the constitutive molecules is needed to focus synthesis efforts on targets of real practical interest.

Whatever its nature, impact, friction, electric spark, or light, any load applied on a material supplies energy that eventually dissipates within the sample. Depending on whether it depends on events that take place before or after this dissipation occurs, the initiation process is athermal or thermal. Athermal processes result from the input energy channeling along specific decomposition pathways, for example, during resonant photoinitiation. Thermal processes result from the rise in temperature following the dissipation of the supplied energy in localized regions (hot spots).

This chapter reviews sensitivity models assuming initiation to be thermal in nature. They fall into two groups. In the first one (direct scenario), the determining step is associated with the decomposition reactions activated by the heat coming from the dissipation in the sample of the energy supplied by the external perturbation. Therefore, the initiation probability

https://doi.org/10.1016/B978-0-12-822971-2.00008-5

depends on the ability of the load to trigger chemical decomposition. In the second group (indirect scenario), the determining step is related to the subsequent decomposition induced by the prior release and dissipation of the chemical energy stored in the material. In the direct scenario, the sensitivity primarily depends on the energy barriers to decomposition, as reflected by an effective activation energy $E^{\dagger}$. In the indirect scenario, the amount of chemical energy E_d released by a molecule upon decomposition and its ability to raise the temperature in the surroundings (inversely proportional to the molar heat capacity of the material) are likely to play a role as well.

Current sensitivity–structure relationships are mostly empirical and do not explicitly assume the initiation to arise as a consequence of an increase in temperature. However, many of them implicitly assume either of the two above scenarios. The direct scenario is assumed in correlations of sensitivities with bond dissociation energy (BDE) of trigger bonds (Section 2). The indirect scenario underpins correlations involving the chemical energy content of the EM possibly along with BDE (Section 3). Finally, a recent two-states model falling under the first scenario but assuming a transient thermal equilibrium between unreacted molecules and activated ones is discussed in Section 4.

2 Direct scenario

This scenario assumes the direct activation of the chemical reactions by the external stimulus to be the determining step in the initiation of an EM. Ignition, i.e., the onset of chemistry, would then ensure initiation, i.e., the emergence of a self-sustained decomposition noticeable on a macroscopic scale. In this picture, it should be possible to neglect any feature associated with late steps in the decomposition, including the amount of energy eventually released. The initiation would be determined by the resilience of a molecule to decay, crudely reflected by the strength of its bonds. The energy needed to break a given bond is the bond dissociation energy (BDE). In a first approximation, it is determined by the nature of the bonded atoms (C, N, O, H, etc.) and by the bond order, i.e., the number of electron pairs contributing to the bond and determining its multiplicity: single, aromatic, double, or triple in order of increasing BDE. This allows the compilation of standard BDE values [1].

In fact, the decomposition of EMs typically involves many elementary steps beyond the early bond breaking and final exothermic formation of the products. In the EM context, the significance of BDEs stems from the trigger bond assumption, according to which the determining step consists in the homolytic cleavage of an especially weak bond, referred to as a trigger bond. Examples of trigger bonds are X–NO_2 bonds in nitro compounds or O–O bonds in peroxides. In this picture, the decomposition rate should primarily depend on the activation energy associated with the bond disruption, loosely identified with the lowest BDE for the molecule and denoted as D_{min}.

2.1 Correlations for thermal stability

Current attempts at predicting thermal stability have mostly resorted to BDE data. A naive idea would be to consider that the decomposition temperature T_d is such that $k_B T_d \simeq E^{\dagger} \simeq D_{min}$

where k_B, $E^{\dagger}$, and $D_{\min}$ are, respectively, the Boltzmann constant, the activation energy of the decomposition, and the BDE of the trigger bond (in principle the lowest BDE value for the compound). This is obviously not the case as this would lead to decomposition temperatures T_d over 10^4 K. In fact, $k_B T_d \ll D_{\min}$ because macroscopic time scales are very large compared to the vibrational period of an oscillation of the trigger bond. Therefore, the latter has many opportunities to break in the course of a real experiment. In the Arrhenius expression for the decomposition rate of a system of N molecules given by $(1/N)(dN/dt) = A\exp(-E^{\dagger}/k_B T)$, the magnitude of the prefactor A is typically $10^{12} - 10^{13}\ \mathrm{s}^{-1}$ [2]. Therefore, observing a sizeable decomposition over a 1-s timescale requires that $\exp(-E^{\dagger}/k_B T) \simeq 10^{-13} - 10^{-12}$, i.e., a temperature between 300 and 650 K for activation energies ranging from 100 to 200 kJ/mol (1–2 eV).

In view of ranking materials on the basis of their respective stability, most authors focus on activation energies ($E^{\dagger}$), prefactors being difficult to estimate while having often a lower impact on the decomposition. Further assuming $E^{\dagger} \simeq D_{\min}$, linear correlations between T_d and $D_{\min}$ might be expected. In spite of numerous exceptions, such correlations were indeed reported for well-studied explosives [3].

A more careful evaluation of $E^{\dagger}$ might lead to better correlations. For instance, it was observed that the disruption of the O–O bond in organic peroxides is often concomitant to other homolytic bond cleavages [4]. Approximating the corresponding $E^{\dagger}$ from suitable decomposition reactions and accounting for the role of the heating rate in DSC experiments, fairly good correlations were obtained to estimate either decomposition temperature (T_d) or self-accelerating decomposition temperatures (T_{SADT}), with the notable exception of hydroperoxides [4].

Nevertheless, such models still rely on the idea that the determining step is associated with the first bond disruptions. In principle, allowing for the possibility of more complex processes as rate-determining steps requires a tedious exploration of the potential energy surface (PES) in search of the predominant decomposition pathways and associated activation energies $E_i^{\dagger}$. Such efforts have been made for ionic liquids, leading to an encouraging correlation between T_d and the lowest activation energy leading to an exothermic process [5]. However, subsequent attempts at extending this correlation proved unsuccessful [6].

An alternative approach to circumvent the costly characterization of PESs is to assign an intrinsic decomposition temperature $T_d^{(i)}$ to each functional group i deemed unstable. The decomposition temperature T_d of any molecule M of interest is then obtained as the minimal $T_d^{(i)}$ taken from the groups i found on the molecule:

$$T_d = \min_{i \in M} \{T_d^{(i)}\} \tag{7.1}$$

The application of this approach to peroxides proves very successful [4]. However, this is probably due to the fact that the compounds studied consist mostly in hydrocarbon parts including peroxide groups in a small number of well-defined environments. Indeed, no such results have been reported to date for more complex energetic materials, in which unstable groups (mostly oxidizing groups) are present in higher proportion. In any case, this approach requires extensive experimental data in order to fix the values of the many parameters involved. This may severely restrict its usefulness in the context of EMs.

2.2 Correlations for mechanical sensitivities

Impact sensitivity is usually characterized by the drop weight impact height h_{50}. In contrast to T_d, this property is specific to EMs due to their ability to quickly release chemical energy through reactions between oxidizing and reducing groups that coexist in the material, eventually leading to gaseous products like H_2O, CO_2, N_2, etc. Typical oxidizing groups are nitro groups (NO_2) in various environments, as encountered in nitroaromatic compounds (NACs) and nitroalkanes ($C–NO_2$), nitramines ($N–NO_2$), and nitric esters ($O–NO_2$). Furazan and furoxan rings are oxidizing moieties of significant interest as well. They are referred to as explosophores because they impart an explosive character to the molecule. Examples of explosophores are shown in Fig. 7.1. A few of them are deprived of oxygen, like $–NF_2$ or the azido group ($–N_3$). In any case, explosophores exhibit especially weak bonds acting as trigger bonds.

Should explosophores determine the explosive character of an EM, then its sensitivities should correlate to their occurrences or to the associated trigger bonds. Actually, standard BDE values do account for the relative impact sensitivities of nitric esters, nitramines, and nitroalkanes, as they prove consistent with the ranking of their corresponding trigger bonds according to their BDE which increase as $O–NO_2 < N–NO_2 < C–NO_2$.

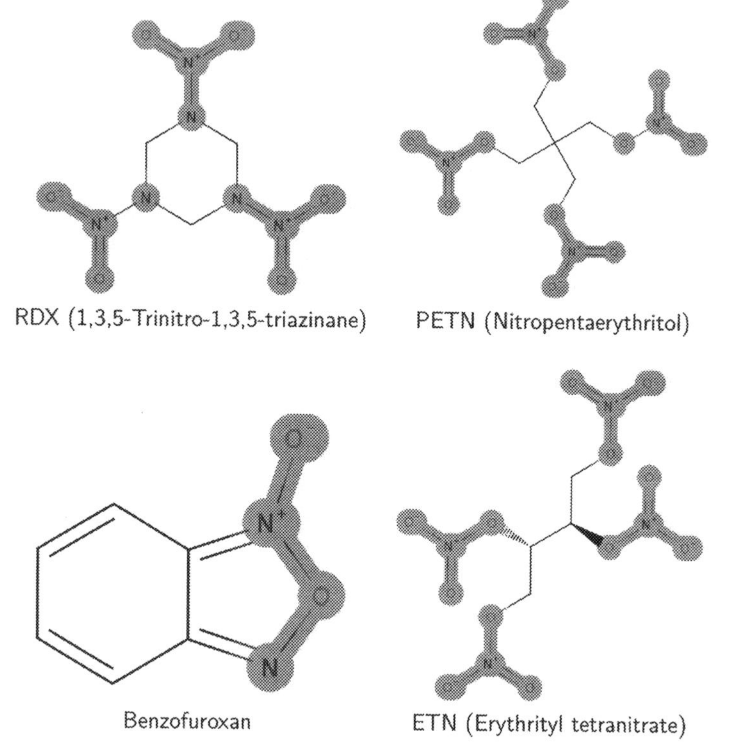

FIG. 7.1 Examples of energetic molecules with explosophores emphasized. RDX exhibits three nitramine groups, benzofuroxan a furoxan group, and both PETN and ETN exhibit four nitric ester groups.

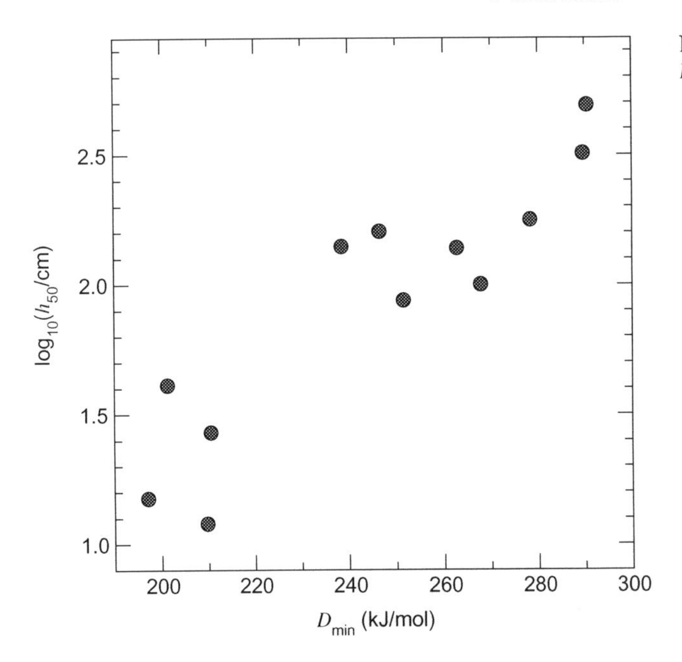

FIG. 7.2 Correlation between log h_{50} and D_{min} reported by Rice et al. [7].

This suggests that fairly accurate BDEs might account for the relative sensitivities within each of those three families of EMs. Further assuming the heat dissipated within a hot spot to be proportional to the energy supplied by the external stimulus, i.e., to h_{50} in the case of the drop weight impact test, we might expect $h_{50} \propto D_{min}$. An early study [7] on nitroaromatic compounds (NACs) rather found h_{50} to be inversely proportional to the rate constant for homolytic disruption of the C–NO_2 bond, i.e., $\log(h_{50}) \propto D_{min}$ (Fig. 7.2). However, further studies for various classes of explosives support the $h_{50} \propto D_{min}$ relationship [8–11].

2.3 Beyond the first decomposition steps

In fact, not only can the early stages of decomposition be more complex than simple homolytic bond scission, but the later stages can also affect sensitivities. To take the latter into account, a search for all relevant products, intermediates, and transition states should be carried out. Typical reaction energy diagrams thus obtained are very complex. The analysis is made easier if the outcome of the reactions depends on a single rate-determining step. As long as transition state theory applies to all forward and reverse steps, an apparent activation energy E_{app} can be derived from the detailed reaction network [12]. The determination of decomposition pathways in terms of related minima and transition states on the PES is the realm of quantum chemistry. While well suited to the elucidation of detailed mechanisms in gas phase, this approach is more difficult to apply to the description of complex reactions in condensed phases.

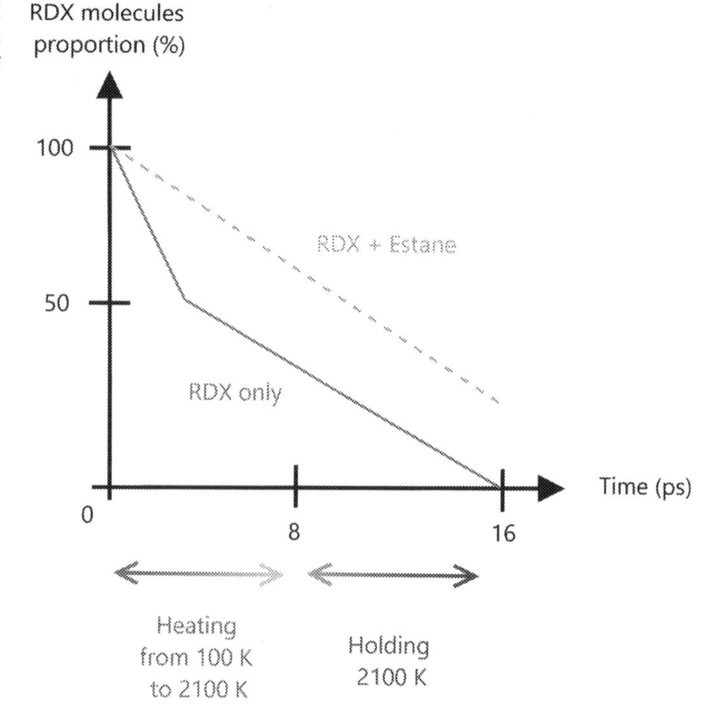

FIG. 7.3 The decomposition of RDX molecules during a thermal load is slowed down by the introduction of estane as a binder [13].

As discussed in Chapter 12, an alternative is provided by reactive molecular dynamics (RMD). Due to the fast chemical kinetics, the decomposition of EMs under extreme conditions, either locally induced by the applied load or inherent to the subsequent detonation regime, provides an unique test bed for this technique.

For instance, it was applied to the dissociation of RDX either pure or associated with an estane binder believed to affect the thermal stability of the EM [13]. The system was first thermalized at $T = 100$ K for 10 ps, then heated up with a ramp of 250 K/ps for 8 ps, and then maintained for another 8 ps at the final temperature of 2100 K. The time required to decompose 50% of the RDX molecules was increased from 4 to 10 ps upon introduction of the binder (Fig. 7.3). This suggests that in spite of the inherent fragility of N–NO_2 bonds, a desensitization may be obtained depending on the overall material composition. Although the binder is mainly introduced to tune mechanical properties, it plays a significant role as well by altering the decomposition mechanism.

This is a striking illustration of the dependence of the decomposition rate and thus thermal stability on complex factors not directly related to the molecular structure of the energetic compound. The decomposition process thus depends not only on early bond breaking reactions but also on intermediate stages which contribute to the rate of energy propagation in the material. Unfortunately, the cost of RMD simulations precludes simulating the response of an explosive over the 250 μs duration of the impact event in a drop weight impact test.

3 Indirect scenario

The indirect scenario assumes that the critical step to initiate an EM is to establish a self-sustaining decomposition regime which propagates through thermal activation. The propagation rate is then given by an Arrhenius expression:

$$k_{\text{pr}} = A\exp(-E^{\dagger}/k_B T_e) \tag{7.2}$$

where $E^{\dagger}$ and A stand for apparent values of the activation energy and prefactor, respectively, and T_e is an effective local temperature in regions of the material adjacent to reactive centers where chemical energy has been released from decomposed molecules. In this picture, initiation depends on k_{pr} being fast enough so that the energy released upon decomposition of a molecule activates further molecular decomposition before it dissipates into the environment. Since it assigns a critical role to the chemical energy released by the material, this approach may be viewed as a physical justification for relationships linking sensitivities to this property. Moreover, it provides a straightforward explanation to the performance–sensitivity contradiction [14]. Finally, it suggests that energy content and BDE should be fruitfully combined into a single hybrid sensitivity index.

3.1 Energy content

Upon decomposition, the exact amount of energy E_d released by an energetic molecule depends on the local conditions prevailing during the process. It may be approximated by the energy content, presently defined the difference in formation enthalpy between the material and its decomposition products derived from the H_2O-CO_2 arbitrary [15]. Clearly, E_d does not play any role if initiation depends only on energy transfers prior ignition (phonon up-pumping model) or on the first homolytic bond scissions (trigger bond hypothesis). However, the emergence of pores in explosives recovered from thermal runaway experiments (Fig. 7.4) shows that decompositions initiated locally do not necessarily propagate to the whole sample. Therefore, the initiation might depend on the propagation step as much as on the creation of hot spots and onset of chemistry. The propagation rate depends on both kinetic (activation energy $E^{\dagger}$ and prefactor A) and thermodynamic (E_d) factors. In fact, since kinetic parameters depend on the chemical bonds to be broken, which are similar for chemically related compounds, E_d might play a major role.

As a molecule releases an energy E_d, it can raise the temperature of a neighboring molecule by a value $\Delta T = E_d/C_p$ where E_d and C_p are, respectively, the decomposition energy and heat capacity, both expressed on a per mole basis. In the lack of reliable estimate for C_p, a crude approximation is to assume ΔT proportional to the heat of explosion reported either on a per mass (Q) or per volume (Q_v) basis. Since ΔT drives the propagation of the decomposition, sensitivity should decrease as Q (or Q_v) increases.

Although not easily seen in the case of impact sensitivity [17], this relationship is clearly observed for the critical explosion shock pressure P_c of a set of CHNO compounds, which is observed to decrease as a function of Q_v, as shown in Fig. 7.5 and summarized by the following equation [16]:

FIG. 7.4 Optical micrograph (polarized light) of an explosive composition heated for 7 h at 250°C, showing voids attributed to decomposed areas along the grain interfaces *(Courtesy of Hervé Trumel and Philippe Lambert).*

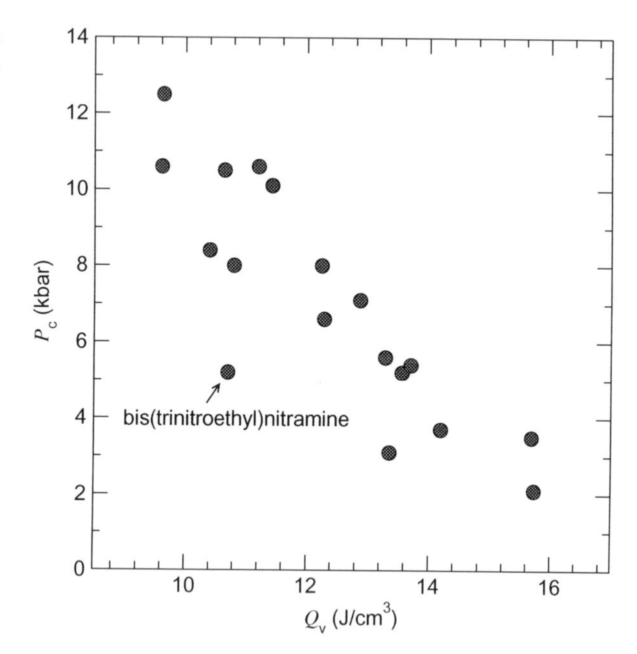

FIG. 7.5 Correlation between critical shock pressure P_c and volumetric energy content Q_v reported by Pepekin et al. [16].

$$P_c = (26.2 \pm 1.7\,\text{kbar}) - (0.15 \pm 0.01) \times Q_v \tag{7.3}$$

The significance of this correlation suggests that the amount of energy released upon decomposition of a molecule is indeed a major determinant of P_c, as expected if propagation (rather than ignition) is the critical step for initiation. This accounts for the well-known fact that EMs fail to provide optimal values for both insensitivity and performance at the same time [18]: low sensitivity requires low energy content, whereas high performance requires the opposite.

3.2 Hybrid sensitivity index

The preceding discussion outlines the fact that in the indirect scenario, impact sensitivity depends primarily on kinetic ($E^{\dagger}$ or simply $D_{\min}$) as well as thermodynamic (Q_v, or the energy content per atom E_d/N_A) factors. Therefore, the ratio of these two factors may be viewed as a hybrid sensitivity index that should be a better predictor of sensitivity than either taken individually [19]. It was extensively investigated as a means to estimate both shock and impact sensitivities [20, 21]. The resulting correlations yield sensitivity estimates on par with state-of-the-art empirical models, using only a single physically motivated descriptor.

In fact, the value of this sensitivity index stems from the fact that it is just a crude measure of $\log(k_{pr})$. In line with the trigger bond assumption, it identifies $E^{\dagger}$ with the lowest BDE, considering only bonds involved in explosophore moieties whose homolytic scission is likely to initiate a self-sustained decomposition. In the case where the lowest BDE is observed for several bonds in the molecule, only one of these bonds is actually taken into account, despite the fact that each new bond of this type introduced into the molecule increases the probability of initiation.

3.3 Inclusion of all explosophores

In view of accounting for the contribution of every explosophore on the initiation probability, a more detained expression is introduced for k_{pr}, along with a semiempirical relationship aimed at linking this quantity to the observed h_{50} values:

$$h_{50} = \left(\frac{k_c}{k_{pr}}\right)^{\omega} \tag{7.4}$$

where k_c and ω are positive empirical parameters [22, 23]. The expression for k_{pr} is obtained by resolving this rate constant into contributions arising from the different decomposition pathways i available:

$$k_{pr} = \sum_i Z_i \exp\left(-\frac{E_i^{\dagger}}{k_B T}\right) \tag{7.5}$$

where $E_i^{\dagger}$ is the thermal activation energy associated with the explosophore whose reactivity triggers the decomposition along the reaction path i, Z_i the corresponding prefactor having the dimension of a frequency, k_B the Boltzmann constant, and T the local temperature close to a reactive center. Then the assumption is made that the local temperature is due to the decomposition of neighboring molecules. Denoting as ηE_d the fraction of this energy effectively contributing to the temperature rise and assuming that a transient thermodynamic equilibrium is reached before the decomposition of additional molecules gets activated, the equipartition theorem yields:

$$\eta E_d = \frac{3N_A}{2} k_B T \tag{7.6}$$

By combining Eqs. (7.4)–(7.6), an explicit expression is obtained for h_{50} as a function of molecular properties, namely the energy content per atom E_d/N_A and the kinetic parameters ($E_i^\dagger$, Z_i) for every decomposition pathway i available to the molecule:

$$h_{50} = k_c^\omega \left[\sum_i Z_i \exp\left(-\frac{3N_A E_i^\dagger}{2\eta E_d} \right) \right]^{-\omega} \tag{7.7}$$

In addition to molecular kinetic features, this approach basically depends on three empirical constant: ω, η, and k_c. For simplicity, current implementations of this approach do not actually calculate the prefactors, but rather assume that they can take only three distinct values depending on the explosophore studied. Similarly, the activation energies $E_i^\dagger$ are identified with standard bond dissociation energies D_i for nitro groups, and treated as empirical parameters otherwise [24].

A comparison on the same test set demonstrates that in spite of its simplicity, such a model outperforms quantitative structure–property relationships, with a determination coefficient R^2 increased from 0.7 to about 0.9 [23]. This suggests that its basic ingredients, i.e., kinetic parameters and energy content, are indeed major determinants of impact sensitivity. Extensive application of the model to large datasets reveals that its performance is especially good for nonaromatic compounds ($R^2 \simeq 0.8$) and significantly drops when considering high nitrogen heteroaromatic compounds. This comes as no surprise as the model relies mostly on nitro compounds for its parameterization.

4 Two-states model of a loaded material

Recently, Ren et al. introduced an alternative approach to the prediction of sensitivities by deriving an expression for the initiation probability as a function of the drop hammer energy, from which h_{50} naturally follows [25]. In their preliminary report, they consider a two-states model, assuming that in a region of the EM equilibrated at temperature T after impact, molecules can be broken down into two categories: activated and nonactivated ones. The former are identified with the intermediary species resulting from the early endothermic stages of the chemical decomposition, while the latter are simply the unreacted energetic molecules. There is no need to consider the final decomposition products and related quantities since activated molecules are supposed to eventually decompose. The partition function of this two-states system is [25]:

$$Q = Q_{mol} + Q_{act} \tag{7.8}$$

Nonactivated and activated molecules are assigned energies E_{mol} and E_{act}, respectively. Assuming they are in number g_{mol} and g_{act}, Q becomes:

$$Q = g_{mol} \exp\left(-\frac{E_{mol}}{k_B T} \right) + g_{act} \exp\left(-\frac{E_{act}}{k_B T} \right) \tag{7.9}$$

The corresponding fraction of activated molecules in the material is then:

$$\rho_{act} = \frac{Q_{act}}{Q} = \left[1 + \exp\left(\frac{\Delta F}{k_B T}\right)\right]^{-1} \tag{7.10}$$

where ΔF denotes the difference in free energy between the activated and nonactivated states. No assumption is made regarding the spatial distribution of the activated molecules, which can be distributed homogeneously or concentrated so as to form hot spots. It is simply assumed instead that the initiation probability P_{init} is proportional to the fraction of activated molecules:

$$P_{\text{init}} = \rho_{act}/\rho_c \tag{7.11}$$

where ρ_c is referred to as a critical density by the authors, who restrict P_{init} to unity in the case where ρ_{act} would be larger than ρ_c. It is further hypothesized that the temperature increase is proportional to the energy W of the drop hammer, leading to a final temperature given by $T = T_0 + \alpha W$, where T_0 denotes the initial (ambient) temperature and α the fraction of the mechanical energy provided by the impact that is locally available and converted into thermal energy. This coefficient is assumed to depend only on the material.

Using Eqs. (7.8)–(7.11) and assuming $g_{act}/g_{mol} = 1$, an explicit expression for P_{init} is obtained. The only unknown are ΔF, α, and ρ_c. Ren et al. fitted their values against experimental initiation probabilities derived from impact test experiments on RDX for various values of the drop weight impact height. As is clear from Fig. 7.6, the model equations enable an excellent fit of experiment. From the curve obtained, it is easy to derive the drop weight impact height h_{50} corresponding to an initiation probability of 50%.

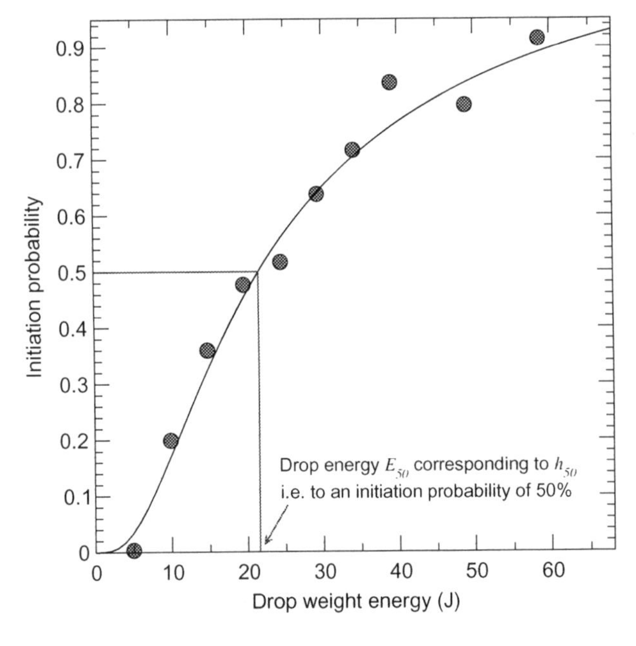

FIG. 7.6 Initiation probability of RDX as a function of the drop weight impact energy. *Red symbols* (*gray* in the print version): experiment. *Black line*: fit according to the Ren et al. model [25].

Regarding the values of the fitted parameters, $\Delta F \approx 30$ kJ/mol is in good agreement with the experimental value of 40 kJ/mol [25]. The critical fraction $\rho_c \approx 0.4$ suggests that initiation takes place in local regions where 40% of the molecules are activated. Finally, the value $\alpha \approx 11,000 \text{J} \cdot \text{K}^{-1}$ is more difficult to interpret as it integrates macroscopic as well as microscopic aspects. Not only is it likely to depend on the material being studied, but it also clearly depends on the size of the sample used for the h_{50} measurement.

As it stands, the Ren et al. model accounts very well for the dependence of P_{init} on the drop weight energy. To turn it into a predictive tool for new compounds, the model parameters ΔF, ρ_c, and α must be expressed on the basis of molecular features. ΔF should roughly correspond to the free energy change on going from the energetic molecules to the radical species resulting from the first steps in the decomposition. The fraction ρ_c of activated molecules required for initiation might depend on E_d: it may be expected that the more energy is released by the decomposition, the fewer activated molecules are needed for initiation. Finally, α might first be assumed constant. Its dependence on the sample size could be ignored as experiments are usually made using samples of standard dimensions.

5 Conclusion

Simple correlations for sensitivities usually involve either kinetic factors such as bond dissociation energies, or thermodynamic ones like energy content. This is consistent with initiation being thermal in nature, with either ignition or propagation of the decomposition as critical step. In the latter case, the ratio of kinetic and thermodynamic factors may be used as a hybrid sensitivity index, since it provides a crude measure of the propagation rate. More elaborate expressions can be developed so as to include the role of all explosophores. Finally, a model to predict the initiation probability as a function of the magnitude of the load is outlined.

References

[1] Y.-R. Luo, Handbook of Bond Dissociation Energies in Organic Compounds, CRC Press, 2002. https://www.routledge.com/Handbook-of-Bond-Dissociation-Energies-in-Organic-Compounds/Luo/p/book/9780849315893.

[2] A.F. Voter, Introduction to the Kinetic Monte Carlo method, in: K.E. Sickafus, E.A. Kotomin, B.P. Uberuaga (Eds.), Radiation Effects in Solids, NATO Science Series, vol. 235, Springer, Dordrecht, The Netherlands, 2007, pp. 1–23 (Chapter 1).

[3] D. Mathieu, S. Beaucamp, Matériaux energétiques, in: Traité "Sciences Fondamentales", Fascicule AF 6710, Encyclopédie "Techniques de l'Ingénieur", Paris, 2004, https://doi.org/10.51257/a-v1-af6710.

[4] D. Mathieu, T. Alaime, J. Beaufrez, From theoretical energy barriers to decomposition temperatures of organic peroxides, J. Therm. Anal. Calorim 129 (1) (2017) 323–337, https://doi.org/10.1007/s10973-017-6114-x.

[5] M.C. Kroon, W. Buijs, C.J. Peters, G.-J. Witkamp, Quantum chemical aided prediction of the thermal decomposition mechanisms and temperatures of ionic liquids, Thermochim. Acta 465 (1-2) (2007) 40–47.

[6] T. Aumond, Modélisation des Températures de Décomposition de Matériaux Organiques, Rapport de stage CEA Le Ripault, Master Chimie, Université de Poitiers, 2018.

[7] B.M. Rice, S. Sahu, F.J. Owens, Density functional calculations of bond dissociation energies for NO2 scission in some nitroaromatic molecules, J. Mol. Struct. THEOCHEM 583 (1) (2002) 69–72, https://doi.org/10.1016/S0166-1280(01)00782-5.

[8] J. Li, A multivariate relationship for the impact sensitivities of energetic N-nitrocompounds based on bond dissociation energy, J. Hazard. Mater. 174 (1) (2010) 728–733, https://doi.org/10.1016/j.jhazmat.2009.09.111.

[9] J. Li, Relationships for the impact sensitivities of energetic C-nitro compounds based on bond dissociation energy, J. Phys. Chem. B 114 (6) (2010) 2198–2202, https://doi.org/10.1021/jp909404f. publisher: American Chemical Society.

[10] Z. Mei, F. Zhao, S. Xu, X. Ju, A simple relationship of bond dissociation energy and average charge separation to impact sensitivity for nitro explosives, J. Serb. Chem. Soc. 84 (2019) 27–40, https://doi.org/10.2298/JSC180404059M.

[11] T.L. Jensen, J.F. Moxnes, E. Unneberg, D. Christensen, Models for predicting impact sensitivity of energetic materials based on the trigger linkage hypothesis and Arrhenius kinetics, J. Mol. Model 26 (4) (2020) 65, https://doi.org/10.1007/s00894-019-4269-z.

[12] Z. Mao, C.T. Campbell, Apparent activation energies in complex reaction mechanisms: a simple relationship via degrees of rate control, ACS Catalysis 9 (10) (2019) 9465–9473, https://doi.org/10.1021/acscatal.9b02761. publisher: American Chemical Society.

[13] L. Zhang, S.V. Zybin, A.C.T. van Duin, S. Dasgupta, W.A. Goddard, Thermal decomposition of energetic materials by ReaxFF reactive molecular dynamics, AIP Conf. Proc. 845 (1) (2006) 589–592, https://doi.org/10.1063/1.2263391. publisher: American Institute of Physics.

[14] D. Mathieu, Sensitivity of energetic materials: theoretical relationships to detonation performance and molecular structure, Ind. Eng. Chem. Res. 56 (29) (2017) 8191–8201, https://doi.org/10.1021/acs.iecr.7b02021.

[15] M.J. Kamlet, S.J. Jacobs, Chemistry of detonations. I. A simple method for calculating detonation properties of C-H-N-O explosives, J. Chem. Phys. 48 (1) (1968) 23–35, https://doi.org/10.1063/1.1667908. publisher: American Institute of Physics.

[16] V.I. Pepekin, B.L. Korsunskii, A.A. Denisaev, Initiation of solid explosives by mechanical impact, Combust. Explos. Shock Waves 44 (5) (2008) 586–590, https://doi.org/10.1007/s10573-008-0089-7.

[17] J. Edwards, C. Eybl, B. Johnson, Correlation between sensitivity and approximated heats of detonation of several nitroamines using quantum mechanical methods, Int. J. Quantum Chem. 100 (5) (2004) 713–719, https://doi.org/10.1002/qua.20235.

[18] H.-H. Licht, Performance and sensitivity of explosives, Propellants Explos. Pyrotech. 1521-4087, 25 (3) (2000) 126–132, https://doi.org/10.1002/1521-4087(200006)25:3<126::AID-PREP126>3.0.CO;2-8.

[19] L.E. Fried, M.R. Manaa, P.F. Pagoria, R.L. Simpson, Design and synthesis of energetic materials, Ann. Rev. Mater. Res. 31 (1) (2001) 291–321, https://doi.org/10.1146/annurev.matsci.31.1.291.

[20] D. Mathieu, Theoretical shock sensitivity index for explosives, J. Phys. Chem. A 116 (2012) 1794–1800.

[21] D. Mathieu, Toward a physically based quantitative modeling of impact sensitivities, J. Phys. Chem. A 117 (10) (2013) 2253–2259, https://doi.org/10.1021/jp311677s. publisher: American Chemical Society.

[22] D. Mathieu, T. Alaime, Predicting impact sensitivities of nitro compounds on the basis of a semi-empirical rate constant, J. Phys. Chem. A 118 (41) (2014) 9720–9726, https://doi.org/10.1021/jp507057r. publisher: American Chemical Society.

[23] D. Mathieu, T. Alaime, Impact sensitivities of energetic materials: exploring the limitations of a model based only on structural formulas, J. Mol. Graph. Model. 62 (2015) 81–86, https://doi.org/10.1016/j.jmgm.2015.09.001. http://www.sciencedirect.com/science/article/pii/S1093326315300449.

[24] D. Mathieu, Modeling Sensitivities of energetic materials using the Python language and libraries–mathieu–2020–propellants, explosives, pyrotechnics–Wiley online library, Propellants Explos. Pyrotech. 45 (2020) 966–973.

[25] G. Ren, Y. Liu, W. Lai, T. Yu, Statistical theory of initiation of explosives by impact, arXiv:1601.03358 [cond-mat] (2016), arXiv: 1601.03358. http://arxiv.org/abs/1601.03358.

PART III

Relationships involving the crystal structure

CHAPTER

8

Some molecular and crystalline factors that affect the sensitivities of explosives

Peter Politzer and Jane S. Murray*

Department of Chemistry, University of New Orleans, New Orleans, LA, United States

***Corresponding author: E-mail: ppolitze@uno.edu**

1 The challenge

A major challenge in designing explosive materials is that improvements in detonation performance are often accompanied by increased and undesirable sensitivity to accidental detonation due to unintended stimuli such as impact, shock, friction, etc. Considerable experience, as well as some systematic studies [1,2], suggest that a high level of detonation performance is generally incompatible with a low level of sensitivity, that the two are fundamentally contradictory [1,2]. However this grim assessment has sometimes been questioned [1–7]. In this chapter, we shall try to provide some guidelines for overcoming the apparent incompatibility between high performance and low sensitivity.

Sensitivities to impact have been the most widely measured and reported, and shall be our focus, although much of the discussion applies to shock sensitivity as well. Impact sensitivity is generally determined by dropping a given mass upon a sample of the explosive and finding the height from which 50% of the drops initiate a reaction [8–10]. This height is labeled h_{50} and its value in centimeters is taken to be a measure of the explosive's impact sensitivity; the smaller is h_{50}, the more sensitive is the explosive. The magnitude of h_{50} depends of course upon the mass being used, and this needs to be specified.

An alternative is to calculate the impact energy with which the mass strikes the sample. By classical physics, the impact energy can be defined either as the potential energy of the mass before it is dropped or as the kinetic energy with which it hits the sample. Before the mass is dropped, it has a certain potential energy due to its height above the sample. This is given by Eq. (8.1):

https://doi.org/10.1016/B978-0-12-822971-2.00009-7

$$\text{impact energy} = \text{potential energy} = mgh_{50} \tag{8.1}$$

where m is the mass being dropped and g is the acceleration due to gravity. The smaller is the impact energy needed to initiate reaction, the more sensitive is the explosive. The apparent advantage of giving impact sensitivity in terms of impact energy is that the mass need not be specified; it has already been included. For a mass of 2.5 kg, a drop height h_{50} of 100 cm corresponds to an impact energy of 24.5 J (or 24.5 Nm).

However Mathieu has shown that the situation is more complicated [11]. For a given explosive, h_{50} is not necessarily twice as great for a 2.5 kg mass as for a 5 kg mass, as Eq. (8.1) would predict; it is frequently less. This can be understood if impact energy is defined in terms of the kinetic energy with which the mass strikes the sample. This is $0.5\ mv^2$, where v is the velocity of the mass at impact. For the 2.5 kg mass to achieve the same impact energy as the 5 kg mass, the 2.5 kg mass must hit the sample with a greater velocity. This may affect the subsequent course of events within the sample.

It is in fact notoriously difficult to obtain reproducible impact sensitivities. They are very dependent upon the physical state of the sample—the sizes, shapes and hardness of the crystals, lattice defects, surface roughness, etc—and the manner in which the sample was prepared—crystallization, grinding, purification, etc. [10,12–16]. Accordingly measured values of h_{50} may differ considerably, for the same compound, from one laboratory to another. For example, three different sources report the h_{50} of 1,3,5-trinitrobenzene to be 100, 71 and 30 cm; for 2,4,6-trinitrotoluene (TNT) they are 160, 98, and 61 cm [17]. However if each laboratory takes great care to ensure that its preparation and testing procedures are scrupulously maintained, then the results from the different laboratories should show similar general trends [17].

There have been numerous attempts to correlate impact and shock sensitivities with a variety of molecular and crystal properties. An incomplete list of these properties would include bond energies [18], bond lengths [19], bond polarities [20], band gaps [21], atomic charges [22,23], electrostatic potentials [24], ratios of bond to total energies [25–28], rates of vibrational energy transfer [29–31] and oxygen balance [13]. Such correlations can be useful for predicting sensitivities, usually for a particular category of compounds (nitroaromatics, nitramines, etc.). However the very diverse and disparate nature of the properties being linked to sensitivity suggests that these relationships cannot all represent basic causes of sensitivity [32,33]; many of them must be symptomatic. They simply reflect some other more fundamental factor. *Correlation does not necessarily imply causation.*

Our present focus will be upon several factors that are believed to be fundamentally related to sensitivity. They do not correlate with it; they correspond to general trends, not correlations. However awareness of these factors can help in designing energetic materials with diminished sensitivities and yet high performance levels. We will begin with a brief overview of the initiation of detonation.

2 Initiation of detonation

The following is a simplified description of the initiation of detonation due to impact or shock [25,33–36]:

(1) When a crystalline explosive undergoes impact or shock, it is compressed. The rate and the extent of the compression depend upon the strength of the impact or shock and the nature of the explosive.
(2) The compression produces changes in the crystal structure of the explosive. These are likely to include shear or slip (lattice planes shifting past each other), disorder in the lattice, collapse of voids and vacancies in the lattice, and other changes in lattice defects.
(3) The resistance to these structural effects may cause local buildups of thermal energy ("hot spots") in small regions of the lattice.
(4) Some of this localized thermal energy may be transferred into molecular vibrational modes that lead to bond breaking and molecular rearrangements. These can eventually result in self-sustaining exothermal chemical decomposition that releases large amounts of energy and gaseous products.
(5) There may consequently form a high-pressure, supersonic shock wave that propagates through the system (detonation).

The relative importance of these steps in determining sensitivity varies from one explosive to another. Any factor that promotes a particular step increases sensitivity. This means that there are a number of potential causes of high sensitivity, and it seems unlikely that a general correlation can be established with any one of them. We will discuss three causative factors that have been identified, no one of which should be viewed as the basis for a general correlation.

3 Some factors related to sensitivity

3.1 Free space in the crystal lattice

Any crystal lattice contains a certain amount of free space. In C,H,N,O energetic compounds, this has been estimated to be 17–29% of the total volume [37]. Baillou et al. suggested that free space promotes sensitivity [38]: "....there seems to be a correlation between shock sensitivity and magnitude of the internal porosity within the explosive grains." We agree that there is a relationship, but not a correlation.

Why would free space in the crystal lattice promote sensitivity? One reasonable explanation is that the more free space there is in the lattice, the greater will be the compression resulting from the initial impact or shock. This in turn will increase the changes in the crystal structure—shear, slip, disorder, collapse of voids, etc—that lead to the formation of hot spots, and thus will promote eventual detonation.

Sensitivity is anisotropic; it depends upon the direction of the impact or shock relative to the crystal lattice. Dick et al. found that pentaerythritol tetranitrate (PETN) is more sensitive to shock that is parallel to the [110, 001] crystallographic directions than when it is parallel to the [101, 100] [39–41]. Piermarini et al. reported that the sensitivity of nitromethane depends upon the relative orientations of the crystal and the applied stress [42].

Kunz subsequently showed that the compressibility of PETN is greater in the more sensitive [001] direction than in the less sensitive [100] [43]. It is also noteworthy that the notoriously insensitive explosive 1,3,5-triamino-2,4,6-trinitrobenzene (TATB) has a very low compressibility [44].

All of this supports the idea of a link between sensitivity and compressibility, and hence free space in the crystal lattice. Rice et al. have indeed proposed that a critical level of compression is needed for detonation [45].

The free space per molecule in a crystal lattice can be quantified in terms of the packing coefficient κ, which is the fraction of the unit cell volume V_{cell} that is occupied by the molecules of the explosive:

$$\kappa = ZV_{int}/V_{cell} \tag{8.2}$$

where Z is the number of molecules per unit cell and V_{int} is the intrinsic volume per molecule. Since V_{cell} can be calculated from the molecular mass M, the crystal density d and Z:

$$V_{cell} = ZM/d \tag{8.3}$$

then,

$$\kappa = V_{int}\, d/M \tag{8.4}$$

The free space per molecule in the unit cell, ΔV, is therefore,

$$\Delta V = (1 - \kappa)\, V_{cell}/Z = (M/d) - V_{int} \tag{8.5}$$

The ratio M/d can be interpreted as the "effective" molecular volume V_{eff} that would correspond to the unit cell being completely filled, i.e., $\kappa = 1$. Thus,

$$\Delta V = \text{Veff} - \text{Vint} \tag{8.6}$$

The issue now becomes assigning a value to V_{int}, the intrinsic molecular volume. Since there is no rigorous definition of molecular volume, various ways of approximating it have been suggested [46–50]. Our preference, following Bader et al. [47], is to define it as the space enclosed by an outer contour of the molecule's electronic density. This has the advantage of directly reflecting the specific features of the particular molecule, such as lone pairs, π electrons, strained bonds and atomic anisotropy.

For some purposes, such as analyzing noncovalent interactions, the 0.001 au contour of the electronic density has been shown to be very effective in defining molecular boundaries [51–54]. However the volume within the 0.001 au contour is quite similar to V_{eff} [55,56], and therefore cannot be used to approximate V_{int}.

After testing several possible contours, we chose the 0.003 au to define V_{int} because this reproduces well, by Eq. (8.4), the range and average value of packing coefficients that Eckhardt and Gavezzotti had found independently for C,H,N,O explosives [37]. We subsequently found that the average distances of the atomic nuclei from the 0.003 au contours are distinctly similar to the atoms' van der Waals radii [57].

When Eq. (8.5) is used to calculate the free space per molecule in 1,3,5-trinitro-1,3,5-triazacyclohexane (RDX), with V_{int} defined by the 0.003 au contour, ΔV is 22% of the unit cell volume. A computational study of the effect of pressure upon RDX showed that a relatively small initial isotropic increase in pressure did rapidly diminish the unit cell volume by about 20% [58]. Further reduction in volume was much more difficult; an additional 20% decrease required roughly a 30-fold increase in pressure. The fact that the region of easy compression closely matches the estimated free space supports the validity of Eq. (8.5) and the use of the 0.003 au contour to determine V_{int}.

Free space in the crystal lattice may also increase sensitivity in another manner, in addition to increasing compressibility. It has been demonstrated computationally that N–NO_2 and C–NO_2 bonds are weaker in the gas phase and at crystal surfaces or by lattice voids (free space) than in the bulk crystals [59–65]. Breaking these bonds is believed to be a key step in initiating the self-sustaining exothermal chemical decomposition that characterizes detonation [12,32,66–69].

A relationship between impact sensitivity and free space per molecule in the crystal lattice was first demonstrated by Pospišil et al. [70], but they defined V_{int} in terms of the 0.002 au contour of the molecule's electronic density. The improved current version, using the 0.003 au contour for V_{int}, came later [57,71].

There is an overall tendency for impact sensitivity to increase as ΔV, the free space per molecule in the crystal lattice, becomes larger. This can be seen in Table 8.1, which lists h_{50} and ΔV for a series of explosives, and in Fig. 8.1, in which $\log(h_{50})$ is plotted against ΔV for these compounds. Note that they encompass a variety of chemical categories: nitroaromatics, nitramines, nitroazoles, and others.

We emphasize that Table 8.1 and Fig. 8.1 simply show a general trend, not a correlation. We do not claim that sensitivity correlates with ΔV because the free space per molecule is not the only factor that governs sensitivity, as shall be seen.

TABLE 8.1 Experimental and computed properties.

Compound	h_{50} [a]	ΔV [b]	Q_{max} [c]	D[d]	P[e]
Bis(trinitroethyl)nitramine	5	73	1.25	8.91	369
Tetraazidoazo-1,3,5-triazine	6[f]	85	1.47	7.85	266
PETN (pentaerythritol tetranitrate)	12	73	1.51	8.65	327
β-CL-20 (hexanitrohexaazaisowurtzitane)	14[g]	86	1.60	9.48	422
Trinitropyridine *N*-oxide	20	47	1.58	8.56	333
RDX (trinitrotriazacyclohexane)	26	46	1.50	8.83	347
TNAZ (1,3,3-trinitroazetidine)	29[h]	37	1.63	9.00	364
HMX (tetranitrotetraazacyclooctane)	29	49	1.50	9.13	381
Tetryl (2,4,6-trinitro-*N*-methyl-*N*-nitroaniline)	32	70	1.44	7.80	264
N,N'-dinitro-1,2-diaminoethane	34	29	1.30	8.28	295
2,3,4,6-Tetranitroaniline	41	57	1.39	8.24	307
2,4,6-Trinitroresorcinol	43	56	1.15	7.66	263
Benzotrifuroxan, **6**	50	55	1.69	8.50	331
2,4,5-Trinitroimidazole	68	44	1.49	8.83	355
FOX-7 (1,1-diamino-2,2-dinitroethylene), **4**	72[i]	23	1.20	8.61	338
Picric acid (2,4,6-trinitrophenol)	87	55	1.28	7.57	251
TNB (1,3,5-trinitrobenzene)	100	48	1.36	7.53	248

Continued

TABLE 8.1 Experimental and computed properties.—Cont'd

Compound	h_{50} [a]	ΔV [b]	Q_{max} [c]	D[d]	P[e]
2,4-Dinitroimidazole	105	37	1.29	7.96	278
2,4,6-Trinitrobenzoic acid	109	61	1.15	7.35	239
TNT (2,4,6-trinitrotoluene)	160	58	1.29	7.01	207
LLM-116 (4-amino-3,5-dinitropyrazole)	165[f]	29	1.28	8.47	328
Picramide (2,4,6-trinitroaniline), **1**	177	50	1.26	7.55	251
NTO (3-nitro-1,2,4-triazole-5-one)	291	21	0.982	8.07	300

[a] *Experimental, in cm, from Ref. [14] unless otherwise indicated.*
[b] *Calculated with Eq. (8.5), in Å^3, Ref. [7] using experimental densities.*
[c] *Calculated with Eqs. (8.8) and (8.9), in kcal/g, Ref. [7] using experimental heats of formation.*
[d] *Calculated with Eq. (8.10), in km/s, Ref. [7] using experimental densities.*
[e] *Calculated with Eq. (8.11), in kbar, Ref. [7] using experimental densities.*
[f] *Ref. [72].*
[g] *Ref. [73].*
[h] *Ref. [74].*
[i] *Ref. 75.*

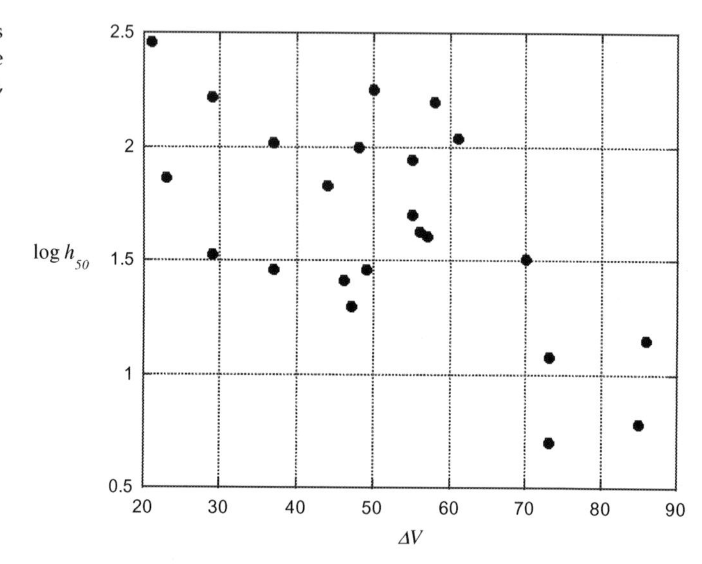

FIG. 8.1 Impact sensitivity, given as $\log h_{50}$ with h_{50} in cm, plotted against free space per molecule in crystal lattice, ΔV in Å^3, for compounds in Table 8.1.

3.2 Molecular electrostatic potential

Any system of nuclei and electrons, such as a molecule, creates an electrostatic potential V(**r**) at each point **r** in the surrounding space; its magnitude is given rigorously by Eq. (8.7):

$$V(\mathbf{r}) = \sum_{A} \frac{Z_A}{|\mathbf{R}_A - \mathbf{r}|} - \int \frac{\rho(\mathbf{r}')d\mathbf{r}'}{|\mathbf{r}' - \mathbf{r}|} \tag{8.7}$$

in which Z_A is the charge on nucleus A, located at $\mathbf{R}_A$, and $\rho(\mathbf{r})$ is the electronic density of the system. The sign of $V(\mathbf{r})$ in any region depends upon whether the positive contribution of the nuclei or the negative one of the electrons is dominant in that region.

The electrostatic potential is a real physical property, an observable, which can be determined experimentally by diffraction techniques [76–78] as well as computationally. It should not be confused with partial atomic charges which are arbitrarily defined (in many different ways) and have no rigorous physical basis [53,79–81].

$V(\mathbf{r})$ directly reflects the *total* charge distribution of a molecule, nuclear, and electronic. It does not in general follow the electronic density, i.e., electron-rich regions do not necessarily have negative potentials [82–85].

The electrostatic potentials on the 0.001 au molecular surfaces of C,H,N,O explosives differ from those of other organic molecules. The molecules of explosives tend to have electron-withdrawing atoms and groups around their peripheries (e.g., aza nitrogen and NO_2 groups) which give rise to strongly positive potentials above the central portions of their molecular surfaces and above any C—NO_2 and N—NO_2 bonds (as well as near any hydrogen atoms) [6,24,86–90]. There are weaker negative potentials around the peripheries of the molecules.

In contrast, typical organic molecular surfaces often have prominent negative regions due to lone pairs or π electrons [77,82]. These may be stronger than their positive potentials.

Compare, for example, the electrostatic potentials on the 0.001 au molecular surfaces of the explosive 2,4,6-trinitroaniline, **1** (Fig. 8.2) and its nonexplosive parent molecule aniline, **2** (Fig. 8.3). In **1**, the potential above nearly the entire molecule is positive, especially above the ring and the C—NO_2 bonds. In **2**, the potential above nearly the entire molecule is negative, particularly above the ring and the nitrogen lone pair. (All calculations have been at the density functional B3PW91/6-31G** level using Gaussian 09 [91] and the WFA-SAS code [92]).

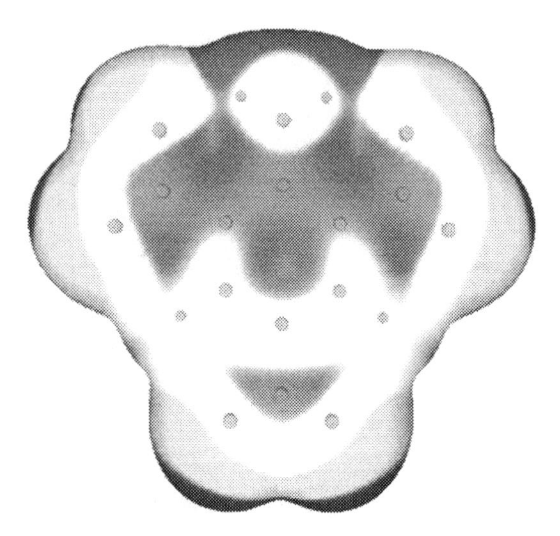

FIG. 8.2 Computed electrostatic potential on 0.001 au molecular surface of 2,4,6-trinitroaniline, **1**. The NH_2 group is at the top. *Gray circles* indicate positions of atoms.Color ranges in kcal/mol: *red*, more positive than 20; *yellow*, from 20 to 0; *green*, from 0 to −15; *blue*, more negative than −15.

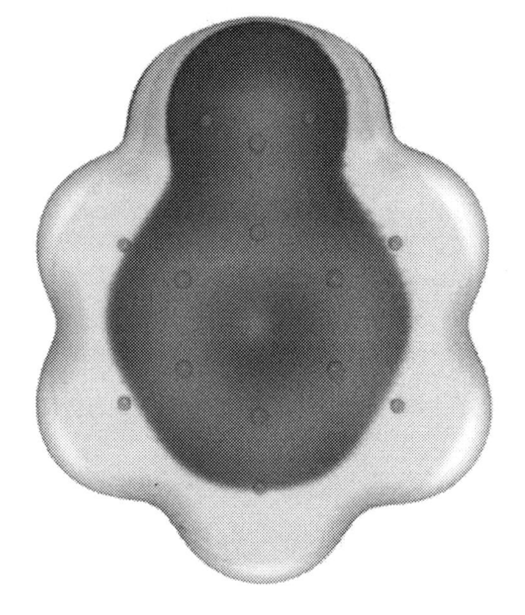

FIG. 8.3 Computed electrostatic potential on 0.001 au molecular surface of aniline, **2**. The NH_2 group is at the top, with the nitrogen lone pair facing the viewer. *Gray circles* indicate positions of atoms. Color ranges in kcal/mol: *red*, more positive than 20; *yellow*, from 20 to 0; *green*, from 0 to −15; *blue*, more negative than −15.

NH_2
O_2N NO_2
NO_2
1

NH_2
2

The strongly positive electrostatic potentials above the central regions of C,H,N,O explosives have been linked to their impact sensitivities, particularly within a given chemical category of compounds (e.g., nitroaromatics) [6,17,24,86–90]. The compounds tend to be more sensitive as these potentials are more positive. See, for example, the comparison of nitroaromatic potentials by Rice and Hare [88] and that of nitropolyazoles by Politzer and Murray [89]. These are again trends, not correlations.

A seemingly reasonable interpretation of these trends is in terms of the shifting of crystal lattice planes past each other (i.e., shear/slip) that accompanies the compression when the explosive is subjected to impact or shock. The repulsion between the strongly positive regions on molecules in neighboring lattice planes, as these planes move past each other, increases the resistance to this movement and may promote the hot spot formation that is a key part of detonation initiation. For example, in RDX and HMX, the crystal directions that show more resistance to shear/slip are also the more sensitive ones [93,94].

A nice demonstration of the electrostatic potential as a guide to relative sensitivity is provided by the 1:1 co-crystal of CL-20 (hexanitrohexaazaisowurtzitane) and TNT (2,4,6-trinitrotoluene) [95]. The co-crystal is less sensitive than CL-20 but more sensitive than TNT. This is consistent with the fact that the positive surface potential of the CL-20 is weaker in the co-crystal than in CL-20 alone while that of TNT is more positive in the co-crystal than in TNT alone.

The strongly positive central region electrostatic potentials of C,H,N,O explosives may also be related to sensitivity in a symptomatic as well as a causative manner. The withdrawal of electronic charge that produces these positive potentials weakens bonds such as $C—NO_2$ and $N—NO_2$, the rupture of which may help to initiate the chemical decomposition leading to detonation [12,32,66–69].

3.3 Detonation heat release

The heat release that accompanies detonation has both chemical and physical components. The chemical portion is the sum of the contributions, initially endothermal but later exothermal, of the various steps that constitute the decomposition of the explosive. The physical portion depends upon factors such as the loading density and the extent of expansion of the gaseous products [34,96–98].

The chemical component of the heat release is commonly estimated by first calculating the enthalpy change is going from the explosive X to the final products; this is given by Eq. (8.8):

$$\Delta H_X = \Sigma n_i \, \Delta H_{f,i} - \Delta H_{f,X} \tag{8.8}$$

In Eq. (8.8), n_i is the number of moles of final product i, having molar heat of formation $\Delta H_{f,i}$, and $\Delta H_{f,X}$ is the molar heat of formation of the explosive X. The quantity ΔH_X is negative, but the detonation heat release (i.e., the heat of detonation) is given as a positive number, either per unit mass of explosive, Q [Eq. (8.9)], or per unit volume of explosive, ρQ, where ρ is the density of the explosive.

$$Q = -\Delta H_X / M_X \tag{8.9}$$

M_X is the molecular mass of the explosive.

The magnitude of Q clearly depends upon the nature of the final detonation products. For most C,H,N,O explosives, these are likely to be almost entirely some combination of N_2(g), CO(g), CO_2(g), H_2O(g), H_2(g), and C(s) [96,99–102]. The proportions of these products that will result in any given case depend upon several factors, including the loading density of the explosive and the temperature. The resulting Q can vary significantly; for instance, its value for RDX with a loading density of 1.00 g/cm^3 is predicted to be 1.32 kcal/g compared to 1.50 kcal/g when the loading density is 1.80 g/cm^3 [103]. Product compositions can be predicted using computer codes [9,100,104] or by applying various sets of rules that have been proposed for this purpose [99,103,105,106].

A particularly useful set of rules was proposed by Kamlet and Jacobs [99], according to which the products are N_2(g), H_2O(g), CO_2(g), and C(s), with oxygen atoms going to H_2O before CO_2, and no CO(g) being produced. This product composition is reasonably accurate for loading densities near the crystal density of the explosive [102,103].

This set of products generally leads to a larger value of Q than do those that involve some CO(g) being produced, because CO_2(g) has a much more negative heat of formation than does CO(g), −94.05 vs. −26.42 kcal/mol [107]. By Eqs. (8.8) and (8.9), this will yield a larger Q. Pepekin referred to this as Q_{max}, and pointed out that it can be regarded as an intrinsic property of the particular compound, the upper limit of its inherent capacity for converting chemical energy into thermal energy [98,108].

It has been noted on a number of occasions that explosives tend to be more impact sensitive as their heats of detonation, particularly Q_{max}, are greater [5,17,25,88,108–111]. Again, this is a general trend, not a correlation. This tendency is observed whether the heat of detonation is expressed on a mass basis, Q_{max}, or a volume basis, ρQ_{max}. In the present discussion we will focus upon Q_{max}.

Table 8.1 contains Q_{max} values for a series of explosives; as pointed out earlier, they include different chemical categories. The relationship of these Q_{max} to impact sensitivity is shown in Fig. 8.4. There is an overall trend for sensitivity to increase (h_{50} to decrease) as Q_{max} is larger; however it is not a correlation.

This link between maximum heat of detonation and sensitivity is reflected in Kamlet's approximate relationships, within groups of chemically similar compounds, between impact sensitivity and oxygen balance [12,13]. The more oxygen the compound contains, the more CO_2 and H_2O is produced upon decomposition. Since these have more negative heats of formation than do the other likely products (CO, H_2, and solid carbon), they will result in greater heats of detonation and thus greater sensitivity.

Can the tendency for impact sensitivity to increase with the maximum heat of detonation be rationalized in some manner? Zeman [111] pointed out that according to the Bell–Evans–Polanyi principle [112,113], which was proposed for reactions that are similar to each other, the more exothermic (endothermic) is the reaction, the lower (higher) is its activation energy. This would imply that a greater heat of detonation is accompanied by a smaller

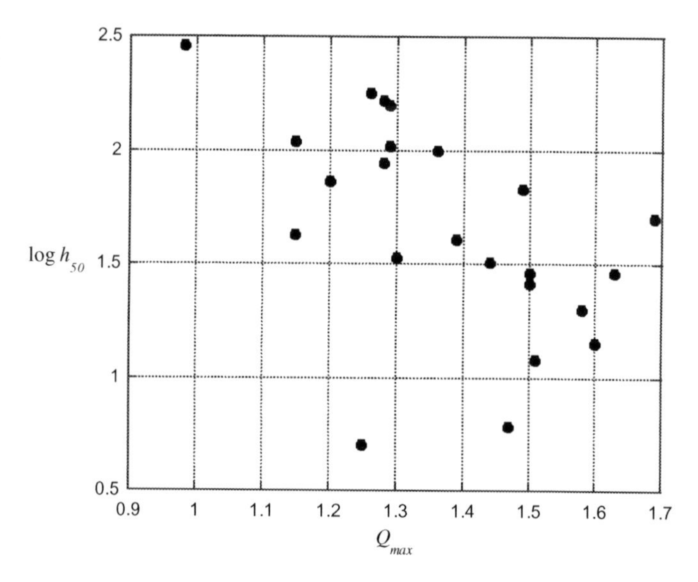

FIG. 8.4 Impact sensitivity, given as log h_{50} with h_{50} in cm, plotted against heat of detonation, Q_{max} in kcal/g, for compounds in Table 8.1.

activation energy for the chemical decomposition, thereby facilitating detonation initiation and increasing sensitivity. However the decomposition reactions are expected to be multi-step, and to differ for the various categories of C,H,N,O compounds (e.g., nitroaromatics and nitramines). Accordingly it is not clear how meaningful is this rationalization.

3.4 General comments

The three factors that have been discussed—free space in the lattice, electrostatic potentials and heat of detonation—all affect sensitivity, and there are likely to be others that do so as well. Accordingly sensitivity cannot be expected to correlate with any one of them alone unless the others are essentially constant for a particular group of molecules. This is sometimes the case for molecules of a given chemical category.

In general, if any one of these factors suggests a high level of sensitivity for a particular compound, then its effect should be expected to dominate and the compound to be sensitive [5–7]. Consider two examples from Table 8.1:

(1) 1,3,3-Trinitroazetidine is quite sensitive ($h_{50}=29$ cm) despite having a low ΔV, because it has a high Q_{max}, 1.63 kcal/g.
(2) Bis(trinitroethyl)nitamine is very sensitive ($h_{50}=5$ cm) even though it has a moderate Q_{max} of 1.25 kcal/g, due to its large ΔV of 73 Å^3.

4 A contradiction?

Before proceeding, it is necessary to address what appears to be a problem, specifically a contradiction, posed by the discussion in Section 3.3. It is well-established that impact sensitivity tends overall to increase as the heat of detonation is larger [5,17,25,88,108–111]. In order to minimize sensitivity, explosives should therefore be designed so that the heat of detonation does not exceed some reasonable limit.

However achieving a large heat of detonation has traditionally been a key objective in designing new explosives. This is a key reason for the past interest in strained or cage-like molecules [8] and for the current focus upon high-nitrogen molecules [4,36]; they often have large, positive heats of formation, and Eqs. (8.8) and (8.9) show that these will increase Q_{max}. The notion of trying to limit the magnitude of Q_{max} has been known to provoke derisive laughter! It seems to contradict the goal of a high level of detonation performance. Is this really the case?

Among the most widely used measures of detonation performance are detonation velocity D (the stable velocity of the shock front that characterizes detonation) and detonation pressure P (the stable pressure developed behind the front) [33,35,36]. While these can be predicted using computer codes [9,100,104], considerable insight can be obtained from equations developed by Kamlet and Jacobs [99], which explicitly identify the properties that determine detonation velocity and detonation pressure:

$$D\,(\text{km/s}) = 1.01\left[N^{0.5}\,{M_{ave}}^{0.25}\,Q^{0.25}\,(1+1.30\,d)\right] \quad (8.10)$$

$$P\,(\text{kbar}) = 15.58\left[N\,{M_{ave}}^{0.5}\,Q^{0.5}d^2\right] \quad (8.11)$$

N is the number of moles of gaseous detonation products per gram of explosive, M_{ave} is their average molecular mass in g/mol, Q is the detonation heat release in cal/g of explosive and d is the loading density of the explosive in g/cm^3. In designing explosives, the loading density is commonly taken to be the crystal density, although in practice it may be significantly less.

The effectiveness of Eqs. (8.10) and (8.11) has been confirmed in numerous studies [99,102,114–117]. These equations show that detonation velocity and detonation pressure depend strongly upon the density but, rather surprisingly, much less upon the amount of gaseous products and the heat release! It follows that a large detonation heat release is not as essential for achieving a high level of detonation performance as has often been believed, and limiting its magnitude can be a very reasonable way to avoid excessive sensitivity.

Is there any evidence to support this conclusion? Consider the following, based upon data in Table 8.1:

(1) Q_{max} for bis(trinitroethyl)nitramine is a moderate 1.25 kcal/g, well below those of TNAZ (1.63 kcal/g) and HMX (1.50 kcal/g). However the D and P of bis(trinitroethyl)nitramine are very similar to those of TNAZ and only slightly less than those of HMX.
(2) The D and P of NTO exceed those of nine of the compounds in Table 8.1 despite NTO having a much lower Q_{max} than any of them.
(3) HMX has much higher D and P than does benzotrifuroxan, even though the Q_{max} of HMX (1.50 kcal/g) is significantly lower than that of benzotrifuroxan (1.69 kcal/g).
(4) The D and P of FOX-7 are only slightly below those of RDX, despite FOX-7 having a Q_{max} of just 1.20 kcal/g, much less than the 1.50 kcal/g of RDX.

These examples confirm the weak dependence of detonation velocity and detonation pressure upon detonation heat release. Limiting the magnitude of Q_{max} in order to keep sensitivity at an acceptable level does not preclude high detonation performance because D and P are much less dependent upon Q_{max} than is sensitivity. This was demonstrated in earlier work on both a percentage basis [5] and graphically [6].

5 Sensitivity and molecular/crystal structure

Section 3.1 has already provided a link between sensitivity and crystal structure, in that impact sensitivity tends to increase as there is more free space per molecule in the crystal lattice. Related to this, and also discussed, are the observations that compressibility and sensitivity vary with crystallographic direction within the lattice.

It has been shown that there is a correlation between ΔV, the free space per molecule, and V_{int}, the intrinsic molecular volume [118]. The free space per molecule increases as the intrinsic molecular volume is larger, suggesting that larger molecules do not pack as well as smaller ones and thus leave more free space. This implies that small molecules are likely to be less sensitive than large ones, although there will surely be many exceptions since free space is not the only factor governing sensitivity (see Section 3).

It is somewhat advantageous for explosive molecules to be planar or near-planar as well as small. Planarity makes it more likely that the crystal lattice will be graphitic, i.e., composed of parallel or near-parallel layers. This should diminish the resistance to the shifting of lattice planes relative to each other (slip/shear) that accompanies the compression following impact or shock. Since this resistance promotes the hot spot formation that is a key feature of detonation initiation, decreasing it should result in lower sensitivity. This may help to explain the remarkable insensitivity of TATB, **3**, which has a crystal lattice of parallel layers [119,120]. The h_{50} of TATB is actually too high (>320 cm) to be determined by conventional means [14].

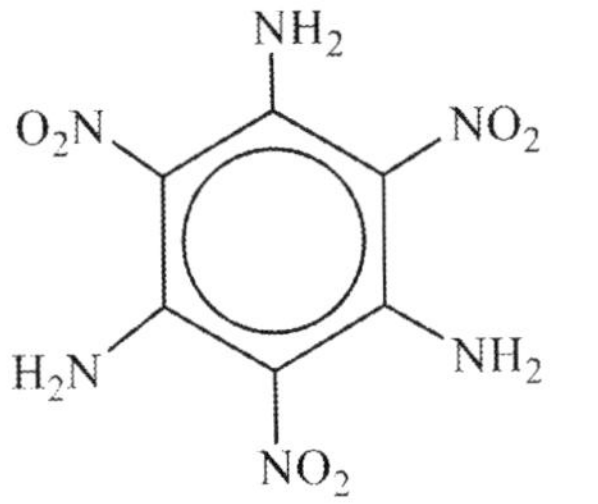

3 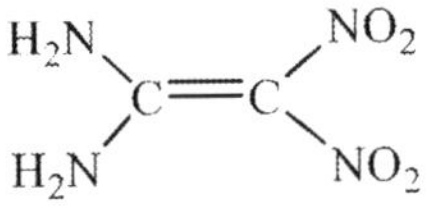 **4**

Intermolecular hydrogen bonding can promote the formation of a layered crystal lattice, especially if the molecules are planar or near-planar. TATB, **3**, and FOX-7, **4**, are examples [119,120]. However the layers in TATB are planar, while those in FOX-7 are zig-zag and may therefore offer more resistance to slip/shear. This may be a factor in FOX-7 not being as insensitive as TATB; their h_{50} are 72 cm [72] and >320 cm [14], respectively.

It has also been suggested that intermolecular hydrogen bonding increases thermal conductivity and thus promotes the diffusion and dissipation of hot spot energy [121], which would diminish sensitivity. Indeed the very insensitive and extensively hydrogen-bonded TATB has "the highest thermal conductivity of the common organic explosive molecules." [122].

It has long been recognized that lattice defects play an important role in facilitating detonation initiation and hence increase sensitivity [33,68,121,123–125]. Such defects can include vacancies, dislocations, voids, cracks, and misalignments. Defects introduce strain in the crystal lattice, which is relieved by localization of some of the external energy coming from shock or impact, thereby forming hot spots. Defect-induced hot spots are in addition to those resulting from resistance to structural effects such as shear or slip that are caused by shock or impact, discussed earlier.

In the context of crystal structure, it is relevant to mention co-crystallization, which has evolved during the past decade into an increasingly active area of explosives research [95,126–131]. The basic concept is that if an explosive A is deficient in a particular property X, then co-crystallization with a second explosive B that has a better X may result in a co-crystal that is improved in X compared to A alone but without excessive detriment to other properties. A particular focus has been upon co-crystallizing a powerful but sensitive explosive with one that is less powerful but also less sensitive. The expectation is that the co-crystal is likely to be intermediate in both respects but it is hoped that the diminished sensitivity will more than offset any decrease in detonation performance.

Some successes have been achieved with co-crystallization [95,126–131], and considerable work is in progress to determine, for example, what are the optimum ratios of a given pair of explosives in a co-crystal. The interactions between the partners in a co-crystal are expected to be noncovalent, and computed electrostatic potentials are frequently used to analyze the interactions and to predict which pairs of compounds will interact most favorably [95,129,130,132,133].

6 Substituent effects upon sensitivity

Amino groups are well known to diminish sensitivity [69,88,134,135]. A frequently cited example is the decrease in sensitivity as NH_2 groups are progressively substituted on 1,3,5-trinitrobenzene (TNB), in going from TNB to its mono-, di- and triamino- derivatives [14].

NH_2 groups on the frameworks of explosive molecules can allow strong H--O or H--N hydrogen bonds with the oxygen and/or nitrogen that are commonly found on the peripheries of explosive molecules. It was pointed out in the previous section that hydrogen bonding is expected to diminish sensitivity.

However one must be cautious in linking cause and effect. There are several reasons why the presence of NH_2 groups might lead to lower sensitivity:

(1) Through intermolecular hydrogen bonding, they promote the formation of a layered crystal lattice [119,120] and thus decrease hot spot formation due to shear/slip.

(2) Again through intermolecular hydrogen bonding, they help to dissipate hot spot energy through thermal conduction [121].

(3) The NH_2 group is electron-donating through conjugation and thereby weakens the positive electrostatic potential above the central portion of an explosive molecule's surface [69,88], this potential being associated with impact sensitivity.

(4) The stabilizing effect of the NH_2 group through hydrogen bonding lowers the heat of formation of the explosive and consequently the detonation heat release, which is also associated with impact sensitivity.

Probably all of the above are involved in the diminished sensitivity resulting from NH_2 substituents.

In assessing possible substituents in the context of sensitivity, it is necessary to keep in mind that an explosive molecule needs some oxygens in order to convert its hydrogens to H_2O and its carbons to CO_2 and CO. In the absence of oxygens, the detonation products will be just N_2, H_2 and C(s). By definition, all three of these have zero heats of formation, and thus contribute nothing to the heat of detonation, which then comes entirely from the heat of formation of the explosive. This is normally not sufficient, even for high-nitrogen compounds, for good detonation performance.

For instance, the compound known as DAAT (**5**) has a very high heat of formation of 206 kcal/mol [136], which is greater than any of the compounds in Table 8.1. However the heat of detonation Q_{max} of **5** is only 936 cal/g, which is *less* than any of the compounds in Table 8.1, and its detonation velocity and detonation pressure, calculated with Eqs. (8.10) and (8.11), are among the lowest: 7.60 km/s and 254 kbar.

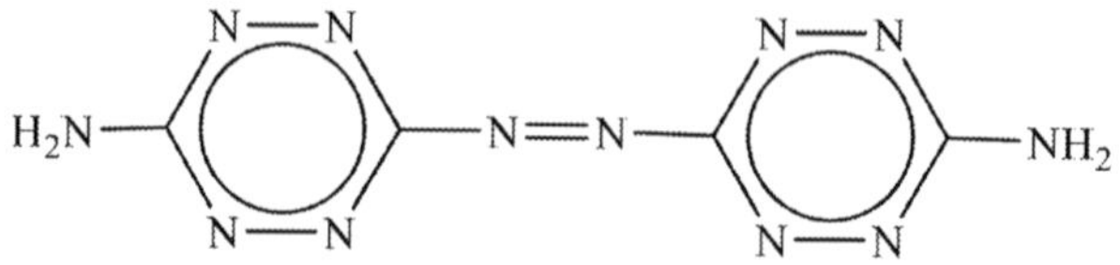

5

On the other hand, as pointed out in Section 3.3, a high heat of detonation is detrimental from the standpoint of sensitivity. This means that the number of oxygens should be restricted so that they are used to form primarily H_2O rather than CO_2; the former has a less negative heat of formation, −57.80 vs −94.05 kcal/mol [107], and will result in a smaller Q_{max}.

Traditionally the most common way of introducing oxygens into explosive compounds has been by means of NO_2 or, less frequently, ONO_2 groups. However these may provide more oxygens than is desirable, resulting in a large Q_{max} and an undesirable level of sensitivity. The idea that high performance and high sensitivity go together may indeed be due to the ubiquity of NO_2 groups in explosive compounds.

ONO_2 is also linked to sensitivity; note PETN in Table 8.1. Of the seven nitrate esters for which Storm et al. report impact sensitivities, only one has $h_{50} > 21$ cm [14].

An increasingly used approach for introducing some oxygens but limiting their number is by using *N*-oxide linkages, $N^+ \rightarrow O^-$, instead of NO_2 groups [136–139]. The *N*-oxide linkage is formally a coordinate covalent bond in which both electrons are provided by the nitrogen. It is inductively electron-attracting [140] but can be either electron-attracting or electron-donating through resonance [140–144].

N-oxide linkages increase both crystal densities and heats of detonation compared to the parent compounds, although to a lesser extent than do NO_2 groups [145]. The more moderate Q_{max} of *N*-oxides is of course in their favor with respect to sensitivity.

The *N*-oxide group is superior to the NO_2 group with respect to the number of moles of gaseous detonation products per gram of explosive [145]. Eqs. (8.10) and (8.11) show that this is actually more important than Q_{max} in determining detonation velocity and detonation pressure.

Another advantage of the *N*-oxide linkage is that it is less electron-attracting than an NO_2 group; thus the electrostatic potential above the central portion of the molecule will be less positive [6] and the compound may be less sensitive. Finally, given the current focus upon high-nitrogen compounds, it is relevant to mention that the *N*-oxide group partially counteracts the destabilizing effect of nitrogen catenation [118,146–148]. For a more extensive comparison of NO_2 groups and $N^+ \rightarrow O^-$ linkages, see Politzer et al., [145].

7 Discussion and summary

Section 3 has provided some general guidelines for trying to minimize the sensitivities of explosives. There should be relatively little free space per molecule in the crystal lattice, a strongly positive electrostatic potential above the central portion of the molecular surface should be avoided, and the detonation heat release should be moderate. Adherence to these guidelines should increase the likelihood of a level of sensitivity low enough to be acceptable, although there is certainly no guarantee of this since there are probably additional factors that also influence sensitivity.

Some specific recommendations seem to be justified. Lower sensitivity tends to be favored by the explosive molecules being small and planar; the latter promotes the desirable feature of the crystal lattice being layered (graphitic). Intermolecular hydrogen bonding does so as well, and also helps to dissipate hot spots through thermal conduction. Lattice defects should be minimized if possible, since they are associated with the formation of hot spots.

The presence of NH_2 groups on explosive molecules is a means of achieving intermolecular hydrogen bonding. It is also directly relevant to two of the general guidelines mentioned above: Since NH_2 is electron-donating through conjugation, it weakens the positive electrostatic potential typically found above the central portion of an explosive molecule. The stabilizing effect of hydrogen bonding also lowers the compound's heat of formation, thereby helping to moderate the detonation heat release.

The *N*-oxide linkage, $N^+ \rightarrow O^-$, seems to be a good candidate as a source of oxygen in an explosive molecule. It serves to moderate both the heat of detonation and the positive electrostatic potential above the central part of the molecular surface, and also counteracts to some extent the destabilization arising from nitrogen catenation.

The idea that maximizing the heat of detonation Q should not be an objective in designing explosives, but that Q should instead be limited to a moderate value, seems counter-intuitive and may be difficult to accept, despite the established general tendency of sensitivity to increase with increasing Q [5,17,25,88,108–111]. Accordingly it may be useful to mention some ways of increasing detonation performance in order to counteract a lower Q.

Eqs. (8.10) and (8.11) show that detonation velocity and detonation pressure depend most strongly upon d and N and least upon Q and M_{ave}. This is actually somewhat misleading, because M_{ave} varies approximately inversely with N [103] and so there is some cancellation. However the fact remains that the effect of an explosive depends upon the volume of gases produced as well as the heat released (and density).

A large value of N is thus highly desirable. This means having more of the lighter gaseous products, H_2O and H_2, and less of the heavier ones, CO, N_2 and especially CO_2. H_2O is most desirable because, unlike H_2, it has a negative heat of formation and thus contributes to the heat release but not excessively, as may be the case for CO_2.

The importance of N is illustrated by the case of benzotrifuroxan, **6**. Its density is very similar to that of HMX, 1.901 and 1.894 g/cm^3, respectively [56], and its Q_{max} is significantly higher, 1.69 vs 1.50 kcal/g (Table 8.1). Yet the detonation velocity and detonation pressure of benzotrifuroxan are markedly inferior to those of HMX, 8.50 vs 9.13 km/s and 331 vs 381 kbar (Table 8.1). This is because benzotrifuroxan has no hydrogens, and thus cannot produce either of the lightest gaseous products, H_2 and H_2O. As a result, it has a very low N value, less than that of any of the other explosives in Table 8.1 [7].

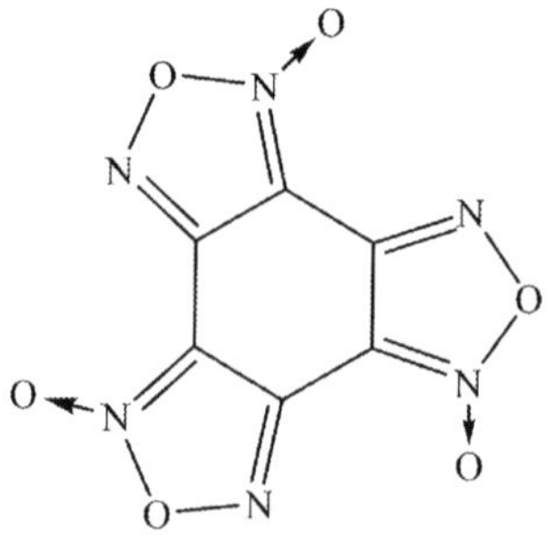

6

This should be kept in mind when replacing C—H units by nitrogen in the quest for high-nitrogen compounds. One consequence is likely to fewer of the lighter H_2O product gases and more of the heavier N_2, which also do not contribute to the heat release.

Finally Eqs. (8.10) and (8.11) establish the loading density *d* as a major determinant of detonation performance, although not quite to the extent that is sometimes believed [102]. For instance RDX has a better detonation velocity and detonation pressure than do LLM-116 and NTO (Table 8.1), even though their densities of 1.90 g/cm^3 [72] and 1.918 g/cm^3 [56] are much greater than the 1.806 g/cm^3 [56] of RDX.

Nevertheless, a high density is definitely desirable. A cautious generalization is that compounds with planar or near-planar molecules are somewhat more likely to have higher densities [49,118], although there are many exceptions to this.

It has often been suggested that intermolecular hydrogen bonding promotes high crystal density, with TATB, **3**, and FOX-7, **4**, serving as examples; their densities are 1.937 g/cm^3 [56] and 1.885 g/cm^3 [72], respectively. Somewhat surprisingly, however, Dunitz et al. found no general relationship between density and hydrogen bonding, using a database of hydrocarbon, C,H,N and C,H,O molecular solids [49]. They pointed out that ice, which is strongly hydrogen-bonded, has an open structure and a low density.

Replacing C—H units in a molecular framework by nitrogen will often (not always [145]) increase the density as well as the compound's heat of formation. But care must be taken that this does not result in a shortage of hydrogens for forming H_2O as a detonation product. Furthermore, the higher heat of formation may produce an excessively large detonation heat release (and consequent high sensitivity). The search for high performance/low sensitivity explosives is always a quest for balance.

References

[1] H.-H. Licht, Performance and sensitivity of explosives, Propell. Explos. Pyrotech. 25 (2000) 126–132.

[2] V. Džingalašević, G. Antić, D. Mlađenović, Ratio of detonation pressure and critical pressure of high explosives with different compounds, Sci. Tech. Rev. 54 (2004) 72.

[3] T.M. Klapötke, C.M. Sabaté, Bistetrazoles: nitrogen-rich, high-performing, insensitive energetic compounds, Chem. Mater. 20 (2008) 3629–3637.

[4] H. Gao, J.M. Shreeve, Azole-based energetic salts, Chem. Rev. 111 (2011) 7377–7436.

[5] P. Politzer, J.S. Murray, Impact sensitivity and the maximum heat of detonation, J. Mol. Model. 21 (2015). 262 (1–11).

[6] P. Politzer, J.S. Murray, High performance, low sensitivity: conflicting or compatible? Propell. Explos. Pyrotech. 41 (2016) 414–425.

[7] P. Politzer, J.S. Murray, High performance, low sensitivity: The impossible (or possible) dream? in: M.K. Shukla, V.M. Boddu, J.A. Steevens, R. Damavarapu, J. Leszczynski (Eds.), Energetic Materials: From Cradle to Grave, Springer, Cham, Switzerland, 2017, pp. 1–22.

[8] S. Iyer, N. Slagg, Molecular aspects in energetic materials, in: J.F. Liebman, A. Greenberg (Eds.), Structure and Reactivity, VCH, New York, 1988, pp. 255–288.

[9] M. Sučeska, Test Methods for Explosives, Springer-Verlag, New York, 1995.

[10] R.M. Doherty, D.S. Watt, Relationship between RDX properties and sensitivity, Propell. Explos. Pyrotech. 33 (2008) 4–13.

[11] D. Mathieu, Sensitivity of energetic materials: theoretical relationships to detonation performance and molecular structure, Ind. Eng. Chem. Res. 56 (2017) 8191–8201.

[12] M.J. Kamlet, The Relationship of Impact Sensitivity with Structure of Organic High Explosives: I. Polynitroaliphatic Exploives, in: Proc. 6th Symp. (Internat.) Detonation, Report No. ACR-221, Office of Naval Research, Arlington, VA, 1976, pp. 312–322.

[13] M. Kamlet, H.G. Adolph, The relationship of impact sensitivity with structure of organic high explosives. II. Polynitroaromatic explosives, Propell. Explos. 4 (1979) 30–34.

[14] C.B. Storm, J.R. Stine, J.F. Kramer, Sensitivity relationships in energetic materials, in: S.N. Bulusu (Ed.), Chemistry and Physics of Energetic Materials, The Netherlands, Kluwer, Dordrecht, 1990, pp. 605–639.
[15] R.W. Armstrong, C.S. Coffey, V.F. DeVost, W.L. Elban, Crystal size dependence for impact sensitivities of Cyclotrimethylenetrinitramine, J. Appl. Phys. 68 (1990) 979–984.
[16] R.W. Armstrong, W.L. Elban, Materials science and technology aspects of energetic (explosive) materials, Mater. Sci. Technol. 22 (2006) 381–395.
[17] P. Politzer, J.S. Murray, Some molecular/crystalline factors that affect the sensitivities of energetic materials: molecular surface electrostatic potentials, lattice free space and maximum heat of detonation per unit volume, J. Mol. Model. 21 (2015) 25(1–11).
[18] B.M. Rice, S. Sahu, F.J. Owens, Density functional calculations of bond dissociation energies for NO_2 scission in some nitroaromatic molecules, J. Mol. Struct. (THEOCHEM) 583 (2002) 69–72.
[19] P. Politzer, J.S. Murray, P. Lane, P. Sjoberg, H.G. Adolph, Shock sensitivity relationships for nitramines and niroaliphatics, Chem. Phys. Lett. 181 (1991) 78–82.
[20] A. Delpuech, J. Cherville, Relation entre la Structure Electronique et la Sensibilité au Choc des Explosifs Secondaires Nitrés-Critère Moléculaire de Sensibilité. I. Cas des Nitroaromatiques et des Nitramines, Propell. Explos. 3 (1978) 169–175.
[21] W. Zhu, H. Xiao, First-principles band gap criterion for impact sensitivity of energetic crystals: a review, Struct. Chem. 21 (2010) 657–665.
[22] C. Zhang, Review of the establishment of nitro group charge method and its applications, J. Hazard. Mater. 161 (2009) 21–28.
[23] G. Anders, I. Borges Jr., Topological analysis of the molecular charge density and impact sensitivity models of energetic molecules, J. Phys. Chem. A 115 (2011) 9055–9068.
[24] J.S. Murray, P. Lane, P. Politzer, Effects of strongly electron-attracting components on molecular surface electrostatic potentials; application to predicting impact sensitivities of energetic molecules, Mol.Phys. 93 (1998) 187–194.
[25] L.E. Fried, M.R. Manaa, P.F. Pagoria, R.L. Simpson, Design and synthesis of energetic materials, Annu. Rev. Mat. Res. 31 (2001) 291–321.
[26] D. Mathieu, Theoretical shock sensitivity index for explosives, J. Phys. Chem. A 116 (2012) 1794–1800.
[27] D. Mathieu, Toward a physically based quantitative model of impact sensitivities, J. Phys. Chem. A 117 (2013) 2253–2259.
[28] T.L. Jensen, J.F. Moxnes, E. Unneberg, D. Christensen, Models for predicting impact sensitivity of energetic materials based on the trigger linkage hypothesis and arrhenius kinetics, J. Mol. Model. 26 (2020). 65(1–14).
[29] L.E. Fried, A.J. Ruggiero, Energy transfer rates in primary, secondary, and insensitive explosives, J. Phys. Chem. 98 (1994) 9786–9791.
[30] K.L. McNesby, C.S. Coffey, Spectroscopic determination of impact sensitivities of explosives, J. Phys. Chem. B 101 (1997) 3097–3104.
[31] S. Ye, M. Koshi, Theoretical studies of energy transfer rates of secondary explosives, J. Phys. Chem. B 110 (2006) 18515–18520.
[32] T.B. Brill, K.J. James, Kinetics and mechanisms of thermal decomposition of Nitroaromatic explosives, Chem. Rev. 93 (1993) 2667–2692.
[33] D.D. Dlott, Fast molecular processes in energetic materials, in: P. Politzer, J.S. Murray (Eds.), Energetic Materials. Part 2. Detonation, Combustion, Elsevier, Amsterdam, 2003, pp. 125–191.
[34] R. Meyer, J. Kohler, A. Homburg, Explosives, sixth ed., Wiley-VCH, Weinheim, Germany, 2007.
[35] S.A. Shackelford, Role of thermochemical decomposition in energetic material initiation sensitivity and explosive performance, Central Eur. J. Energ. Mater. 5 (2008) 75–101.
[36] T.M. Klapötke, Chemistry of High-Energy Materials, fourth ed., de Gruyter, Berlin, 2017.
[37] C.J. Eckhardt, A. Gavezzotti, Computer simulations and analysis of structural and energetic features of some crystalline energetic materials, J. Phys. Chem. B 111 (2007) 3430–3437.
[38] F. Baillou, J.M. Dartyge, C. Spyckerelle, J. Mala, Tenth Symposium (International) on Detonation, ADA 304862, Office of Naval Research, Arlington, VA, 1993, pp. 816–823.
[39] J.J. Dick, Effect of crystal orientation on shock initiation sensitivity of pentaerythritol tetranitrate explosive, Appl. Phys. Lett. 44 (1984) 859–861.

[40] J.J. Dick, R.N. Mulford, W.J. Spencer, D.R. Pettit, E. Garcia, D.C. Shaw, Shock response of pentarythritol tetranitrate single crystals, J. Appl. Phys. 70 (1991) 3572–3587.
[41] C.S. Yoo, N.C. Holmes, P.C. Souers, C.J. Wu, F.H. Ree, J.J. Dick, Anisotropic shock sensitivity and Detonation temperature of pentaerythritol tetranitrate single crystal, J. Appl. Phys. 88 (2000) 70–75.
[42] G.J. Piermarini, S. Block, P.J. Miller, Effects of pressure on the thermal decomposition kinetics and chemical reactivity of nitromethane, J. Phys. Chem. 93 (1989) 457–462.
[43] A.B. Kunz, An *ab initio* investigation of crystalline PETN, Mater. Res. Soc. Symp. Proc. 418 (1996) 287–292.
[44] C. Zhang, Investigation of the slide of the single layer of the 1,3,5-triamino-2,4,6-trinitrobenzene crystal: sliding potential and orientation, J. Phys. Chem. B 111 (2007) 14295–14298.
[45] B.M. Rice, W. Mattson, S.F. Trevino, Molecular-dynamics investigation of the desensitization of detonable material, Phys. Rev. E 57 (1998) 5106–5111.
[46] A.I. Kitaigorodski, Organic Chemical Crystallography, Consultants Bureau, New York, 1961.
[47] R.F.W. Bader, M.T. Carroll, J.R. Cheeseman, C. Chang, Properties of atoms in molecules: atomic volumes, J. Am. Chem. Soc. 109 (1987) 7968–7979.
[48] J.R. Stine, On predicting properties of explosives, J. Energ. Mater. 8 (1990) 41–73.
[49] J.D. Dunitz, G. Filippini, A. Gavezzotti, A statistical study of density and packing variations among crystalline isomers, Tetrahedron 56 (2000) 6595–6601.
[50] H.L. Ammon, New Atom/Functional Group volume additivity data bases for the calculation of the crystal densities of C-, H-, N-, O-, F-, S-, P-, Cl-, and Br-containing compounds, Struct. Chem. 12 (2001) 205–212. and references cited.
[51] P. Politzer, J.S. Murray, Statistical analysis of the molecular surface electrostatic potential: An approach to describing noncovalent interactions in condensed phases, J. Mol. Struct. (THEOCHEM) 425 (1998) 107–114.
[52] P. Politzer, J.S. Murray, Computational prediction of condensed phase properties from statistical characterization of molecular surface electrostatic potentials, Fluid Phase Equil. 185 (2001) 129–137.
[53] J.S. Murray, P. Politzer, The electrostatic potential: An overview, WIREs Comput. Mol. Sci. 1 (2011) 153–163.
[54] P. Politzer, J.S. Murray, T. Clark, Halogen bonding and other σ-hole interactions: a perspective, Phys. Chem. Chem. Phys. 15 (2013) 11178–11189.
[55] P. Politzer, J. Martínez, J.S. Murray, M.C. Concha, A. Toro-Labbé, An electrostatic interaction correction for improved crystal density prediction, Mol. Phys. 107 (2009) 2095–2101.
[56] B.M. Rice, E.F.C. Byrd, Evaluation of electrostatic descriptors for predicting cyrstalline density, J. Comput. Chem. 34 (2013) 2146–2151.
[57] P. Politzer, J.S. Murray, Impact sensitivity and crystal lattice compressibility/free space, J. Mol. Model. 20 (2014) 2223(1–8).
[58] M.M. Kuklja, A.B. Kunz, *Ab initio* simulation of defects in energetic materials. II. Hydrostatic compression of cyclotrimethylene trinitramine, J. Appl. Phys. 86 (1999) 4428–4434.
[59] S. Roszak, P.B. Keegstra, D.W. O'Neal, P.C. Hariharan, J.J. Kaufman, *Ab- initio* MRD-CI calculations for breaking a chemical bond in a molecule in a crystal or other solid environment. II. H_3C-NO_2 decomposition of nitromethane in a nitromethane crystal with voids, Int. J. Quantum Chem. 36 (1989) 353–368.
[60] M.M. Kuklja, Thermal decomposition of solid cyclotrimethylene trinitramine, J. Phys. Chem. B 105 (2001) 10159–10162.
[61] D. Tsiaousis, R.W. Munn, Energy of charged states in the RDX crystal: trapping of charge-transfer pairs as a possible mechanism for initiating detonation, J. Chem. Phys. 122 (2005) 184708.
[62] O. Sharia, M.M. Kuklja, Rapid materials degradation induced by surfaces and voids: *ab initio* modeling of β-octatetramethylene tetranitramine, J. Am. Chem. Soc. 134 (2012) 11815–11820.
[63] O. Sharia, M.M. Kuklja, Surface-enhanced decomposition kinetics of molecular materials illustrated with cyclotetramethylene-tetranitramine, J. Phys. Chem. C 116 (2012) 11077–11081.
[64] C. Zhang, Stress-induced activation of deomposition of organic explosives: a simple way to understand, J. Mol. Model. 19 (2013) 477–483.
[65] K. Nomura, R.K. Kalia, A. Nakano, P. Vashishta, Reactive nanojets: Nanostructure-enhanced chemical reactions in a defected energetic crystal, Appl. Phys. Lett. 91 (2007) 183109(1–3).
[66] M.J. Kamlet, H.G. Adolph, Proc. 7th Symp. (Internat.) Detonation, Report No. NSWCMP-82-334, Naval Surface Warfare Center, Silver Springs, MD, 1981, pp. 60–67.

[67] P. Politzer, J.S. Murray, C-NO_2 dissociation energies and surface electrostatic potential maxima in relation to the impact sensitivities of some nitroheterocylic molecules, Mol. Phys. 86 (1995) 251–255.
[68] P. Politzer, J.S. Murray, Detonation performance and sensitivity: a quest for balance, Adv. Quantum Chem. 69 (2014) 1–30.
[69] P. Politzer, J.S. Murray, Some perspectives on sensitivity to initiation of detonation, in: T. Brinck (Ed.), Green Energetic Materials, Wiley, Chichester, UK, 2014, pp. 45–62.
[70] M. Pospišil, P. Vávra, M.C. Concha, J.S. Murray, P. Politzer, A possible crystal volume factor in the impact sensitivities of some energetic compounds, J. Mol. Model. 16 (2010) 895–901.
[71] M. Pospišil, P. Vávra, M.C. Concha, J.S. Murray, P. Politzer, Sensitivity and the available free space per molecule in the unit cell, J. Mol. Model. 17 (2011) 2569–2574.
[72] P.F. Pagoria, G.S. Lee, A.R. Mitchell, R.D. Schmidt, A review of energetic materials synthesis, Thermochim. Acta 384 (2002) 187–204.
[73] M.-H.V. Huynh, M.A. Hiskey, E.L. Hartline, D.P. Montoya, R. Gilardi, Polyazido high-nitrogen compounds: Hydrazo- and Azo-1,3,5-triazine, Angew. Chem. Int. Ed. 43 (2004) 4924–4928.
[74] R.L. Simpson, P.A. Urtiew, D.L. Ornellas, G.L. Moody, K.J. Scribner, D.M. Hoffman, CL-20 performance exceeds that of HMX and its sensitivity is moderate, Propell. Explos. Pyrotech. 22 (1997) 249–255.
[75] D.S. Watt, M.D. Cliff, TNAZ Based Melt-Cast Explosives: Technology Review and AMRL Research Directions, DSTO-TR-0702, Sect. 4.1, Defense Science and Technology Organization, Melbourne, Australia, 1998, p. 13.
[76] R.F. Stewart, On the mapping of electrostatic properties from Bragg diffraction data, Chem. Phys. Lett. 65 (1979) 335–342.
[77] P. Politzer, D.G. Truhlar (Eds.), Chemical Applications of Atomic and Molecular Electrostatic Potentials, Plenum Press, New York, 1981.
[78] C.L. Klein, E.D. Stevens, Experimental measurements of electronic denstiy distributions and electrostatic potentials, in: J.F. Liebman, A. Greenberg (Eds.), Structure and Reactivity, VCH Publishers, New York, 1988, pp. 25–64.
[79] S.M. Bachrach, Population analysis and electron densities from quantum mechanics, in: K.B. Lipkowitz, D.B. Boyd (Eds.), Reviews in Computational Chemistry, Vol. 5, VCH Publishers, New York, 1994, pp. 171–227.
[80] S.L. Price, Applications of realistic electrostatic modelling to molecules in complexes, solids and proteins, J. Chem. Soc. Faraday Trans. 92 (1996) 2997–3008.
[81] K.C. Gross, C.M. Hadad, P.G. Seybold, Charge competition in halogenated hydrocarbons, Internat. J. Quantum Chem. 112 (2012) 219–229.
[82] P. Politzer, J.S. Murray, Molecular electrostatic potentials and chemical reactivity, in: K.B. Lipkowitz, D.B. Boyd (Eds.), Reviews in Computational Chemistry, Vol. 2, VCH Publishers, New York, 1991, pp. 273–312.
[83] S.E. Wheeler, K.N. Houk, Through-space effects of substituents dominate molecular electrostatic potentials of substituted arenes, J. Chem. Theory Comput. 5 (2009) 2301–2312.
[84] J.S. Murray, Z.P.-I. Shields, P.G. Seybold, P. Politzer, Intuitive and counterintuitive noncovalent interactions of aromatic π regions with the hydrogen and nitrogen of HCN, J. Comput. Sci. 10 (2015) 209–216.
[85] P. Politzer, J.S. Murray, σ-Hole interactions: perspectives and misconceptions, Crystals 7 (2017) 212(1–14).
[86] J.S. Murray, P. Lane, P. Politzer, Relationships between impact sensitivities and molecular surface electrostatic potentials of Nitroaromatic and Nitroheterocyclic molecules, Mol. Phys. 85 (1995) 1–8.
[87] P. Politzer, J.S. Murray, Relationships between dissociation energies and electrostatic potentials of C-NO_2 bonds; applications to impact sensitivities, J. Mol. Struct. 376 (1996) 419–424.
[88] B.M. Rice, J.J. Hare, A quantum mechanical investigation of the relation between impact sensitivity and the charge distribution in energetic molecules, J. Phys. Chem. A 106 (2002) 1770–1783.
[89] P. Politzer, J.S. Murray, Sensitivity correlations, in: P. Politzer, J.S. Murray (Eds.), Energetic Molecules. Part 2. Detonation, Combustion, Elsevier, Amsterdam, 2003, pp. 5–23.
[90] J.S. Murray, M.C. Concha, P. Politzer, Links between surface electrostatic potentials of energetic molecules, impact sensitivities and C-NO_2/N-NO_2 bond dissociation energies, Mol. Phys. 107 (2009) 89–97.
[91] M.J. Frisch, G.W. Trucks, H.B. Schlegel, G.E. Scuseria, M.A. Robb, J.R. Cheeseman, G. Scalmani, V. Barone, B. Mennucci, G.A. Petersson, et al., Gaussian 09, Revision A1, Gaussian Inc, Wallingford, CT, 2009.
[92] F.A. Bulat, A. Toro-Labbé, T. Brinck, J.S. Murray, P. Politzer, Quantitative analysis of molecular surfaces: areas, volumes, electrostatic potentials and average local ionization energies, J. Mol. Model. 16 (2010) 1679–1691.

[93] T. Zhou, S.V. Zybin, Y. Liu, F. Huang, W.A. Goddard III, Anisotropic shock sensitivity for β-octahydro-1, 3,5,7-tetranitro-1,3,5,7-tetrazocine energetic material under compressive-shear loading from ReaxFF-lg reactive dynamics simulations, J. Appl. Phys. 111 (2012) 124904(1–11).
[94] Q. An, Y. Liu, S.V. Zybin, H. Kim, W.A. Goddard III, Anisotropic shock sensitivity of cyclotrimethylene trinitramine (RDX) from compress-and-shear reactive dynamics, J. Phys. Chem. C 116 (2012) 10198–10206.
[95] H. Li, Y. Shu, S. Gao, L. Chen, Q. Ma, X. Ju, Easy methods to study the smart energetic TNT/CL-20 co-crystal, J. Mol. Model. 19 (2013) 4909–4917.
[96] M.J. Kamlet, J.E. Ablard, Chemistry of detonations. II. Buffered equilibria, J. Chem. Phys. 48 (1968) 36–42.
[97] D.L. Omellas, The heat and products of detonation of cyclotrimethylenetrinitramine, 2,4,6-trinitrotoluene, nitromethane, and Bis[2,2-dinitro-2-fluoroethyl]formal, J. Phys. Chem. 72 (1968) 2390–2394.
[98] V.I. Pepekin, S.A. Gubin, Heat of explosion of commercial and Brisant high explosives, Combust. Explos. Shock Waves 43 (2007) 212–218.
[99] M.J. Kamlet, S.J. Jacobs, Chemistry of detonations. I. A simple method for calculating detonation properties of C, H, N, O explosives, J. Chem. Phys. 48 (1968) 23–35.
[100] C.L. Mader, Numerical Modeling of Explosives and Propellants, second ed., CRC Press, Boca Raton, FL, 1998.
[101] B.M. Rice, J. Hare, Predicting heats of detonation using quantum mechanical calculations, Thermochim. Acta 384 (2002) 377–391.
[102] P. Politzer, J.S. Murray, Some perspectives on estimating detonation properties of C,H,N,O compounds, Central Eur. J. Energ. Mater. 8 (2011) 209–220.
[103] P. Politzer, J.S. Murray, The role of product decomposition in determining detonation velocity and detonation pressure, Central Eur. J. Energ. Mater. 11 (2014) 459–474.
[104] S. Bastea, L.E. Fried, K.R. Glaesemann, W.M. Howard, P.C. Sovers, P.A. Vitello, CHEETAH 5.0, User's Manual, Lawrence Livermore National Laboratory, Lawrence, CA, 2006.
[105] J. Akhavan, The Chemistry of Explosives, second ed., Royal Society of Chemistry, Cambridge, UK, 2004.
[106] H. Muthurajan, A. How Ghee, Software development for the Detonation product analysis of high energetic materials—part I, Central Eur. J. Energ. Mater. 5 (3–4) (2008) 19–35.
[107] P. J. Linstrom and W. G. Mallard (Eds.), NIST Chemistry Webbook, NIST Standard Reference Database Number 69, National Institute of Standards and Technology, Gaithersburg, MD, http://www.nist.gov.
[108] V.I. Pepekin, B.L. Korsunskii, A.A. Denisaev, Initiation of solid explosives by mechanical impact, Combust. Explos. Shock Waves 44 (2008) 586–590.
[109] C. Wu, L.E. Fried, Proc. 6th Symp. (Internat.) Detonation, 2000, p. 490.
[110] J. Edwards, C. Eybl, B. Johnson, Correlation between sensitivity and approximated heats of detonation of several nitroamines using quantum mechanical methods, Internat. J. Quantum Chem. 100 (2004) 713–719.
[111] S. Zeman, Sensitivities of high energy compounds, Struct. Bond. 125 (2007) 195–271.
[112] R.P. Bell, The theory of reactions involving proton transfers, Proc. Royal Soc. London A 154 (1936) 414–429.
[113] M.G. Evans, M. Polanyi, Further considerations on the thermodynamics of chemical equilibria and reaction rates, J. Chem. Soc. Faraday Trans. 32 (1936) 1333–1360.
[114] M.J. Kamlet, H. Hurwitz, Chemistry of detonations. IV. Evaluation of a simple predictional method for detonation velocities of C-H-N-O explosives, J. Chem. Phys. 48 (1968) 3685–3692.
[115] M.J. Kamlet, C. Dickinson, Chemistry of detonations. III. Evaluation of the simplified calculational method for chapman-jouguet detonation pressures on the basis of available experimental information, J. Chem. Phys. 48 (1968) 43–50.
[116] T. Urbánski, Chemistry and Technology of Explosives, Vol. 4, Pergamon Press, Oxford, 1984.
[117] H. Shekhar, Studies on empirical approaches for estimation of detonation velocity of high explosives, Central Eur. J. Energ. Mater. 9 (2012) 39–48.
[118] P. Politzer, J.S. Murray, Perspectives on the crystal densities and packing coefficients of explosive compounds, Struct. Chem. 27 (2016) 401–408.
[119] M.M. Kuklja, S.N. Rashkeev, Shear-strain induced chemical reactivity of layered molecular crystals, Appl. Phys. Lett. 90 (2007) 151913(1–3).
[120] C. Zhang, X. Wang, H. Huang, π-Stacked interactions in explosive crystals: buffers against external mechanical stimuli, J. Am. Chem. Soc. 130 (2008) 8359–8365.
[121] C.M. Tarver, S.K. Chidester, A.L. Nichols III, Critical conditions for impact- and shock-induced hot spots in solid explosives, J. Phys. Chem. 100 (1996) 5794–5799.

[122] C.M. Tarver, P.A. Urtiew, T.D. Tran, Sensitivity of 2,6-diamino-3,5-dinitropyrazine-1-oxide, J. Energ. Mater. 23 (2005) 183–203.
[123] D.H. Tsai, R.W. Armstrong, Defect-enhanced structural relaxation mechanism for the evolution of hot spots in rapidly compressed crystals, J. Phys. Chem. 98 (1994) 10997–11000.
[124] J. Sharma, C.S. Coffey, A.L. Ramaswamy, R.W. Armstrong, Atomic force microscopy of hot spot reaction sites in impacted RDX and laser heated AP, Mater. Res. Soc. Symp. Proc. 418 (1996) 257–264.
[125] C.T. White, J.J.C. Barrett, J.W. Mintmire, M.L. Elert, D.H. Robertson, Effects of nanoscale voids on the sensitivity of model energetic materials, Mater. Res. Soc. Symp. Proc. 418 (1996) 277–280.
[126] K.B. Landenberger, A.J. Matzger, Cocrystal engineering of a prototype energetic material: supramolecular chemistry of 2,4,6-TNT, Cryst. Growth Des. 10 (2010) 5341–5347.
[127] O. Bolton, L.R. Simke, P.F. Pagoria, A.J. Matzger, High power explosive with good sensitivity: a 2:1 cocrystal of CL-20:HMX, Cryst. Growth Des. 12 (2012) 4311–4314.
[128] C. Zhang, Y. Cao, H. Li, Y. Zhou, J. Zhou, T. Gao, H. Zhang, Z. Yang, G. Jiang, Toward low-sensitive and high-energetic cocrystal I: evaluation of the power and the safety of observed energetic cocrystals, Cryst. Eng. Comm. 15 (2013) 4003–4014.
[129] W. Qian, Y. Shu, H. Li, Q. Ma, The effect of HNS on the reinforcement of TNT crystal: a molecular simulation study, J. Mol. Model. 20 (2014) 2461(1–7).
[130] K.-P. Song, F.-D. Ren, S.-H. Zhang, W.-J. Shi, Theoretical insights into the stabilities, detonation performance, and electrostatic potentials of cocrystals containing α- or β-HMX and TATB, fox-7, NTO, or DMF in various molar ratios, J. Mol. Model. 22 (2016) 249(1–15).
[131] C. Zhang, F. Jiao, H. Li, Crystal engineering for creating low sensitivity and highly energetic materials, Cryst. Growth Des. 18 (2018) 5713–5726.
[132] D. Musumeci, C.A. Hunter, R. Prohens, S. Scuderi, J.F. McCabe, Virtual cocrystal screening, Chem. Sci. 2 (2011) 883–890.
[133] T. Grecu, C.A. Hunter, E.J. Gardiner, J.F. McCabe, Validation of a computational Cocrystal prediction tool: comparison of virtual and experimental cocrystal screening results, Cryst. Growth Des. 14 (2014) 165–171.
[134] J.P. Agrawal, Past, present and future of thermally stable explosives, Central Eur. J. Energ. Mater. 9 (2012) 273–290.
[135] X. Cao, Y. Wen, B. Xiang, X. Long, C. Zhang, Are amino groups advantageous to insensitive high explosives (IHES)? J. Mol. Model. 18 (2012) 4729–4738.
[136] D.E. Chavez, M.A. Hiskey, D.L. Naud, Tetrazine explosives, Propell. Explos. Pyrotech. 29 (2004) 209–215.
[137] D. Fischer, T.M. Klapötke, D.G. Piercey, J. Stierstorfer, Synthesis of 5-aminotetrazole-1 N-oxide and its azo derivative: a key step in the development of new energetic materials, Chem. A Eur. J. 19 (2013) 4602–4613.
[138] P. Politzer, P. Lane, J.S. Murray, Computational characterization of two Di-1,2,3,4-tetrazine tetraoxides, DTTO and iso-DTTO, as potential energetic compounds, Central Eur. J. Energ. Mater. 10 (2013) 37–52.
[139] O.V. Larionov (Ed.), Heterocyclic *N*-Oxides, Springer, Cham, Switzerland, 2017.
[140] A.R. Katritzky, R. Taylor, Heteroaromatics containing one six-membered ring, Adv. Heterocycl. Chem. 47 (1990) 277–323.
[141] P. Lane, J.S. Murray, P. Politzer, A computational analysis of the structural features and reactive behavior of some heterocyclic aromatic N-oxides, J. Mol. Struct. (THEOCHEM) 236 (1991) 283–296.
[142] A. Albini, S. Pietra, Heterocyclic *N*-Oxides, CRC Press, Boca Raton, FL, 1991.
[143] J.S. Poole, A computational study of the chemistry of substituted 3-nitrenopyridine 1-oxides, J. Mol. Struct. (THEOCHEM) 894 (2009) 93–102.
[144] P. Politzer, P. Lane, J.S. Murray, Some interesting aspects of N-oxides, Mol. Phys. 112 (2014) 719–725.
[145] P. Politzer, J.S. Murray, Nitro groups vs. N-oxide linkages: effects upon some key determinants of detonation performance, Central Eur. J. Energ. Mater. 14 (2017) 3–25.
[146] K.J. Wilson, S.A. Perera, R.J. Bartlett, J.D. Watts, Stabilization of the pseudo-benzene N_6 ring with oxygen, J. Phys. Chem. A 105 (2001) 7693–7699.
[147] A.M. Churakov, V.A. Tartakovsky, Progress in 1,2,3,4-tetrazine chemistry, Chem. Rev. 104 (2004) 2601–2616.
[148] P. Politzer, J.S. Murray, in: A. Greer, J.F. Liebman (Eds.), N-Oxides of Polyazoles and Polyazines, in the Chemistry of Nitrogen-Rich Functional Groups, Wiley, Hoboken, NJ, 2020, pp. 231–249. ch. 8.

CHAPTER

9

Interplay between chemical and mechanical factors

*Sergey V. Bondarchuk**

Department of Chemistry and Nanomaterials Science, The Bohdan Khmelnytsky National University of Cherkasy, Cherkasy, Ukraine

*Corresponding author: E-mail: bondchem@cdu.edu.ua

Abbreviations

14DNI	1,4-dinitroimidazole
ABT	1,1′-azobistetrazole
AGTZ	aminoguanidinium-1*H*-tetrazolate
ATZ	5-amino-1*H*-tetrazole
bicyclo-HMX	1,3,4,6-tetranitrooctahydroimidazo-[4,5-d]imidazole
BTF	benzotrifuroxan
CL-20	hexanitrohexaazaisowurtzitane
DAAF	3,3′-diamino-4,4′-azoxyfurazan
DAAzF	diaminoazofuraz
DATB	1,3-diamino-2,4,6-trinitrobenzene
DDNP	diazodinitrophenol
DGTZ	diaminoguanidinium-1*H*-tetrazolate
DNDP	diazodinitrophenol
DNIT	4,5-dihydro-5-nitrimino-1*H*-tetrazole
DNTF	3,4-bis(3-nitrofurazan-4-yl)furoxan
ETN	erythritol tetranitrate
FOX-7	1,1-diamino-2,2-dinitroethene
GTZ	guanidinium-1*H*-tetrazolate
HBT	5,5′-hydrazinebistetrazole
HMX	1,3,5,7-tetranitro-1,3,5,7-tetrazocane
HNB	hexanitrobenzene
HNS	hexanitrostilbene
LLM-105	2,6-diamino-3,5-dinitropyrazine-1-oxide
MATB	1-amino-2,4,6-trinitrobenzene

Theoretical and Computational Chemistry, Volume 22
https://doi.org/10.1016/B978-0-12-822971-2.00015-2

MATNB	methylamino-2,4,6-trinitrobenzene
MNT	mononitrotoluene
NQ	nitroguanidine
NTO	nitrotriazolone
PETN	pentaerythritol tetranitrate
RDX	1,3,5-trinitro-1,3,5-triazinane
TATB	1,3,5-triamino-2,4,6-trinitrobenzene
TEX	diazatetracyclo[5.5.0.0^{5,9}.0^{3,11}]-dodecane
TKX-50	dihydroxylammonium 5,5′-bistetrazole-1,1′-diolate
TNAZ	1,3,3-trinitroazetidine
TNB	1,3,5-trinitrobenzene
TNE	tetranitratoethane
TNT	2,4,6-trinitrotoluene

1 Definition of the problem

Impact sensitivity (IS) is a complex phenomenon, which cannot be defined analytically due to its stochastic nature. The known unit of IS is h_{50}, which determines a minimum height needed to initiate explosion of 50% of samples. A quantitative model of ignition probability was recently developed for HMX [1]. Indeed, each sample has own unique distribution of separate crystals and impact energy (IE) dissipation is always different. In other words, we can never repeat two identical IS tests at microscopic level. As a result, IS of the same material depends on density of the sample, shape, size, and arrangement of crystals, etc. [2–7]. If one assumes an ideal crystal of a mechanically isotropic material, which is impacted by a hammer with an ideally flat surface in a strictly parallel direction to the sample surface, the IE needed to initiate explosion can be defined precisely. In reality, mechanical anisotropy of a material, defects in crystals and irregular hammer surface cause some divergence in mechanical forces acting on each single crystal in the sample leading to stochastic nature of the explosion event.

Meanwhile, a common feature of any impact event is crystal compression. Thus, in a general case, mechanical energy (ME) must be transformed into activation energy (AE) of explosion. The latter can be either purely thermic ("dark") process [8] or can involve electronically excited states (ExSs) [9,10]. In the first case, decomposition reaction occurs on the potential energy surface of high vibrational levels of the ground electronic state. In the second case, thermally and mechanically induced electron transfer (ET) can take place that leads to the involvement of ExSs. Ab initio calculations showed that compression of a crystal to at least 30 GPa causes an electronic excitation (EE) equivalent to 2–5 eV [11]. A principal scheme of these two processes is presented in Fig. 9.1. The left side illustrates how a ME imparted to a crystal can be transformed into valence vibrations in a molecule.

Each fundamental frequency creates a set of resonant frequencies above, which are multiples of its fundamental and are called *overtones*. Overtones resonantly interact with higher fundamentals of valence vibrations and release energy from the phonon manifold to valence vibrational manifold. Therefore, the smaller the difference in energy between the phonon overtones and vibrational fundamentals (Δ) is, the better the energy transfers [12,13]. With the rise of Δ, rate constant of the energy transfer decreases sharply since it is proportional to $\text{sech}^2(\Delta E/kT)$. This process usually occurs in so-called "hot spots," instead of the whole crystal. This follows from the assumption that in the case of a uniform distribution of IE, a 2 J

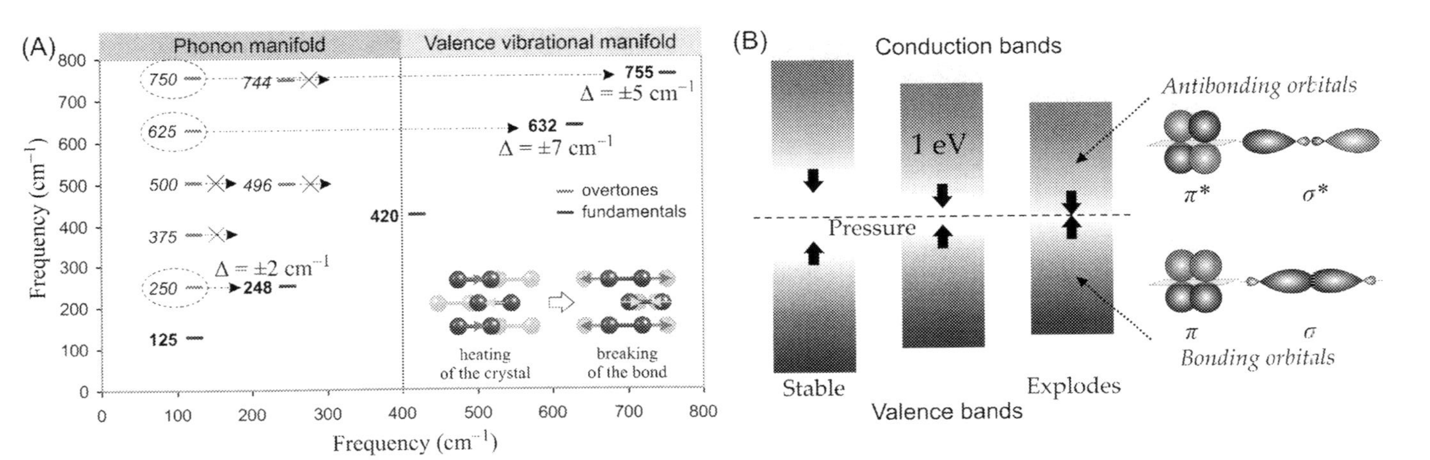

FIG. 9.1 Two mechanisms of initiation involving vibrational "hot" states (A) and thermally excited states (B).

impact of a 50 mg sample at room temperature similar to RDX ($C_P = 1.3$ J/g K) heats the sample approximately by 31 K [14,15].

Analytically, this energy transfer rate between the phonon manifold and valence vibrational manifold can be expressed as in Eq. (9.1) [12,13]:

$$\kappa = \frac{j\hbar\Omega}{\tau_1(0)\theta_e} \tag{9.1}$$

where κ is the energy transfer rate, j is the number of vibrational modes in the higher frequency region, Ω is the vibrational frequency of these modes, θ_e is the equivalence temperature, and $\tau_1(0)$ is the lifetime of the higher frequency mode at $T = 0$ K.

In practice, it is more convenient to estimate the κ value using scheme in Fig. 9.1A. Assuming a harmonic coupling process with $\Delta < 10\ \text{cm}^{-1}$, a new parameter ζ is introduced. The latter is the total number of phonon–vibron coupling modes. Thus, the κ value can be simply done as in Eq. (9.2) [16]:

$$\kappa = \mu\zeta, \tag{9.2}$$

where μ is assumed to be constant. This model was applied for TATB, PETN, FOX-7, TEX, 14DNI, and β-HMX and a correlation of ζ with IS was found.

However, for a more precise calculation of the ζ values, one needs to take into account the effect of crystal compression on vibrational frequencies. The increase in pressure leads to a shift in the vibrational frequencies of the material, which is then followed by a temperature effect. In the case of adiabatic compression of a crystal this effect can be expressed as the following [17]:

$$\frac{T}{T_0} = \left(\frac{V}{V_0}\right)^{-\Gamma}, \tag{9.3}$$

where Γ is the Grüneisen parameter and V and T stands for the volume and temperature, respectively. Thus, the rate of energy up-conversion with account of Eq. (9.3) was performed for a series of metal azides (NaN_3, NH_4N_3, LiN_3, HN_3, $Ba(N_3)_2$, $Zn(N_3)_2$, $Sn(N_3)_2$, and AgN_3), ABT, DNIT, HNB, MNT, β-HMX, HBT, α-FOX-7, NTO, TATB, ATZ, GTZ, AGTZ, and DGTZ [17–20].

Another mechanism of ME transformation assumes involving of thermally excited states of explosives [21]. In the absence of any external energetic perturbations, molecular crystal of an arbitrary organic compound has ExSs lying too high to be effectively occupied at ambient temperature. Indeed, according to the Boltzmann distribution, the number of excited (N_u) and ground (N_l) state molecules can be expressed as the following (Eq. 9.4):

$$\frac{N_u}{N_l} = e^{-\frac{\Delta E}{kT}}. \tag{9.4}$$

If one assumes the $\Delta E = 200$ kJ/mol, then, at 293 K, $N_u/N_l \approx 10^{-36}$, which is negligibly small and becomes noticeable only near 800 K ($N_u/N_l \approx 10^{-15}$). It is obvious that such extremely high temperatures are not achievable under average impact loading (IL) that we have stressed above. Moreover, ExSs of most explosives lay much higher, ca. 500–800 kJ/mol as an average.

From the mechanochemical point of view, mechanically induced decomposition of crystals is a kinetic phenomenon, in which the thermal motion of atoms gives rise to energy fluctuations. The latter may cause breaking of some bonds or rearrangement of atoms in the molecule. If an IL causes *stress* (σ), then each crystal possesses *endurance* (τ) period before decomposition. In this case, AE $U(\sigma)$ can be expressed as the following (Eq. 9.5) [22]:

$$U(\sigma) = U_0 - \gamma\sigma = kT \ln\left(\frac{\tau}{\tau_0}\right), \tag{9.5}$$

where k is the Boltzmann constant, T is the absolute temperature, and γ is the material-specific coefficient; $\tau_0 = (10^{-12} \div 10^{-14})$ s. For polymeric materials $U(\sigma)$ is about 120–200 kJ/mol, which is significantly lower than the energy of a C—C bond (350 kJ/mol). Thus, it becomes clear that ME imparted to a crystal spends for decreasing of AE $U(\sigma)$. Apart from the external mechanical loading, there can be an intrinsic stress caused by bending of the valence angles or stretching (shrinking) of bonds. Such compounds usually demonstrate high chemical reactivity, which can be related to the low-lying ExSs.

According to the band structure theory of solids, appearance of free charge carriers (electrons) means that the band gap (ΔE_{gap}) is narrowing under external mechanical loading. In the lowest lying ExSs the electron–hole (radical ion) pairs appear. In the case of molecular crystals, ΔE_{gap} can be expressed as in Eq. (9.6) [23].

$$E_{gap} = I - A - 2P \tag{9.6}$$

Herein, I and A stand for ionization energy and electron affinity; $P = P_e = P_h$ is the polarization energy of electron and hole.

Thus, Fig. 9.1B demonstrates compression-induced narrowing of ΔE_{gap} from an insulator (left) to a conductor (right). Intermediate position has $\Delta E_{gap} = 1$ eV (96.5 kJ/mol) indicating conditional energetic boundary between "dark" reactions in the ground and ExS. In the case of reactions in solution, thermal EE above this boundary is less probable according to Eq. (9.4); therefore, involvement of low-lying ExSs is possible only when AE is lower 1 eV [24]. This limits the range of reactants to polymers and hydrogen donors [25–27]. Involvement of compounds with higher lying ExSs (unsaturated hydrocarbons) requires paramagnetic catalysts [28].

Slightly different situation is in the case of IL of crystalline materials. Although most explosives have wide ΔE_{gap}, crystal compression accompanied with heating narrows ΔE_{gap} and makes possible occupation of the conduction band. So, what happens when an electron occupies conduction band? First of all, a strong redistribution of the electron density takes place. Excited electrons occupy molecular orbitals (constituents of the conduction band), which have antibonding character with respect to most of chemical bonds in the molecule (Fig. 9.1B). This leads to breaking of the latter and formation of reactive radical species which can trigger further decomposition of the crystal. Thus, different molecular and crystalline properties affecting mechanically induced thermal electron occupation of the conduction band of an explosive will be discussed below.

For many explosives, crystal properties can be simply neglected since vacuum isolated molecules are good representatives of the whole material. However, the properties of separate *polymorphs*, which often have significant differences in sensitivity, can be described only via consideration of their crystalline properties. For example, HMX demonstrates increase in sensitivity (up to 80%) when the β-HMX (1.96 J)→δ-HMX (0.39 J) transition takes place [29,30]. Another example of such shortcomings is ionic explosives (metal azides, aryl diazonium salts, etc.). In this case, description of vacuum isolated ions or contact ions in vacuum is not correct. And the only way to describe such chemical systems properly is to model their crystalline state. However, a model based on molecular electrostatic potential is developed for ionic explosives [31].

2 Band gap compressibility

Theoretical analysis of the possible role of EEs in the problem of initiation was put forward long ago [32]. It is well known that crystal compression leads to the insulator–metal transition. This happens when the concentration of matrix atoms reaches a critical value predicted by the criteria of Herzfeld, Mott, Hubbard, Edwards-Sienko, and others [33]. Additionally, decreasing of ΔE_{gap} yields the increased sensitivity according to principle of the easiest electron transition, PET [34]. Thus, a number of theoretical studies were performed to combine ΔE_{gap} and IS, which is well-reviewed in the literature [35–40]. Usually, crystal compression immediately leads to a narrowing of ΔE_{gap}, like in the case of DDNP [41], RDX [42], HMX [43,44], TATB [45,46], CH_3NO_2 [47], etc. In some cases, however, ΔE_{gap} can demonstrate flat regions, like in AgN_3 [48] or even rise (e.g., in NH_4ClO_4) [49] with increased external pressure. Recently, pressure dependence of ΔE_{gap} was studied for a series of 24 C—H—N—O—Cl explosives of various families and was found that the response of ΔE_{gap} to hydrostatic compression has generally linear character with different slopes. However, in some cases (TNE, NQ, etc.) this dependence has nonlinear character (Fig. 9.2) [50].

Generally, a simple correlation between ΔE_{gap} and IS is possible only when energetic crystals have similar structure or similar thermal decomposition mechanism [40]. Comparison of the ΔE_{gap} values for different families of compounds results in significant divergences and does not applicable. For example, ΔE_{gap} (eV) and h_{50} (cm) values for MATB, ε-CL-20, and $Sr(N_3)_2$ are the following: 1.89 and 177, 3.63 and 27, and 3.71 and not explosive [40]. However, according to Cartwright and Wilkinson [51], $Sr(N_3)_2$ has impact sensitivity 9.10 N cm^{-2}.

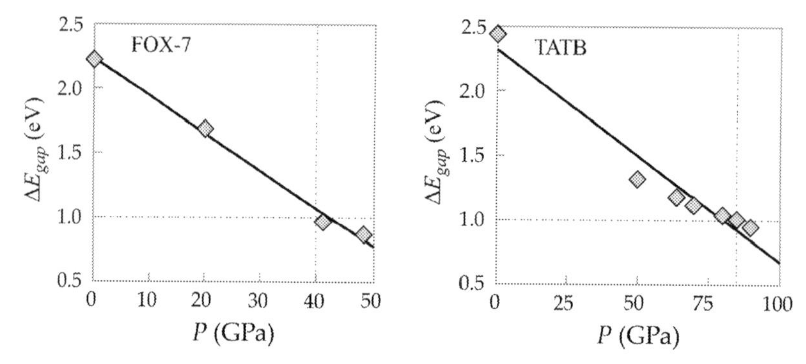

FIG. 9.2 ΔE_{gap} as a linear (FOX-7) and nonlinear (TATB) function of hydrostatic compression.

Thus, it is crucial to know how ΔE_{gap} is changed with the rise of pressure. In this aspect, ideal crystals and those containing defects behave differently. It has been found that the RDX crystals with vacancies reach metallic state at pressure being 30% lower than the prefect ones [42]. Meanwhile, the edge dislocations in the RDX and FOX-7 crystals produce local electronic states in the optical gap whereas the external pressure moves these states deep within ΔE_{gap} [52,53].

Obviously, applied ME should be transformed into vibrational and then into electronic energy. This process can proceed via the known vibronic coupling mechanism, which is essential in such nonadiabatic processes, like crystal compression [37]. Otherwise speaking, ΔE_{gap} corresponds to a part of IE (or h_{50}) and can be considered as AE of the explosion reaction. Assuming the rate constant of the latter is inversely proportional to h_{50}, one can express ΔE_{gap} as the following [50]:

$$h_{50} \propto A \exp\left(\frac{\Delta E_{gap}}{\beta_E k_B T}\right), \tag{9.7}$$

where $A = \prod_i^{3N-3} \nu_i^R \Big/ \prod_i^{3N-4} \nu_i^{TS}$, k_B is the Boltzmann constant; ν_i^R and ν_i^{TS} are the positive normal mode frequencies of the reactant minimum and transition state, respectively [54]. A similar approximation was shown to be successful for bimolecular reactions between electrophiles and nucleophiles [55]. Herein, β_E is a function, which reflects the *band gap compressibility* and, in the finite differences approximation, this can be expressed as the following [50]:

$$\frac{1}{\beta_E} = \left(\frac{P_{metal}}{\Delta E_{gap}}\right) = \tan\alpha, \tag{9.8}$$

where, P_{metal} is the pressure at $\Delta E_{gap} = 0$ (metallic state). In general case the function β_E is unknown and is peculiar for each single explosive; therefore, it is convenient to use pressure at $\Delta E_{gap} = 1$ eV as we have described above. This value can be called *triggering pressure* (P_{trigg}), above which the thermal electron occupation of the conduction band becomes possible. Substituting Eq. (9.8) in Eq. (9.7) and taking into account that ΔE_{gap} is a part of IE, one can obtain the following correlation (Eq. 9.9) [50].

$$h_{50} \propto \exp\left(P_{trigg}\right) \tag{9.9}$$

The search of triggering pressure is a time-consuming and computationally expensive process. Usually, a few cell relaxation and band structure calculations are absolutely required. In this context, it is more convenient to find crystal compressibility β (or bulk modulus $K = 1/\beta$) instead of the β_E, which will be described below.

3 Bulk modulus

Isothermal bulk modulus (K_T) determines how a material resists to a uniform (isotropic or hydrostatic) compression at constant temperature. There are two accepted notations of bulk modulus in the literature, K and B. Henceforth, the quantity K_T will be simply denoted as K. The reciprocal of K is *compressibility* $\beta = 1/K$. These quantities are important characteristics of an energetic material (EM), which reflect how the latter responses to an IL (Eq. 9.10) [56].

$$K = -V\left(\frac{\partial P}{\partial V}\right)_T \quad (9.10)$$

Thus, hard materials have high bulk moduli, while soft materials have small ones. Often, K is expressed via *volumetric thermal expansion coefficient* (α), *isobaric* and *isochoric heat capacities* (C_P and C_V) as the following (Eq. 9.11) [57]:

$$K = \frac{C_P - C_V}{\alpha^2 VT} \quad (9.11)$$

To calculate the K values, a set of cell relaxations must be performed at different pressures followed by fitting these data to one of the equations of state (EOS) (Tait, Murnaghan, Vinet, Sun, or Birch-Murnaghan). The latter EOS, in the form of the third-order expansion (Eq. 9.12), is the most popular approximation [56].

$$P = \frac{3K_0}{2}\left[\left(\frac{V_0}{V}\right)^{7/3} - \left(\frac{V_0}{V}\right)^{5/3}\right]\left[1 + \frac{3}{4}(K_0' - 4)\left(\left(\frac{V_0}{V}\right)^{2/3} - 1\right)\right] \quad (9.12)$$

Herein, subscript 0 stands for standard temperature and K_0' is the pressure derivative of the bulk modulus.

An alternative approach for obtaining K is energy-vs-strain method. In this case, elastic stiffness $\{C_{ij}\}$ and compliance $\{S_{ij}\} = \{C_{ij}\}^{-1}$ tensors are calculated, from which one can extract various elastic moduli. This can be done using three popular approximations due to Voigt, Reuss, and Hill. Bulk moduli obtained with the first two approximations are expressed in Eqs. (9.13) and (9.14), while the Hill bulk modulus is $K_{Hill} = 1/2$ $(K_{Voigt} + K_{Reuss})$ [58].

$$K_{Voigt} = \frac{1}{9}[C_{11} + C_{22} + C_{33} + 2(C_{12} + C_{13} + C_{23})] \quad (9.13)$$

$$K_{Reuss} = \frac{1}{S_{11} + S_{22} + S_{33} + 2(S_{12} + S_{13} + S_{23})} \quad (9.14)$$

The calculation of a $\{C_{ij}\}$ tensor allows obtaining other mechanical properties, namely, shear (G) and Young's (E) moduli, which are also important [58]. For example, it has been found that the K/G ratio correlates with IS [59–61]. Actually, many explosives demonstrate orientation-dependent sensitivity, which is caused by steric hindrance to shear flow [62]. This phenomenon is well-established for CH_3NO_2, β-HMX, bicyclo-HMX, α-RDX, ε-CL-20, HNS, TNT, TATB, and LLM-105 [63,64]. The predicted Hugoniot elastic limits for TKX-50 show anisotropic IS with [010] as the most sensitive direction and [100] as the least sensitive [65]. We should stress that TKX-50 has the biggest K value (21.9 GPa) ever reported for high explosives [66]. The latter EM, along with TATB ($K = 17.1$ GPa) [67], are one of the most insensitive explosives. Thus, the link between IS and bulk modulus is obvious, which was recently shown for energetic nitrate esters [68]. Bulk modulus was also applied for calculations of the integral $(P - P_c)^2 dt$, which is constant for each explosive [69]. Herein, P_c is a critical pressure above which the explosive is no longer elastic; it is equal to Hugoniot elastic limit.

Recently, we have shown that K is directly related to metallization point (assuming the P_{trigg} values too) [70]. For example, in a covalently bound solid, which possesses the sp^3 hybrid orbitals (like diamond), the energy W needed to close the gap can be expressed as in Eq. (9.15) [33]:

$$W = k_\theta/4, \tag{9.15}$$

where the force constant k_θ is then found as in Eq. (9.16):

$$k_\theta = \frac{3l^3}{8}(C_{11} - C_{12}). \tag{9.16}$$

Herein, l is the bond length (Å); C_{11} and C_{12} are the corresponding elastic stiffness constants (GPa). It is obvious that K is directly proportional to IE; however, as we have recently shown, the K value cannot serve as a single parameter in the correlation [70]. For the low-symmetry molecular or ionic crystals of the most explosives, analogous analytical expression is expected to be very complex. Therefore, empirical correlations are much more convenient in this case.

Thus, compression of an explosive's crystal leads to numerous physical changes (collapse of voids [71], dislocations [52,53], decreasing of the lattice free space [72–74], static electric charge formation [75], etc.) that are difficult to take into account within a simple theoretical model. Even despite this, many processes still remain to be incomprehensible to the end [76,77].

4 Crystal packing and growth morphology

Crystal packing is an important factor influencing sensitivity of EMs. Typical packing modes of explosives are illustrated in Fig. 9.3 [78,79]. Recently, a straightforward method was offered for a quick identification of the packing mode by the shape of Hirshfeld surface and the distribution of red dots on the molecular surface. As a result, the packing mode with high shear sliding probability possesses low IS, in combination with a high molecular stability

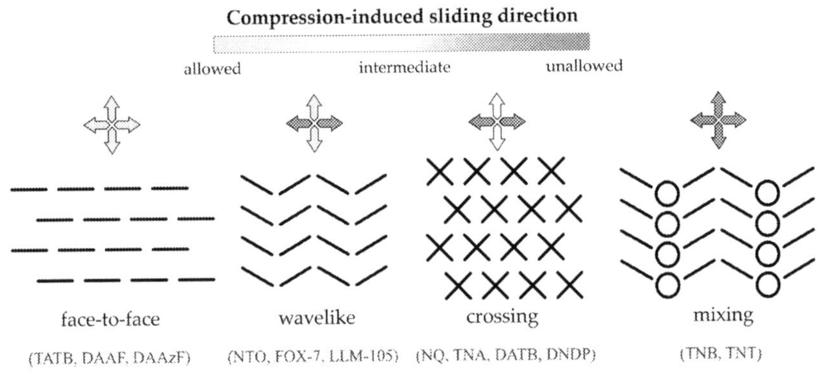

FIG. 9.3 Typical packing modes of explosives with indication of sliding direction probability.

[80]. This is consistent with previously described model of steric hindrance to shear flow [62]. Thus, one can conclude that insensitive EM must [78–81]:

(a) demonstrate a perfect face-to-face π-stacking;
(b) be constructed with big π-bonded molecules;
(c) possess strong intra- and intermolecular hydrogen bonding.

In addition to compounds listed in Fig. 9.3, the same packing-sensitivity relationships were found for a series of furazan-triazole energetic compounds [82].

QTAIM analysis of impact sensitive materials revealed a convergent intramolecular bond bundle with a low electron count that serves as a trigger linkage differing them from insensitive ones [83]. The presence of intermolecular hydrogen bonds effectively buffers the system from heating, thereby delaying the decomposition reaction [84]. Along with molecular polarizability, hydrogen bonding causes an increase of binding energy of energetic cocrystals (BTF/MATNB, BTF/MATB, BTF/TNAZ, BTF/TNB, BTF/TNT, and BTF/CL-20) by 19%–41% over the pristine crystals [85]. This reduces the sensitivity to external stimuli and improves the detonation performance by 5%–10% [85]. Indeed, more tightly packed crystals demonstrate higher detonation performances. The natural upper limit of the density of cyclic and polycyclic organic compounds can be considered the diamond density 3.51 g cm^{-3} [86]. Thus, the methods for obtaining more tightly packed crystals of EM are of great importance. A promising method is trapping of HMX, formed under compression in solvent, into the triaminoguanidine-glyoxal polymer (TAGP) layers [87]. As a result, so-called qy-HMX crystals are obtained with increased density (2.13 g cm^{-3}), enthalpy of formation and stability [88]. Recently, a correlation between *crystal packing coefficient*, which quantifies the fraction of the volume in a crystal that is occupied by hard sphere atoms, and thermostability was shown for 67 EM including cocrystals [89,90]. For thermostable explosives, the crystal packing coefficient should hit in the narrow range 73%–77% [89].

Also, crystal growth morphology is an important factor affecting IS. However, this influence is often hard to estimate due to simultaneous contribution of size, purity, and arrangement of crystals in the sample [2–7,91]. On one hand, plate-like crystals of DNTF (H_2O/EtOH) are less sensitive than rod-like (H_2O/AcOH) [92]. Also, bar-shaped RDX/CL-20 cocrystals demonstrate lower IS than polyhedral crystals of raw CL-20 [93]. On the other hand, for RDX [94,95] and ε-CL-20 [96] spherical shape was found to be the least sensitive. Finally, morphology can cause a little effect on IS, like in the case of ETN [5].

Thus, to quantify crystal habits, a few measures were applied, which characterize their deviation from an ideal spherical shape in 1D-, 2D-, and 3D-space. The simplest measure is the *length/diameter ratio* (L/D) [95,96], which is the most appropriate for determination of the habit ellipsoidization degree. For strongly irregular shapes, however, the L/D ratio is a poor descriptor; therefore, another quantity was introduced to characterize shape of the habit in 2D, namely, *circularity* (C) (Eq. 9.17) [97]:

$$C = 4\pi \frac{S}{P^2}, \quad (9.17)$$

where S is the area of the crystal projection on a selected plane; P is the crystal projection perimeter (Fig. 9.4).

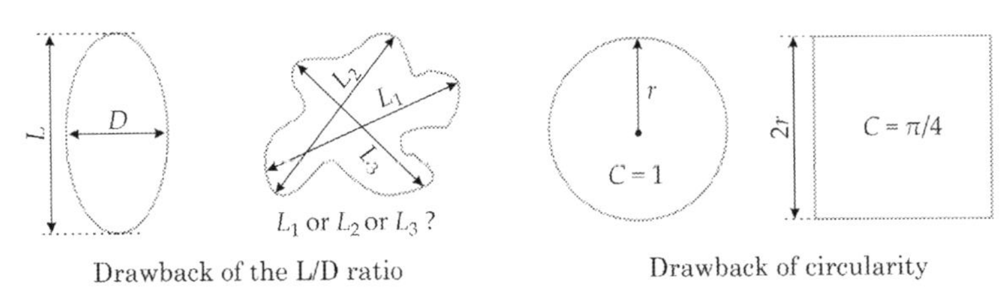

FIG. 9.4 Presentation of the L/D ratio and circularity along with their drawbacks.

Taking into account drawbacks presented in Fig. 9.4, it is important to introduce a quantity, which describes the habit in 3D. The latter can be expressed via sphericity (Ψ) as presented in Eq. (9.18) [98].

$$\Psi = \frac{S_{cryst}}{6^{2/3} V_{cryst}^{2/3} \pi^{1/3}}, \tag{9.18}$$

where S_{cryst} and V_{cryst} are surface and volume of a crystal habit, respectively. For an ideal sphere, $\Psi = 1$ and for any other shapes, $\Psi > 1$. The stronger the crystal expands in one or two dimensions, the greater the Ψ value is.

The influence of Ψ on IS can be graphically described as in Fig. 9.5 [99]. During consolidation, ME dissipates throughout the sample and transforms into heat due to friction of crystal edges. As a result, no significant pressure rise is observed (Fig. 9.5B). At the next step, separate crystals compress, which leads to a sharp rise of pressure and change in the band structure. Of course, as we have discussed above, anisotropy of elasticity is an important factor affecting IS. Compression of a real sample, however, can be thought as isotropic, since separate crystals are distributed randomly (Fig. 9.5C). Different consolidation/compression ratios, naturally occurring due to different crystal habit sphericity, can, probably, be responsible for *dead-pressing* of explosives [100]; this is easily seen in Fig. 9.5D and E. Indeed, pressurizing of the rod-shaped crystals with higher consolidation/compression ratio (Fig. 9.5E) means local heating will prevail over the pressure rise. Thus, since the achievement of ΔE_{gap} narrowing is easier when the consolidation/ compression ratio is smaller, the crystals are expected to be more sensitive with high rather than low sphericity [99].

Another important factor affecting IS is melting point, which is well discussed in statistical crack mechanics of initiation of explosives [101]. The formation of "hot spots" due to frictional heating [102] of the crystal edges as a result of an IL can be schematically viewed in Fig. 9.5F [99]. Once the thermally excited molecules decompose into free radicals (or other reactive intermediates) these cannot propagate chain reaction because of the restriction of their translational motion within the hot spot. As a result, if local heating is not enough, the vibrational energy dissipates throughout the crystal and molecules return to their ground state (quenching of the hot spots). Thus, low-melting crystals are expected to be more sensitive than the high-melting ones.

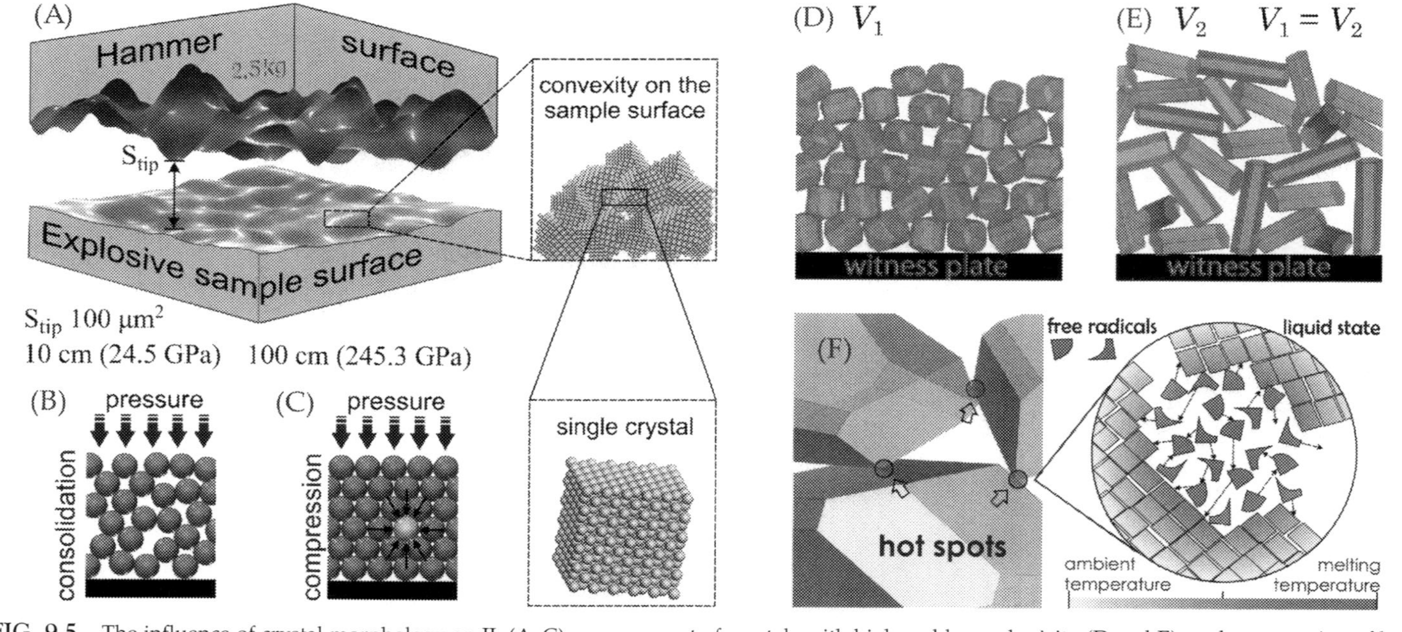

FIG. 9.5 The influence of crystal morphology on IL (A–C), arrangement of crystals with high and low sphericity (D and E), and presentation of hot spot (F).

5 Empirical models

Using quantities described above, a few empirical models for prediction of IS were developed covering aromatic, aliphatic, and heterocyclic nitro and nitrato compounds (Eq. 9.19), metal azides (Eq. 9.20), bistetrazole-based energetic salts (ESs) (Eq. 9.21), and diazonium salts (Eqs. 9.22 and 9.23) [50,70,99,103].

$$\Omega_1 = \frac{\Psi T_m^2}{N_F^7} \exp\left(\frac{P_{trigg}}{1000}\right) \exp\left(-\frac{E_c}{100}\right) \tag{9.19}$$

$$\Omega_2 = \frac{B\Psi\eta}{N_F A^2} \exp\left(\frac{\Delta E_{gap} - E_c}{1000}\right) \tag{9.20}$$

$$\Omega_3 = \frac{\Psi T_{dec}^2 \eta}{N_F} \exp\left(-\frac{E_c}{100}\right) \tag{9.21}$$

$$\Omega_4 = \frac{l_1 q_{NN}}{l_2 \nu_1 A^7} \tag{9.22}$$

$$\Omega_5 = \frac{B^3 T_{dec}^2 \eta^{1/2}}{N_F} \exp\left(-\frac{E_c}{100}\right) \tag{9.23}$$

Herein, Ω is the dimensionless empirical sensitivity function; N_F is the number of electrons in the valence shell; T_m and T_{dec} are melting and decomposition temperatures; E_c is the enclosed energy content, which is the negative of the decomposition enthalpy normalized to a certain unit [104]; A is the electron affinity; $\eta = I - A$ is chemical hardness; l_1 and l_2 are the N≡N and C—N bond lengths; ν_1 is the N≡N bond frequency; q_{NN} is the sum of partial charges on the diazonium group. Plots of the obtained functions Ω vs the corresponding experimental data on IS are illustrated in Fig. 9.6.

Among the presented empirical models, Ω_4 is based on purely molecular features, while the other models are compiled either with solid-state characteristics (Eq. 9.19) or their combination (Eqs. 9.20, 9.21, and 9.23). In all these cases, the described solid-state properties, along with E_c, have general character and can be applicable in a similar manner with different weights. It is important to note that extension of a sensitivity model on predicted crystal structures (Fig. 9.6B, cross-hair markers) assumes avoiding of use of strongly polymorph-dependent properties (T_m, Ψ, etc.), since these values may become the reason of a significant error. Meanwhile, the molecular properties (η and A) reflect peculiarities of the decomposition mechanism. For example, in the case of metal azides, an electron transfers from the azide anion to the metal cation followed by the secondary reactions of the $N_3^{\bullet}$ radical. In terms of the η values, this means that extremely hard species, for example, a Li^+ cation, must be reduced to much softer ones (a lithium atom). Conversely, the A values determine how much energy is released after reduction meaning IE is directly proportional to η, but is inversely proportional to A.

Another example is decomposition of aryl diazonium salts. It is known that a key step in this process is a compression-induced ET to a diazonium cation yielding an unstable aryl diazenyl radical [103]. Thus, this process can be quantified via electron affinity of the given

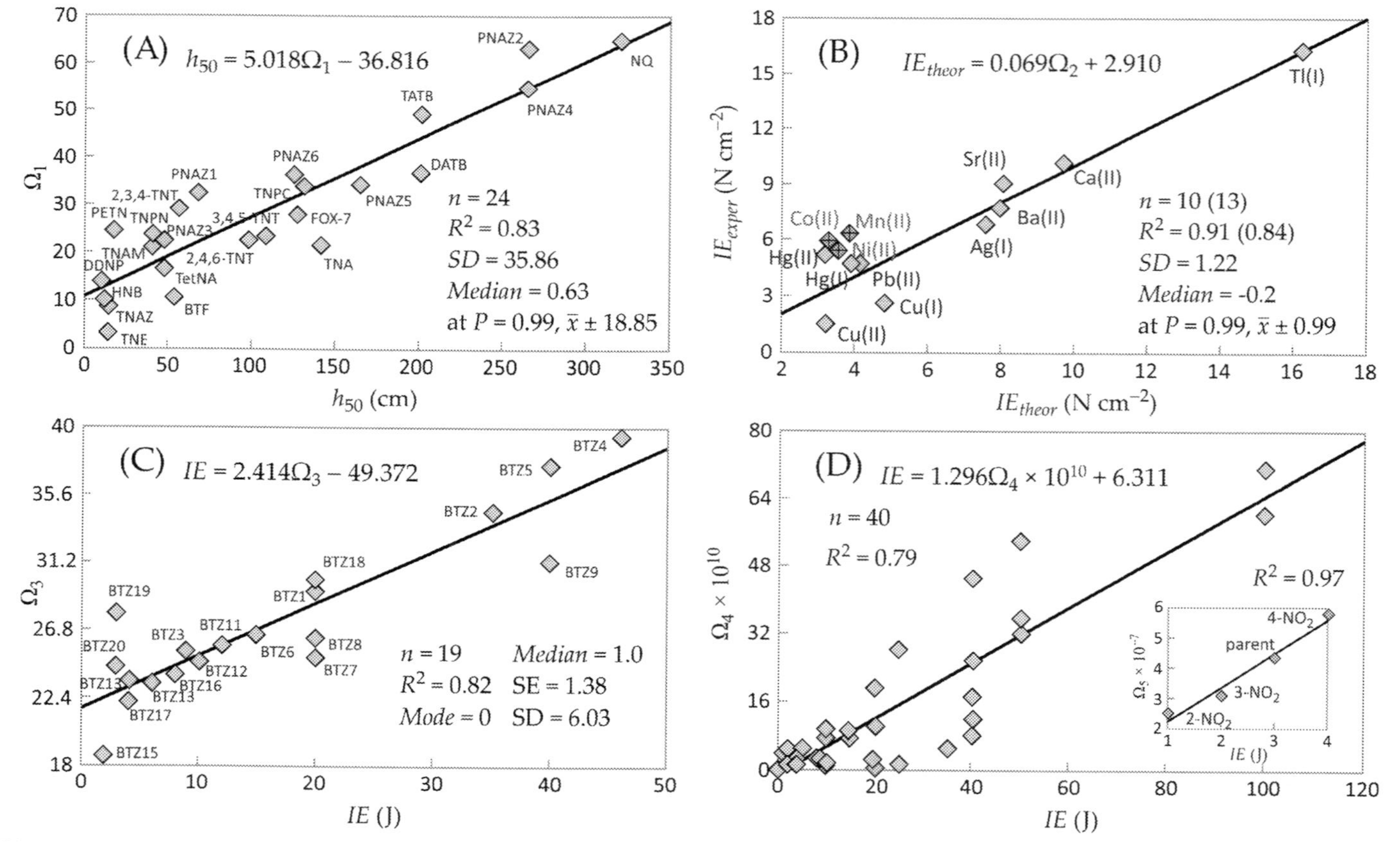

FIG. 9.6 The plots of sensitivity functions Ω_1 (A), Ω_2 (B), Ω_3 (C), and Ω_4–Ω_5 (D) vs the experimental ISs.

cation. Moreover, total charge of the diazonium group changes to some degree, which can be reflected in the q_{NN} values. Finally, there are known geometrical changes leading to a decrease of the C—N and increase of the N≡N bond lengths as well as opposite change in the valence vibrational frequencies of these bonds. All these, parameters can be easily quantified (Eq. 9.22) and contribute to the sensitivity model.

6 Conclusions

In summary, IS is a complex phenomenon with stochastic nature, which cannot be simply described by an analytical expression or an arbitrary empirical correlation. This chapter describes how the conception of pressure-induced thermal EEs can be applied for detonation in solids and how to reach these ExSs using ME. It is probable that the other sensitivities (spark, shock, friction) have a similar nature but differ in the mechanisms of energy transformation leading to EE. The results of theoretical analysis of the factors affecting IS agree with the hypothesis about the role of conduction band filling in the initialization of decomposition. In this chapter, we have provided a theoretical background for criteria of IS in terms of the hot spot formation mechanism assuming the decomposition reaction should take place in the liquid state in the vicinity of a hot spot.

The obtained correlation coefficients are generally good to justify application of the developed sensitivity models, which cover a wide range of EM. Meanwhile, some initial guess about the decomposition mechanism is often essential. If the mechanism is known, molecular features are usually enough to compile a good sensitivity model since the difference in IS between polymorphs is small compared to that of distinct molecules. General chemical reactivity theory, however, cannot always explain IS, which means the crystalline properties must be considered in the model.

Finally, the described empirical models often require crystal structures to be known. This may be challenging when modeling novel explosives, since it needs a very computationally expensive crystal structure prediction, which, nevertheless, does not guarantee the predicted structure will be true. Thus, the use of strongly polymorph-dependent properties must be avoided in such sensitivity models.

Acknowledgments

This work was supported by the Ministry of Education and Science of Ukraine, Research Fund (Grant No. 0118U003862).

References

[1] H.-F. Guo, Y.-Q. Wu, F.-L. Huang, A hot spots ignition probability model for low-velocity impacted explosive particles based on the particle size and distribution, Math. Probl. Eng. 2017 (2017) 7421842.
[2] H. Czerski, W.G. Proud, Relationship between the morphology of granular cyclotrimethylene-trinitramine and its shock sensitivity, J. Appl. Phys. 102 (11) (2007) 113515.
[3] Y. Li, P. Wu, C. Hua, J. Wang, B. Huang, J. Chen, Z. Qiao, G. Yang, Determination of the mechanical and thermal properties, and impact sensitivity of pressed HMX-based PBX, Cent. Eur. J. Energ. Mater. 16 (2) (2019) 299–315.

[4] F. Ma, P. Chen, K. Dai, Q. Zhou, Specimen size effect of explosive sensitivity under low velocity impact, J. Phys. Conf. Ser. 500 (2014), 052026.
[5] V.W. Manner, B.C. Tappan, B.L. Scott, D.N. Preston, G.W. Brown, Crystal structure, packing analysis, and structural-sensitivity correlations of erythritol tetranitrate, Cryst. Growth Des. 14 (11) (2014) 6154–6160.
[6] X. Song, F. Li, Y. Wang, C. An, J. Wang, J. Zhang, A fractal approach to assess the risks of nitroamine explosives, J. Energ. Mater. 30 (1) (2012) 1–29.
[7] C. Zhang, Q. Peng, L. Wang, X. Wang, Thermal sensitivity of HMX crystals and HMX-based explosives treated under various conditions, Propellants Explos. Pyrotech. 35 (6) (2010) 561–566.
[8] R.V. Tsyshevsky, O. Sharia, M.M. Kuklja, Molecular theory of detonation initiation: insight from first principles modeling of the decomposition mechanisms of organic nitro energetic materials, Molecules 21 (2) (2016) 236.
[9] A. Bhattacharya, Y. Guo, E.R. Bernstein, Nonadiabatic reaction of energetic molecules, Acc. Chem. Res. 43 (12) (2010) 1476–1485.
[10] G. Chu, Z. Yang, T. Xi, J. Xin, Y. Zhao, W. He, M. Shui, Y. Gu, Y. Xiong, T. Xu, Relaxed structure of typical nitro explosives in the excited state: observation, implication and application, Chem. Phys. Lett. 698 (2018) 200–2005.
[11] M.M. Kuklja, B.P. Aduev, E.D. Aluker, V.I. Krasheninin, A.G. Krechetov, A.Y. Mitrofanov, Role of electronic excitations in explosive decomposition of solids, J. Appl. Phys. 89 (7) (2001) 4156–4166.
[12] D.D. Dlott, M.D. Fayer, Shocked molecular solids: vibrational up pumping, defect hot spot formation, and the onset of chemistry, J. Chem. Phys. 92 (6) (1990) 3798–3812.
[13] A. Tokmakoff, M.D. Fayer, D.D. Dlott, Chemical reaction initiation and hot-spot formation in shocked energetic molecular materials, J. Phys. Chem. 97 (9) (1993) 1901–1913.
[14] C.S. Coffey, S.J. Jacobs, Detection of local heating in impact or shock experiments with thermally sensitive films, J. Appl. Phys. 52 (11) (1981) 6991–6993.
[15] K.L. McNesby, C.S. Coffey, Spectroscopic determination of impact sensitivities of explosives, J. Phys. Chem. B 101 (16) (1997) 3097–3104.
[16] J. Bernstein, Ab initio study of energy transfer rates and impact sensitivities of crystalline explosives, J. Chem. Phys. 148 (8) (2018) 084502.
[17] A.A.L. Michalchuk, M. Trestman, S. Rudić, P. Portius, P.T. Fincham, C.R. Pulham, C.A. Morrison, Predicting the reactivity of energetic materials: an ab initio multi-phonon approach, J. Mater. Chem. A 7 (33) (2019) 19539–19553.
[18] A.A.L. Michalchuk, S. Piggott, I. Lapsanska, C. Pulham, C. Morrison, The big bang theory: predicting energetic materials properties, Acta Crystallogr. A73 (2017) C775.
[19] A.A.L. Michalchuk, S. Rudić, C.R. Pulham, C.A. Morrison, Vibrationally induced metallisation of the energetic azide α-NaN_3, Phys. Chem. Chem. Phys. 20 (46) (2018) 29061–29069.
[20] A.A.L. Michalchuk, P.T. Fincham, P. Portius, C.R. Pulham, C.A. Morrison, A pathway to the athermal impact initiation of energetic azides, J. Phys. Chem. C 122 (34) (2018) 19395–19408.
[21] K.K. Kalninsh, Significance of electron excitation in chemical reactions, J. Chem. Soc. Faraday Trans. 1 77 (2) (1981) 227–238.
[22] V.R. Regel', A.I. Slutsker, E.E. Tomashevskh, The kinetic nature of the strength of solids, Sov. Phys. Usp. 15 (1) (1972) 45–65.
[23] E.A. Silinsh, Organic Molecular Crystals: Their Electronic States, Springer, Berlin, 1980.
[24] K.K. Kalninsh, E.F. Panarin, On the physical meaning of the activation energy of a chemical reaction, Dokl. Chem. 456 (2014) 103–106.
[25] K.K. Kalninsh, Thermal-electron transfer in crystalline complexes with hydrogen bonding, J. Chem. Soc. Faraday Trans. 2 78 (2) (1982) 327–337.
[26] K.K. Kalninsh, Dark catalytic electron transfer in solid-state hydrogen-bonded electron-donor–electron-acceptor complexes, J. Chem. Soc. Faraday Trans. 2 80 (12) (1984) 1529–1538.
[27] K.K. Kalninsh, Joint experimental and theoretical study of the poly(styryl sodium) and poly(α-methylstyryl sodium) polymerization/depolymerization, Phys. Chem. Chem. Phys. 3 (20) (2001) 4542–4546.
[28] S.V. Bondarchuk, B.F. Minaev, Thermally accessible triplet state of π-nucleophiles does exist. Evidence from first principles study of ethylene interaction with copper species, RSC Adv. 5 (15) (2015) 11558–11569.
[29] B.W. Asay, B.F. Henson, L.B. Smilowitz, P.M. Dickson, On the difference in impact sensitivity of beta and delta HMX, J. Energ. Mater. 21 (4) (2003) 223–235.
[30] M. Herrmann, W. Engel, N. Eisenreich, Thermal expansion, transitions, sensitivities and burning rates of HMX, Propellants Explos. Pyrotech. 17 (4) (1992) 190–195.

[31] P. Politzer, P. Lane, J.S. Murray, Sensitivities of ionic explosives, Mol. Phys. 115 (5) (2017) 497–509.
[32] F. Williams, Electronic states of solid explosives and their probable role in detonations, Adv. Chem. Phys. 21 (1971) 289–302.
[33] J.J. Gilman, Shear-induced metallization, Philos. Mag. B 67 (2) (1993) 207–214.
[34] H.M. Xiao, Y.F. Li, Banding and electronic structures of metal azides—sensitivity and conductivity, Sci. China Ser. B Chem. 38 (5) (1995) 538–545.
[35] Z.-X. Chen, X.M. Xiao, Quantum chemistry derived criteria for impact sensitivity, Propellants Explos. Pyrotech. 39 (4) (2014) 487–495.
[36] J.J. Kay, Mechanisms of shock-induced reactions in high explosives, AIP Conf. Proc. 1793 (1) (2017), 030023.
[37] M.M. Kuklja, On the initiation of chemical reactions by electronic excitations in molecular solids, Appl. Phys. A Mater. Sci. Process. 76 (3) (2003) 359–366.
[38] Q.-L. Yan, S. Zeman, Theoretical evaluation of sensitivity and thermal stability for high explosives based on quantum chemistry methods: a brief review, Int. J. Quantum Chem. 113 (8) (2013) 1049–1061.
[39] H. Zhang, F. Cheung, F. Zhao, X.-L. Cheng, Band gaps and the possible effect on impact sensitivity for some nitro aromatic explosive materials, Int. J. Quantum Chem. 109 (7) (2009) 1547–1552.
[40] W. Zhu, H. Xiao, First-principles band gap criterion for impact sensitivity of energetic crystals: a review, Struct. Chem. 21 (3) (2010) 657–665.
[41] Y. Liu, X. Gong, L. Wang, G. Wang, Effect of hydrostatic compression on structure and properties of 2-diazo-4,6-dinitrophenol crystal: density functional theory studies, J. Phys. Chem. C 115 (23) (2011) 11738–11748.
[42] M.M. Kuklja, A.B. Kunz, Ab initio simulation of defects in energetic materials: hydrostatic compression of cyclotrimethylene trinitramine, J. Appl. Phys. 86 (8) (1999) 4428–4434.
[43] Q. Peng, Rahul, G. Wang, G.-R. Liu, S. De, Structures, mechanical properties, equations of state, and electronic properties of β-HMX under hydrostatic pressures: a DFT-D2 study, Phys. Chem. Chem. Phys. 16 (37) (2014) 19972–19983.
[44] W.H. Zhu, X.W. Zhang, T. Wei, H.M. Xiao, DFT studies of pressure effects on structural and vibrational properties of crystalline octahydro-1,3,5,7-tetranitro-1,3,5,7-tetrazocine, Theor. Chem. Acc. 124 (2–3) (2009) 179–186.
[45] Y. Kohno, K. Mori, R.I. Hiyoshi, O. Takahashi, K. Ueda, Molecular dynamics and first-principles studies of structural change in 1,3,5-triamino-2,4,6-trinitrobenzene (TATB) in crystalline state under high pressure: comparison of hydrogen bond systems of TATB versus 1,3-diamino-2,4,6-trinitrobenzene (DATB), Chem. Phys. 472 (2016) 163–172.
[46] C.J. Wu, L.H. Yang, L.E. Fried, Electronic structure of solid 1,3,5-triamino-2,4,6-trinitrobenzene under uniaxial compression: possible role of pressure-induced metallization in energetic materials, Phys. Rev. B 67 (23) (2003) 235101.
[47] E.J. Reed, J.D. Joannopoulos, L.E. Fried, Electronic excitations in shocked nitromethane, Phys. Rev. B 62 (24) (2000) 16500–16509.
[48] W.H. Zhu, H.M. Xiao, First-principles study of structural and vibrational properties of crystalline silver azide under high pressure, J. Solid State Chem. 180 (12) (2007) 3521–3528.
[49] W.H. Zhu, X.W. Zhang, W. Zhu, H.M. Xiao, Density functional theory studies of hydrostatic compression of crystalline ammonium perchlorate, Phys. Chem. Chem. Phys. 10 (48) (2008) 7318–7323.
[50] S.V. Bondarchuk, Quantification of impact sensitivity based on solid-state derived criteria, J. Phys. Chem. A 122 (24) (2018) 5455–5463.
[51] M. Cartwright, J. Wilkinson, Correlation of structure and sensitivity in inorganic azides I. Effect of non-bonded nitrogen nitrogen distances, Propellants Explos. Pyrotech. 35 (4) (2010) 326–332.
[52] M.M. Kuklja, S.N. Rashkeev, F.J. Zerilli, Shear-strain induced decomposition of 1,1-diamino-2,2-dinitroethylene, Appl. Phys. Lett. 89 (7) (2006) 071904.
[53] M.M. Kuklja, A.B. Kunz, Compression-induced effect on the electronic structure of cyclotrimethylene trinitramine containing an edge dislocation, J. Appl. Phys. 87 (5) (2000) 2215–2218.
[54] A.F. Voter, Introduction to the kinetic Monte Carlo method, in: K.E. Sickafus, E.A. Kotomin, B.P. Uberuaga (Eds.), Radiation Effects in Solids, NATO Science Series, vol. 235, Springer, Dordrecht, 2007, pp. 1–23.
[55] S.V. Bondarchuk, B.F. Minaev, The singlet-triplet energy splitting of π-nucleophiles as a measure of their reaction rate with electrophilic partners, Chem. Phys. Lett. 607 (2014) 75–80.
[56] D.M. Dattelbaum, L.L. Stevens, Equations of state of binders and related polymers, in: S.M. Peiris, G.J. Piermarini (Eds.), Static Compression of Energetic Materials. Shock Wave and High Pressure Phenomena, Springer, Berlin, Heidelberg, 2009.

[57] P. Gillet, P. Richet, F. Guyot, G. Fiquet, High-temperature thermodynamic properties of forsterite, J. Geophys. Res. 96 (B7) (1991) 11805–11816.
[58] G. Schubert (Ed.), Treatise on Geophysics, second ed., Elsevier, Amsterdam, 2015.
[59] Y. Lu, Y. Shu, N. Liu, X. Lu, M. Xu, Molecular dynamics simulations on ε-CL-20-based PBXs with added GAP and its derivative polymers, RSC Adv. 8 (9) (2018) 4955–4962.
[60] S. Zeman, M. Jungová, Q.-L. Yan, Impact sensitivity in respect of the crystal lattice free volume and the characteristics of plasticity of some nitramine explosives, Chin. J. Energ. Mater. 23 (12) (2015) 1186–1191.
[61] S. Zeman, M. Jungová, Sensitivity and performance of energetic materials, Propellants Explos. Pyrotech. 41 (3) (2016) 426–451.
[62] J.J. Dick, Orientation-dependent explosion sensitivity of solid nitromethane, J. Phys. Chem. 97 (23) (1993) 6193–6196.
[63] J.C. Gump, High-pressure and temperature investigations of energetic materials, J. Phys. Conf. Ser. 500 (2014), 052014.
[64] Z. Zheng, J. Zhao, Unreacted equation of states of typical energetic materials under static compression: a review, Chin. Phys. B 25 (7) (2016), 076202.
[65] Q. An, T. Cheng, W.A. Goddard III, S.V. Zybin, Anisotropic impact sensitivity and shock induced plasticity of TKX-50 (dihydroxylammonium 5,5′-bis(tetrazole)-1,1′-diolate) single crystals: from large-scale molecular dynamics simulations, J. Phys. Chem. C 119 (4) (2015) 2196–2207.
[66] Z.A. Dreger, A.I. Stash, Z.-G. Yu, Y.-S. Chen, Y. Tao, High-pressure structural response of an insensitive energetic crystal: dihydroxylammonium 5,5′-bistetrazole-1,1′-diolate (TKX-50), J. Phys. Chem. C 121 (10) (2017) 5761–5767.
[67] L.L. Stevens, N. Velisavljevic, D.E. Hooks, D.M. Dattelbaum, Hydrostatic compression curve for triamino-trinitrobenzene determined to 13.0 GPa with powder X-ray diffraction, Propellants Explos. Pyrotech. 33 (4) (2008) 286–295.
[68] V.W. Manner, M.J. Cawkwell, E.M. Kober, T.W. Myers, J.W. Brown, H. Tian, C.J. Snyder, R. Perriot, D.N. Preston, Examining the chemical and structural properties that influence the sensitivity of energetic nitrate esters, Chem. Sci. 9 (15) (2018) 3649–3663.
[69] M. Kornhauser, Correlation of explosive sensitivity to compressional inputs, in: Proceedings Ninth Symposium (International) on Detonation, vol. 2, Red Lion Inn, Columbia River Portland, Oregon, August 28–September 1, 1989, 1989, p. 1451.
[70] S.V. Bondarchuk, A unified model of impact sensitivity of metal azides, New J. Chem. 43 (3) (2019) 1459–1468.
[71] H. Czerski, M.W. Greenaway, W.G. Proud, J.E. Field, Links between the morphology of RDX crystals and their shock sensitivity, AIP Conf. Proc. 845 (1) (2006) 1053–1056.
[72] P. Politzer, J.S. Murray, Impact sensitivity and crystal lattice compressibility/free space, J. Mol. Model. 20 (5) (2014) 2223.
[73] M. Pospíšil, P. Vávra, M.C. Concha, J.S. Murray, P. Politzer, A possible crystal volume factor in the impact sensitivities of some energetic compounds, J. Mol. Model. 16 (5) (2010) 895–901.
[74] S. Zeman, N. Liu, M. Jungová, A.K. Hussein, Q.-N. Yan, Crystal lattice free volume in a study of initiation reactivity of nitramines: impact sensitivity, Def. Technol. 14 (2) (2018) 93–98.
[75] K. Raha, J.S. Chhabra, Static charge development and impact sensitivity of high explosives, J. Hazard. Mater. 34 (3) (1993) 385–391.
[76] D.E. Hooks, K.J. Ramos, C.A. Bolme, M.J. Cawkwell, Elasticity of crystalline molecular explosives, Propellants Explos. Pyrotech. 40 (3) (2015) 333–350.
[77] D.E. Hooks, M.J. Cawkwell, K.J. Ramos, Plasticity in crystalline molecular explosives—a key to unraveling "unpredictable" responses, Propellants Explos. Pyrotech. 41 (2) (2016) 203–204.
[78] Y. Ma, A. Zhang, X. Xue, D. Jiang, Y. Zhu, C. Zhang, Crystal packing of impact-sensitive high-energy explosives, Cryst. Growth Des. 14 (11) (2014) 6101–6114.
[79] Y. Ma, A. Zhang, C. Zhang, D. Jiang, Y. Zhu, C. Zhang, Crystal packing of low-sensitivity and high-energy explosives, Cryst. Growth Des. 14 (9) (2014) 4703–4713.
[80] B. Tian, Y. Xiong, L. Chen, C. Zhang, Relationship between the crystal packing and impact sensitivity of energetic materials, CrystEngComm 20 (6) (2018) 837–848.
[81] C. Zhang, X. Wang, H. Huang, π-Stacked interactions in explosive crystals: buffers against external mechanical stimuli, J. Am. Chem. Soc. 130 (26) (2008) 8359–8365.

[82] Y. Liu, Y. Xu, Q. Sun, M. Lu, Energetic furazan–triazoles with high thermal stability and low sensitivity: facile synthesis, crystal structures and energetic properties, CrystEngComm 21 (40) (2019) 6093–6099.
[83] T.E. Jones, Role of inter- and intramolecular bonding on impact sensitivity, J. Phys. Chem. A 116 (46) (2012) 11008–11014.
[84] L. Zhang, Y. Yu, M. Xiang, A study of the shock sensitivity of energetic single crystals by large-scale ab initio molecular dynamics simulations, Nanomaterials 9 (9) (2019) 1251.
[85] L. Zhang, S.-L. Jiang, Y. Yu, J. Chen, Revealing solid properties of high-energy-density molecular cocrystals from the cooperation of hydrogen bonding and molecular polarizability, Sci. Rep. 9 (1) (2019) 1257.
[86] S.S. Novikov, The natural upper limit of the ideal detonation velocity of the condensed explosives, Dokl. Chem. 425 (1) (2009) 57–59.
[87] Z.-H. Xue, X.-X. Zhang, B.-B. Huang, X. Bai, L.-Y. Zhu, S. Chen, Q.-L. Yan, The structural diversity of hybrid qy-HMX crystals with constraint of 2D dopants and the resulted changes in thermal reactivity, Chem. Eng. J. 390 (2020) 124565.
[88] Q.-L. Yan, Z. Yang, X.-X. Zhang, J.-Y. Lyu, W. He, S. Huang, P.-J. Liu, C. Zhang, Q.-H. Zhang, G.-Q. He, F.-D. Nie, High density assembly of energetic molecules under the constraint of defected 2D materials, J. Mater. Chem. A 7 (30) (2019) 17806–17814.
[89] H. Li, L. Zhang, N. Petrutik, K. Wang, Q. Ma, D. Shem-Tov, F. Zhao, M. Gozin, Molecular and crystal features of thermostable energetic materials: guidelines for architecture of "bridged" compounds, ACS Cent. Sci. 6 (1) (2020) 54–75.
[90] C. Li, H. Li, H.-H. Zong, Y. Huang, M. Gozin, C.Q. Sun, L. Zhang, Strategies for achieving balance between detonation performance and crystal stability of high-energy-density materials, iScience 23 (3) (2020), 100944.
[91] J. Liu, W. Jiang, Q. Yang, J. Song, G.-Z. Hao, F.-S. Li, Study of nano-nitramine explosives: preparation, sensitivity and application, Def. Technol. 10 (2) (2014) 184–189.
[92] N. Liu, Y.-N. Li, S. Zeman, Y.-J. Shu, B.-Z. Wang, Y.-S. Zhou, Q.-L. Zhao, W.-L. Wang, Crystal morphology of 3,4-bis(3-nitrofurazan-4-yl)furoxan (DNTF) in a solvent system: molecular dynamics simulation and sensitivity study, CrystEngComm 18 (16) (2016) 2843–2851.
[93] H. Gao, W. Jiang, J. Liu, G. Hao, L. Xiao, X. Ke, T. Chen, Synthesis and characterization of a new co-crystal explosive with high energy and good sensitivity, J. Energ. Mater. 35 (4) (2017) 490–498.
[94] D.-X. Wang, S.-S. Chen, Y.-Y. Li, J.-Y. Yang, T.-Y. Wei, S.-H. Jin, An investigation into the effects of additives on crystal characteristics and impact sensitivity of RDX, J. Energ. Mater. 32 (3) (2014) 184–198.
[95] Y. Wang, X. Li, S. Chen, X. Ma, Z. Yu, S. Jin, L. Li, Y. Chen, Preparation and characterization of cyclotrimethylenetrinitramine (RDX) with reduced sensitivity, Materials 10 (8) (2017) 974.
[96] H. Chen, L. Li, S. Jin, S. Chen, Q. Jiao, Effects of additives on ε-HNIW crystal morphology and impact sensitivity, Propellants Explos. Pyrotech. 37 (1) (2012) 77–82.
[97] J.K. Miller, J.O. Mares, I.E. Gunduz, S.F. Son, J.F. Rhoads, The impact of crystal morphology on the thermal responses of ultrasonically-excited energetic materials, J. Appl. Phys. 119 (2) (2016) 024903.
[98] S.V. Bondarchuk, Impact sensitivity of crystalline phenyl diazonium salts: a first-principles study of solid-state properties determining the phenomenon, Int. J. Quantum Chem. 117 (21) (2017) e25430.
[99] S.V. Bondarchuk, Significance of crystal habit sphericity in the determination of the impact sensitivity of bistetrazole-based energetic salts, CrystEngComm 20 (38) (2018) 5718–5725.
[100] R. Meyer, J. Köhler, A. Homburg, Explosives, seventh, completely revised and updated edition, Wiley-VCH, Weinheim, 2015.
[101] J.K. Dienes, Q.H. Zuo, J.D. Kershner, Impact initiation of explosives and propellants via statistical crack mechanics, J. Mech. Phys. Solids 54 (6) (2006) 1237–1275.
[102] T. Hussain, L. Yan, Single and double shock initiation modelling for high explosive materials in last three decades, IOP Conf. Ser. Mater. Sci. Eng. 146 (2016), 012041.
[103] S.V. Bondarchuk, Impact sensitivity of aryl diazonium chlorides: limitations of molecular and solid-state approach, J. Mol. Graph. Model. 89 (2019) 114–121.
[104] D. Mathieu, Sensitivity of energetic materials: theoretical relationships to detonation performance and molecular structure, Ind. Eng. Chem. Res. 56 (29) (2017) 8191–8201.

CHAPTER

10

From lattice vibrations to molecular dissociation

Adam A.L. Michalchuk[a] and Carole A. Morrison[b,*]

[a]Federal Institute for Materials Research and Testing (BAM), Berlin, Germany [b]The School of Chemistry, University of Edinburgh, Edinburgh, United Kingdom

*Corresponding author: E-mail: c.morrison@ed.ac.uk

1 Introduction

The ease with which an energetic material can be initiated by mechanical impact is a critical parameter directing material safety and application. While impact sensitivity metrics are traditionally derived experimentally, recent developments have highlighted that the phenomenon is amenable to first principles simulation. In this chapter, we will outline a fully *ab initio* approach to predict the relative impact sensitivities of energetic materials based on the mechanochemical principles that link the impact event to vibrational energy transfer. This mechanism is key to rationalizing how a mechanical impact—which deposits energy into the low-frequency lattice vibrations—results in a molecular response. By simulating the vibrational energy levels (the so-called phonon density of states, PDOS) using first-principles computational methods (typically dispersion-corrected plane-wave density functional theory, PW-DFT) we can calculate the relative rate of energy propagation from the delocalized low-energy lattice vibrations through to the localized molecular modes. The latter traps the energy, which eventually results in bond rupture through heightened vibrational excitation. This method, based on vibrational up-pumping, offers a route toward predicting the impact sensitivities of a broad range of energetic materials, provided the crystal structure of the compound (or salt or co-crystal) is known. While it does not offer insight into the sensitizing roles undoubtedly played by crystal defects or grain boundaries, it does provide a level of understanding at the molecular and crystal packing levels. Correspondingly, this approach offers a feedback mechanism to chemists and materials scientists to guide the design of new materials with desired impact sensitivity behavior.

Theoretical and Computational Chemistry, Volume 22
https://doi.org/10.1016/B978-0-12-822971-2.00010-3

The chapter is constructed as follows. The background condensed matter physics and mechanochemistry are presented first, using energetic azides as explanatory examples. We then illustrate the fundamental understanding that the energy propagation model provides by predicting the impact sensitivities of five well-documented energetic molecular crystals, namely, hexanitrobenzene (HNB), the β-polymorph of 1,3,5,7-tetranitro-1,3,5,7-tetrazocane (known as β-HMX), the α-polymorph of 1,1-diamino-2,2-dinitroethene (known as α-FOX-7), nitrotriazolone (NTO) and 2,4,6-triamino-1,3,5-trinitrobenzene (TATB). This test set represents not only a structurally diverse series of molecules and crystal structures but also encapsulates a broad spectrum of material impact sensitivity, from the highly sensitive primary energetic HNB through to the highly insensitive secondary energetic TATB. As the predictive model is based purely on the vibrational up-pumping of the PDOS, its success testifies to the mechanical shock initiation process being dominated by phonon scattering effects. Consideration of phonon scattering alone is enough to predict this important metric for energetic materials. Finally, the discussion and future outlook sections highlight how this model offers a route toward *ab initio* materials design of new energetic materials with tailored impact sensitivity behavior.

2 Background theory

2.1 Vibrations in crystal lattices

Atoms in molecules and crystals are in constant motion. As atoms separate, the electronic–nuclear interactions pull them back together. This perpetual tug-of-war leads to the dynamic nature of matter that underpins much of chemistry and physics. To understand this phenomenon more clearly, we begin with a one-dimensional chain of atoms, separated by distance r, bound by a periodic pattern of repeat distance a, and restricted to movement along the chain by some small perturbation u. The total energy of the 1D chain is given by the standard Taylor expansion,

$$E = E_0 + \frac{dE}{du} + \frac{1}{2}\frac{d^2E}{du^2}u^2 + \frac{1}{3!}\frac{d^3E}{du^3}u^3 \ldots \tag{10.1}$$

where E_0 describes the energy of the ideal chain of atoms at rest, i.e., at their equilibrium positions r_0. All subsequent terms denote the energy correction associated with distorting the chain away from this minimum energy configuration. Note the first derivative term describes the forces acting on the atoms, which must equal zero if $r = r_0$. It is convenient at this stage to truncate the total energy after the second derivative term. This defines the harmonic approximation and is valid for the small atomic perturbations induced by vibrations being considered here. The higher order derivative terms are the anharmonic terms; we will return to these latter to demonstrate their importance in describing vibrational (phonon) scattering processes. For now, within the confines of the harmonic approximation, the energy of our 1D chain of equivalent atoms can be recast accordingly,

$$E = E_0 + \frac{1}{2}\frac{d^2E}{du^2}u^2 = \frac{1}{2}j\sum_{n=1}^{N}(u_n - u_{n+1})^2 \tag{10.2}$$

where the symbol j represents the force constant acting between neighbors in the chain of n atoms. With the ultimate aim of identifying the vibrational motion of atoms within crystal lattices, we are tasked with identifying the direction and velocity of each atom, with the underlying motion described by u_n. For this, we must determine the force F acting on each atom of mass m, which in turn is simply.

$$F_n = -\frac{\partial E_n}{\partial u_n} = m\frac{\partial^2 u_n}{\partial t^2} \tag{10.3}$$

This, of course, yields Newton's classical equations of motion. Solving this equation for u_n gives the expected result that the oscillating motions of the atoms along the chain conform to sinusoidal (traveling) waves. Introducing a time-dependence t allows the frequencies of those motions to be deduced, as.

$$u_n(t) = \sum\nolimits_K \tilde{u}_k \exp\left[i(kna - \omega t)\right] \tag{10.4}$$

where ω is the angular frequency, $\tilde{u}_k$ is the amplitude of motion, and $\mathbf{k}$ is any wave vector (i.e., $2\pi/\lambda$, where $\lambda =$wavelength). Substitution of Eq. (10.4) into Eq. (10.3) yields,

$$-\frac{\partial E_n}{\partial u_n} = -4ju_n \sin^2\left(\frac{ka}{2}\right) \tag{10.5}$$

And the following expression for ω as a function of $\mathbf{k}$ can be derived,

$$\omega_k = \left(\frac{4j}{m}\right)^{\frac{1}{2}} \left|\sin\left(\frac{ka}{2}\right)\right| \tag{10.6}$$

It therefore follows that ω depends on (i) the interatomic force constant, j, (ii) the mass of the vibrating atom, m, and (iii) the wave vector of the vibration, $\mathbf{k}$. This last dependency results in the so-called phonon dispersion relationship. Note only discrete values of $\mathbf{k}$ are permitted: owing to the translational symmetry of crystals (i.e., the real space distance a in our 1D chain), any wave vectors that differ by an integer of a reciprocal lattice vector, $\mathbf{G}$ (which is a vector perpendicular to any given crystallographic Miller Index plane, of length inverse to the plane separation, and defined as $n \times \frac{2\pi}{a}$) are equivalent. It follows that all unique wave vectors must exist within the limiting region of $-\pi/a < k < \pi/a$. This range defines the boundary of the first Brillouin zone. Any value of $\mathbf{k}$ that lies on this boundary has the property that $\partial\omega/\partial \mathbf{k} = 0$, meaning these wave vectors must be standing waves.

The discussion for a 1D chain gives a general result that is readily extended to three dimensions. Now every atom in the chain can move in three directions, meaning that a system of N atoms displays $3\,N$ degrees of freedom. In the long wavelength limit, three of the $3\,N$ motions correspond to simple translations of the chain along the x, y, or z directions. These are referred to as the acoustic modes as their pattern of oscillation is a longitudinal wave that mimics the motion of a sound wave through matter. Acoustic modes conform to the property that $\omega \to 0$ as $\mathbf{k} \to 0$ (i.e., the frequency tends to zero as the mode adopts an increasingly longer wavelength). The remaining $3N$-3 modes are combinations of $\pm$x, $\pm$y, and $\pm$z motions; these are referred to as optical modes, as their motions are longitudinal and transverse waves which mimic the oscillating dipole pattern of an electromagnetic wave. The name (wrongly) suggests excitation through optical light adsorption, but every chemist knows it is infrared (thermal) adsorption that leads to population excitation of the permitted vibrational states.

The constraint of confining permitted $\boldsymbol{k}$ vectors to lie within the first Brillouin zone reminds us to consider the implication of the real space (crystallographic) lattice in describing vibrational motion, for one is a Fourier transformation of the other. We take as an example the stretching motion of a covalent bond. This vibrational mode occurs in every unit cell throughout the entire crystal, although it does not necessarily occur with the same relative phase in each case. The phase shift is defined by the wave vector, as shown by Eq. (10.6). At the long wavelength limit, $\boldsymbol{k} \to 0$ (and hence $\lambda \to \infty$), the covalent bond in all unit cells oscillates in perfect phase with each other. In three dimensions this corresponds to the coordinate point (0,0,0) in the Brillouin zone. This special k-point position is referred to as the Γ-point or zone-centred point, and expresses the permitted modes of vibration under long-range order (or coherent) conditions. This corresponds to the description of the phonon modes determined experimentally by, e.g., infrared and Raman spectroscopies. At the other extreme, in the short wavelength limit, $\boldsymbol{k} = \pm\frac{\pi}{a}$ (*i.e.*, corresponding to a real-space wavelength of $\lambda = 2a$), the stretching mode has a positive phase in one unit cell and a negative phase in the immediate neighbor unit cell. Hence, each unit cell vibrates out of phase with the next. By convention, each point in the first Brillouin zone is denoted by its fractional translation along the reciprocal lattice vector ($\boldsymbol{G} = \frac{2\pi}{a}$). In this notation, a k-point on the surface of the first Brillouin zone (i.e., at $\boldsymbol{k} = \frac{\pi}{a}$) is denoted as $\boldsymbol{k} = \frac{1}{2}$. This notation offers a convenient real-space interpretation where the k-point (½,½,½) thus corresponds to a real-space supercell representation of $2\boldsymbol{a} \times 2\boldsymbol{b} \times 2\boldsymbol{c}$, where $\boldsymbol{a}$, $\boldsymbol{b}$, and $\boldsymbol{c}$ are the crystallographic unit cell axes. Hence the general point $k = (1/h, 1/k, 1/l)$ is defined by real space supercells of $h\boldsymbol{a} \times k\boldsymbol{b} \times l\boldsymbol{c}$. To measure the modes of vibration experimentally at these non Γ-point positions requires specialized techniques, such as inelastic neutron or X-ray scattering.

Time for an example. The vibrational spectrum for the energetic material HN_3 is shown for the range 0–700 cm^{-1} in Fig. 10.1 in the form of the phonon dispersion curves (i.e., showing

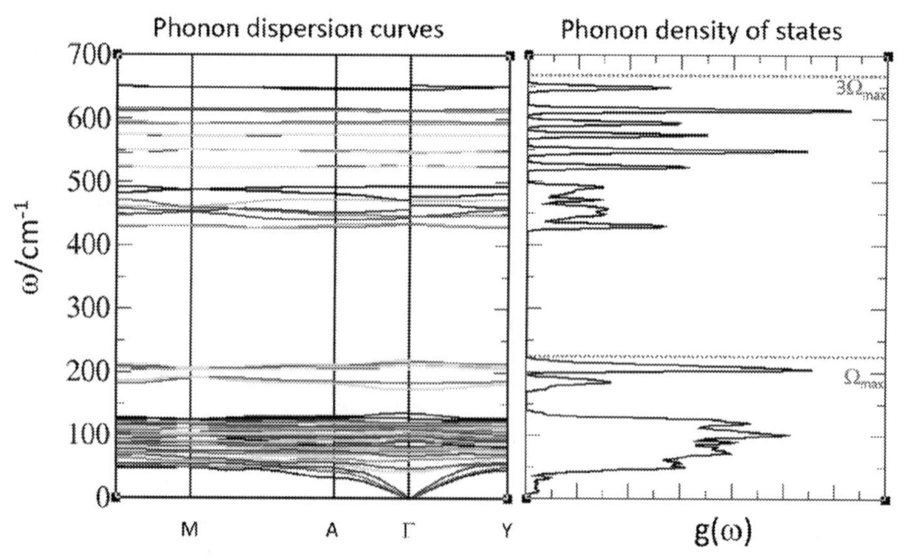

FIG. 10.1 Calculated phonon dispersion curves and density of states (0–700 cm^{-1}) for the energetic material HN_3.

the dependency of ω on $\boldsymbol{k}$) and the corresponding phonon density of states (obtained as an integration of ω over $\boldsymbol{k}$). Considering the dispersion curves first, the Latin characters along the x-axis denote the k-point positions on the surface of the Brillouin zone, whereas Greek letters are used to denote special positions within the Brillouin zone. Point M refers to the k-point (½,½,½) discussed above; A = (½,0,0), and therefore captures the phonons expressed by a $2\boldsymbol{a} \times \boldsymbol{b} \times \boldsymbol{c}$ supercell representation, while Y equates to (0,½,½), i.e., $\boldsymbol{a} \times 2\boldsymbol{b} \times 2\boldsymbol{c}$ supercell representation. Note for all k-points on the Brillouin zone surface, all vibrational bands conform to $\partial\omega/\partial k = 0$. This confirms the standing wave status at these special positions. The three acoustic bands, for which $\omega \rightarrow 0$ as $\boldsymbol{k} \rightarrow 0$ (i.e., as they approach the Γ-point) are also readily identifiable.

The phonon density of states (PDOS) show one very important feature: phonon bands become sharper (i.e., $\boldsymbol{k}$ invariant) as ω increases. This implies that the low-frequency bands are much more dependent on the crystal packing than the high frequency bands. These low-frequency bands, denoted below the dotted line labeled Ω_{max}, are termed the lattice (or external, or bath) modes. Here molecules move as semi-rigid bodies, and it is the variation in the space *between* the molecules that induces the oscillations that are expressed on the graph. These modes are therefore, by definition, delocalized. The sharper modes that appear above Ω_{max} are the molecular modes. Here, the molecules twist, bend, and stretch, and the length scale of the vibration reduces until ultimately it resides on just pairs of atoms described by the high frequency bond stretching motions. Any dependency on molecular orientation or crystal packing is now lost, and the phonon dispersion curves flat-line.

2.2 The mechanochemistry of impact excitation

As a mechanical impact travels through a crystalline material, it induces dynamical compression that ultimately leads to crystal fracturing. These events are relatively slow, however; the first response is felt by the vibrational modes in the crystal lattice, and occurs on the order of picoseconds [1,2]. The impact energy is initially adsorbed by the Γ-point acoustic modes as a compression wave, and is rapidly equilibrated across all bands below Ω_{max}. As a result, the phonon bath reaches quasi-temperatures of the order of thousands of degrees prior to thermalization.

If the material behaved as an harmonic solid the impact energy would remain locked in the acoustic branches indefinitely; we will see in the next section that anharmonic effects provide pathways for energy dissipation, and that the direction of energy travel must be up through the vibrational states. The critical window for the so-called vibrational up-pumping effect is the Ω_{max} to $3\Omega_{max}$ window, as defined by the first-order anharmonic approximation limit (see Fig. 10.2). The modes of vibration which lie at the bottom of this limit (i.e., below $2\Omega_{max}$) are generally comprised of hybrid modes of mixed molecular and external nature. Correspondingly, they act as excellent conduits for the conversion of lattice vibrational energy into molecular vibrational energy at the upper edge of this limit (from $2\Omega_{max}$ to $3\Omega_{max}$). Modes ranging from Ω_{max} to $2\Omega_{max}$ are aptly termed the doorway modes, denoted Q_D. Within the region of molecular vibrational modes ($2\Omega_{max}$ to $3\Omega_{max}$) a special subset requires particular attention; these are the target frequencies, denoted Q_T. For energetic materials these will likely involve oscillations of N-N or C-N type bonds. As we will discuss in a later section, under

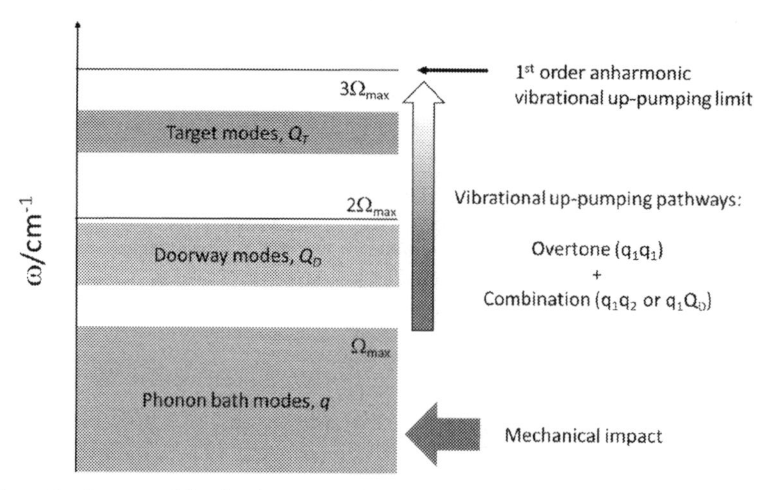

FIG. 10.2 Schematic diagram of the shock wave excitation and the vibrational up-pumping process.

heightened vibrational excitation these modes induce changes in electronic structure that ultimately trigger bond rupture. Note for the schematic diagram shown in Fig. 10.2 the Q_T modes sit below the $3\Omega_{max}$ upper limit; any crystal lattice displaying a PDOS conforming to this vibrational distribution is likely to be sensitive to impact initiation. The reverse phenomenon is also noteworthy: should Q_T sit above $3\Omega_{max}$ the molecular crystal will likely be insensitive to impact initiation.

2.3 Phonon scattering

While restricted by the constraints of the harmonic model, the normal vibrational modes remain entirely orthogonal and untouched by the effects of temperature. This means that excitation of one mode has no influence on the others. Introducing vibrational anharmonicity permits phonons to interact (scatter) with one another, and in doing so transfer energy from one vibrational state to another. These interactions change the wave vectors and energies of the phonon states, and ultimately leads to equilibration of the thermal energy in the system. If we think of phonons as quanta of energy then it is not too much of a conceptual stretch to appreciate that new phonon states can be created from old phonon states, and in that process the old phonon states are destroyed. Phonons, therefore, have finite lifetimes. From the perspective of impact energy dissipation, the more phonon states that exist in the lattice phonon bath and doorway regions, the more efficient the crystal lattice is at absorbing the shock wave energy and channeling it toward the molecular modes that trigger bond rupture and induce initiation. Thus, a correlation between the coupling of phonon states and impact sensitivity can be drawn.

From Eq. (10.1) it is clear that the anharmonic contributions to the total energy of the crystal become increasingly negligible as each higher order term is added to the Taylor expansion. As such, it is common to discuss only first-order anharmonicity, which accounts for all three-phonon collisions (i.e., the combination of two phonon states to create a third, or *vice versa*). This can be described by a general Hamiltonian:

$$H = H_q + H_Q + H_{q.Q} \tag{10.7}$$

where Eq. 10.7 describes the Hamiltonian for the low energy external lattice modes (H_q) and the higher energy internal molecular modes (H_Q) as unique entities. The third term refers to the conversion of energy between the two, and can be expanded by the Fermi golden rule formalism to:

$$H_{q.Q} = \left|V^{(3)}\right|\delta(\omega_1(\mathbf{k}) - \omega_2(\mathbf{k}') - \omega_3(\mathbf{k}'')) \tag{10.8}$$

Here, $V^{(3)}$ is the cubic anharmonic coupling constant, which describes the strength of interactions between the three vibrational modes. While this is a system specific term, recent work using high-resolution Raman spectroscopy and computational modeling has yielded broadly similar values for a range of energetic materials, including HMX, TATB, RDX and PETN [3–5]. It is therefore a reasonably safe assumption to omit $V^{(3)}$ in any further discussion. The remainder of Eq. (10.8) describes the so-called two-phonon density of states (2phonDOS) and offers a count of the number of coupling pathways that link the higher frequency mode (ω_1) to the two lower frequency modes (ω_2 and ω_3). The scattering pathways that are available are restricted by conservation of energy and momentum, such that $\Delta \mathbf{k} = 0$ and $\Delta\hbar\omega = 0$. Correspondingly, two pathways become available for phonon scattering, namely (i) a high-energy phonon decomposing into two lower energy phonons or (ii) two low-energy phonons coupling to form a higher energy phonon. The first process is termed vibrational cooling, and is a phenomenon that was well established both experimentally and theoretically in the late 20th century [6]; the second is vibrational up-pumping, and is the mechanism by which the impact energy, absorbed by the acoustic phonon modes, will be transferred to the localized molecular modes to trigger the initiation event. This was first proposed by Coffey and Toton [7] in an attempt to describe the processes occurring immediately behind a shock wave propagating through a material and was subsequently developed further by Dlott and Fayer [2,8].

Through Eq. (10.8), we make the distinction between two different scattering processes. In the first, ω_2 and ω_3 originate from the same phonon branch. We borrow the spectroscopic term overtone to describe this process. Similarly, where ω_2 and ω_3 do not come from the same branch index—i.e., are two different vibrational modes—we borrow the spectroscopic term combination.

2.4 From vibrational excitation to electronic excitation

Before we explore the vibrational up-pumping model in more detail, let us skip ahead and assume that the impact energy has been successfully channeled through the PDOS by phonon scattering until it reached the target frequencies. How does vibrational energy illicit a chemical (electronic) response, which for energetic molecules will lead to initiation? As mentioned above, initiation is likely to follow rupture of N-N or C-N bonds, and so presumably follows excitation of normal modes comprising oscillation of these bonds. These oscillations reduce the energy separation of the frontier electronic states, such that bond order is reduced and athermal bond dissociation follows [9].

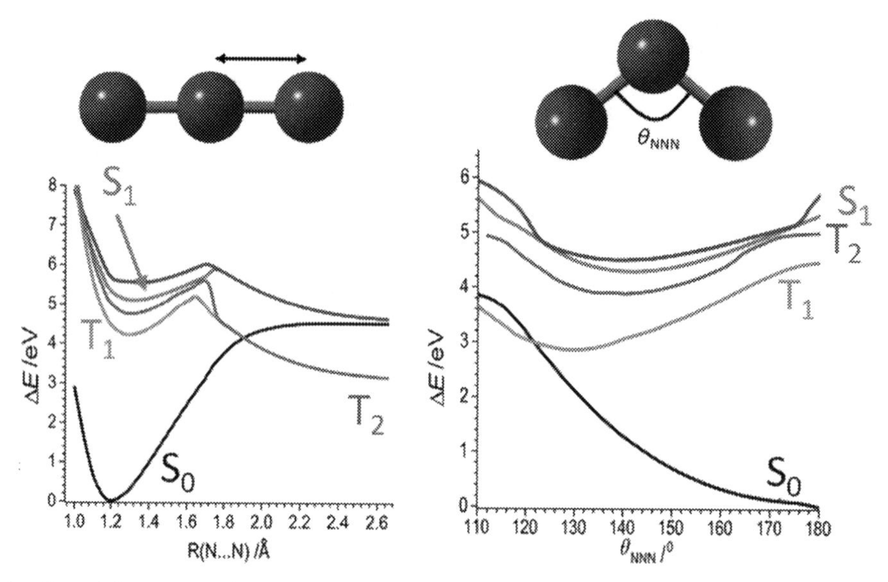

FIG. 10.3 Potential energy surfaces (S_0 black, S_1 red, S_2 blue, T_1 pink and T_2 green) for the N_3^- anion (linear geometry on the left, bent geometry on the right), derived from MRCI calculations. *Adapted from Ref. Michalchuk, A. A. L.; Fincham, P. T.; Portius, P.; Pulham, C. R.; Morrison, C. A. A pathway to the Athermal impact initiation of energetic Azides. Journal of Physical Chemistry C 2018, 122 (34), 19395–19408. https:/doi.org/10.1021/acs.jpcc.8b05285, with permission American Chemical Society.*

To illustrate how vibrational excitation leads to bond breaking, we turn to the structural simplicity of the azide anion, N_3^-. Calculated potential energy surfaces (determined by multireference configuration interaction (MRCI) methods, reported in full elsewhere) are shown in Fig. 10.3 [9]. In the ground (S_0) state, N-N bond dissociation for the equilibrium linear geometry is achieved at $R > 2$ Å, but the barrier to dissociation (*ca.* 4.5 eV) is significant and the process is endothermic, meaning there is an absence of energetic driving force to achieve dissociation. Within the confines of the harmonic approximation, a potential well with such high curvature will have widely spaced vibrational quantum energy levels, and it is highly unlikely that the higher vibrational levels, producing the most pronounced vibrational amplitudes $\tilde{u}_k$, will ever be reached to allow the dissociation barrier to be breached.

If direct bond dissociation from the S_0 surface is unlikely, is dissociation from an electronically excited potential energy surface more likely? Promoting N_3^- to the first electronically excited state (the T_1 surface) does appear to stack the odds more in our favor: bond dissociation occurs at a shorter distance ($R = 1.75$ Å) and the barrier drops to *ca.* 1 eV. In fact, the same trend holds for any of the excited state surfaces. However, the energy required to leave the ground state surface, at *ca.* 4.2 eV, is still considerable and remains unaccounted for. As all covalent bond potential energy surfaces tend to have high curvatures and steep sides, by inference we can deduce that pure bond stretching modes in general make unlikely candidates for Q_T modes. Note also that pure bond stretching vibrations are generally the higher frequency modes, and are therefore more likely to fall outside the Ω_{max} to $3\Omega_{max}$ first-order anharmonic vibrational up-pumping window, further compounding their unsuitability for Q_T assignment.

If bond dissociation cannot be achieved by direct vibrational excitation of the N_3^- stretching modes, we must look to other, lower energy, molecular vibrations to bypass the dissociation barrier. The lowest frequency molecular vibration is the most susceptible to mechanical perturbation. For N_3^- this is the θ_{NNN} angle bend. Deviating θ_{NNN} from 180° causes the ground state energy S_0 to increase until a plateau is reached at *ca.* 110° (see Fig. 10.3). In contrast, all excited state potential energy surfaces stabilize as the molecule bends. This effect is most pronounced for the lowest energy T_1 state, which in fact forms a conical intersection with the S_0 surface when $\theta_{NNN} =$ *ca.* 120°. Thus, once the molecule is bent the conical intersection provides a route to access the T_1 surface; from there dissociation of the N-N bond is energetically favored and is accompanied by a significantly lower energy barrier.

Up to this point, the discussion has focused on the isolated N_3^- anion, but the same effect has also been verified in the condensed phase by density functional theory (DFT) simulations [10]. For the energetic azide NaN_3 the conventional unit cell setting supports a total of 24 (16 q and 8 Q) vibrational modes. This unit cell setting permits study of the edge of the first Brillouin zone, where the motions of the atoms in the neighboring primitive unit cells are out-of-phase standing waves, as well as the centre of the Brillouin zone where the motions of the atoms are in-phase standing waves. Distorting (or "mode-walking") along each of the eigenvectors, and calculating the associated electronic band structures, revealed that the θ_{NNN} angle bend vibrations are the only viable modes to close the electronic band gap (see Fig. 10.4). The implication is clear: if the vibrational energy transfer from an adsorbed compressive wave excites

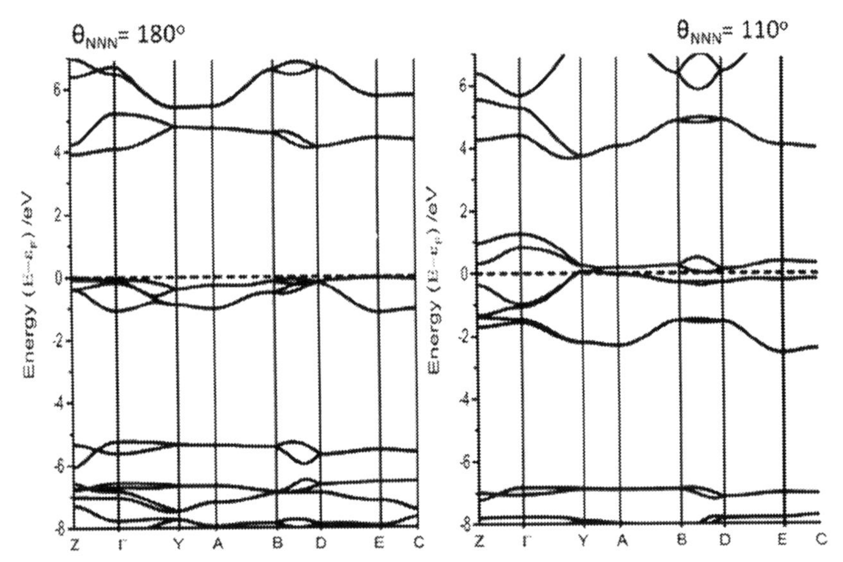

FIG. 10.4 Simulated electronic band structures for NaN_3, with $\theta_{NNN} = 180°$ and 110°, demonstrating band gap closing. *Adapted with from Ref. Michalchuk, A. A. L.; Rudić, S.; Pulham, C. R.; Morrison, C. A. Vibrationally induced metallisation of the energetic Azide α-NaN3. Physical Chemistry Chemical Physics 2018, 20 (46), 29061–29069. https:/doi.org/10.1039/C8CP06161K, with permission from Royal Society of Chemistry.*

the θ_{NNN} low energy modes, to the extent that $\tilde{u}_k$ distorts the linear geometry by 70°, NaN_3 will change from being an insulator to a metal. The effect is transient, lasting a matter of femtoseconds, until the eigenvector oscillation restores the linear geometry and the bandgap reopens. But this short time period is more than enough for electrons to flow and unstable reactive intermediates to form, which will trigger the initiation process. The term "dynamical metallization" has been adopted to describe this phenomenon [10].

Note the electronic structure of the N_3^- anion upon perturbation in the crystal lattice contrasts markedly with that observed for the isolated (gas phase) anion in one crucial respect. The band structures shown in Fig. 10.4 arise from closed-shell simulations, and therefore suggest a S_0/S_1 metallization channel. The multireference calculations reported for the isolated anion could only locate a S_0/T_1 conical intersection, i.e., a spin "forbidden" channel. Thus the periodicity of the crystalline lattice has opened up a more probable singlet channel for electronic excitation. The MRCI calculations reported in Fig. 10.3 suggest the barrier to N-N bond dissociation on the S_1 surface is very similar to the barrier on the T_1 surface, and so a low energy route to bond rupture on this alternative excited state surface can still be accessed.

In the case of the energetic azides, the structural simplicity of the energetic N_3^- moiety meant that it was a relatively straightforward process to identify the Q_T modes as the θ_{NNN} angle bends. Moreover, they are likely to occupy a distinctly separate band of vibrational modes in the PDOS. For HN_3 shown in Fig. 10.1 these are the bands in the 425–500 cm^{-1} region. For other energetic molecules, such as ring compounds, however, it is highly unlikely that the Q_T modes will be vibrationally distinct. Moreover, in more complex systems it is probable that the molecular deformations that induce band gap closing will be manifest through a number of vibrational modes. As such, with increasing molecular complexity, the task of identifying which of the Q modes should be labeled as Q_T becomes an increasingly challenging and futile process. To that end, the task can be simplified enormously by considering all internal modes within the Ω_{max} to $3\Omega_{max}$ 1st order anharmonic approximation limit to be acting as target frequency modes. Correspondingly, from here on the term Q_T will be dropped in favor of vibrational excitation of all of the more broadly defined internal modes.

2.5 The vibrational up-pumping process

The probability of phonon scattering events is proportional to the number of phonon states that are available to scatter. While the anharmonic coupling constant $V^{(3)}$ introduced in Eq. (10.8) is itself omitted from any numerical calculation, it is known to decreases across the three-phonon collision series as $qqq > qqQ > qQQ > QQQ$, where the nomenclature denotes the first two phonons combining to create the third, higher energy phonon [11]. The qqq collisions therefore represent the effectively instantaneous equilibration of the phonon bath, following energy deposition from the impact into the three acoustic modes. The vibrational up-pumping process will therefore be dominated by qqQ collisions. Scattering from qQQ events will also have an important role to play, for while their probabilities are lower, any scattering will result in an energy transfer "boost" as the newly created Q state will be of higher frequency than can be reached by q mode scattering alone. The probability of QQQ collisions is too low to consider any further. Given the observation that Ω_{max} generally sits around 200 ± 50 cm^{-1} for molecular crystals, the limit imposed by the first-order anharmonic approximation sets the Ω_{max} to $3\Omega_{max}$ up-pumping window at *ca.* 200 ± 50 to 600 ± 150 cm^{-1}.

At this stage, it is timely to outline briefly the historical narrative that has shaped our understanding to this point. Much of the early work in this field (performed in this 1990s by Fried and Ruggiero [12], McNesby and Coffey [13]) relied upon vibrational spectra obtained from inelastic neutron scattering or Raman spectroscopy. A universal value of $\Omega_{max} = 200$ cm^{-1} was commonly adopted [12]. Despite the limitations in data resolution and maximum measurable energy transfer (initially limited to 600 cm^{-1}, but subsequently pushed out to 700 cm^{-1}) [13], promising correlations were observed. While variations in the details of the vibrational up-pumping models and the phonon scattering pathways were explored, the importance of energy transfer from the q-manifold to the Q_D region was clear, a phenomenon echoed by Koshi [14] who reported the first attempts to employ computational modeling in this field. Using classical (force field) methods, Koshi simulated PDOS and 2phonDOS for a range of energetic molecules and established a clear connection between the rate of energy transfer through the vibrational states with impact sensitivity. Later work also highlighted the importance of the Q_D region; by simulating the gas-phase (i.e., isolated molecule) vibrational spectra using DFT methods, and simply counting the vibrational frequencies that fell in the 200–700 cm^{-1} window, a correlation with measured impact sensitivities was observed [5]. Bernstein recently expanded this work by calculating the Γ-point PDOS for the crystal lattice directly using DFT, which greatly improved the data resolution as compared to the earlier experimental studies [15]. Bernstein's model considered overtone phonon scattering only, but included projections up to the 10th order in order to purposefully capture all overtone states that fall within the window $200 < \omega < 700$ cm^{-1}. Our own contributions in the field builds on these developments, and the basis for our vibrational up-pumping model is as outlined below:

1. Calculate a PDOS using DFT that is representative of the whole of the first Brillouin zone through multiple k-point sampling.
2. Use the resulting PDOS to set system-specific values for Ω_{max}. To achieve this we look for a gap in the PDOS over the region 200 ± 100 cm^{-1}, which marks the onset of the first set of Q_D modes. Visualization of the eigenvectors as expressed at the Γ-point also assist in deducing the switch-over between q and Q mode behavior.
3. Introduce shock temperature heating to thermally populate the q modes to 1000s of degrees and the Q modes to room temperature. This is achieved using the Bose-Einstein distribution, where the phonon population p at a given temperature T is expressed as:

$$p(\omega, T) = \frac{1}{\exp\left(\hbar\omega / k_B T\right) - 1} \tag{10.9}$$

4. Explore vibrational up-pumping processes according to first-order anharmonic principles, i.e., into the window Ω_{max} to $3\Omega_{max}$, *via* a two-stage process that is normalized with respect to the number of q modes and unit cell volume. In the first stage, the 2phonDOS arising from q_1q_1Q (*i.e.*, the overtone scattering pathways, capped at the physically realistic second-order limit) are calculated and projected onto the fundamental PDOS. This forms the set of phonons that subsequently scatter through the combination pathways, i.e., q_1q_2Q and q_1QQ.

5. Calculate the resulting 2phonDOS, project it onto the underlying fundamental PDOS bands, and integrate. This provides a numerical quantification of the transfer of vibrational energy through the crystalline lattice that can be compared directly with the experimental impact sensitivity measurement.

In this form, the up-pumping model takes only a crystallographic structure as input; no further experimental values are needed. The predictive power is derived solely from the comparison of the relative rates of vibrational energy transfer, which in turn depends on the molecular and crystallographic structure of the energetic material being considered. Hence, this model captures system-specific information at both the solid state and molecular levels. While we have shown that there is a direct link to be drawn between vibrational excitation and the electronic excitations that drive chemical decomposition, in the next section we will demonstrate that explicit consideration of the electronic excitations are probably not needed to gauge the relative mechanical reactivity of energetic materials. Consideration of the phonon scattering pathways alone is (probably) enough.

3 Predicting the impact sensitivities for a diverse range of energetic molecular crystals

In this section, we apply the theoretical model outlined above to predict the impact sensitivities of the five well-known energetic molecules HNB, β-HMX, α-FOX7, NTO and TATB, as

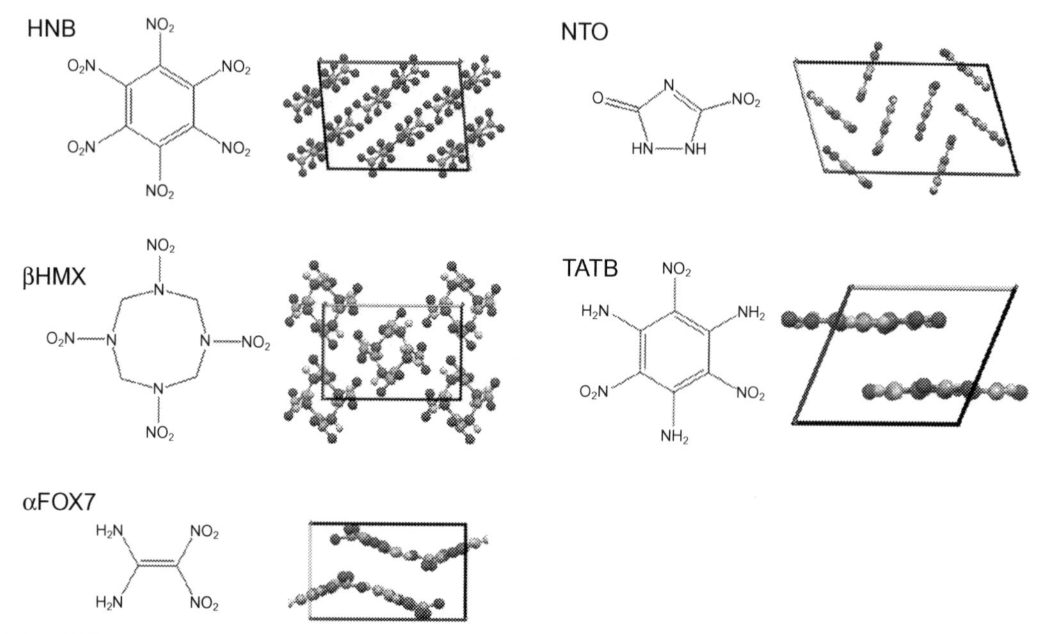

FIG. 10.5 Chemical structures and crystal structure packing diagrams of five structurally diverse energetic compounds.

shown in Fig. 10.5. This test set is not only structurally diverse, but also encapsulates a broad spectrum of material property response to mechanical impact. This ranges from the highly sensitive primary energetic HNB, through to the highly insensitive secondary energetic TATB. Reliable experimental impact sensitivity (E_{50} values) are available for all five compounds, allowing ranking against computational predictions. It therefore provides a suitable and challenging test set for the vibrational up-pumping model.

The process begins with calculating the full PDOS, across the first Brillouin zone, using first-principles simulation methods (typically dispersion-corrected plane-wave DFT). These simulations have been reported elsewhere for the five molecular crystal structures shown in Fig. 10.5 and the results are reproduced here in Fig. 10.6 [16]. Several important features of the PDOS plots warrant discussion first. The plots run (from top to bottom) in order of decreasing sensitivity to impact. The lower frequency vertical dotted lines denote Ω_{max}. Crystallographic unit cells that contain Z molecules will have three acoustic bands and $6Z-3$ external modes, meaning it should in principle be a straightforward process to simply count the expected number of external modes, and then draw the line to mark the external/internal distinction (this is known as the "rigid body approximation"). It is also reasonable to expect a gap to appear on the PDOS as the mode behavior changes from the broad ($\boldsymbol{k}$-dependent) q modes to the sharper ($\boldsymbol{k}$ -invariant) Q modes. However, floppy organic molecules may have low-frequency Q mode vibrations that fall at energies below those of the highest-frequency q modes, i.e., below Ω_{max}. These vibrations are termed amalgamated modes, and for energetic molecules will often involve $-NO_2$ motions. Within the context of the model developed here, Ω_{max} is therefore defined as the highest frequency band above the $6Z$ external mode limit that cannot be distinguished from the other lattice modes by a gap in the PDOS.

The values obtained for Ω_{max} vary over a range from 135 cm^{-1} (HNB) to 200 cm^{-1} (NTO) (see Table 10.1 and Fig. 10.6). For all molecular crystals bar TATB the top of the phonon bath comprises $-NO_2$ amalgamated motions. For TATB these motions appear at a higher wavenumber in the doorway region (*ca.* 290 cm^{-1}), which suggests a functional group flexibility that is not shared by the least sensitive energetic compound. No obvious correlation between Ω_{max} and the resulting impact sensitivity order is observed. This appears counter-intuitive, as a higher Ω_{max} value by definition defines a larger phonon bath region, which in turn extends the second overtone population limit and increases the size of the vibrational up-pumping window (both given by $3\Omega_{max}$); Ω_{max} should therefore be classed as an indirect design parameter to control the impact sensitivity response.

While a clear correlation between Ω_{max} and impact sensitivity ordering is not readily forthcoming, what is much more visually apparent from Fig. 10.6 is a clear connection between the density of Q_D states and the impact sensitivity response. This observation has been noted several times in the literature already [14,15]. While HNB has a narrow doorway region (directed by its low value for Ω_{max}) it is densely populated. Contrast this with the sparsely populated, albeit wider, doorway region for the structurally similar TATB, and a sense for why one structure can rapidly transfer the impact energy, while the other cannot, becomes clear. For HNB the doorway density is dominated by $-NO_2$ wagging motions, ring deformation modes and $C-NO_2$ stretching modes; the corresponding modes for TATB all occur at higher frequencies as the $-NO_2$ groups are pinned into the plane of the aromatic ring through internal hydrogen bond interactions with the neighboring $-NH_2$ groups (see Fig. 10.5). This molecular stiffening places these modes beyond the doorway region, and consequently they are less likely to be

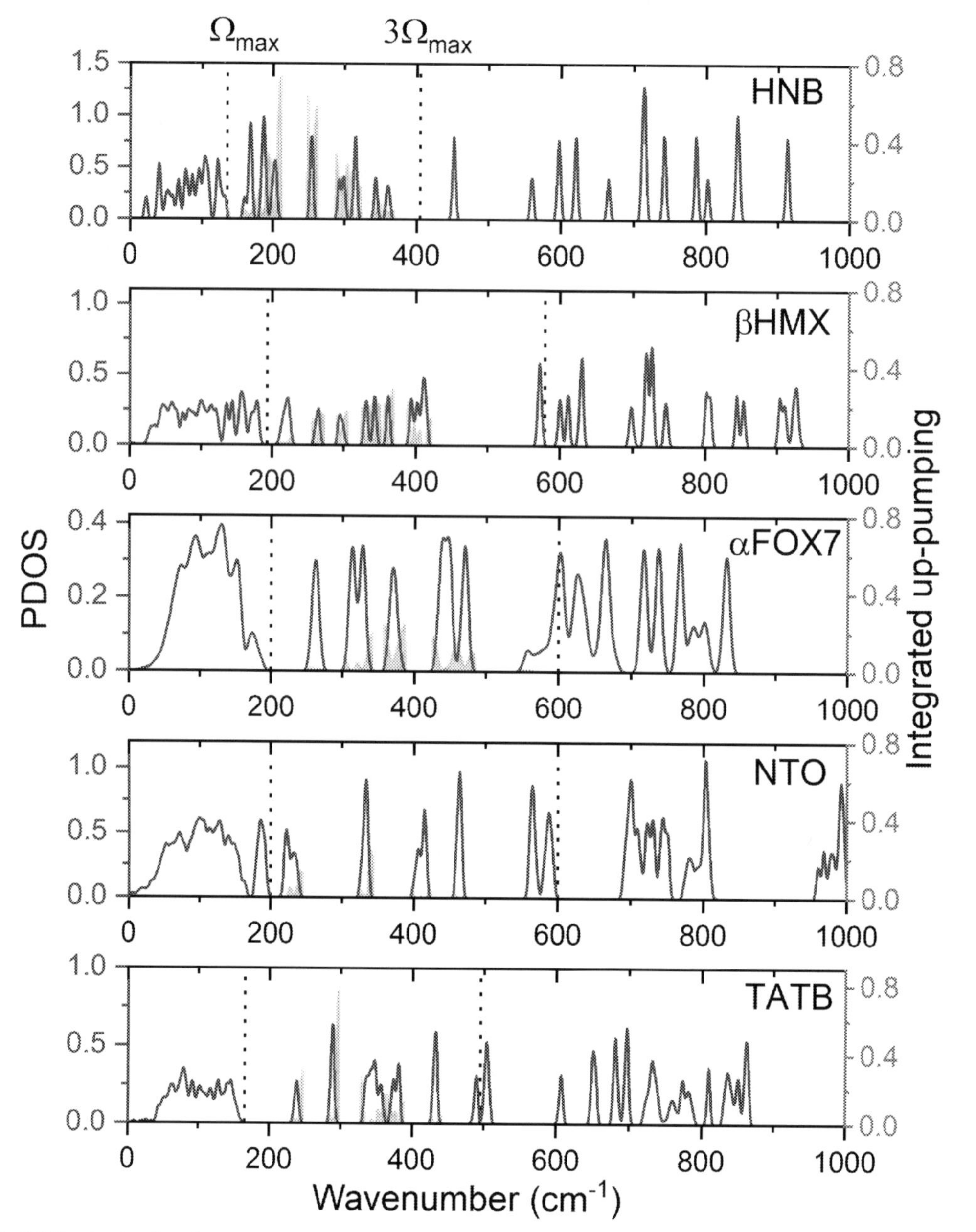

FIG. 10.6 Stack plot showing calculated phonon density of states (PDOS, black) and integrated vibrational up-pumping (red shading) for HNB, β-HMX, α-FOX7, NTO and TATB, arranged in order of decreasing sensitivity to impact. The top of the phonon bath (Ω_{max}) is marked by a dotted vertical line (left hand side marker). The vibrational up-pumping region lies between Ω_{max} and $3\Omega_{max}$ (vertical dashed line).

TABLE 10.1 Parameters employed in the two-stage vibrational up-pumping model.

Energetic crystal	Ω_{max}/cm^{-1}	Z	Y	No. External modes (< Ω_{max})	*Scaled* C_{ph}[a]	Scaled T_{shock}/K[b]
HNB	135	2	18	30	0.6	5000
β-HMX	193	4	10	34	0.68	4410
α-FOX7	200	4	10	34	0.68	4410
NTO	200	8	16	64	1.28	2343
TATB	160	2	12	24	0.48	6250

[a] *Arbitrary scale 50 external modes: 1.0* C_{ph}.
[b] *Arbitrary scale 1.0* C_{ph}*: 3000 K* T_{shock}.

reached by the vibrational up-pumping energy. Thus, while structural rigidity acts to pull amalgamated modes into the doorway region, it shifts the other low-lying internal Q_D modes to higher wavenumbers. This overall depopulation of the doorway density region results in a breakdown in the channeling of vibrational energy toward the molecular modes. For the other three more structurally diverse molecules represented in Fig. 10.6 the correlation holds: a lower Q_D vibrational density of states results in a less sensitive molecular crystal.

This qualitative assessment of the PDOS plots can be converted into a quantitative assessment through application of the vibrational up-pumping mode described earlier in this chapter. Key parameters and other data utilized in the up-pumping model are summarized in Table 10.1. Alongside the values adopted for Ω_{max}, a value for the q mode heating must be deduced. The phonon heat capacities (C_{ph}) quoted are derived from the Einstein model, which equates to roughly $6k_B$ per molecule in the primitive unit cell. Due to the amalgamation of some molecular modes (Y) into the phonon bath the approximation $C_{ph} \approx 6k_B Z + Y$ holds. Defining the heat capacity in this way reflects the number of modes that should be considered to reside within the phonon bath. As it is reasonable to assume that the compressibility and anharmonicity of the test set systems will be similar, it is appropriate to use the number of external modes to express a scaled C_{ph} value, and in turn use this to deduce a scaled shock temperature, T_{shock}. In our previous report that encompassed a larger test set of compounds, we selected the arbitrary scaling that 50 phonon bath modes would equate to a scaled C_{ph} of 1.0, which we set to a temperature equivalent of $T_{shock} = 3000$ K. This temperature was chosen as a roughly representative value of phonon bath heating following impact excitation. Note the actual values used are not critically important; rather, it is the spread of values that will act to differentiate the impact response for the different crystal lattices. The remaining molecular modes are then populated according to the Bose Einstein distribution at room temperature (i.e., 300 K).

The results obtained from the two-stage vibrational up-pumping model are shown in Fig. 10.6 (as the up-pumped density) and Fig. 10.7 (as the integrated results). Considering the up-pumped density first, represented as the red shaded regions, the role played by the Q_D modes is clear, as they trap the 2phonDOS density arising from the qqQ (and qQQ) couplings. The 2phonDOS decays rapidly after the $2\Omega_{max}$ limit, as this represents the upper boundary for qq scattering, which is massively activated by the Bose Einstein T_{shock} population levels. Low lying Q_D modes, sitting just above Ω_{max} therefore channel more impact energy

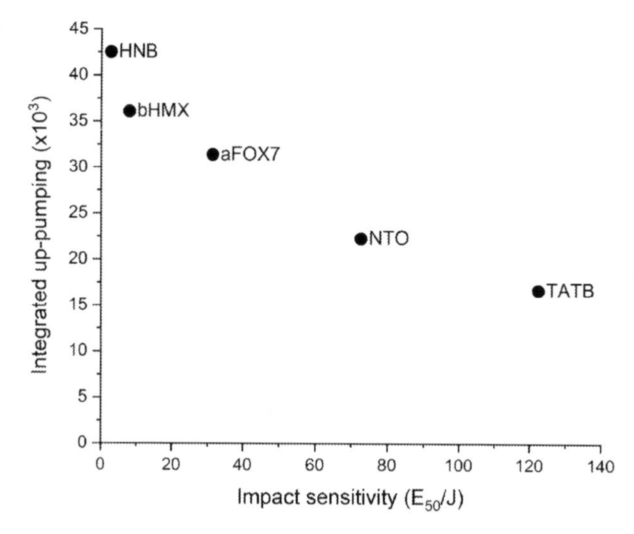

FIG. 10.7 Plot of (experimental) impact sensitivity vs (simulated) integrated vibrational up-pumping, derived from the first-principles two-stage vibrational up-pumping model.

than higher energy Q_D modes, and therefore directly influence the impact sensitivity behavior of the molecular crystal. The vibrational sparsity of TATB in this region, and to a lesser extent NTO and α-FOX7 therefore render these materials less amenable to initiation by mechanical impact. Integrating the red shaded areas over the Ω_{max} to $3\Omega_{max}$ up-pumping window gives the final outcome shown in Fig. 10.7, where the computed up-pumping data is plotted directly alongside the accepted literature values for experimental impact sensitivities. All five compounds are unambiguously ranked in order of increasing impact sensitivity, paying testimony to the two-stage vibrational up-pumping model capturing enough of the underlying condensed matter physics to predict the correct outcome, without recourse to any fitting parameters.

4 Discussion and future outlook

Making new energetic materials is extremely hazardous work. Aside from the standard risks of chemical synthesis, the product the chemist has made in the bottom of their flask has the added dimension that it may (or may not) explode on the slightest touch. The safety tests required to ascertain whether the crystals are (or are not) safe to handle require considerable scale up, or multiple repeats, of the synthetic procedure. In short, this is not work for the faint-hearted. Would it not be better if we could design and test energetic materials on a computer, and then only make the compounds we want?

The vibrational up-pumping model outlined in this chapter has the potential to render this thought a reality. Impact sensitivity is a predictable quantity, and as the predictive model is free from any fitting parameters, it can be applied with just as much confidence to any energetic material, provided the crystal structure is known. This last step can be overcome too, without recourse to returning to the laboratory, by making use of the increasingly robust tools being developed by the crystal structure prediction community. The reality of predicting a material property response, starting with nothing more than a molecule drawn on a page, is with us now.

The up-pumping model goes beyond simply reproducing known experimental data: it also provides a feedback loop to rationalize energetic material design. We have seen in the previous section the importance of the Ω_{max} to $3\Omega_{max}$ up-pumping window. Populating that window with a high density of low-lying doorway modes creates an efficient vibrational transfer conduit to trap the energy arising from the many lattice mode collisions into the molecular vibrations. The ensuing large amplitude vibrations distort the molecular structure to the degree that electronic changes occur. Band gaps narrow, electrons flow, and unstable species emerge on the time scale of a molecular vibration. Thus, a crowded doorway phonon density of states signifies a molecular crystal that is highly shock sensitive. The flip side, having a vibrationally sparse doorway region, or pushing the doorway modes present up to higher wavenumbers, acts to break up the vibrational energy transfer. The efficiency with which the impact energy is transferred into the molecular vibrations is now compromised; thus crystal lattices with sparsely populated doorway regions are insensitive to shock wave initiation.

The doorway modes of vibration are molecular vibrational modes, and so influencing their vibrational distribution is firmly in the control of the chemist. Floppier molecules, with e.g., dangling -NO_2 groups, will have lower lying doorway modes. Stiffer molecules, constrained by internal hydrogen bonds or more double bonds, will have higher lying doorway modes. While this is a very simplistic interpretation of what remains a complex phenomenon (the factors driving the location of Ω_{max}, and the influences of polymorphism still need to be explored, for example), having a predictable model gives the chemist a powerful new angle to explore and rationalize energetic material design. These developments are welcomed by a community dealing with inherently hazardous materials.

References

[1] H. Kim, D.D. Dlott, Theory of Ultrahot molecular solids: Vibrational cooling and shock-induced multiphonon up pumping in crystalline naphthalene, The Journal of Chemical Physics 93 (3) (1990) 1695–1709. https://doi.org/10.1063/1.459097.

[2] D.D. Dlott, M.D. Fayer, Shocked molecular solids: Vibrational up pumping, defect hot spot formation, and the onset of chemistry, The Journal of Chemical Physics 92 (6) (1990) 3798–3812. https://doi.org/10.1063/1.457838.

[3] S.D. McGrane, A.P. Shreve, Temperature-dependent Raman spectra of Triaminotrinitrobenzene: Anharmonic mode couplings in an energetic material, The Journal of Chemical Physics 119 (12) (2003) 5834–5841. https://doi.org/10.1063/1.1601601.

[4] S.D. McGrane, J. Barber, J. Quenneville, Anharmonic vibrational properties of explosives from temperature-dependent Raman, The Journal of Physical Chemistry. A 109 (44) (2005) 9919–9927. https://doi.org/10.1021/jp0523219.

[5] Ye; Koshi, M., Theoretical studies of energy transfer rates of secondary explosives, The Journal of Physical Chemistry. B 110 (37) (2006) 18515–18520. https://doi.org/10.1021/jp062815l.

[6] J.R. Hill, E.L. Chronister, T. Chang, H. Kim, J.C. Postlewaite, D.D. Dlott, Vibrational relaxation and vibrational cooling in low temperature molecular crystals, The Journal of Chemical Physics 88 (2) (1988) 949–967. https://doi.org/10.1063/1.454175.

[7] C.S. Coffey, E.T. Toton, A microscopic theory of compressive wave-induced reactions in solid explosives, The Journal of Chemical Physics 76 (2) (1982) 949–954. https://doi.org/10.1063/1.443065.

[8] A. Tokmakoff, M.D. Fayer, D.D. Dlott, Chemical reaction initiation and hot-spot formation in shocked energetic molecular materials, The Journal of Physical Chemistry 97 (9) (1993) 1901–1913. https://doi.org/10.1021/j100111a031.

[9] A.A.L. Michalchuk, P.T. Fincham, P. Portius, C.R. Pulham, C.A. Morrison, A pathway to the Athermal impact initiation of energetic Azides, Journal of Physical Chemistry C 122 (34) (2018) 19395–19408. https://doi.org/10.1021/acs.jpcc.8b05285.
[10] A.A.L. Michalchuk, S. Rudić, C.R. Pulham, C.A. Morrison, Vibrationally induced metallisation of the energetic Azide α-NaN $_3$, Physical Chemistry Chemical Physics 20 (46) (2018) 29061–29069. https://doi.org/10.1039/C8CP06161K.
[11] D.D. Dlott, Multi-phonon up-pumping in energetic materials. In *advanced series in physical chemistry*, World Scientific 16 (2005) 303–333. https://doi.org/10.1142/9789812775283_0010.
[12] L.E. Fried, A.J. Ruggiero, Energy transfer rates in primary, secondary, and insensitive explosives, The Journal of Physical Chemistry 98 (39) (1994) 9786–9791. https://doi.org/10.1021/j100090a012.
[13] K.L. McNesby, C.S. Coffey, Spectroscopic determination of impact sensitivities of explosives, The Journal of Physical Chemistry. B 101 (16) (1997) 3097–3104. https://doi.org/10.1021/jp961771l.
[14] S. Ye, Energy transfer rates and impact sensitivities of crystalline explosives, Combustion and Flame 132 (1–2) (2003) 240–246. https://doi.org/10.1016/S0010-2180(02)00461-3.
[15] J. Bernstein, *Ab initio* study of energy transfer rates and impact sensitivities of crystalline explosives, The Journal of Chemical Physics 148 (8) (2018), 084502. https://doi.org/10.1063/1.5012989.
[16] A.A.L. Michalchuk, M. Trestman, S. Rudić, P. Portius, P.T. Fincham, C.R. Pulham, C.A. Morrison, Predicting the reactivity of energetic materials: An *ab initio* multi-phonon approach, Journal of Materials Chemistry A 7 (33) (2019) 19539–19553. https://doi.org/10.1039/C9TA06209B.

CHAPTER

11

Role of electronic excited states in the initiation of explosives

Didier Mathieu[a,*], *Romain Claveau*[a,b], *and Julien Glorian*[b]
[a]CEA, DAM, Le Ripault, Monts, France [b]French-German Research Institute of Saint-Louis, Saint-Louis, France
*Corresponding author: E-mail: didier.mathieu@cea.fr

1 Introduction

Throughout this chapter devoted to the relevance of electronic excited states regarding the initiation of explosives, terms like "state" or "excitation" will refer to electrons, unless mentioned otherwise. Besides the term "initiation," which refers to the occurrence of an explosion, the term "ignition" will be used to refer to the onset of decomposition reactions [1]. Although the two terms are sometimes used interchangeably, this distinction is crucial in the present context.

There is little doubt that excited electrons are involved in the explosion of an energetic material (EM), as evidenced by the visible glow reflecting their relaxation to lower states. A more difficult question, widely debated over decades of research, is the extent to which they contribute to EM initiation and affect sensitivities to various stimuli. In typical high explosives, breaking a trigger bond requires an energy of about 2 eV, whereas promoting an electron across the gap requires even more energy, usually well above 3 eV. Therefore, focusing on the ground state sounds a reasonable approximation, supported by the widespread belief that initiation is thermal in origin [2–6]. In this case, excited electrons should play no significant role under most stimuli, which primarily heat up the nuclei rather than the electrons. This is the case not only for mechanical or thermal loads but also for laser initiation, unless the photon energies are so high as to promote electrons into excited states. However, such stimuli do not just heat up the nuclei. For instance, the primary effect of a mechanical impact or shock is to squeeze the material. This could hamper chemical reactions so that electronic excitations

Theoretical and Computational Chemistry, Volume 22
https://doi.org/10.1016/B978-0-12-822971-2.00018-8

might precede chemical decomposition as the first step of the mechanical response. Accordingly, the possible role of excited electrons on ignition and initiation has been extensively studied [7–9].

Nonetheless, the literature on the subject may still seem rather confusing to newcomers. Some authors consider that experimental data already provide abundant evidence for an important role of electronic excitations [9]. However, although the results cited in support of this view confirm that electrons may get excited in loaded materials even before the onset of the first reactions, they say little regarding their actual significance in the initiation process [10]. Experimental studies are scarce due to the obvious challenges associated with the characterization of fast and localized chemical processes in opaque condensed phases, while a theoretical description of the excited states and their dynamics in condensed matter is notoriously challenging. Furthermore, there are various types of loads to consider.

Therefore, much work is still needed to definitely establish the relative impact of excited and ground-state processes on observed sensitivities, including theoretical and numerical studies taking advantage of current progress in the simulation of the dynamics of excited states in materials. Beyond their fundamental interest, such simulations might support the development of improved techniques to initiate high explosives using selectively induced electronic transitions (i.e., resonant photoinitiation). Indeed, current initiation methods are either thermal, mechanical, or electrical. These broad classifications include techniques using high-energy lasers in the red to infrared region, and operating via either thermal (through direct heating) or mechanical (through an indirect approach involving a shock wave) initiation of the target EM. These techniques are not inherently safe because coincidental thermal or mechanical stimuli could result in unwarranted detonation. In contrast, direct optical initiation involving a photochemical pathway could provide improved security and safety. The ignition would be controlled through laser irradiation of the sample at a specific wavelength consistent with excitation energies. These prospects warrant investigations to get further insight into photochemical initiation and to guide the synthesis of new EMs with optimal properties in view of direct optical initiation [11–13].

This chapter first provides some background on the electronic states of molecules (Section 2) and materials (Section 3), as well as on the transitions between these states (Section 4). Finally, Section 5 gets to the heart of the matter by reviewing published initiation mechanisms involving electronic excitations and preliminary numerical simulations of their interplay with external loads and chemistry.

2 Electronic structure and excited states

Because atoms move slowly compared to electrons, electronic wavefunctions are usually well described within the clamped nuclei approximation, i.e., neglecting the nuclear kinetic energy operator in the molecular Hamiltonian. The eigenstates Ψ_i and eigenvalues E_i of the resulting electronic Hamiltonian are, respectively, the stationary (adiabatic) states and energies of the electrons for the current configuration $\mathbf{R}$ of the atoms. Each function $E_i(\mathbf{R})$

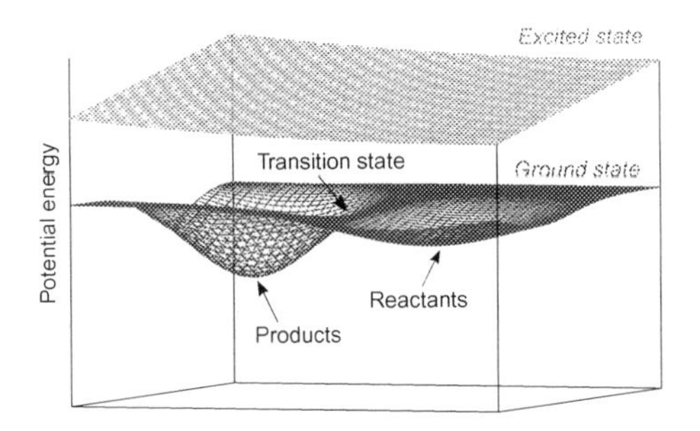

FIG. 11.1 Schematic sketch of a PES.

is represented by a hypersurface over the atomic configuration space, referred to as a potential energy surface (PES) since it determines the interatomic forces in the state Ψ_i. Hypothetical PESs are sketched in Fig. 11.1 as 2D surfaces. The lower one exhibits two minima corresponding to equilibrium atomic configurations, separated by a saddle point representing a transition state (TS).

The electronic structure refers to the set of PESs and associated states Ψ_i. It is a static description for electrons in that they remain in a stationary state, made challenging in practice due to the high dimensionality of the configuration space.

Typical high explosives are wide-gap insulators so that near the equilibrium configuration, the ground state Ψ_0 lies lower in energy than any excited state Ψ_i $(i > 0)$ by several eV, corresponding to a characteristic frequency in the femtosecond range. Therefore, most perturbations that may be applied to such a system will primarily affect the nuclei. In the adiabatic approximation, the electronic wavefunction adjusts instantaneously to any change in the atomic configuration $\mathbf{R}$ so that a system initially in a state Ψ_i never departs from the corresponding PES. This description holds for a system evolving on the ground-state surface $E_0(\mathbf{R})$ as long as it does not enter a region in the configuration space where the energy gap $E_1 - E_0$ gets too small, in which case the transition probability $\Psi_0 \rightarrow \Psi_1$ becomes significant.

The ground electronic state Ψ_0 and corresponding PES are routinely computed on the basis of the Hartree–Fock (HF) approximation or density functional theory (DFT) [14]. However, accurate first-principles estimates of the excited states energies E_i are more difficult to obtain, requiring higher levels of theory such as time-dependent DFT (TD-DFT) or complete active space perturbation theory (CASPT2) [15].

Nevertheless, the one-particle picture based on HF or DFT approaches provides a convenient description of excited states. In this picture, a specific state is associated with a distribution of the electrons over one-particle energy levels (orbitals) consistent with the Pauli principle which states that two electrons may occupy the same orbital provided they exhibit opposite spins. Depending on the value of the total spin, electronic states are classified as singlet, doublet, triplet…Starting with the lowest energies, they are denoted S_0, S_1, S_2…for singlet states, T_1, T_2…for triplet states, and so on.

Fig. 11.2 sketches the lowest electronic states of a hypothetical system along with the corresponding PESs. Horizontal lines in every potential well represent vibrational energy

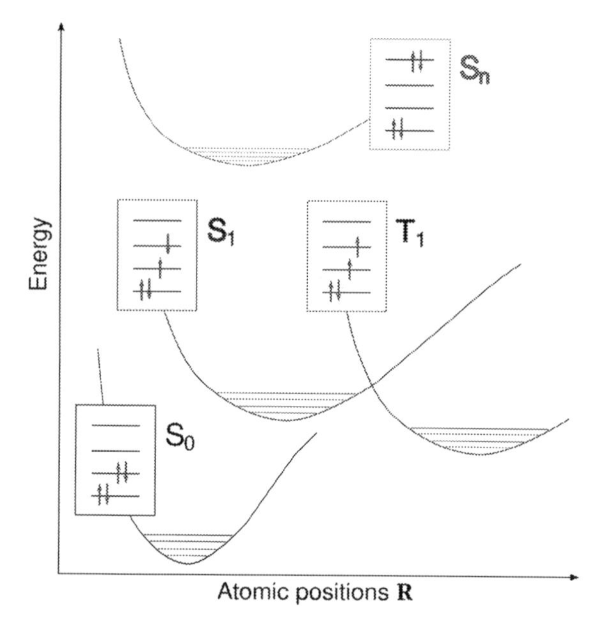

FIG. 11.2 Adiabatic states of a hypothetical molecule: ground electronic state S_0, first singlet and triplet excited states S_1 and T_1, and arbitrary singlet excited state S_n. For every PES, electrons are shown in a box as *up* and *down arrows* depending on their spin states, lying on horizontal scales representing orbital energies. *Black*, *blue(gray* in print version), and *red(dark gray* in print version) colors indicate, respectively, ground, monoexcited, and polyexcited states.

levels. The ground state (S_0) corresponds to all electrons paired into the orbital with lowest energies. Excited states may be viewed as those obtained from S_0 as electrons are promoted into higher (initially empty) orbital energy levels.

Such a description is restricted to finite systems like isolated species. However, since high explosives are made of energetic molecules held together by weak interactions, their electronic structure under ambient conditions is similar to that of a single constitutive molecule. Molecular excited states provide the information needed to optimize fluorescence-based detection methods [16]. Accordingly, electronic excited states have been computed for many model and real explosive molecules, including $H_2N–NO_2$ [17], *N,N*-dimethylnitramine [18], RDX [19], and many others.

3 Band structure and defects

Although the electronic structure of isolated molecules allows to rationalize and estimate such properties of molecular crystals as optical and electrical ones, this is not the case regarding EM sensitivities. This is because ignition occurs in locally compressed regions or near defects associated with specific electronic features rather than in bulk crystalline phases where molecules undergo little strain from their environment.

The electronic energy spectrum of a large collection of *M* noninteracting molecules is the superposition of *M* times the spectrum of a single one. As these molecules are put together to form a crystal, the *M*-fold degeneracy of every molecular level is lifted by intermolecular interactions, giving rise to a quasi-continuum of crystal energy levels (crystal orbitals) spread over a finite energy range called a band. By virtue of the crystal periodicity, every crystal

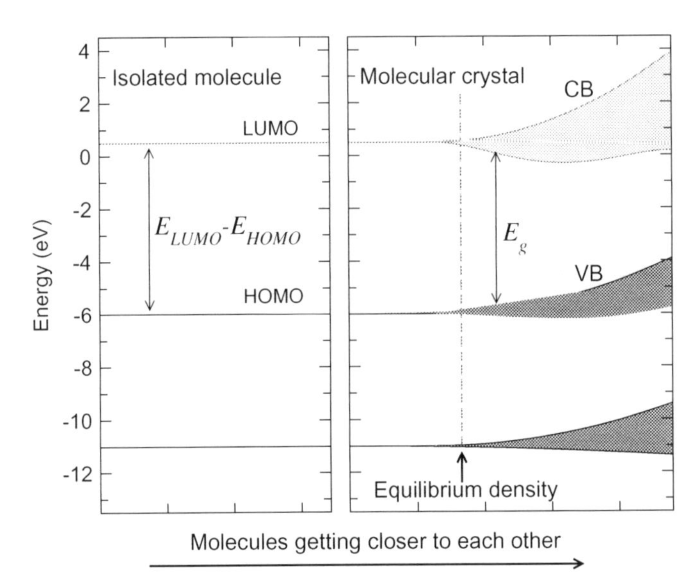

FIG. 11.3 Evolution of the electronic structure of a large collection of molecules from infinite separation (*left*) to forming a dense crystal (*right*). Occupied and unoccupied energy levels shown, respectively, in *dark* and *light gray*.

orbital (CO) is delocalized over the whole crystal. This is shown in Fig. 11.3, along with the widening of the bands and the resulting closure of the forbidden gaps upon further compression. The HOMO and LUMO of the isolated molecules give rise to the highest occupied and lowest unoccupied (at $T = 0$ K) energy bands, referred to, respectively, as the valence and conduction bands (VB and CB) by analogy with semiconductors. The CB–EB energy gap is referred to as the band gap E_g of the material. It is typically $\geq$5 eV for organic crystals under ambient conditions and theoretically decreases to zero at very high compression ratio.

In contrast to perfectly periodic models considered in solid state physics, any real crystal exhibits defects. Point defects give rise to local electron states that may be involved in electronic transitions. In fact, local states occur even in perfect crystals owing to transient electron excitations. Any electron promoted across the gap into the CB leaves a hole in the VB, which may be viewed as a positively charged particle. In the lack of lattice distortions, the excited electron and the hole are delocalized over the whole crystal. However, they can stabilize by inducing lattice distortions (polarization) in their surroundings. They behave then like a negatively (resp. positively) charged entity called an electron polaron (resp. a hole polaron). Alternatively, the excited electron and the resulting hole can remain bound together by virtue of their Coulomb attraction, thus forming an electrically neutral hydrogenic quasiparticle called an exciton. Polarons and excitons are best described as quasiparticles since they are mobile excitations moving in the crystal along with the surrounding crystal distortions.

The energy required to create such quasiparticles equals the energy required to promote an electron across the gap (E_g) minus the relaxation energy, which is usually small compared to E_g. However, in the case of organic materials, it may be significant, leading to tightly bound excitons (large electron–hole attraction). In that case, the excitonic levels lie significantly lower than the bottom of the conduction band so that the threshold energy for photon absorption (optical band gap) is lower than E_g (electrical band gap).

4 Electronic transitions

Should excited states be involved in an EM response, then this means that the system does not evolve on the ground state, but instead makes transitions between electronic states. As shown in Fig. 11.4, such transitions basically arise from two mechanisms, depending on whether they involve photons or atomic motions (phonons).

Transitions involving photons take place over extremely short periods of time ($\lesssim$1 fs) so that they do not involve atomic motions and are represented on such plots by vertical arrows joining initial and final states. For this reason, they are referred to as vertical transitions. Vertical relaxations are called radiative since they are associated with photoemission. Conversely, vertical excitations (T_V in Fig. 11.4) involve the absorption of photons providing the energy required to promote an electron in an excited state. In the case of one-photon absorption (OPA), this requires an incident radiation typically in the UV or visible range to overcome the gap of several eV between the final and initial states. However, any electromagnetic radiation may induce electronic excitations through multiphonon processes, especially two-photon absorption (TPA).

Transitions involving phonons are called nonradiative or nonadiabatic (T_{NA}) because they stem from the breakdown of the adiabatic approximation, with electrons unable to rearrange themselves in time to the fast atomic moves. They are significant when the energy difference between electronic states is not significantly larger than frequencies associated with atomic motions. Nonadiabatic excitations from the ground state typically arise as the system goes through a transition state (TS), as shown in Fig. 11.4.

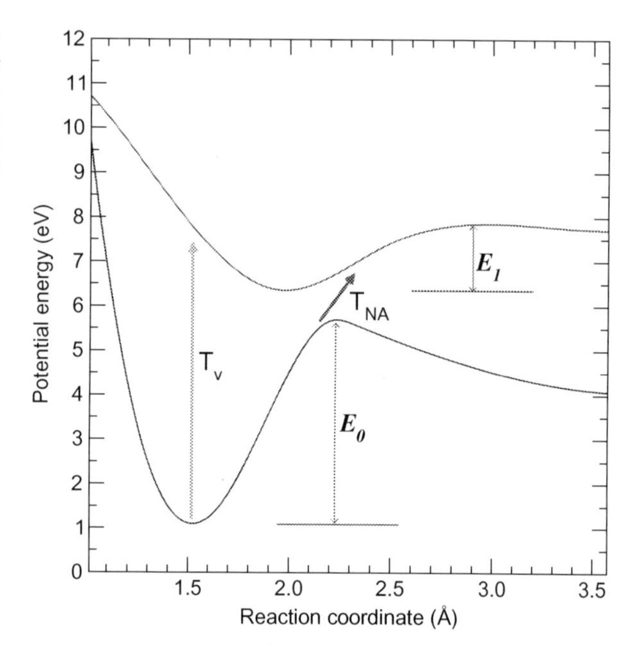

FIG. 11.4 Schematic view of electronic ground state and first excited state of a molecular system plot along an arbitrary reaction coordinate to illustrate vertical (V) and nonadiabatic (NA) transitions. E_0 and E_1 denote energy barriers to decomposition for the ground and excited states.

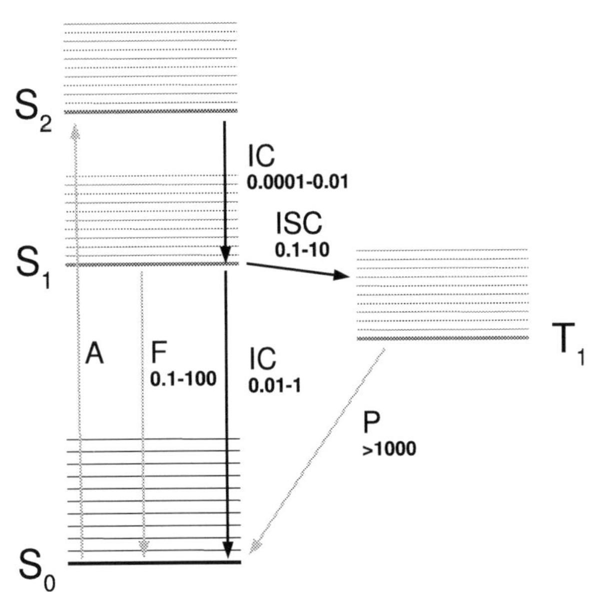

FIG. 11.5 Jablonski diagram showing electronic (*thick lines*) and vibrational (*thin lines*) energy levels of a hypothetical system and transitions between them, shown using *orange(light gray* in print version) and *black arrows*, respectively, for radiative and nonradiative ones, with corresponding lifetimes indicated in ns units.

The distribution of the electrons over the various states at any time is determined by the set of all transition rates. Fig. 11.5 summarizes significant types of transitions: photon absorption (A), fluorescence (F), phosphorescence (P), internal conversion (IC), and intersystem crossing (ISC). To set the ideas down, typical lifetimes obtained as the reciprocals of the corresponding transition rates are reported. Not surprisingly, ISC and P occur at relatively low rates as they involve a change in spin state, in contrast to IC and F. On the other hand, nonradiative transitions (IC and ISC) are more efficient than radiative ones (F and P); therefore, relaxation of excited states is often driven by internal conversion IC.

While these relaxation rates are inherent to the material, the excitation rates depend on the external perturbation applied. For electronic excitations to be significant, they must overcome the global deexcitation rate resulting from the IC, ISC, F, and P processes. Photon absorption (A) is the most straightforward excitation mechanism. However, the initiation of explosives, whether accidental or voluntary, is most often caused by mechanical stress. In order to promote electrons into excited states, such a stress must occur at high rate and bring the atoms to configurations where an excited PES lies close in energy to the ground PES. As suggested by Fig. 11.4, such configurations are mostly encountered in the vicinity of transition states associated with chemical reactions. Depending on the rate at which the transition state is approached, the system will remain in the ground state or hop to the excited state. Another nonadiabatic excitation pathway is possible for highly compressed materials, owing to the closing of the band gap shown in Fig. 11.3.

5 Initiation mechanisms and excited states

As mentioned in Section 1, there is little doubt that electronic excitations are involved in the initiation of an EM. This comes as no surprise as even the simple act of rubbing a material causes the emission of visible photons reflecting the radiative relaxation of excited electrons, according to the well-known but not fully understood phenomenon known as triboluminescence. The question is whether these excitations must be taken into account for a satisfactory description of the initiation process, or whether they can reasonably be neglected.

The answer to this question is a long-term goal in view of the lack of suitable experimental probes, the complexity of the physics involved, the various kinds of loads that may be applied, and the fact that excited electrons are likely to be involved differently in the subsequent steps of the initiation process. Although the latter depend on the nature of the external stress applied to the material, three subsequent steps may be roughly distinguished (Fig. 11.6). Step 1 is a prereactive one, with the energy provided by the external load getting trapped in local regions and transferred to high-frequency vibrations through phonon up-pumping. Step 2 is the ignition, i.e., the onset of chemistry with the cleavage of so-called trigger bonds. Finally, Step 3 is the release of chemical energy and the propagation of a self-sustained decomposition. This step is unstable as the release of gaseous species increases the local pressure, thus causing a continuous acceleration of the chemical reactions (deflagration) until a steady state (detonation) is possibly achieved.

With regard to the particularly studied case of impact sensitivity, the most convincing models currently available do not take electronic excitations into account, which might suggest that the latter play no major role. However, as emphasized in Fig. 11.6, they rely on contradictory assumptions regarding the determining step: prereactive phenomena for the phonon up-pumping model [20], endothermic onset of chemistry for correlations based on bond dissociation energy (BDE) of trigger bonds [21], and propagation of the decomposition for a more recent thermokinetic approach [22, 23]. This clearly confirms that correlation does not imply causation. Therefore, a possible critical role of excited states cannot be definitely ruled out, either in the first, second, or third step, or a combination of them.

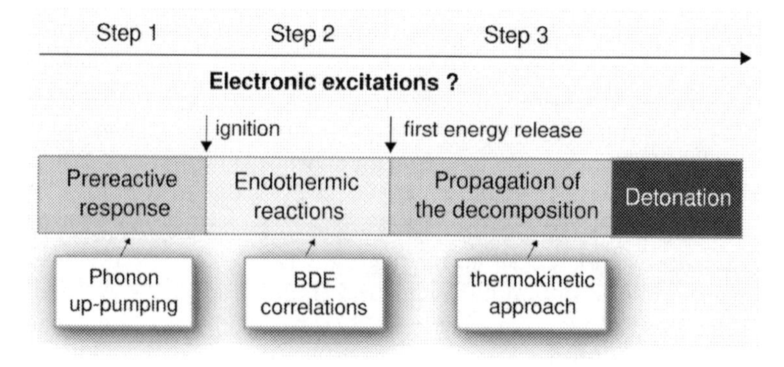

FIG. 11.6 Subsequent steps in the initiation process and assumptions of current models regarding the determining one for impact sensitivity. Electronic excitations might be involved at any stage.

5.1 Detonation regime

Researchers addressing the role of electronic excitations on EM sensitivities do not necessarily focus on a specific initiation step among those shown in Fig. 11.6. However, extreme conditions are mostly considered, e.g., compression ratios close to 50% [24] and pressures above tens of GPa [25, 26]. Such conditions could be relevant either in step 1, within localized regions trapping energy supplied by a mechanical load [27], or in step 3, in front of a detonation wave.

In fact, early studies of electronic excitations in EMs were primarily motivated by the fast decomposition rate observed in this regime [28–33]. As pointed out by Gilman [31], the time of about 10^{-13} s taken for a detonation wave with velocity close to 10^4 m/s to pass through a molecule is too short for reactions to be thermally activated in the process. Therefore, it might be reasonable to assume that electrons provide the driving force. According to Dremin, the shock energy is first absorbed only by translational degrees of freedom (DOFs), resulting in a transient translational temperature overheat up to several tens of thousands of degrees before spreading among all DOFs [32, 33]. Such extreme transient conditions led the authors to envision three mechanisms by which the shock could trigger chemical decomposition, one of them involving electronic excitation. However, the model derived on this basis exhibits an unphysical linear increase of the temperature overheat with the number of atoms per molecule [33].

Similarly, vertical electronic excitation induced by the oscillating field arising from excited lattice phonons is highly unlikely as it would involve no less than about 30 phonons, as pointed out by Hooper [34]. This is especially true at rest in view of the gap between vibrational and electronic transition frequencies. However, such transitions should be enhanced under shock conditions in view of the broad frequency range contained in a shock wave. In an attempt to estimate the corresponding rates for RDX using first-order time-dependent perturbation theory, Zwitter et al. obtained very significant values (of the order of 1 transition per picosecond) for transitions in the far UV range [35]. Such magnitudes could reflect a loss of validity of the perturbational treatment. Furthermore, they critically depend on arbitrary hypotheses regarding the amplitude of the electromagnetic field induced by the fluctuating charges. Finally, they assume a purely coherent evolution of the electronic wavefunction.

In the lack of incoming photons or electrons, electronic excitations can only be produced by atomic motions, i.e., either by temperature or by fast coherent atomic displacements. However, this requires a significant decrease of the band gap E_g so that E_g/h (with h the Planck constant) matches the vibrational frequencies. In the early 2000s, Kuklja and coworkers investigated conditions enabling a significant reduction of E_g and estimated the corresponding fraction η of excited electrons from Maxwell–Boltzmann statistics [26, 36, 37]. Under ambient conditions, they obtained a gap value of 5.25 eV for a defect-free crystal of a typical high explosive (RDX). They attributed the lower value of 3.4 eV derived from optical absorption experiments to the imperfections inherent to real-life materials. Under a pressure of 30 GPa, they estimated that E_g decreases by 2 to 3.25 eV. Even under such a large pressure, η remains only $\simeq 5 \times 10^{-28}$ at $T = 600$ K.

In the next step, the authors considered defective crystals. While they observed no significant change in electronic structure upon introduction of cracks and vacancies in the RDX crystal [26, 36], a significant decrease of E_g was noted for a one-dimensional model of an edge

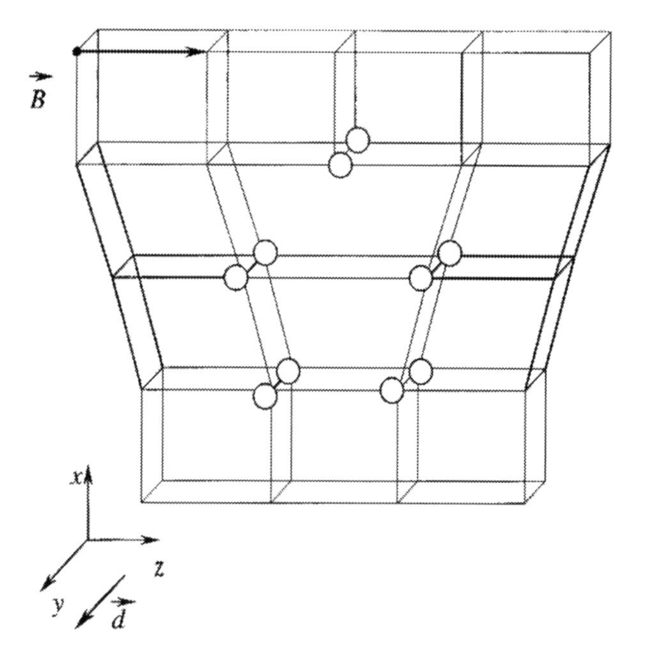

FIG. 11.7 One-dimensional atomistic model introduced by Kuklja et al. [37] for an edge dislocation in the RDX crystal. The Burger vector **B** is along the *z* direction and the *slip plane* is orthogonal to **d**. RDX molecules are located at the nodes of the represented mesh. Calculations are restricted to the molecules shown as *open circles. Adapted from M.M. Kuklja, E.V. Stefanovich, A.B. Kunz, An excitonic mechanism of detonation initiation in explosives, J. Chem. Phys. 112 (7) (2000) 3417, https://doi.org/10.1063/1.480922. With permission from AIP.*

dislocation (Fig. 11.7). For this system, the authors estimated that the gap should be reduced to about 1 eV under pressures of 15–20 GPa [37]. However, notwithstanding potential artifacts due to the "rigid molecule approximation" [38], this conclusion should be considered with caution as details are missing regarding how exactly volume and pressure are defined for such a 1D polymer model.

Likewise, Reed et al. calculated the electronic structure of energetic crystals for even higher hydrostatic pressures (up to 180 GPa) and under uniaxial compression and shear, considering also the role of vacancies and stacking faults [25, 39]. They concluded that even huge static pressures do not allow to promote a significant fraction of the electrons into excited states. For nitromethane, they estimated that dynamic stimuli leading to molecular velocities in excess of 6 km/s (corresponding to a temperature $\simeq 10^4$ K) are necessary for this purpose [25], and speculated that electronic excitations might arise from high-velocity collisions between nitro groups [39]. Such results point to the significance of dynamical electronic effects demonstrated in a subsequent study [40].

5.2 Hot spot formation

It is well established that the amount of energy required to initiate an EM is much less than that required to heat the sample uniformly up to the ignition temperature [41]. This implies that the input energy be concentrated into specific regions (hot spots) within the sample [4]. Therefore, the external stimulus must be strong enough so that a sufficient amount of energy is locally available to break chemical bonds before dissipating in the surroundings. In the extreme case, the use of a detonator to generate a strong shock is especially likely to produce electronic excitations. This is because the resulting compression may hinder chemical decomposition up to high temperatures, while making electronic excitations easier through band gap reduction. At the same time, high velocities are imparted to atoms, thus favoring nonadiabatic transitions. As such a shock traveling at more than 6000 m/s enters an explosive sample, mechanisms such as those discussed in Section 5.1 may be envisioned. Therefore, shockwave initiation will not be considered further in this section.

There is little doubt that electrostatic discharges and electron beams produce excited states as incoming electrons get trapped into previously empty orbitals or knock other electrons into them. Regarding light beams, including lasers, their ability to promote electrons into excited states obviously depends on their wavelength. In resonant photo-initiation, UV or visible light is likely to produce excitations quite efficiently if the photon energy matches the band gap of the material. This sudden supply of energy is all the more likely to induce fast chemical decomposition as photoexcited states may be dissociative or decompose with a small activation barrier, as suggested by Fig. 11.4.

However, laser initiation is currently most commonly performed using IR lasers that produce localized heating through vibrational excitation, i.e., by stimulating atomic motion rather than electron transfers to higher orbitals. Likewise, thermal loads heat up the atoms through collisions with surroundings at higher temperatures, while mechanical stimuli impart coherent velocities to the atoms through collisions with a moving solid, such as the falling hammer in the drop weight impact test.

At first sight, such stimuli are unlikely to promote electrons to excited states. This is especially the case for thermal stimuli. A rising temperature ramp will trigger the decomposition of a typical EM between 150°C and 300°C, i.e., well before promoting electrons into the excited states lying several eV's higher in energy. Likewise, while even modest mechanical stimuli generate hot spots and sometimes observable luminescence, they do not produce the high molecular velocities of thousands of km/s required to initiate the material through electronic excitation [25] so that most electrons are able to follow adiabatically the atomic displacements. In drop weight impact experiments, the hammer hits the sample with a velocity of about 10 m/s, to be compared with thermal atomic velocities under ambient conditions (>400 m/s). Under both thermal/mechanical loads, kinetic energy is transferred from external atoms (those that make up the hammer or surrounding material) to the atoms in the EM until a final equilibrium is reached for all nuclei. In view of the relative weakness of nonadiabatic couplings under mild conditions, the electrons gradually gain energy from the nuclei and the whole system is expected to get equilibrated only in a second step, leaving time for the ignition to take place in the ground electronic state (GS). This may justify that many studies of thermal/mechanical initiation focus on the GS [42].

In view of their ability to trigger thermo- or triboluminescence, this theoretical lack of efficiency of mild thermal/mechanical loads in exciting electrons may seem paradoxical. This raises the question of the minimum number of excited electrons needed to form a hot spot and ignite the material. As a tentative explanation to the fact that explosive samples do not detonate below a critical diameter, Kuklja et al. have suggested that initiation is made impossible as the number of electronically excited molecules becomes less than 1 [26]. As an excited energetic species relax to the ground state, the released electronic excitation energy of typically >5 eV is deposited into internal molecular vibrations and may lead to bond cleavage and generation of hot reactive species that could propagate the decomposition process.

However, in a condensed phase, this energy might spread out long before it allows the decomposition of a molecule, which is a slower process involving significant atomic rearrangements.

In fact, experimental and theoretical work showed that in order to initiate a thermal explosion, typical hot spots must exhibit temperatures of at least 500°C, be about 1 μm in size and last for at least 1 μs [43]. For a typical energetic compound like RDX, this corresponds to at least a few billion molecules with an average excess energy of 1 eV. Therefore, the very small concentrations of electronic excitations derived from equilibrium distributions suggest that they do not contribute significantly to hot spot formation [42].

Another possibility envisioned earlier is that hot spots arise as a consequence of the diffusion of preexisting thermal excitons into localized regions of reduced band gap [44]. However, assuming a typical value of 10^6 Hz for the exciton hopping frequency and an intermolecular spacing of 10–100 Å, the time needed for an exciton to move over a distance of 1 μm (minimal hot spot size) is about 100–1000 μs, i.e., much too long to account for the fast hot spot generation. Thus, it is quite unlikely that hot spots could arise merely as a result of exciton migration.

To conclude this section, it must be stressed that the concept of a thermal origin of initiation [2–6] is just a convenient approximation. The term "hot spot" suggests a region where energy has first accumulated and reached a transient equilibrium state, before the local temperature overshoot activates chemical decomposition in a second step. In fact, the energy put into a hot spot may trigger various events, including chemical reactions, even before thermalization is completed, i.e., before vibrational (and perhaps electronic) transitions lead to a transient equilibrium distribution. However, it might still be possible to qualitatively describe the propagation of the decomposition as a thermally activated process as long as the corresponding rate increases with the amount of energy released by exothermic reactions. Thus, it must be emphasized that the term "hot spot" does not necessarily imply a clear separation between thermal equilibrium and the onset of chemical reactions.

5.3 Growth of reactive centers

According to the preceding discussion in Section 5.2, while mild thermal/mechanical loads do produce excited states leading to the observation of luminescence phenomena, they appear unlikely to produce critical hot spots so as to initiate an explosion. In this section, we consider excited states in reactive centers. Although the latter are just hot spots where the ignition has started, it is important to make a distinction in the present context. Indeed, electronic transitions are naturally much more prevalent in reactive centers as chemical reactions imply species passing through transition states where the energy gap between excited and ground states markedly decreases.

As pointed out elsewhere by Bondarchuk [27], even mild loads could produce extreme conditions in very localized regions of a sample, i.e., near microscopic defects. For instance, the shock-induced collapse of a void may result in areas combining extreme stresses, including shear, with high local temperature arising from the acceleration of atoms through the pore (see Chapter 12). These are ideal conditions to promote electrons into excited states, and at the same time to trigger chemical reactions.

What is clear is that the chemical reactions observed under these conditions are most certainly accompanied by electronic transitions. The latter could operate in synergy with vibrational excitations, as pointed out by Michalchuk et al. [45, 46]. In simple words, an increase in temperature or a mechanical load could knock the system from a near equilibrium geometry to a region of high nonadiabatic coupling (e.g., near a transition state) where it could then hop to an excited state (channel T_{NA} in Fig. 11.4). In this picture, vibrational excitation is a prerequisite for electronic excitation. This reaction path is in competition with ground-state decomposition mechanism, made probable as well as soon as the system gets close to a dissociative transition state. In this case, even if they contribute to the decomposition, electronic transitions are not necessarily critical to it. Their actual significance depends on the relative rates associated with the nonadiabatic and ground-state mechanisms. Furthermore, it must be kept in mind that since excited states are likely to relax very quickly, they do not necessarily contribute the decomposition even if they exhibit small/zero barriers to dissociation.

5.4 Relaxation of electronically excited species

Although electron populations in excited states are easily obtained from Fermi–Dirac or even Maxwell–Boltzmann statistics as done in many studies, this approach does not provide insight into short time phenomena that take place before a transient electronic thermal equilibrium may be assumed. As a system makes a transition from an initial state i to a new state j, the atoms experience forces leading to fast changes in their positions. Within a few femtoseconds, the system may enter a region where state j is strongly coupled with other states. In particular, configurations with PES degeneracy (conical intersections) play a ubiquitous role in the dynamics of excited states due to strong nonadiabatic coupling allowing efficient transfers of electron population from one state to the other.

Getting insight into this nonequilibrium dynamics requires characterizing the relevant PESs and evaluating the transition probabilities. In contrast to vertical transitions, the probability of a nonadiabatic transition between two electronic states i and j cannot be estimated just from the corresponding wavefunctions Ψ_i and Ψ_j because they depend on the atoms trajectories (in a semiclassical picture where nuclei are treated classically) or on vibrational states (in a full quantum picture). This means that the relevant parts of the corresponding energy surfaces $E_i(\mathbf{R})$ and $E_j(\mathbf{R})$ must be determined.

5.4.1 *Static calculations*

Most current findings regarding the decay of electronic excitations in EMs were obtained from static electronic structure calculations. These studies mostly focus on a single molecule initially assumed to be electronically excited. This option is needed to check the validity of the theoretical method employed because only under such conditions do the experimental measurements and theoretical calculations directly correspond [8].

Excited states are typically calculated for several significant configurations of the atoms, sampling potential energy surfaces to identify prominent relaxation channels, especially conical intersections. A detailed characterization of the PESs features provides valuable insight into deactivation mechanisms. For instance, following the determination of the electronic spectrum of nitroethylene, a model nitro compound with formula $H_2C{=}CHNO_2$ [47], a photochemical nonradiative deactivation pathway from the $\pi \rightarrow \pi^*$ excited S_5 state down to the ground state through a set of conical intersections was proposed [48]. This process involves conical intersections associated with various distortions of the nitro group, and the possibility of NO_2 release in the ground state was pointed out. Such calculations are often associated with time-resolved experiments on isolated molecules, like time-of-flight mass spectroscopy and laser-induced fluorescence spectroscopy, which provides additional information regarding the early decomposition products. Taking dimethylnitramine (DMNA) as an example, such techniques indicate that the decomposition of the electronically excited molecule yields NO as a major decomposition product [49]. This observation is rationalized with the help of CASSCF calculations pointing to the prominent role of a $(S_2/S_1)_{CI}$ conical intersection along the nitro–nitrite isomerization reaction coordinate. This CI isomerization is hindered by a $(S_1/S_0)_{CI}$ accounting for the predominance of NO_2 elimination in the thermal decomposition of DMNA.

Many studies of decomposition pathways of energetic molecules have been reported by Bernstein and coworkers, as summarized in a couple of reviews [7, 8]. Not surprisingly, excited electrons dramatically affect the nature of the intermediate species as well as the overall kinetics of the process [9]. However, these authors pointed out a more intriguing conclusion from their extensive studies. Although virtually all nonenergetic compounds considered were found to decompose in excited states, this is not the case for the energetic molecules. Instead, the latter undergo an ultrafast cascade relaxation from the initial photoexcited state S_n to the ground state S_0. In the process, the electronic excitation energy (typically $>$5 eV) provided by the laser is transferred to molecular vibrations in the ground state and used to break chemical bonds, thus generating hot reactive radicals for further decomposition [8].

It would be interesting to explore in depth the implications of these observations for molecules in solid or liquid states. In a solid, one might expect dissociation to be hindered by surrounding molecules, in contrast to relaxation processes that could be even faster due to the additional pathways made possible by the increased number of interacting species. However, a direct correlation between dynamics following electron excitation and energetic character is not to be expected, since EMs are primarily characterized by a high chemical energy content, as required to propagate a self-sustained decomposition.

5.4.2 *Dynamic simulations*

Dynamical simulations could potentially provide deeper insight into the fate of excited electrons in isolated molecules. In the Ehrenfest approach, nuclear trajectories are propagated on an effective PES, namely the average of the various energy surfaces weighted according to their respective populations. Coupled to a classical radiation field in view of describing responses to ultrafast laser pulses, this approach is known as the semiclassical electron-radiation-ion dynamics or SERID method [50]. In a recent application of this method to the photophysical deactivation of excited states in TATB [51], 201 trajectories were simulated, 68% of which led to molecular fragmentation due to the very high fluence of the incoming photons. The authors estimated a value of 2400 fs the lifetime of the excited state resulting from HOMO → LUMO transition, in fair agreement with the experimental value of 6000 fs [51]. With regard to the above-mentioned Bernstein correlation [8], this result indicates that such a lifetime remains compatible with energetic character.

A serious drawback of the Ehrenfest approach is the fact that the use of a single effective PES implies deterministic trajectories, whereas electronic transitions must split them into various branches as interatomic forces depend on the current PES. More advanced approximations to a full quantum dynamics expand the nuclear wavefunction linearly in term of coupled traveling Gaussian basis functions built around trajectories [52]. New trajectory basis functions (TBFs) are spawned whenever necessary (allowing for branching). With further approximations, the approach can be made compatible with on-the-fly nonadiabatic dynamics with the PESs computed at each time step as needed. The resulting scheme is known as the Ab Initio Multiple Spawning (AIMS) method [52]. It was used by Ghosh et al. to study the decay to the ground state of a number of compounds initially promoted into the S_1 excited state [53]. The relaxation processes usually involve internal conversion via $(S_1/S_0)_{CI}$ conical intersections. The corresponding decay time constant was found to be over 100 ns for 1,2,4,5-tetrazine, about 400 fs for 1,2,4,5-tetrazine-2,4-di-N-oxide, and below 50 fs for compounds with nitro (with $C–NO_2$, $N–NO_2$, and $O–NO_2$ bonds) or azido (N_3) groups. The large difference in time constants for the two former molecules is attributed to the fact that the conical intersection lies higher in energy compared to the Franck–Condon geometry for the S_1 state of 1,2,4,5-tetrazine. For nitro compounds, the ultrafast decay of the S_1 state through IC does not involve any dissociation. In contrast, for methyl azide, it leads to N_2 elimination. Based on the fact that the only compound whose relaxation takes place on a timescale of more than 500 fs (namely 1,2,4,5-tetrazine) exhibits poor detonation performance, the authors speculate that shorter excited state lifetimes could be a necessary condition for a compound to be of value as energetic material, in line with the Bernstein assumption [8].

The relaxation of photoexcited methyl azide was also studied by Peters et al. [54], albeit starting from the manifold of highly excited states produced via a 10 fs UV pulse at 8 eV. After excitation, the molecules undergo immediate structural changes leading to nonadiabatic relaxation to lower excited states on the extremely fast timescale of 25 fs. Using the AIMS method, it was shown that this ultrafast process was a consequence of the strong nonadiabatic couplings between the excited surfaces involved, leading to a transfer of the wavepacket to a lower excited state before it reaches the minimal energy region along the conical intersection seam.

5.5 Coupled electron-nuclear dynamics in materials

No matter how well they approximate the exact quantum dynamics, the above-mentioned approaches are not fully satisfactory for electrons in bulk matter. This is because the huge number of DOFs provided by the environment makes their dynamics appears irreversible. Such open quantum systems are commonly described by master equations for the density matrix.

However, a semiclassical description in term of electronic wavefunctions evolving according to a nonunitary (irreversible) fashion along nuclear trajectories may prove more intuitive [55, 56]. In view of accounting the branching of the trajectories, the latter are propagated classically on the current PES, except in regions of strong coupling where stochastic electron excitation can occur and the trajectories may split into several independent branches, resulting in a swarm of trajectories evolving on various PESs. Such approaches are known for about 50 years as trajectory surface hopping (SH) methods. There are usually implemented with electronic structures described using semiempirical models. Nevertheless, they remain difficult to apply to realistic systems due to the need to keep track of a large swarm of trajectories.

However, the problem becomes much simpler if we are only interested in excitation probabilities and not in the fate of the system after electronic excitation. Provided nonadiabatic couplings remain sufficiently small, the population decay of the ground state is mostly determined by the transition probabilities along the trajectory. Electronic relaxations may be neglected as long as excited states populations remain small. Therefore, it may be sufficient to simulate the GS trajectory while neglecting the influence of electronic excitations. This approach is known as the neglect of backpropagation approximation (NBRA) method.

It was used to assess the fraction η of electrons promoted into excited states by the sudden acceleration of the atoms under shock [57]. For this purpose, a compression wave (loosely referred to as a shock) was propagated in a linear chain of atoms interacting via a simple analytic potential, while the electronic wavefunction was propagated according to a time-dependent Schrödinger equation accounting for atomic motions [44, 58–60]. This approach allows a qualitative description of the excited states population as a compression wave passes over the material [57, 58], as shown in Fig. 11.8 in the case of a sustained compression.

However, various deficiencies are obvious. In particular, preliminary simulations of the system at rest proved unable to achieve correct thermal equilibrium populations [58]. It was subsequently pointed out that simulating the evolution of the electronic wave function in a coherent way along ground-state nuclear trajectory overestimates the population in excited states by several orders of magnitude [44]. This issue proved inherent to the NBRA approximation, which cannot describe an equilibrium state due to the lack of feedback between the classical and quantum subsystems [61, 62]. For a specific simulation, equilibrium

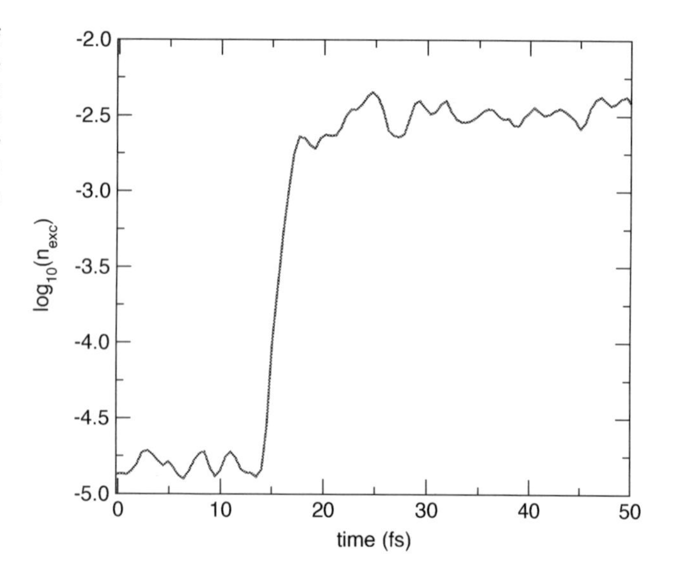

FIG. 11.8 Evolution of the fraction of excited electrons as a sustained shock wave passes through the material. *Adapted from D. Mathieu, P. Martin, P. Simonetti, Nonadiabatic simulation of valence electrons in solids under sustained shockwaves, AIP Conf. Proc. 505 (2000) 409–412, https://doi.org/10.1063/1.1303504. With permission from AIP.*

populations can be obtained by introducing empirical decoherence times and/or relaxation rates to account for the role of the macroscopic environment. However, this cannot be done in a consistent manner under various conditions [58–60].

Real energetic materials under shock are typically studied using the multiscale shock simulation technique (MSST) which allows one to focus on a small unit cell. A new modification of this scheme for simulation of the coupling between electrons and ions allowed a very interesting MD simulation of the shock-induced decomposition of a primary explosive (HN_3) in detonation regime, including thermal electron transfers to excited states [63]. This simulation was focused on the evolution of the system following the onset of chemistry. During the reactions, the fraction of excited electrons was dramatically enhanced, up to one electronic excitation for every eight molecules. This might be expected in view of the significant band gap reduction observed under dynamic compression, so much so in nitromethane that a transient semimetallic layer is formed [40]. As a consequence, the decomposition kinetics of HN_3 is enhanced by about 30% [63]. Depending on the material, this excitation-induced enhancement of the reaction rate could be critical to the propagation of the detonation. In other words, excited electrons might contribute not only to the sensitivity to strong shocks but also to the sensitivity under milder mechanical loads through their role in the propagation of the detonation.

The approaches described in this section are complementary to those described in Section 5.4.2. While more approximate descriptions of the excited states are used, realistic condensed phase conditions and environmental effects are included.

6 Conclusion

It is difficult to imagine that the initiation of an energetic material is not accompanied by transitions to excited states, if only during chemical reactions. Nevertheless, this is not enough to conclude that these excitations play a fundamental role in the overall process and in measured sensitivity values. Although even mild mechanical loads may induce such excitations, it is not clear that these are in sufficient concentration to induce the initiation. Certainly, the different types of loads and materials must be considered on a case-by-case basis. Furthermore, the successive stages of the initiation process should be considered since each of them might contribute to the observed sensitivities. Ultimately, much remains to be learned about the role of excited states on initiation. This could be achieved with the help of nonadiabatic simulation methods. Although their application to energetic materials has barely started, steady progress enhances the ability of these techniques to describe realistic systems.

References

[1] N.K. Bourne, On the laser ignition and initiation of explosives, Proc. R. Soc. Lond. A Math. Phys. Eng. Sci. 457 (2010) (2001) 1401–1426, https://doi.org/10.1098/rspa.2000.0721. https://royalsocietypublishing.org/doi/abs/10.1098/rspa.2000.0721.

[2] S.A. Shackelford 2, A general concept concerning energetic material sensitivity and initiation [1], J. Phys. IV France 05 (C4) (1995) C4-485-C4-499. https://doi.org/10.1051/jp4:1995439.

[3] S.A. Shackelford, Role of thermochemical decomposition in energetic material initiation sensitivity and explosive performance, Central Eur. J. Energ. Mater. 5 (2008) 75–101.

[4] J.E. Field, Hot spot ignition mechanisms for explosives, Acc. Chem. Res. 25 (11) (1992) 489–496, https://doi.org/10.1021/ar00023a002.

[5] J.F. Field, D.M. Williamson, S.M. Walley, S.J.P. Palmer, Hot spot ignition of explosives and their thermal properties and experimental measurement of the thermal properties of a PBX and its Binder system, in: R.B. Hetnarski (Ed.), Encyclopedia of Thermal Stresses, Springer Netherlands, Dordrecht, 2014, pp. 2293–2306, ISBN: 978-94-007-2739-7, https://doi.org/10.1007/978-94-007-2739-7_709.

[6] S.R. Ahmad, M. Cartwright, Laser Ignition of Energetic Materials, Wiley and Sons, 2015.

[7] A. Bhattacharya, Y. Guo, E.R. Bernstein, Nonadiabatic reaction of energetic molecules, Acc. Chem. Res. 43 (12) (2010) 1476–1485, https://doi.org/10.1021/ar100067f.

[8] E.R. Bernstein, Chapter Two–On the release of stored energy from energetic materials, in: J.R. Sabin (Ed.), Advances in Quantum Chemistry, Energetic Materials, vol. 69, Academic Press, 2014, pp. 31–69, https://doi.org/10.1016/B978-0-12-800345-9.00002-7 (January).

[9] M.M. Kuklja, Chapter Three–Quantum-chemical modeling of energetic materials: chemical reactions triggered by defects, deformations, and electronic excitations, in: J.R. Sabin (Ed.), Advances in Quantum Chemistry, Energetic Materials, vol. 69, Academic Press, 2014, pp. 71–145, https://doi.org/10.1016/B978-0-12-800345-9.00003-9. (January).

[10] J. Sharma, B.C. Beard, Electronic excitations preceding shock initiation in explosives, MRS Online Proc. Library Arch. 296 (1992) 189–198, https://doi.org/10.1557/PROC-296-189. https://www.cambridge.org/core/journals/mrs-online-proceedings-library-archive/article/div-classtitleelectronic-excitations-preceding-shock-initiation-in-explosivesdiv/EA3456D84528EEE351E3284C27A6DFAD.

[11] A.E. Sifain, L.F. Tadesse, J.A. Bjorgaard, D.E. Chavez, O.V. Prezhdo, R.J. Scharff, S. Tretiak, Cooperative enhancement of the nonlinear optical response in conjugated energetic materials: a TD-DFT study, J. Chem. Phys. 146 (11) (2017) 114308, https://doi.org/10.1063/1.4978579.

[12] L. Belau, Y. Haas, S. Zilberg, Formation of the cyclo-Pentazolate N5–anion by high-energy dissociation of phenylpentazole anions, J. Phys. Chem. A 108 (52) (2004) 11715–11720, https://doi.org/10.1021/jp0469057.

[13] Y. Haas, Conical intersections leading to chemical reactions in the gas and liquid phases, Adv. Chem. 2014 (2014) 1–10, https://doi.org/10.1155/2014/419102. https://www.hindawi.com/archive/2014/419102/.

[14] J. Kohanoff, Electronic Structure Calculations for Solids and Molecules: Theory and Computational Methods, Cambridge University Press, Cambridge, 2006, ISBN: 978-0-521-81591-8, https://doi.org/10.1017/CBO9780511755613. https://www.cambridge.org/core/books/electronic-structure-calculations-for-solids-and-molecules/0C0AF2B01A380912FC13816A9A0C350F.

[15] M.A. Robb, Theoretical Chemistry for Electronic Excited States, Royal Society of Chemistry, 2018, ISBN: 978-1-78262-864-4, https://doi.org/10.1039/9781788013642. https://pubs.rsc.org/en/content/ebook/978-1-78262-864-4. (March).

[16] J.K. Cooper, C.D. Grant, J.Z. Zhang, Experimental and TD-DFT study of optical absorption of six explosive molecules: RDX, HMX, PETN, TNT, TATP, and HMTD, J. Phys. Chem. A 117 (29) (2013) 6043–6051, https://doi.org/10.1021/jp312492v.

[17] I. Borges, Excited electronic and ionized states of the nitramide molecule, H2NNO2, studied by the symmetry-adapted-cluster configuration interaction method, Theor. Chem. Acc. 121 (5) (2008) 239–246, https://doi.org/10.1007/s00214-008-0469-9.

[18] I. Borges, Excited electronic and ionized states of N,N-dimethylnitramine, Chem. Phys. 349 (1) (2008) 256–262, https://doi.org/10.1016/j.chemphys.2008.02.043.

[19] I. Borges, A.J.A. Aquino, M. Barbatti, H. Lischka, The electronically excited states of RDX (hexahydro-1,3,5-trinitro-1,3,5-triazine): Vertical excitations, Int. J. Quantum Chem. 109 (11) (2009) 2348–2355, https://doi.org/10.1002/qua.22043. https://onlinelibrary.wiley.com/doi/abs/10.1002/qua.22043.

[20] A.A.L. Michalchuk, J. Hemingway, C.A. Morrison, Predicting the impact sensitivities of energetic materials through zone-center phonon up-pumping, J. Chem. Phys. 154 (6) (2021) 064105, https://doi.org/10.1063/5.0036927.

[21] B.M. Rice, S. Sahu, F.J. Owens, Density functional calculations of bond dissociation energies for NO_2 scission in some nitroaromatic molecules, J. Mol. Struct. (Theochem) 583 (2002) 69–72.

[22] D. Mathieu, T. Alaime, Predicting impact sensitivities of nitro compounds on the basis of a semi-empirical rate constant, J. Phys. Chem. A 118 (2014) 9720–9726.

[23] D. Mathieu, Modeling sensitivities of energetic materials using the Python language and libraries, Propellants Explos. Pyrotech. 45 (6) (2020) 966–973, https://doi.org/10.1002/prep.201900377.

[24] M.M. Kuklja, A.B. Kunz, Ab initio simulation of defects in energetic materials: hydrostatic compression of cyclotrimethylene trinitramine, J. Appl. Phys. 86 (8) (1999) 4428–4434, https://doi.org/10.1063/1.371381.

[25] E.J. Reed, J.D. Joannopoulos, L.E. Fried, Electronic excitations in shocked nitromethane, Phys. Rev. B 62 (24) (2000) 16500–16509, https://doi.org/10.1103/PhysRevB.62.16500.

[26] M.M. Kuklja, An excitonic mechanism of detonation initiation in explosives, J. Chem. Phys. 112 (2000) 3417–3423.

[27] S.V. Bondarchuk, Quantification of impact sensitivity based on solid-state derived criteria, J. Phys. Chem. A 122 (24) (2018) 5455–5463, https://doi.org/10.1021/acs.jpca.8b01743.

[28] F. Williams, Electronic states of solid explosives and their probable role in detonations, in: Advances in Chemical Physics, John Wiley & Sons, Ltd, 1971, pp. 289–302, ISBN: 978-0-470-14369-8, https://doi.org/10.1002/9780470143698.ch20. https://onlinelibrary.wiley.com/doi/abs/10.1002/9780470143698.ch20.

[29] A. Delpuech, J. Cherville, Relation entre la structure electronique et la sensibilité au Choc des explosifs secondaires nitrés-critère moléculaire de sensibilité. I. Cas des nitroaromatiques et des nitramines, Propellants, Explos. Pyrotech. 3 (1978) 169–175.

[30] A. Delpuech, J. Cherville, Relation entre La structure electronique et la sensibilité au Choc des explosifs secondaires nitrés. Critère moléculaire de sensibilité II. Cas des esters nitriques, Propellants, Explos. Pyrotech. 4 (1979) 121–128.

[31] J.J. Gilman, Chemical reactions at detonation fronts in solids, Philosoph. Mag. B 71 (6) (1995) 1057–1068, https://doi.org/10.1080/01418639508241895.

[32] A.N. Dremin, V.Y. Klimenko, O.N. Davidova, T.A. Zoludeva, Multiprocess detonation model, in: Proceedings of the International Symposium on Detonation (9th) Held in Portland, Oregon, American Institute of Physics, 1989, pp. 724–729.

[33] A.N. Dremin, Towards detonation theory, J. Phys. IV France 05 (C4) (1995) C4–C4-276. https://doi.org/10.1051/jp4:1995420.

[34] J. Hooper, Vibrational energy transfer in shocked molecular crystals, J. Chem. Phys. 132 (1) (2010) 014507, https://doi.org/10.1063/1.3273212.

[35] D.E. Zwitter, A computation of the frequency dependent dielectric function for energetic materials, in: AIP Conference Proceedings, vol. 505, AIP, Snowbird, Utah (USA), 2000, pp. 405–408, https://doi.org/10.1063/1.1303503.

[36] M.M. Kuklja, B.P. Aduev, E.D. Aluker, V.I. Krasheninin, A.G. Krechetov, A.Y. Mitrofanov, Role of electronic excitations in explosive decomposition of solids, J. Appl. Phys. 89 (2001) 4156–4166.

[37] A. Barry Kunz, M.M. Kuklja, T.R. Botcher, T.P. Russell, Initiation of chemistry in molecular solids by processes involving electronic excited states, Thermochim. Acta 384 (1) (2002) 279–284, https://doi.org/10.1016/S0040-6031(01)00804-8.

[38] D. Margetis, E. Kaxiras, M. Elstner, T. Frauenheim, M.R. Manaa, Electronic structure of solid nitromethane: effects of high pressure and molecular vacancies, J. Chem. Phys. 117 (2) (2002) 788–799, https://doi.org/10.1063/1.1466830.

[39] E.J. Reed, M.R. Manaa, J.D. Joannopoulos, L.E. Fried, Electronic excitations, vibrational spectra, and chemistry in nitromethane and HMX, AIP Conf. Proc. 620 (1) (2002) 385–390, https://doi.org/10.1063/1.1483559.

[40] E.J. Reed, M. Riad Manaa, L.E. Fried, K.R. Glaesemann, J.D. Joannopoulos, A transient semimetallic layer in detonating nitromethane, Nat. Phys. 4 (1) (2008) 72–76, https://doi.org/10.1038/nphys806. https://www.nature.com/articles/nphys806.

[41] W. Taylor, A. Weale, The mechanism of the initiation and propagation of detonation in solid explosives, Proc. R. Soc. Lond. A Containing Papers Math. Phys. Charact. 138 (1932), https://doi.org/10.1098/rspa.1932.0173.

[42] N.V. Garmasheva, V.P. Filin, B.G. Loboiko, A.N. Averin, D. Mathieu, P. Simonetti, R. Belmas, Modeling and prediction of sensitivity in energetic materials, AIP Conf. Proc. 620 (1) (2002) 439–441, https://doi.org/10.1063/1.1483572.

[43] F.P. Bowden, A.D. Yoffe, Initiation and Growth of Explosion in Liquids and Solids, Cambridge University Press, United Kingdom, 1952.

[44] D. Mathieu, P. Martin, J.-P. La Hargue, Simulation of the electron dynamics in shockwaves and implications for the sensitivity of energetic materials, Phys. Scr. T 118 (2005) 171–173.

[45] A.A.L. Michalchuk, P.T. Fincham, P. Portius, C.R. Pulham, C.A. Morrison, A pathway to the athermal impact initiation of energetic azides, J. Phys. Chem. C 122 (34) (2018) 19395–19408, https://doi.org/10.1021/acs.jpcc.8b05285.

[46] A.A.L. Michalchuk, S. Rudić, C.R. Pulham, C.A. Morrison, Vibrationally induced metallisation of the energetic azide α-NaN3, Phys. Chem. Chem. Phys. 20 (46) (2018) 29061–29069, https://doi.org/10.1039/C8CP06161K. https://pubs.rsc.org/en/content/articlelanding/2018/cp/c8cp06161k.
[47] I. Borges, M. Barbatti, A.J.A. Aquino, H. Lischka, Electronic spectra of nitroethylene, Int. J. Quantum Chem. 112 (4) (2012) 1225–1232, https://doi.org/10.1002/qua.23080. https://onlinelibrary.wiley.com/doi/abs/10.1002/qua.23080.
[48] I. Borges, A.J.A. Aquino, H. Lischka, A multireference configuration interaction study of the photodynamics of nitroethylene, J. Phys. Chem. A 118 (51) (2014) 12011–12020, https://doi.org/10.1021/jp507396e.
[49] A. Bhattacharya, Y.Q. Guo, E.R. Bernstein, Experimental and theoretical exploration of the initial steps in the decomposition of a model nitramine energetic material: dimethylnitramine, J. Phys. Chem. A 113 (5) (2009) 811–823, https://doi.org/10.1021/jp807247t. publisher: American Chemical Society.
[50] Y. Dou, B.R. Torralva, R.E. Allen, Semiclassical electron-radiation-ion dynamics (SERID) and cis-trans photoisomerization of butadiene, J. Mod. Opt. 50 (15-17) (2003) 2615–2643, https://doi.org/10.1080/09500340308233589.
[51] W. Zhang, J. Sang, J. Cheng, S. Ge, S. Yuan, G.V. Lo, Y. Dou, A photophysical deactivation channel of laser-excited TATB based on semiclassical dynamics simulation and TD-DFT calculation, Molecules 23 (7) (2018) 1593, https://doi.org/10.3390/molecules23071593. https://www.mdpi.com/1420-3049/23/7/1593.
[52] B.F.E. Curchod, T.J. Martínez, Ab initio nonadiabatic quantum molecular dynamics, Chem. Rev. 118 (7) (2018) 3305–3336, https://doi.org/10.1021/acs.chemrev.7b00423.
[53] J. Ghosh, H. Gajapathy, A. Konar, G.M. Narasimhaiah, A. Bhattacharya, Sub-500 fs electronically nonadiabatic chemical dynamics of energetic molecules from the S1 excited state: ab initio multiple spawning study, J. Chem. Phys. 147 (20) (2017) 204302, https://doi.org/10.1063/1.4996956.
[54] W.K. Peters, D.E. Couch, B. Mignolet, X. Shi, Q.L. Nguyen, R.C. Fortenberry, H.B. Schlegel, F. Remacle, H.C. Kapteyn, M.M. Murnane, W. Li, Ultrafast 25-fs relaxation in highly excited states of methyl azide mediated by strong nonadiabatic coupling, Proc. Natl. Acad. Sci. U.S.A 114 (52) (2017) E11072–E11081, https://doi.org/10.1073/pnas.1712566114. https://www.pnas.org/content/114/52/E11072.
[55] B. Smith, A.V. Akimov, Modeling nonadiabatic dynamics in condensed matter materials: some recent advances and applications, J. Phys. Condens. Matter 32 (7) (2020) 073001, https://doi.org/10.1088/1361-648X/ab5246.
[56] L. Wang, J. Qiu, X. Bai, J. Xu, Surface hopping methods for nonadiabatic dynamics in extended systems, WIREs Comput. Mol. Sci. 10 (2) (2020) e1435, https://doi.org/10.1002/wcms.1435.
[57] D. Mathieu, P. Martin, Preliminary investigations of weak non-adiabatic effects in materials from simulations on model clusters, Comput. Mater. Sci. 10 (1) (1998) 235–239, https://doi.org/10.1016/S0927-0256(97)00132-8.
[58] D. Mathieu, P. Simonetti, P. Martin, A model to study the electronic response to an impact in energetic materials, AIP Conf. Proc. 429 (1) (1998) 309–312, https://doi.org/10.1063/1.55538.
[59] D. Mathieu, P. Martin, Molecular dynamics simulation of shockwaves including some nonadiabatic effects, Comput. Mater. Sci. 17 (2000) 347–351.
[60] D. Mathieu, P. Martin, P. Simonetti, Nonadiabatic simulation of valence electrons in solids under sustained shockwaves, AIP Conf. Proc. 505 (1) (2000) 409–412, https://doi.org/10.1063/1.1303504.
[61] J.R. Schmidt, P.V. Parandekar, J.C. Tully, Mixed quantum-classical equilibrium: surface hopping, J. Chem. Phys. 129 (4) (2008) 044104, https://doi.org/10.1063/1.2955564.
[62] B. Smith, A.V. Akimov, A comparative analysis of surface hopping acceptance and decoherence algorithms within the neglect of back-reaction approximation, J. Chem. Phys. 151 (12) (2019) 124107, https://doi.org/10.1063/1.5122770.
[63] E.J. Reed, Electron-ion coupling in shocked energetic materials, J. Phys. Chem. C 116 (2012) 2205–2211.

PART IV

Insight from numerical simulations

CHAPTER

12

Molecular dynamics simulation of hot spot formation and chemical reactions

Didier Mathieu[a,*] *and Itamar Borges, Jr*[b]

[a]CEA, DAM, Le Ripault, Monts, France [b]Departamento de Química, Instituto Militar de Engenharia (IME), Rio de Janeiro, RJ, Brazil

*Corresponding author: E-mail: didier.mathieu@cea.fr

1 Introduction

In different technological areas, chemical reactions play an essential role and have fundamental and practical interest. For instance, a detailed knowledge of chemical reaction mechanisms at the reaction fronts for the high pressures and temperatures involved in combustion conditions is useful for designing thrusters for rocket engines and satellites. Regardless of the application, materials must generally withstand temperature variations not only without untimely phase change but also without chemical degradation, which is of paramount importance for energetic compounds since their exothermic decomposition can lead to a runaway reaction and possibly cause an explosion, i.e., a deflagration or a detonation. Such events can result from various external stimuli. Accordingly, a range of criteria reflecting the sensitivity of a given material to standard loads have been defined (Chapter 2). For any new energetic material (EM) of potential interest, they are measured using standard testing equipment to assess its safe storage and transportation.

For instance, the probability that a given EM is initiated under shock is characterized by a sensitivity derived from gap test experiments involving pressures typically between 3 and 20 GPa [1]. This is in sharp contrast with the drop-weight impact test, which is the most commonly used sensitivity criterion and for which relevant pressures are below 1.5 GPa. Likewise, the time constants for reactions may differ dramatically between various tests, ranging from 0.05 to 2 μs in a typical shock experiment, to 200–250 μs in a drop-weight impact test [1] and about 10^3 s in thermal analyses for the determination of decomposition temperatures [2]. In view of the challenges inherent to *in situ* monitoring of very fast and poorly controlled molecular processes occurring in opaque condensed phases, it comes as no surprise

https://doi.org/10.1016/B978-0-12-822971-2.00012-7

that the origin of EM sensitivities remains largely controversial, despite steady efforts and significant progress in instrumentation and experimental resolutions [3].

The initiation of an EM is a multiscale process, a macroscopic load applied to a sample causing chemical reactions and energy release at the molecular level and finally leading to an explosion revealed at the continuum scale. Describing this process is a twofold challenge. We must first establish how the applied perturbation affects the local molecular environment, e.g., by modifying the local pressure and temperature. Then, we must find out how this affects the molecular dynamics and the outcome of chemical reactions. Except maybe for thermal loads, describing how the macroscopic insult affects the lower spatial scales down to that of molecules requires linking various disciplines, such as material mechanics or heat transfer [4, 5]. This goes beyond the scope of this book. The present chapter deals with the second challenge, assuming that molecules are subjected to given local conditions (e.g., stress, temperature, initial material structure, either homogeneous or defective, and so on) and determining the corresponding outcome from molecular simulation.

The main focus is on molecular dynamics (MD), which is by far the most commonly used simulation technique in the EM community. MD simulations of materials solve the classical equations of motion for the atoms in a representative volume of microscopic scale. This volume must be small enough to limit the number of atoms to handle, and large enough to minimize artifacts that may for instance prevent the formation of shear bands. For a crystal, such a simulation box is typically obtained by repeating the unit cell along each axis. For liquids or amorphous solids, a large cubic simulation box is used so that the system is approximately isotropic. Periodic boundary conditions (PBCs) are applied to this simulation box so as to minimize edge effects. MD simulations of materials based on this approach fall into two groups:

- Nonequilibrium molecular dynamics (NEMD) is used to simulate the transient response of a material to an external load. This can be done by integrating the classical equations of motion for the atoms, which leave the total energy E constant, in addition to the number of atoms N and simulation cell volume V (NVE ensemble).
- Equilibrium MD is used to simulate a system in thermal/mechanical equilibrium with its surroundings. This entails modifying the classical equations of motion of the atoms so as to fix the average value of temperature T (NVT ensemble) using a thermostat, and possibly pressure P (NPT ensemble) using a barostat.

In the context of EMs, MD is extensively used to get insight into the microscopic details of the initiation process. Regardless of the nature of the insult, the latter can usually be broken down into three stages loosely labeled in this chapter as follows:

- Physical step: describes prereactive phenomena, like the localization of the external energy provided by the applied load, leading to the formation of hot spots where enhanced P-T conditions prevail, enabling vibrational and/or electronic excitation of molecules;
- Chemical step: describes the onset of chemistry and subsequent reactions until the formation of the final products;
- Propagation step: describes how the intermediates and hot products evolved from early decompositions activate further reactions, leading to a self-sustained reaction front.

Simulations of the initiation process including the self-sustained regime (propagation step) are mostly restricted to model compounds whose evolution into final products takes place in very few steps and therefore on time scales accessible to simulations [6]. Therefore, MD is mostly limited to the physical and chemical steps.

Simulations including the physical step are often NEMD simulations applied to shocked materials, with the goal to understand nonreactive processes prior to ignition: appearance of shear bands under plastic deformation following mechanical stress, nucleation and growth of hot spots arising from defects such as voids, cracks, and dislocations. They require simulating very large systems (e.g., several millions atoms) over significant periods of time (several nanoseconds), especially if the dynamic compression is to be included in the description. This calls for highly efficient computers and potentials, as well as special algorithms to describe shocked samples.

Applications to the chemical step usually consist in equilibrium MD simulations in the *NV T* or *NPT* ensemble. They are used to get insight into decomposition mechanisms, explain empirical findings regarding sensitivities, or estimate reactions rates.

Pioneering applications of MD to shocked materials were reviewed by White et al. [7]. Using simple models such as 2D atomic arrays, they yielded valuable background information on interactions of molecules with shockwaves. A more recent review focuses on equilibrium MD simulations of chemically realistic EMs under static compression [8]. The present chapter provides an update in this area, including both equilibrium and nonequilibrium MD simulations of heated or mechanically loaded EMs. Section 2 provides a brief overview of the special MD techniques introduced to deal with shocked materials and reviews simulations aimed at understanding the basic physics of shock initiation. Section 5 reviews reactive molecular dynamics (RMD) simulations aimed at determining decomposition pathways. Since ab initio molecular dynamics (AIMD) and density functional tight-binding (DFTB) dynamics simulations are reviewed in Chapter 13, emphasis is put here on simulations based on the reactive force field known as ReaxFF.

2 Simulation techniques for initiation

2.1 Reactive molecular dynamics

In contrast to conventional quantum chemical studies restricted to gas-phase mechanisms, MD accounts for the dependence of the reaction network on the molecular surroundings. However, the approach implies an accurate description of the molecular dynamics, including bond stretching preceding bond dissociation. This requires an integration time step even smaller than the 1 fs value often used in classical MD. In addition, chemical transformations occur infrequently compared to this time step. Under thermal activation, a given reaction step takes place with a characteristic time τ such that $1/\tau = A\exp(-E^{\dagger}/k_BT)$ where $E^{\dagger}$ and A are the activation energy and associated prefactor, k_B is the Boltzmann constant, and T the temperature. The value of $k_BT/E^{\dagger}$ required for a reaction to occur with a characteristic period τ is plotted in Fig. 12.1 for realistic values of $1/A$ of 1, 5, and 10 ps. This shows for instance that in the case where $1/A = 5$ ps, k_BT must be such that $k_BT/E^{\dagger} = 0.16$ for a reaction with

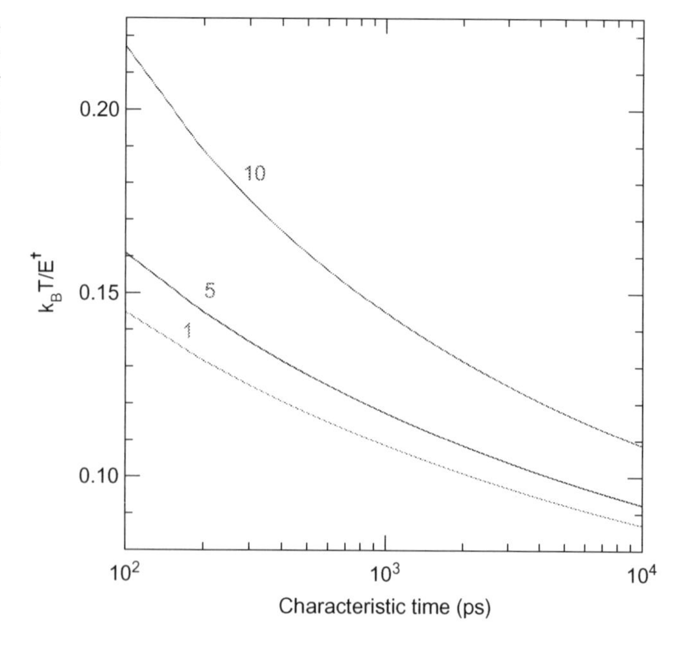

FIG. 12.1 Ratio $k_BT/E^†$ between temperature and apparent activation energy required so that the characteristic time scale τ for decomposition is equal to the value on the x-axis. The *three curves* are for inverse prefactors $1/A$ of respectively 1, 5, and 10 ps.

activation energy $E^†$ to take place over a characteristic time period of 100 ps. For a typical value of $E^†$ of 1 eV, i.e., about 100 kJ/mol, this implies that T should be about 2000 K. For instance, a ReaxFF study shows that hydrazine molecules just begin to decompose at 50 ps when temperature is set to 1000 K, whereas they decompose after about only 7 ps above 2000 K [9].

For lower temperatures, the large gap in time scales between chemical reactions and molecular vibrations would require the use of techniques specially designed for rare events that are not necessarily implemented along with the reactive potentials needed to efficiently derive the forces between atoms involved in reacting groups. To ensure that reactions will occur on the time scale of a simulation, applications of MD to the decomposition of EMs are mostly carried out under extreme P-T conditions ($P > 10$ GPa and $T \sim 10^3$ K) experimentally achieved in static regime by laser heating a sample confined in a diamond anvil cell (DAC). Such conditions are typical of the Chapman–Jouguet (CJ) state in front of a detonation front. However, assuming ignition to be thermal in nature, the onset of chemistry calls for the occurrence of high temperatures upstream in the initiation process. In fact, even under mild mechanical loads, extreme conditions are assumed to prevail in localized regions of high stress concentration [10]. Therefore, RMD applies either to the first reactions following loading or to the decomposition induced as a detonation front propagates in the material.

Even under such extreme conditions, MD simulations of decomposition processes remain computationally demanding. Twenty years ago, a simulation of HMX at 3500 K for up to 55 ps took over 1 year of simulation time [11]. Despite dramatic progress in computer hardware and software, the difference in time scales between molecular dynamics and chemical reactions remains an obstacle to the direct simulation of thermal decomposition, especially under mild conditions, whereas the extrapolation of kinetic data from high temperature simulations to lower temperatures is risky at least.

This is why much effort has been put into the development of more efficient alternative to AIMD, especially DFTB-based MD and ReaxFF. AIMD studies tend to focus on early reactive events, whereas DFTB and ReaxFF allows for more comprehensive descriptions of the decomposition paths, with the latter being suited to large-scale simulations focused on the physics involved in the interaction between shock, molecular structure, and overall decomposition process. Such studies are invaluable as they provide plausible explanations to sensitivity differences between materials. However, they are hardly amenable to detailed experimental investigations. Even the final products derived from the simulations cannot be unambiguously validated since experimental determinations of the gaseous decomposition products are done after their isentropic expansion and subsequent cooling.

2.2 Nonequilibrium molecular dynamics

NEMD is a natural approach to simulate the transient processes following mechanical loading, before or during the build-up of a shock. Shock waves can be produced within a NEMD simulation, thus providing insight into their structure. In addition, an explosion arising from the exothermic decomposition of an EM can only be observed in the NVE ensemble as the use of a thermostat would hamper the temperature rise inherent to such a thermal explosion.

Including the shock within the simulation obviously requires that 3D periodicity be dropped. PBCs are applied only to directions xy orthogonal to the shock direction z. The resulting model is actually a slab orthogonal to z. A shock can be produced using another slab (flyer plate) impacting the front surface of the sample, by impacting the sample with a rigid piston moving at constant velocity, or by launching the sample against a fixed rigid wall. However, as soon as the shock reaches the downstream surface, a rarefaction wave propagates rapidly back into the compressed material so that it relaxes prematurely from the Hugoniot state. Therefore, the simulation cell must extend over a very long distance in the z direction so as to observe the shock over a significant period of time. This makes the computation prohibitive and prevents the study of interesting shock-induced processes taking place over long time periods, including plastic deformation or chemical reactions.

To avoid the need to simulate a tremendous number of atoms, one may focus on a moving computational cell surrounding the shock front, with planes of unshocked material introduced downstream of the shock, while slices of shocked matter are removed upstream [12]. Such simulations provide nonequilibrium quantities as a function of the distance from the shock through time-averaging procedures.

The shock-front absorbing boundary condition (SFABC) method extents the Hugoniot state by preventing the emission of rarefaction waves [13]. It starts as a standard NEMD run where the distance between the downstream surface and the piston is monitored so as to detect the point when the maximum compression is reached. At this point, the molecules at the end of the simulation cell are frozen by setting the corresponding velocities and interatomic forces to zero. The sample is thus maintained at constant volume between the two frozen pistons, which mimics the compressed state of matter behind a sustained shock front. In the field of EMs, this method was successfully applied to the study of processes occurring over time scales beyond the reach of standard NEMD simulations, like the nucleation and growth of shear bands over $\sim$150 ps [14] or hot spot growth for 0.5 ns after the shock-induced collapse of a cylindrical void 20 nm in diameter [15].

2.3 Compress-and-shear reactive dynamics

The compress-and-shear reactive dynamics (CS-RD) is a special NEMD technique specifically designed to quickly (within a few ps of MD) assess shear-induced sensitivities [16]. It acknowledges the major role of shear in the chemical response of EMs to compression, which was already anticipated from static simulations [17, 18]. Under hydrostatic pressure, an organic crystal is forced to occupy a smaller volume. Such constraints do not favor bond rupture and molecular dissociation, but rather promote polymerization reactions through the formation of new covalent bonds as the molecules are forced closer to each other. Decomposition could be easier under uniaxial compression, as a consequence of an increased proximity of the molecules in the direction of compression. However, in static high pressure simulations of molecular crystals, mechanically-induced decompositions are mostly observed upon application of shear in addition to uniaxial compression, as illustrated in Fig. 12.2 for nitromethane.

In the CS-RD procedure, a piston is first applied to precompress the crystal uniaxially, for instance up to a 10% decrease of the size of the simulation cell in the direction of the shock direction. After reorientation of the system to make the slip plane parallel to a new xz-plane containing the slip direction, the system is equilibrated under ambient conditions, and finally sheared at a constant rate by changing the periodic cell parameters and rescaling atomic fractional coordinates at every time step [16].

This method designed so as to promote chemical reactions immediately proved very promising, accounting for the observed anisotropy of the shock sensitivity in PETN single crystals [16]. Even more interestingly, they show this anisotropic response to be due to shear. Indeed, despite the inherent anisotropy of the crystal, uniaxial compression without shear (at a rate sufficient to generate a shock velocity close to the detonation threshold) yields similar responses for various shock directions, in contrast to experimental observations. In subsequent articles, CS-RD yielded anisotropic sensitivities consistent with experimental data for other energetic crystals beyond PETN [16], including RDX [19], β-HMX [20], and the TNT/CL-20 cocrystal [21].

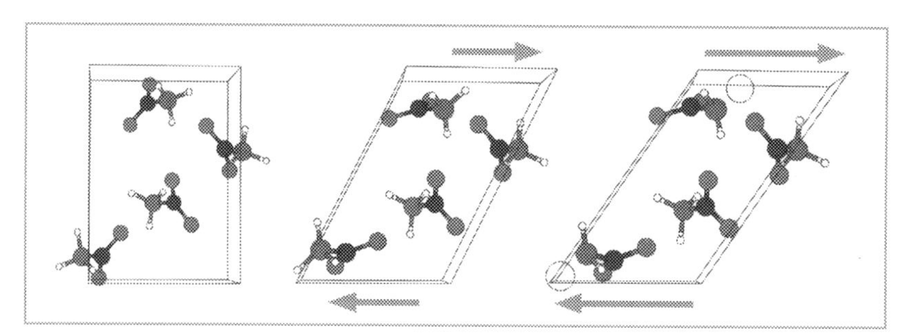

FIG. 12.2 Shear applied to a nitromethane crystal tears the molecules with departure of the hydrogen atoms, represented on the rightmost image by *white spheres* of large diameter. No such reaction is observed under hydrostatic compression. *Data from V. Guilbaud, J.-P. Dognon, D. Mathieu, C. Morell, A. Grand, P. Maldivi, Chemistry is everywhere, CLEFS CEA 60 (2011) 17–20.*

2.4 Equilibrium molecular dynamics

Although EM initiation is a nonequilibrium process, valuable insight is gained using equilibrium MD. First, the response of the material to loading may be viewed as a relaxation toward a new equilibrium state consistent with altered external conditions. For instance, the decomposition thermally induced in nitromethane, e.g., by flash heating the compound in a DAC under constant volume or propagating a strong shock into it, can be viewed as a natural evolution of the system to match extreme *P-V* conditions consistent with those prevailing in the surroundings. This allows the decomposition process to be identified with the reactions observed in *NVT* simulations under such conditions [22]. Likewise, equilibrium MD is used to simulate cook-off experiments, which involve lower heating rates. After a preliminary equilibration step under ambient conditions, a ramp of temperature is applied to stimulate the heating process [9].

Much of the value of equilibrium MD stems from the fact that many processes of practical interest, including any change in the macroscopic thermodynamic variables, occur over long time scales by comparison with the relaxation of a microscopic volume toward a transient local equilibrium state. Therefore, equilibrium MD may be used to describe such a state, or the evolution of a system through a series of equilibrium states. For instance, the kinetic parameters describing the shock-induced decomposition of TATB were obtained from a series of constant pressure RMD simulations for *P* ranging from 0 to 30 GPa [23].

2.5 Hugoniostats

The relaxation of an anisotropically compressed state in a shocked solid may require some plastic flow, i.e., strong enough shear stresses to overcome the corresponding energy barriers. In this context, there can be a residual shear behind the shock that cannot be considered using standard *NVT* simulations. The uniaxial Hugoniostat method is a procedure enforcing the Hugoniot conditions and allowing for such anisotropic states. Starting from an initial P_0, T_0 equilibrium state, it involves applying a uniaxial compression to the simulation cell followed by an equilibrium MD simulation coupled with a thermostat ensuring the relaxation of the internal energy to a final value satisfying the Hugoniot relationships [24]. This approach is referred to as the *NV*Hug formulation to emphasize this constraint, as well as the fact that *N* and *V* are prescribed [25].

The abrupt compression involved in this scheme might induce high stresses leading to a final state different from that obtained in a full NEMD simulation. This could be avoided by introducing an extra dynamical variable into the Hugoniostat equations, acting as a piston to equilibrate to a preset value the component of the stress tensor in the *z* direction [25]. This variable was introduced along with damping rates aimed at eliminating large overshoots in stress and temperature that might lead to artifacts, like plasticity or phase transitions not observed in corresponding NEMD simulations. The procedure is referred to as NP_{xx}Hug. This emphasizes the fact that unlike *NV*Hug, it is a diagonal component P_{xx} of the pressure which is fixed in place of the volume *V*. Such simulations are used to study various aspects of shock responses, including the anisotropic sensitivity and plasticity of the TKX-50 energetic salt [26] and the detailed chemistry involved in the shock-induced decomposition of polyvinyl nitrate [27].

2.6 Multiscale shock technique

To simulate shocked materials over long time periods using only small simulation cells, an alternative to Hugoniostats is the so-called multiscale shock technique (MSST) [28]. This method combines MD with the Euler equations for a compressible flow. The volume V of the simulation cell is a dynamical variable associated with a mass like parameter Q. By varying the shock velocity, arbitrary values of the final pressure P may be obtained. This method allowing computational speedups of at least 10^5 with respect to standard NEMD simulations is used to study reactions in shocked materials [27, 29] and their orientation dependence [30]. A potential drawback might be long-lived volume oscillations not observed in NEMD and making it more difficult to determine phase changes along the Hugoniot [25]. MSST simulations of shock-induced reactions in various crystals yield valuable rules of thumb to design insensitive explosives [31].

2.7 Thermal versus athermal initiation

Depending on the external load applied, a wide range of mechanisms might explain its ability to initiate an EM, possibly involving direct mechanical disruption of bonds, phonon up-pumping in the crystal phase (Chapter 10) or electronic excitations (Chapter 11), without necessarily requiring prior thermal equilibrium (athermal mechanisms). In fact, whatever the nature of the insult, an induction time is usually observed between loading and chemical response. This suggests that reactions are not a direct consequence of the load, whose role could be just to heat the material up to the decomposition temperature.

Accordingly, the standard model of initiation is based on the notion that the process is thermal in origin [32]. It assumes a localization of the energy supplied by any external stress into so-called "hot spots" where mechanical/electrical energy is converted to heat and nucleates self-propagating reaction fronts. Critical hot spots (i.e., those able to initiate an explosion) must be typically 0.1–10 μm in size, reach ~700 K in temperature, and last for $10^{-5} - 10^{-3}$ s. Different types of loads produce different hot spots leading to distinct ignition mechanisms. For instance, shock-induced hot spots are short lived (μs) but exhibit high temperatures (~1000 K), while impact scenarios produce hot spots larger in size (mm) and duration (ms) but lower in temperature (few hundred Kelvin).

This thermal approach assumes that the system achieves equilibrium before the transition state is reached. In such equilibrium reactions, the exact nature of the initiating stimulus is irrelevant as far as chemistry is concerned. This treatment is more or less valid depending on the circumstances. In a RMD study of nonequilibrium reaction kinetics in NM, HMX, and PETN under direct heating at various rates and irradiation with electric fields of various frequencies and strengths, the equilibrium picture was found to hold only for HMX [33]. For this compound, decomposition does not significantly deviates from statistical behavior owing to fast internal relaxation. For NM, which involves a bimolecular mechanism, insults targeting specific modes do lead to strong nonequilibrium states. However, their relaxation takes place over time scales comparable to those of the initial decomposition reactions so that they do not affect the overall kinetics. Finally, athermal decomposition stimulated by loads targeting specific modes is best demonstrated for PETN, whose decomposition involves O–NO_2 as a trigger bond [33].

3 Void collapse and hot spot formation

3.1 Early steps

Voids in EMs, including cavities in solids or bubbles in liquids, are well-known hot spot precursors [32]. MD simulations are well suited to get insight into the formation and subsequent evolution of such hot spots. A number of mechanisms may explain how voids in shocked materials leads to local overshoots in temperature, including adiabatic heating through *PV* work of gaseous molecules trapped into the collapsing void, local energy dissipation upon viscoplastic collapse of the pore, or local heating at the rear surface as it is impacted by a hydrodynamic jet accelerated through the void (Fig. 12.3). The relative significance of such mechanisms depends on many factors. For instance, *PV* work is expected to be especially significant for the largest voids, whereas the hydrodynamic jetting regime prevails for strong shocks.

Direct spectroscopic monitoring of pore collapse in real explosives is currently not possible. At higher scales, experimental shock studies on model systems such as transparent explosive emulsions with millimeter size cavities show asymmetrical void collapse and formation of material jets accelerated across the cavity [34–37]. At lower scales, MD simulations of shocked materials provide additional insight.

By introducing a planar gap into a Lennard–Jones (LJ) crystal, a simple 1D crush-up model demonstrates local temperature overshoots despite the lack of focusing effects, in addition to accounting for the role of void size [38]. The hot spot arises from the vaporization of ejected molecules from upstream into the gap, followed by their stagnation as a vapor and finally, recompression upon collision with the far side of the void. Additional details are obtained considering a cubic nanovoid instead of a planar gap [39]. Under shock, some molecules are ejected into the void with a characteristic spall velocity u_p and accelerated up to $\sim 2.6u_p$ until they collide with the far side of the void. This acceleration is attributed to momentum transfer from the collapsing wall, eventually leading to jetting. A detailed analysis shows that local temperatures estimated from molecular kinetic energies do not reflect the reaction rate and that nonequilibrium effects must be taken into account.

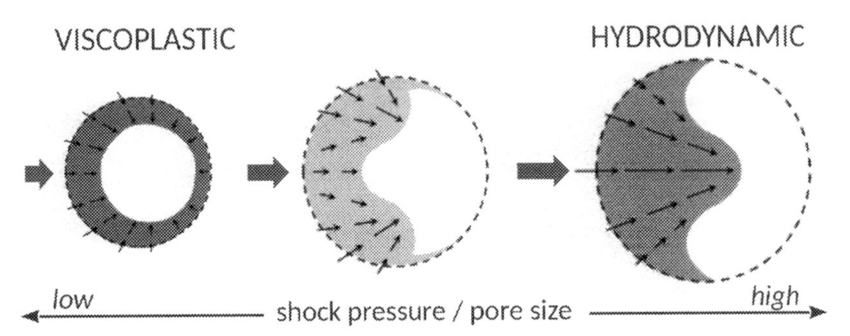

FIG. 12.3 Going from viscoplastic to hydrodynamic collapse regime as pore size and/or shock strength is increased. The *arrow* represents the shock coming from the *left*. *Adapted from M.A. Wood, D.E. Kittell, C.D. Yarrington, A.P. Thompson, Multiscale modeling of shock wave localization in porous energetic material, Phys. Rev. B 97 (1) (2018) 014109, https://doi.org/10.1103/Phys-RevB.97.014109 with permission from APS.*

3.2 Chemically realistic systems

Following early simulations of shock-induced void collapse on model systems, similar simulations are made for chemically realistic systems. For the collapse of an 8 nm diameter spherical void in RDX, ReaxFF simulations yield results similar to those mentioned above for LJ solids. For an impact velocity of V_p, molecules from the upstream wall are ejected into the void with a velocity of $2V_p$, forming a nanojet with a maximum velocity of $\sim 3V_p$ [40]. The jet acceleration is attributed to a lensing effect due to interactions with the void walls along the shock direction. Upon void closure, chemical reactions in RDX first produce NO_2 fragments. Then, after the molecular jet strikes the downstream wall, a wider variety of products is observed, including N_2, H_2O, and HONO. This chemistry might differ from that induced in larger pores. However, similar simulations on CL-20 crystals reveal that early bond-breaking reactions are little affected as shock strength or void size increases, although they occur at a faster rate due to enhanced hot spot temperatures [41].

Using a nonreactive force field for RDX, a larger cylindrical pore 35 nm in diameter is simulated under shock pressures ranging from 9.7 to 42.3 GPa [42]. Material flow during collapse is visualized through the spread and mixing of sets of initially contiguous molecules and evolution of local velocity fields, which proves to be a very visual way of apprehending what is going on. From the results, a clear distinction can be made between the weak and strong shock cases. For a weak shock, the pore collapses via a viscoplastic flow of material without jet formation and with little penetration of material into the downstream region. For a strong shock, the collapse is hydrodynamic with a jet that penetrates significantly into the downstream pore wall, causing a large deposition of mechanical energy leading to much enhanced temperatures in the collapse zone.

These results are reminiscent of experiment [36] and in full agreement with continuum simulations showing how voids can lead to significant local overshoots in temperature under dynamic compression [43]. This is summarized in Fig. 12.4 for a cylindrical pore. The primary effect of the shock is to start a void collapse via plastic deformation (Fig. 12.4A). A continuum of behaviors is observed upon increasing shock strength [44], with the ejection of upstream

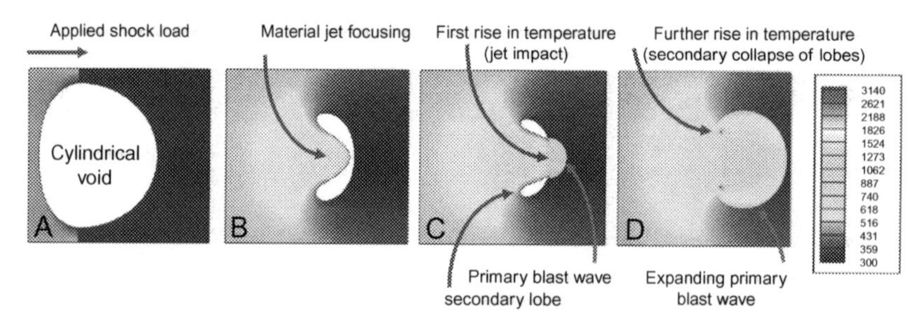

FIG. 12.4 Shock-induced collapse of a cylindrical pore compressed by a shock coming from the *left* as viewed using temperature maps (K): 1. compression of the pore as the shock front comes into contact; 2. jetting of molecules inside the pore; 3. impact of the ejected molecules on the rear surface of the pore whose temperature is increased; 4. collapse of the residual lobes. See text for details. *Adapted from N.K. Rai, M.J. Schmidt, H.S. Udaykumar, High-resolution simulations of cylindrical void collapse in energetic materials: effect of primary and secondary collapse on initiation thresholds, Phys. Rev. Fluids 2 (4) (2017) 043202, https://doi.org/10.1103/PhysRevFluids.2.043202. With permission from APS.*

material coming primarily from an angle of ~45 degree to the shock propagation and flowing toward the centerline of the pore (Fig. 12.4B). The focusing and acceleration of upstream molecules (lensing effect) increases with shock strength until they form a hydrodynamic jet that strongly impacts the downstream surface, producing a primary blast wave (Fig. 12.4C) and hot spot that may develop further. For a moderate shock, ignition might fail to start from this primary hot spot, taking place instead at secondary hot spots created during the implosion of the residual lobes, giving rise to a pinching process within vertices resulting in highly compressed regions where the temperature can exceed that reached in the primary hot spot (Fig. 12.4D).

This shows that an initiation resulting from hydrodynamic pore implosion may involve very different decomposition regimes depending on many factors including shock strength, pore size, and geometry or chemical nature of the material, which can render the modeling of sensitivities highly challenging.

3.3 Simulations showing self-sustained decomposition

The simplicity of a diatomic reactive system $2AB \rightarrow A_2 + B_2$ allows to simulate the initiation of a self-sustained decomposition and its dependence on various parameters [45]. In the perfect AB crystal described with the REBO reactive potential, a shock initiates a detonation provided it compresses the material by about 25% (~10 GPa). The corresponding impact velocity is close to ~4 km/s, i.e., three orders of magnitude larger than impact velocities that ignite real samples in drop weight impact tests [7]. Small defects like vacancies, isolated radicals, or thin planar 1D gaps tend to have no effect on this threshold. In contrast, larger voids or planar gaps of sufficient width lower this threshold through the formation of hot spots [7, 38, 45]. For ellipsoidal voids, this effect depends on their orientation with respect to the shock. Such simulations show that void-induced initiation is not a direct mechanical process, but a thermal one taking place long after the first ejecta impacts.

Further insight is obtained in simulations of shocks in solid nitrogen cubane (N_8) with rectangular nanovoids of various aspect ratios [46]. An analytic model for the detonation threshold as a function of the void dimensions is put forward based on the simulations. Most results are reminiscent of experimental observations at larger scales, including enhanced sensitivity relative to the void-free crystal, asymmetric collapse, and jetting, with a temperature overshoot at the jet impact site on the downstream wall where initiation begins. Furthermore, a transition from single to double jetting is observed with increasing transverse void length above ~15 nm, similar to the observation of dual jets during the collapse of macroscopic voids with a flat upstream wall [34].

Interestingly, high temperatures (2030–4642 K) decompose N_8 into mostly N_2 dinitrogen molecules, with a small fraction of short linear oligomers (up to N_6). Because N_8 decompositions take place randomly within the volume of the sample in this thermal decomposition regime under heat loads, the resulting fragments have few opportunity to polymerize. By contrast, strong shocks decompose N_8 into longer chain intermediates. This is because neighboring molecules at the shock front decompose almost simultaneously, so the resulting fragments readily polymerize. This suggests significant differences between a shocked material and the same system subject to similar temperatures and pressures.

3.4 Shock to deflagration transition in RDX

Large-scale ReaxFF simulations make it possible now to study the onset of deflagration in chemically realistic materials, as shown recently for RDX [47]. In this study, a shock pressure of ~11 GPa, unable to ignite a defect-free RDX sample, is shown to ignite a sample with a cylindrical pore whose axis is perpendicular to the shock direction. As the shock encounters the upstream wall of the pore, molecules are ejected into the void until they impact the far wall. This impact heats up the downstream region, giving rise to a crescent-shaped hot spot reaching local temperatures of ~1500 K. While pores ≤20 nm in diameter get quenched in less than 40 ps, a 40 nm void leads to a self-sustained deflagration wave.

A detailed analysis of the simulation reveals that the transition to deflagration takes place in three stages after the impact of the ejecta against the downstream wall (Fig. 12.5). The first step corresponds to the growth of the hot spot sustained by the complete decomposition of some molecules up to the exothermic formation of early final products in less than 10 ps. The chemical energy thus released contributes to sustain the hot spot which otherwise might get quenched due to dissipation and rarefaction waves.

In the second step illustrated in Fig. 12.5, the rate of exothermic product formation increases and two reaction fronts may be observed: one that propagates forward into the crystalline RDX and another one moving backward into the material pushed into the pore. A steady-state temperature ~4000 K is achieved at the end of this stage. Finally, the third stage is the propagation of the reaction fronts at speeds of ~250 m/s.

3.5 Hot spot growth and relaxation

Continuum simulations of collapsing cylindrical pores show that they generate approximately radial hot spots on a time scale of a few tens of picoseconds, which develop further over times on the order of nanoseconds [43]. The use of large-scale multimillion MD simulations based on ReaxFF and the SFABC technique provides insight into the atomistic details of this process, as shown recently for a shocked cylindrical void 20 nm in diameter in a PETN crystal [15].

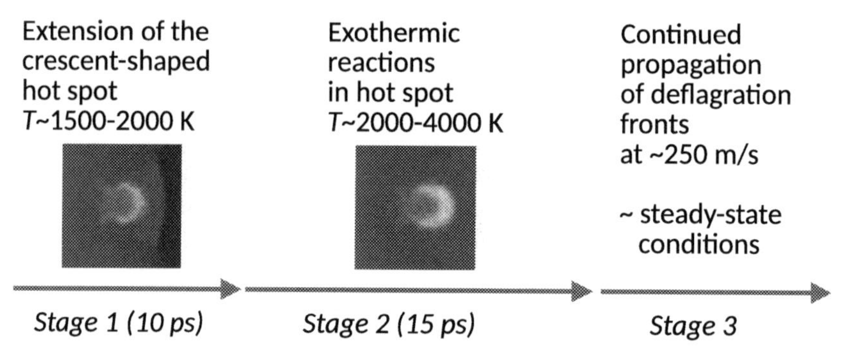

FIG. 12.5 Three stages observed in the shock to deflagration transition induced by a cylindrical nanopore compressed by a shock coming from the *left*. The colors representing the vibrational temperature highlight the crescent-shaped hot spot growing from the downstream region of the pore. *Adapted from M.A. Wood, M.J. Cherukara, E.M. Kober, A. Strachan, Ultrafast chemistry under nonequilibrium conditions and the shock to deflagration transition at the nanoscale, J. Phys. Chem. C 119 (38) (2015) 22008–22015, https://doi.org/10.1021/acs.jpcc.5b05362 with permission from ACS.*

The shock compression of the pore induces a hot spot that is roughly the area of the original void in about 30 ps. Its subsequent growth continues slowly for about 0.2 ns. From 0.2 to 0.4 ns, the hot spot grows more and more rapidly, while remaining radially symmetric. After 0.4 ns, it grows extremely rapidly, filling most of the simulation cell about 0.5 ns after the start of the collapse. Between 0.3 and 0.5 ns, the hot spot section evolves from a disc ~46 nm in diameter into a diamond with longitudinal × transverse dimensions of 110 × 90 nm.

A detailed analysis shows that the very fast increase in size of the hot spot at ~0.4 ns stems from its coalescence with a secondary triangular hot zone which coincides with the double-shocked region formed during the first tens of picoseconds as the primary (planar) shock wave is overtaken by the secondary (cylindrical) shock wave. The successful initiation of the EM depends not only on hot spot formation and growth, but also on its relaxation rate. In addition to continuum thermal simulations, the details of the relaxation processes can be studied using MD, as done for TATB using 1D models [48].

4 Influence of extended defects

4.1 Steric hindrance model and π-stacked buffers

By analogy with metals, simple atomistic models of dislocations in EMs have been studied to assess how they could modify the band structure (Chapter 11). However, the primary incentive for considering dislocations in EM initiation lies in the fact that shocks above a critical strength (called the Hugoniot elastic limit or HEL) induce plastic deformation. Dick explains the anisotropy of shock sensitivity in PETN [49] as a consequence of dislocation-mediated plasticity in the less sensitive directions, facilitating the relaxation of shear stresses that would otherwise distort the molecules and induce bond-breaking events [50, 51]. This approach, widely known in the EM community as the steric hindrance model (SHM), implies that shock directions with high sensitivity correspond to a high value of the HEL, and vice versa.

To estimate sensitivity on the basis of the SHM, the plastic deformation mechanisms observed under quasistatic loading must be extrapolated over many orders of magnitude in strain rate so that the anisotropy in shock sensitivity and HEL data should correlate with the slip systems identified in the crystal [50, 51]. However, it is established that the slip systems of RDX obtained from quasistatic measurements do not correlate with HEL data from shock experiments, a finding rationalized with the help of MD simulations showing that shock-induced deformations involve slip systems that are not readily activated under quasistatic loading at ambient conditions [52]. In this respect, due to its highly symmetric crystal structure, PETN might prove an exception rather than the rule.

In the same spirit as the SHM, it was suggested that any EM with molecules arranged in graphitic planes within crystalline phases (as observed in TATB) would have its sensitivity decreased due to the fact that molecular planes can slide over each other during shearing. This is attributed to the fact that π-stacked interactions between layers would act as buffers temporarily absorbing part of the shock energy instead of letting it immediately excite vibrations leading to molecular decomposition [53]. Although attractive, this approach is difficult to verify since the compounds forming graphitic structures are generally aromatic molecules that are naturally insensitive [54].

4.2 Dislocations as hot spot precursors

Beyond their role in plastic deformation, dislocations could act as precursors of linear hot spots in shocked materials, akin to the way voids act as precursors of hot spots. For instance, the role of dislocations on the shock sensitivity of RDX was studied through MSST simulations based on ReaxFF-lg [55]. It was observed that the decay rate of a perfect RDX crystal is hardly increased as screw dislocations are introduced, whereas it is significantly enhanced by edge dislocations, which could indicate an increased sensitivity. Interestingly, RDX decay rates do not correlate with shear stress barriers. In contrast, they correlate with density, a reduction in density being accompanied by an increase in sensitivity. The authors assume that this stems from the larger activation volume available for decomposition. The direct role of edge dislocations was also studied in MSST/ReaxFF-lg simulations by Deng and coworkers along with shock and/or preheating [56].

4.3 Shear bands

In addition to the activation of dislocation slip systems [52], plastic deformation may be associated with micron-scale adiabatic shear bands where the temperature could become high enough for initiation [57]. Molecular simulations are well suited to address the complexity of plastic deformation, as done in large-scale NEMD simulations of HMX subjected to shocks of increasing strength [58]. A gradual transition was observed from usual dislocation-mediated plasticity to a regime involving nanoscale shear bands. Weak shocks induce plastic deformation associated with gliding dislocations and resulting crystallographic slip. Strong shocks induce plastic deformation through the formation of amorphous nanoscale shear bands where part of the shock energy gets trapped, leading to local overshoots in temperature. However, in contrast to what is observed in the macroscopic shear bands, the molecules are only slightly displaced with respect to their neighbors. Such extended regions might act as some kinds of extended hot spots for initiation, provided the high temperature is maintained long enough to ignite the material [58].

SFABC simulations of shocks along [100] in RDX crystals for long periods of time (>100 ps) showed the formation of such liquid-like shear bands and their propagation to the entire system, whereas their structure evolves to a steady-fluctuating state [14]. The intense viscous-flow-driven heating within them suggests that they should be considered as homogeneously nucleated hot spots. Therefore, while the SHM suggests that the associated deformation mechanisms promote low sensitivity through shear stress relaxation, molecular decomposition is clearly promoted through thermal (due to the enhanced temperature in shear bands) as well as mechanical (due to viscous flow assisted bond breaking) processes.

Reactive simulations of the formation of shock-induced nanoscale shear bands through plastic failure including the onset of chemistry were demonstrated for TATB [59]. This required a scale bridging technique since only the use of classical force fields enables to simulate the millions of atoms needed to resolve the shock front, whereas a quantum method like DFTB is most appropriate for a reliable description of chemical reactions. More specifically, using a nonreactive force field for TATB, the authors first carried out large-scale (>13 M atoms) classical MD simulations of a perfect crystal under an ~30 GPa shock, i.e., conditions close to the von Neumann shock pressure. They observed the formation and growth of hot shear band regions behind the shock and their amorphization through a ps-scale shear failure process distinct from the dislocation-mediated shear banding discussed above. These bands are typically ~10 nm wide and relax in roughly 20 ps to a final temperature of ~1200 K.

The decomposition kinetics was studied in a second step, starting from representative shock-compressed bulk 3D periodic configurations extracted from both the crystalline and shear band regions [59]. These configurations were simulated past their characteristic reaction time at the DFTB level and in the *NVT* ensemble, with *V* corresponding to the average postshock density of 2.96 g·cm^{-3} and *T* ranging from 2400 to 3200 K. Fitting the decay rates of TATB molecules to the Arrhenius expression $k = A\exp(-E_a/k_B T)$ reveals that the activation energy E_a decreases from 285 ± 30 to 213 ± 38 kJ/mol on going from bulk crystal to shear bands. Accordingly, shear bands prove to be very reactive, with decomposition of the TATB molecules two orders of magnitude faster than in the bulk crystal. They could ignite during the reported reaction zone time of ~100–300 ns [59]. Finally, this shear banding ignition mechanism departs from the usual hot spot concept where the onset of reactions stems from local heating but does not assume any change in activation energy [59].

5 Simulation of decomposition mechanisms

5.1 Nitromethane

As a prototypical energy compound, nitromethane (NM) is extensively studied. Most investigations consider neat NM, although the role of water and dioxygen molecules from the surroundings is addressed as well [60]. Liquid under ambient conditions, NM forms a simple crystal with four molecules per unit cell upon cooling. In an early DFT study near CJ conditions, i.e., compressed by a factor of 1.75–1.95 and at temperatures of 3000–4000 K, the formation of H_2O is observed as a first stable product within the very short simulation time of about 1 ps [22]. This demonstrates the role of the condensed phase in favoring proton extraction as the first chemical event, since gas-phase decomposition is expected to start with the C–N bond rupture. The transferred proton bonds to an oxygen atom of a surrounding molecule, which becomes less tightly bound to its nitrogen neighbor. The net outcome of this process is an intermolecular proton transfer uniquely associated with the condensed fluid:

$$2CH_3NO_2 \rightarrow CH_3NO_2H^+ + CH_2NO_2^-$$

The reaction products are a protonated NM molecule and the aci anion $CH_2NO_2^-$ [22]. More details are provided by a subsequent AIMD study of four NM molecules at 2200 K under ambient pressure, up to the formation of water after about 200 ps [61]. This long period of time can be investigated due to the small system size. The results reveal a complex decomposition mechanism involving about 75 species and 100 elementary reactions. The first step observed involves both the C–N bond cleavage (as observed in gas phase) and an intramolecular proton transfer in a single step:

$$CH_3NO_2 \rightarrow CH_2O + HNO$$

Four distinct stages are distinguished in the process, associated with the sequential involvement of the four initial NM molecules in the simulation box. Water acts as a catalyst transporting hydrogen atoms. The obtained final products are H_2O, CO_2, N_2, and linear –(–N=C–N=C=N–)– chains which do not decompose further due to the constraints associated with the PBCs.

The high efficiency of ReaxFF allows a relief of these constraints through the use of much larger simulation boxes [60, 62, 63]. The application of ReaxFF to liquid NM by Rom et al. consider 240 molecules in the simulation cell, simulated in the *NVT* ensemble for densities varying from ambient to 44% compression and temperatures ranging from 2500 to 4500 K [63]. Three steps are clearly observed in all cases: initiation, intermediate reactions, and recombination of unstable species into stable products. Given that the ReaxFF model used was fitted against gas-phase data, it cannot describe the formation of ions in condensed phases. Then, according to these simulations, the initiation of NM mainly consists in unimolecular C–N bond cleavage at lower densities, and in H-transfer and/or N–O bond rupture producing CH_3NO (nitrosomethane) as density rises approaching the CJ state. The corresponding rate therefore exhibits a nonmonotonic variation as density is increased, as unimolecular C–N cleavage is impeded whereas bimolecular CH_3NO production is enhanced. This transition is observed at lower temperatures (<3500 K). At temperatures above 4000 K, the decomposition rate increases monotonically with density as NM molecules decompose very fast and cooling by thermal bath is ineffective [63].

Finally, using the DFTB lanl31 scheme, Perriot et al. simulate NM at lower temperatures (1450–1850 K), high pressures (14–28 GPa), and on large periods of time up to 1500 ps [64]. Series of MD simulations with the same initial conditions highlight the stochastic character of individual runs in terms of reaction paths, with different random seeds for thermal equilibration leading to times to runaway spanning over 1000 ps. This implies that many simulations must be carried out to obtain relevant information regarding the dominant reaction pathways and associated kinetics. For every considered temperature–density (T–ρ) combination, the authors carry out between 17 and 58 simulations, as required to obtain satisfactory statistics. The temperature profiles are used to derive a two-step model based on the Frank–Kamenetskii equation, which suggests initiation to be dominated by a single main reaction pathway associated with the formation of the aci-anion $CH_2NO_2^-$, as observed in earlier short-time quantum simulations [22].

While the decomposition of NM near the CJ state is extensively investigated, relatively few studies consider shocked NM. In a recent one [65], distinct ReaxFF parameterizations are shown to predict shock states and final products in agreement with experiment. Bimolecular reactions play a major role and CH_3NO formation is the dominant initiation pathway, in agreement with ReaxFF simulations of Rom et al. under near-CJ conditions [63]. This finding is consistent with the tendency of ReaxFF schemes fitted against gas-phase data to favor radical rather than ionic mechanisms, thus precluding the formation of the aci-anion. The overall results clearly demonstrate the potential of ReaxFF simulations to gain insight into shock-induced chemistry.

5.2 Nitramines: RDX, HMX, BCHMX

Nitramines include high explosives like RDX, HMX, and bicyclo-HMX (BC-HMX). In 2002, a DFTB simulation over 55 ps provides early insight into the decomposition of HMX at 3500 K (CJ temperature) and slightly compressed by 0.5% (by comparison, the CJ state is compressed by about 10%). In this study, the first reactions are the breaking of the N–NO_2 bond, followed by the dissociation of the NO_2 fragments and C–N bond cleavage.

Nitramines should be well described by ReaxFF since its first extension beyond hydrocarbons was developed for RDX [66] and applied to its decomposition over 10^4 ps and under various density and temperature conditions, including an extreme simulation close to CJ conditions, i.e., $T = 3000$ K and material compressed by about 10% [67]. For this compound, higher densities lead to faster reactions. The release of NO_2 through N–N bond breaking is the dominant early reaction in spite of the similar energy barrier for HONO elimination. The simulations yield kinetic parameters in good agreement with experimental data compiled from a variety of sources [67]. They reveal two distinct regimes: one leading to large carbon clusters (promoted by high density and low temperature) and another one where most C atoms end up in small molecules (for low density and high temperature).

AIMD studies of the decomposition of nitramines are scarce. A recent one focuses on the constant volume decomposition of HMX at high temperatures [68]. At the studied temperatures of 2700 and 3000 K, the breaking of the N–NO_2 bond is not found to be the dominant early reaction, in contrast to what is observed for RDX in the above-mentioned DFTB and ReaxFF simulations.

Another especially interesting study is the simulation of the decomposition of BCHMX [69]. High-temperature MD simulations at a low level of DFT are used to identify the early reaction pathways. In a second step, the molecules involved in the reactions are extracted from the resulting trajectories and the reaction paths are refined on the finite cluster models thus obtained. The results obtained reveal that HONO elimination exhibits a lower energy barrier than N–NO_2 bond breaking. Moreover, an intermolecular hydrogen transfer significantly decreases the reaction barrier for NO_2 elimination. These observations may explain the higher sensitivity of BCHMX in comparison with other nitramines.

Finally, MSST simulations of shocked β-HMX using ReaxFF-lg were used to investigate the role of defects including void (VH), entrained oxygen (OH), and entrained amorphous carbon (CH) by comparison with the perfect crystal [70]. The type of defect was found to impact the sensitivity through its effect on the temperature distribution following the shock (Fig. 12.6). Local high-temperature areas were observed in defective HMX but not in perfect HMX. Both OH and VH crystals have much higher shock sensitivities than in CH. The initial reaction is a N–NO_2 bond rupture in perfect-HMX and N–NO_2 and N–O in defective-HMX crystals. Defective-HMX generates more products during shock simulations.

In contrast, perfect-HMX produces fewer intermediate products and no final product. During the propagation of the shock wave, the perfect crystal is always at a lower temperature and displays a uniform temperature distribution. On the other hand, in the defective systems, local high-temperature areas are formed (see Fig. 12.6C). During the simulation, average values of temperature, pressure, and a gradual decrease of the potential energy are much higher in defect-HMX than perfect-HMX.

5.3 Nitrate esters: PETN and PVN

Pentaerythritol tetranitrate (PETN) is extensively studied since early experiments on single crystals demonstrated the dependence of shock sensitivity on the orientation of the sample with respect to the shock front [49, 71]. To study this dependence using MD simulations,

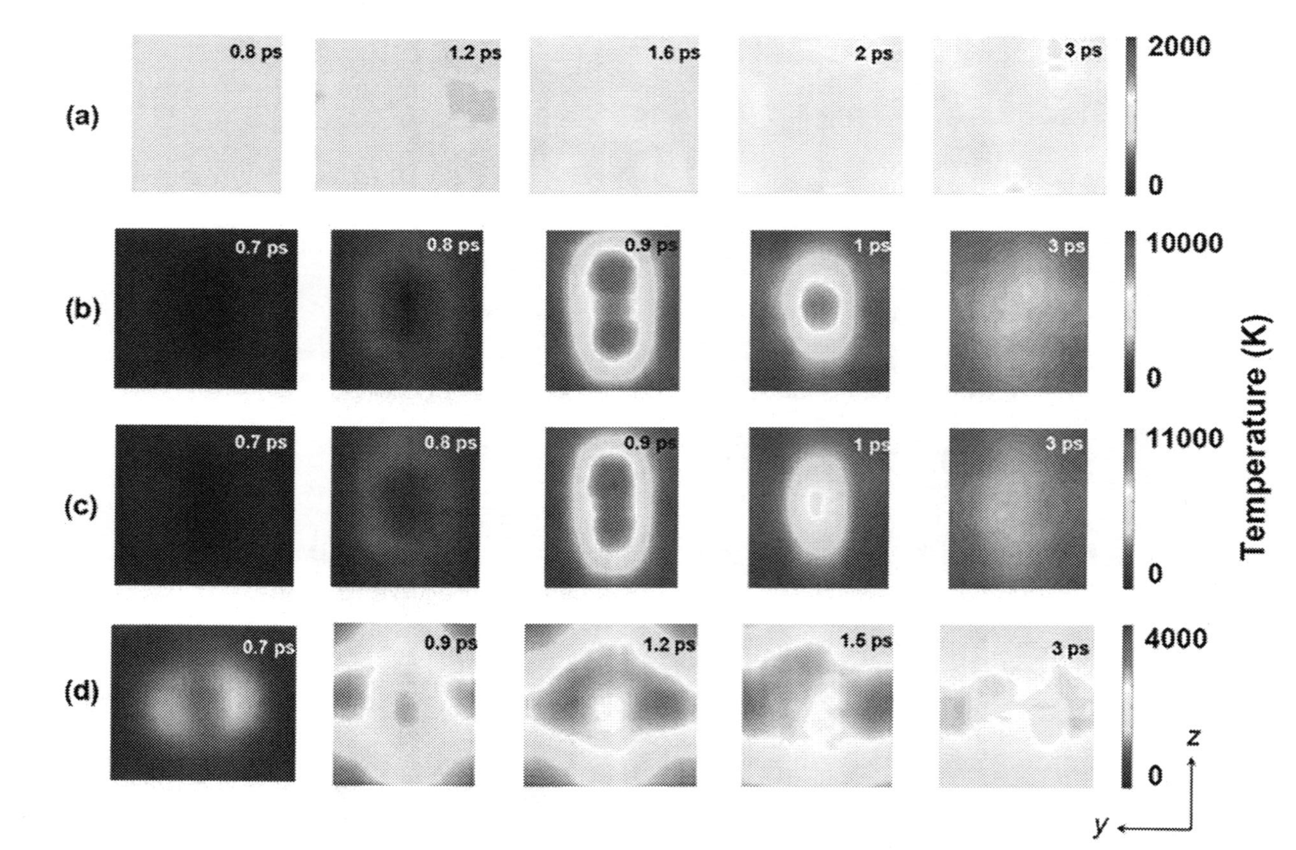

FIG. 12.6 Temperature distributions at different times with different color scales for each system. The shock wave propagates in the *y*-direction. (A) Perfect crystal; (B) crystal with void; (C) crystal with entrained oxygen; (D) entrained amorphous carbon. *Adapted from X. Huang, Z. Qiao, X. Dai, K. Zhang, M. Li, G. Pei, Y. Wen, Effects of different types of defects on ignition mechanisms in shocked β-cyclotetramethylene tetranitramine crystals: a molecular dynamics study based on ReaxFF-lg force field, J. Appl. Phys. 125 (19) (2019) 195101, https://doi.org/10.1063/1.5086916.*

uniaxial shocks are generated in [100] and [110] slabs of PETN. No difference in the rate of NO_2 production is observed between the sensitive [110] and insensitive [100] directions. By contrast, in a CS-RD study, including shear in addition to uniaxial stresses, large stress overshoots and fast temperature increases resulting in early bond-breaking processes are observed for sensitive crystal directions, and vice versa for insensitive directions [16]. This suggests that while sufficiently strong uniaxial stresses initiate the material, they do not account for the orientation dependence of sensitivity, in contrast to CS-RD simulations. Hence, mechanical shear is required to account for the anisotropy of the initiation process.

A subsequent MSST study reveals that the anisotropic shock sensitivity of PETN correlates with a dependence of the first decomposition reactions on the shock direction, with NO_2 group dissociation being predominant along [001] but mixed with ONO_2 dissociation for the [110] and [001] orientations [30]. The results are rationalized on the basis of the amplitudes of vibrational excitation of the C–ONO_2 vs CO–NO_2 bonds.

No matter the material studied, the experimental validation of simulation results is challenging in view of the gap in time and space scales between simulation and experiment. Ultrafast infrared spectroscopic studies on laser-shocked poly(vinyl nitrate) (PVN) films showed that the resulting chemistry occurs on a time scale of tens of picoseconds, involving the nitro group as a primary reactant. The corresponding loss of infrared absorption required an induction time of tens of picoseconds after shock passage, supporting reaction mechanisms that require vibrational energy transfer rather than prompt reaction [72].

Being an amorphous polymer, PVN is well suited to a detailed comparison between such experiments and simulations as it is free of the grain boundaries and other defects that are known to affect shock sensitivities of crystalline materials. Therefore, the shock-induced chemistry observed in this experiment [72] was simulated using the Hugoniostat approach [27]. Theoretical spectroscopic data were extracted from the simulations as Fourier transforms of the power spectra of the MD trajectories (Fig. 12.7), thus enabling a one-to-one comparison with experiment of the reaction initiation time scales as revealed by the rate of change in the N–O stretching modes.

As clear from Fig. 12.7, using a ReaxFF parameter set denoted as ReaxFF-2014, the original NO_2 antisymmetric modes disappears at ~150 ps, in agreement with experiment [27]. In fact, ReaxFF-2014 yields rapid chemical decomposition for shocks of at least 18 GPa, corresponding to a decrease in density of 10% and an increase in temperature of 2000 K. The induction time for exothermic reactions is ~180 ps at 18 GPa. Both this threshold

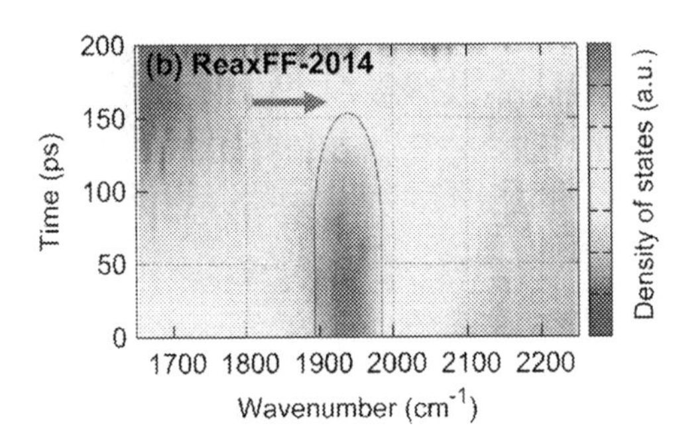

FIG. 12.7 Time-resolved vibrational spectrum of a PVN sample undergoing shock-induced decomposition at 18 GPa. The initial density of states is especially high in the 1900–2000 cm^{-1} frequency range corresponding to the NO_2 antisymmetric stretching mode (shifted from the experimental value of 1624 cm^{-1}) that disappear at ~150 ps. *Adapted from M.M. Islam, A. Strachan, Decomposition and reaction of polyvinyl nitrate under shock and thermal loading: a ReaxFF reactive molecular dynamics study, J. Phys. Chem. C 121 (40) (2017) 22452–22464, https://doi.org/10.1021/acs.jpcc.7b06154. with permission from ACS.*

shock pressure and reaction time scale agree with experiment [72]. Not surprisingly, shock and volumetric compression hinder NO_2 production as observed in other explosives.

To get further insight into mechanochemistry effects, the shock-induced decomposition of PVN was compared to heat-induced decomposition, using isothermal MD simulations with the same ReaxFF force fields [27]. The nature of the loading was found to have a significant effect on the evolution of the intermediate species, which a shock dissociating on PVN into a wider variety of small fragments.

5.4 TATB

The nitroaromatic compounds (NACs) used as explosives are mainly nitroarenes, such as TNT or TATB. They are usually oxygen deficient, so they tend to produce soot on detonation, in contrast to nitramines. This was confirmed by comparative ReaxFF simulations of TATB and HMX [73]. In addition, this study showed that the thermal decomposition of the former at moderate temperature is an order of magnitude slower than for HMX, which may be related to the respective sensitivities of both compounds. The thermal decomposition of TATB was found to produce a carbon-rich phase of polyaromatic rings leading to graphitic regions. AIMD simulations have been used to get further details into the decomposition of NACs [74]. Although equilibrium MD simulations are extensively used to study chemical reactions under CJ conditions, they can also provide insight into detonation wave profiles, as illustrated by a detailed study of the shock response of TATB [23]. The authors simulated the nonreactive state of TATB using a classical nonreactive force field, thus obtaining a MD-based equation of state (EoS) and mechanical properties in good agreement with previous experimental data. Afterward, they used ReaxFF to simulate the decomposition of TATB. They considered a periodic supercell with $\sim 10^5$ atoms and carried out a series of constant pressure simulations for P ranging from 0 to 30 GPa. For each pressure, the simulation was first carried out in the NPT ensemble for T ranging from 600 to 1400 K, and in the NPH ensemble for $T > 1400$ K. These extensive simulations allow the authors to identify 5592 elementary reactions involving 452 intermediate species, and to derive the corresponding Arrhenius parameters describing the reaction rate, i.e., activation energies, activation volumes, and prefactors.

Finally, a complete detonation wave profile was obtained, using the MD-validated JWL EoS for the leading shock wave, the kinetic parameters derived from the ReaxFF simulations for the fast reaction zone, a chemical equilibrium assumption associated with the CJ condition for the CJ state, and finally, a carbon clustering kinetics in the final slow reaction zone [23]. Some intermediate species including H_2 and NH_3 are found to be produced in significant amount, although they have virtually disappeared at the end of the fast reaction zone.

In another study [75], TATB was used as a test case to investigate the influence of nuclear quantum effects on the initiation threshold under shock and thermal loading. The effect of the energy distribution on initiation was investigated with MSST simulations using ReaxFF with quantum and classical thermostats. This showed that nuclear quantum effects increase the temperature rise during dynamical loading and lower the shock temperature, which is the threshold for initiating chemical reactions. These effects have similar contributions of lower specific heat and zero-point energy effects. Nuclear quantum effects not included in the classical simulations lead to a lower activation barrier for the decomposition reaction.

5.5 TNT and other NACs

Rom and coworkers pointed out that some NACs like TNT and TATB exhibit dramatically decreased activation energies in condensed phases compared to gas-phase data, whereas nonaromatic explosives (including RDX, HMX, and PETN) and some NACs (e.g., TNB, i.e., 1,3,5-trinitrobenzene) exhibit similar values in both gas and condensed phases [76]. To understand this difference, they conducted extensive ReaxFF-lg simulations of the decomposition of NACs in condensed phases, focusing on TNT as a representative example [76, 77]. For this compound, the major decomposition pathway in gas phase was found to change from C–H α–attack at lower temperatures to C–NO_2 homolysis above ~1250–1500 K. In condensed phases, the initial decomposition stage involves a more complex set of reactions, including bimolecular reactions leading to the formation of TNT dimers, while higher densities hamper the initial formation of gaseous fragments and promote carbon-clustering. The kinetics of the overall decomposition process may be described as a two-stages process involving a first endothermic stage followed by an exothermic one. In contrast to what might be expected on the basis of the popular picture (derived from gas-phase quantum chemical calculations and experiment) of a decomposition triggered by an initial unimolecular rate-determining step, both stages are enhanced by pressure, as clear from the decrease of the activation volume derived from the fit of the simulated decomposition kinetics.

On the basis of their simulations, the authors find the initial steps to be associated with the formation of a wide range of radical fragments with bimolecular pathways playing a significant role. For TNT and probably many other NACs, this endothermic step involves intermolecular hydrogen transfer, which explains why it is enhanced by pressure, in contrast to unimolecular fragmentation which triggers the decomposition of nonaromatic nitro explosives through X–NO_2 homolysis. On the other hand, this mechanism naturally accounts for the similar activation energies for TNB decomposition in both gas and condensed phases. Indeed, in contrast to TNT, no intermolecular hydrogen transfer can take place as all hydrogen atoms are tightly bound to the aromatic ring.

The MSST technique combined with ReaxFF was also applied to simulate TNT under shock velocities ranging from 6 to 10 km/s [78]. The primary reaction observed is the formation of TNT dimers. However, above a shock velocity of 7 km/s, such dimers quickly decompose to $C_7H_5O_5N_3$, NO_2, and NO. According to the authors, TNT-dimer formation plays an important role in the insensitivity of shocked TNT. This supports the general picture for nitroarenes that carbon clusters arising from the polymerization of aromatic rings delay the release of chemical energy, thus contributing to their insensitivity.

5.6 Liquid mixtures and hypergolic bipropellants

While pure NM has been extensively studied, its mixtures with the recently synthesized $NCNO_2$ molecule were recently investigated using MD simulations at the DFT and ReaxFF levels [79]. A 1:1 mixture was found to significantly improve the detonation properties of NM, including CJ temperature and pressure. The reason for these results is twofold: (1) the number of nitrogen atoms increases the gaseous final products while producing fewer carbon clusters and (2) the increased initial density leads to higher gas expansion capability. Therefore, $NCNO_2$ is an excellent additive to NM or even a replacement for it.

Among liquid mixtures, hypergolic bipropellants consist of fuel and oxidizers that can ignite rapidly upon mixing within tens of million seconds without the assistance of external ignition devices. This self-ignition property improves the reliability of ignition in rocket engines. Reactive MD simulations compared the hypergolic bipropellant mixture of monomethylhydrazine (MMH) and dinitrogentetroxide (NTO) with an ethanol (EtOH) and NTO mixture that is reactive but nonhypergolic [80]. Compared to the nonhypergolic EtOH/NTO mixture, the hypergolic MMH/NTO mixture releases more energy at a higher reaction rate and displays a lower energy barrier. Moreover, it has more reaction channels and events. Upon mixing MMH and NTO, hydrogen abstractions and N–N bond scissions were found to be critical early chemical reactions. The search for novel effective propellants should then focus on molecules with lower energy barriers of H abstractions and bond scissions. Additionally, selecting small oxidizer molecules with considerable diffusion mobility can prompt efficient physical mixing that facilitates chemical reactions and increases the chance of turning the chemical system hypergolic.

5.7 Cocrystals

The performances and sensitivities of cocrystals tend to fall between those of their components [54]. MD simulations may provide additional insight. One possible combination is the TNT:CL-20 cocrystal, which was investigated along with the pure components employing the CS-RD technique and ReaxFF-lg force field [21]. Considering the number of produced NO_2 fragments, TNT:CL-20 has 70% lower shear-induced sensitivity under atmospheric pressure and 46% under high pressure (5 GPa) as compared to the pristine CL-20 crystal, in line with drop-weight experiments.

In another paper, the thermal decomposition of the energetic benzotrifuroxan:1,3,3-trinitroazetidine (BTF:TNAZ) cocrystal is studied using high-temperature AIMD simulations [81]. Many distinct decomposition pathways were observed, including four initial steps and three kinds of subsequent channels, namely BTF-chain isomerization, C–NO_2 homolysis, and ring opening. In addition to oxygen-containing groups, hydrogen atoms released from the decomposing species play a major role in the propagation of the decomposition, as observed in pure crystals.

5.8 Salts and miscellaneous compounds

Beyond simple explosives classically investigated as case studies, molecular simulation is applied to less known compounds, including newly synthesized molecules of potential practical interest like FOX-7 (first reported in 1998) or TKX-50 (reported in 2012). Novel materials exhibit a wide structural diversity, including energetic salts, zwitterions, and high nitrogen compounds. Standard ReaxFF parameterizations should be applied with caution to unusual structures since they were derived from classical explosives like nitramines and NM. Furthermore, novel high-energy compounds include many energetic salts beyond TKX-50. As mentioned above, although ReaxFF can be applied to salts through careful parameterization [82], the fact that standard parameter sets extensively rely on QM calculations of isolated species makes them ill-suited to the study of ionic compounds.

In this context, it comes as no surprise that salts and novel high-energy compounds are most often simulated using quantum potentials, i.e., DFT or DFTB schemes. The initiation and decomposition mechanisms of crystalline FOX-7 at 504 K at pressures in the 1–5 GPa interval, 604 K at 5 GPa, and 704 K at 5 GPa were investigated with AIMD at the DFT level [83]. It was found that the initial decomposition mechanism of FOX-7 is independent of the pressure but depends on the temperature, in agreement with experiment [84]. The initial decomposition step is the bimolecular intermolecular hydrogen transfer, whereas the subsequent decomposition of FOX-7 is sensitive to both temperature and pressure.

In fact, many new compounds of current interest are quite complex and exhibit large unit cells. Therefore, many studies focus on the early steps of the decomposition. For instance, An et al. investigated the initial steps in the decomposition of TKX-50, an energetic salt made of bistetrazole dianions and hydroxylammonium ($HO–NH_3^+$) cations [85]. The authors followed a methodology similar to that described above for BCHMX. In a first step, they carried out high-temperature MD simulations at the DFTB level, which revealed the decomposition to be initiated by proton transfers from cations to dianions and to proceed through the dissociation of the protonated (or diprotonated) bistetrazole into N_2 and N_2O products. Afterward, the authors determined the energy barriers associated with the dissociation reactions using finite cluster calculations at a higher (B3LYP) theoretical level.

Beyond TKX-50, another energetic salt of interest is dihydrazinium 3,3′-dinitro-5,5′-bis-1,2,4-triazole-1,1-diolate (DBTD). Like TKX-50, this salt is made of extended dianions and small monocations (in this case $H_2N–NH_3^+$). The initial steps of its decomposition mechanism were recently investigated through AIMD simulations, using small ($2 \times 1 \times 1$ or $3 \times 1 \times 1$) supercells due to the high cost inherent to this technique [86]. The authors carried out NPT simulations over 100 ps, under high pressure (34.2 GPa) and various temperatures ranging from 298 to 3842 K. In all cases, they reported the decomposition to be triggered by a proton transfer from the cation to the dianion. However, the subsequent steps were found to occur according to a variety of pathways for the different simulations done. The wide range of reactions observed may reflect either the stochastic character of individual runs [64] or the major role of temperature.

6 Granular and composite systems

Perfect or defective monocrystals are obviously limited in their ability to represent real EMs, which consist of a grain of powder, usually bound together using typically 5–10 wt% of a polymer matrix. The resulting polymer-bonded explosive (PBX) can be made insensitive through the use of a rubbery matrix that absorbs shocks. The mechanism underlying hot-spot formation and the relation to detonation combines a complex coupling of thermal, mechanical, and chemical degrees of freedom, which is even more complicated in PBXs due to the introduction of additional parameters. Early MD simulations of model PBXs revealed two mechanisms depending on the shock impedance mismatch between the binder and EM, namely shock focusing and local compression of the crystalline facets, that could explain the relative insensitivity of PBXs with near-spherical crystal shapes [87]. In recent years, a growing number of MD studies were reported on granular materials and chemically realistic PBXs, as illustrated here.

6.1 Granular materials

6.1.1 Shock-to-detonation transition threshold on hexanitrostilbene

Hexanitrostilbene (HNS) is often pressed neatly without the use of binders or additives. This makes it an ideal material for investigating mesoscale phenomena produced by impact initiation. A grain-scale model of HNS was developed to investigate the response of the microstructure to shock loading [88]. A multi-material, large deformation, solid mechanics package was used to describe a high-velocity impact of a fully dense HNS with local porosity included explicitly in the model. DFT-based MD was used to predict the unreacted principal Hugoniot of a fully dense HNS and develop a new equation of state necessary for use in the solid mechanics package. Initial reactive simulations of the threshold velocities for shock-to-detonation transition using both approaches are in good agreement with experiment. Trends seen when pore size distribution is changed were also reproduced: more extensive pore size distribution results in longer run distances [88].

6.1.2 Granularity impact in nanostructured RDX

The initial granularity of an explosive may change the chemistry involved during the reaction process and modify the chemical pathways, resulting in granularity-dependent products. Moreover, granularity can also change the sensitivity of energetic materials. Reactive as well as nonreactive MD simulations of RDX were carried out to investigate different nanogranular structures [89]. It was shown that global shock properties, especially the temperature, are sensitive to the porosity magnitude but not to the nanograin size. The latter plays a role only in local properties, such as local temperature and local chemistry. For a given porosity value, the larger the nanograins, the larger and hotter the hot spots. In these regions, the local chemistry is significantly modified—the result is a more prominent reactivity with a faster formation of final products. It was suggested that a quicker consumption of H, O, and N in regions with higher local temperatures probably affects the formation process of solid carbonaceous phases [89].

6.1.3 Modeling hot spot formation in granular materials

While MD is essential to rationalize the underlying chemistry of hot spot formation, the technique is still impractical to make quantitative predictions at the device scale. To overcome this limitation, a new approach considering hot spots as rare events in a complex dynamical system was recently introduced [90]. The probability of their appearance was quantified using temperature distributions and corresponding reaction rates computed via MC simulations. The latter uses a two-phase (solid/gas) five-equation dynamic compaction model, supplemented with a mesoscale model of the thermal localization at the solid–gas interface. A robust nonlinear dependence of the probability of reaction initiation via hot spots on the initial pore size distribution using a fluctuating initial microstructure was found. Given that the amplitude of the initial porosity fluctuations affected the pore-surface temperature distribution, a probabilistic approach to hot spot formation was used. The reaction rate and the initiation probability prove sensitive to the fluctuation amplitude.

6.2 Polymer-bonded explosives (PBXs)

6.2.1 *RDX-based PBXs*

Fundamental processes of shock-induced instabilities of the polymer-bonded explosive (PBX) N-106 were simulated using 10 ps MD with the ReaxFF reactive force field [91]. The N-106 material consists of RDX crystals bound together using hydroxyl-terminated polybutadiene (HTPB) and isophorone diisocyanate (IPDI)-based polyurethane rubber. Hot spots were formed at the nonuniform interface due to shear relaxation, which results in shear along the interface leading to a significant temperature increase persisting after the shock front has passed the surface. For EMs, this temperature increase is coupled to chemical reactions leading to detonation. It was found that decreasing the binder's density eliminates the hot spot. Thus, it is suggested that in developing insensitive energetic materials for propulsion and explosives, a binder with much lower density is recommended—about one-third of the explosive. Under these conditions, no hot spot develops under the same shock conditions.

It is believed that the formation of hot spots at high temperatures that accelerate reactive energy-releasing events has a critical role in detonation. However, the hot spot formation mechanism in PBXs is still controversial, given that several different ones have been proposed. For investigating this issue, the effect of shock waves in heterogeneous materials upon decomposition and reactive processes was investigated with MD/ReaxFF simulations employing 1.6 million atoms to represent the nonplanar interface. The material, investigated before as discussed above [91], is the prototypical polymer-bonded explosive made up of RDX bonded to HTPB. It is shown that shock propagation from high-density RDX to the low-density polymer (RDX→Poly) across a nonplanar periodic interface (sawtooth) results in a hot spot at the initial asperity but no additional hotspot at the second asperity [3]. This hot spot is due to shear along the interface induced by relaxation of the stress at the asperity.

In contrast, the MD/ReaxFF simulations of shock propagation from the low-density polymer to the high-density RDX (Poly→RDX) indicated the formation of a hot spot at the initial asperity and a second much higher temperature hot spot at the second asperity. The latter is likely a source of detonation in polymer-bonded explosive systems. The hot spots depend on the density mismatch between the RDX and the polymer; in particular, reducing the density by a factor of 2 dramatically reduces the hot spot. Therefore, to produce polymer-bonded explosives less sensitive for use in propellants and explosives, the binder should have a low density at the asperity in contact with RDX. In other words, "coating a low-density polymer between the normal polymer binder and the energetic material dramatically reduces sensitivity" [3]. This effect was verified by simulating the propagation of shock waves in the two directions, namely, RDX→Poly and Poly→RDX.

6.2.2 *PETN-based PBXs*

Reactive MD/ReaxFF simulations were applied to models of PBXs with nonplanar interfaces of PETN, a common PBX energetic material (EM), and the very sensitive silicon pentaerythritol tetranitrate (Si-PETN), whose shock properties could not be characterized experimentally [92]. In both cases, the EM embedded in HTPB-based polymer matrix made up a model of PBX bearing a periodic sawtooth nonplanar interface. When the shock wave propagates from the EM to the polymer (EM→Poly), a hot spot arises from shear localization at the convex polymer asperity. In the opposite situation (Poly→EM), a hot spot is initiated at the concave polymer asperity, and a second more significant hot spot forms at the convex polymer asperity due to the interactions between the shock waves with the nonplanar surface. Those preferential locations of the hot spots at the explosive–PBX interface exhibit little dependence on the energetic compound considered, since similar results were obtained for an RDX-based PBX as discussed above [3]. However, the fate of the hot spot clearly depends on the nature of this compound. For Si-PETN, the decomposition starts with the Si–C–O–X rearrangement to Si–O–C–X through a five-centered transition state on the Si that releases energy (188 kJ/mol), leading to a continuous increase of temperature and pressure in the hot-spot region until detonation. By contrast, the decomposition of PETN starts with the endothermic (163 kJ/mol) NO_2 release, resulting in the attenuation of the hot spot by the polymer binder, which reaches a steady temperature involving the NO_2 dissociation and HONO formation.

6.2.3 Hot spot interactions with polymer binder

To quantify binder properties, reactive MD/ReaxFF simulations were used to investigate the commonly used HTPB binder in a model composite of γ-RDX, shown in Fig. 12.8A [93]. Given that hot spots can be formed near or away from the interface, three different types of hot spots were simulated (Fig. 12.8B). It was found that the binder phase creates a safety buffer by extending the time required for the hot spot to suffer an ignition-to-deflagration transition. This behavior is due mainly to the higher heat capacity of HTPB as compared to RDX. The rate that HTPB absorbs heat from a hot spot is at least four times higher than RDX. A deflagration wave created in RDX moves in HTPB three or more times slower than in RDX, thus inducing high temperatures ($>$2000 K) and pressures ($>$10 GPa) in HTPB. Under these conditions, thermal degradation of the HTPB binder involves chain-branching bimolecular reactions.

6.2.4 Desensitization using a nonpolar additive

Sensitivity can be reduced using desensitizing agents, such as inorganic salts, stearic acids, or fluoropolymers. Polar desensitizing agents induce stabilizing molecule–ion or H-bonding interactions. However, the desensitizing mechanism of nonpolar desensitizing agents is not clear. Examples of the latter are the fluoropolymers used in PBXs. They contain the strongly electronegative F atom, which allows molecular interactions between the polar C—F bond of the nonpolar additive agent and the $-NH_2$, $-OH$, and $-CH_3$ groups in the explosive. To understand how they work, Xue et al. [94] carried out 3 ns MD simulation aimed at investigating binding energies and mechanical properties of various β-NQ/polytetrafluoroethylene PBXs. Weak interactions between NQ and $F_2C{=}CF_2$ barely affect the strength of the trigger bond or the energy barrier for the intramolecular hydrogen-transfer isomerization of NQ. In contrast, the mechanical properties of the β-NQ/polytetrafluoroethylene PBXs significantly differ from those of pristine β-NQ because PBXs reduce rigidity and brittleness, provide greater elasticity and plasticity, and especially improve the ductility. Therefore, β-NQ-based PBXs composed of polytetrafluoroethylene are less sensitive to external stimuli, which results in reduced explosive sensitivity.

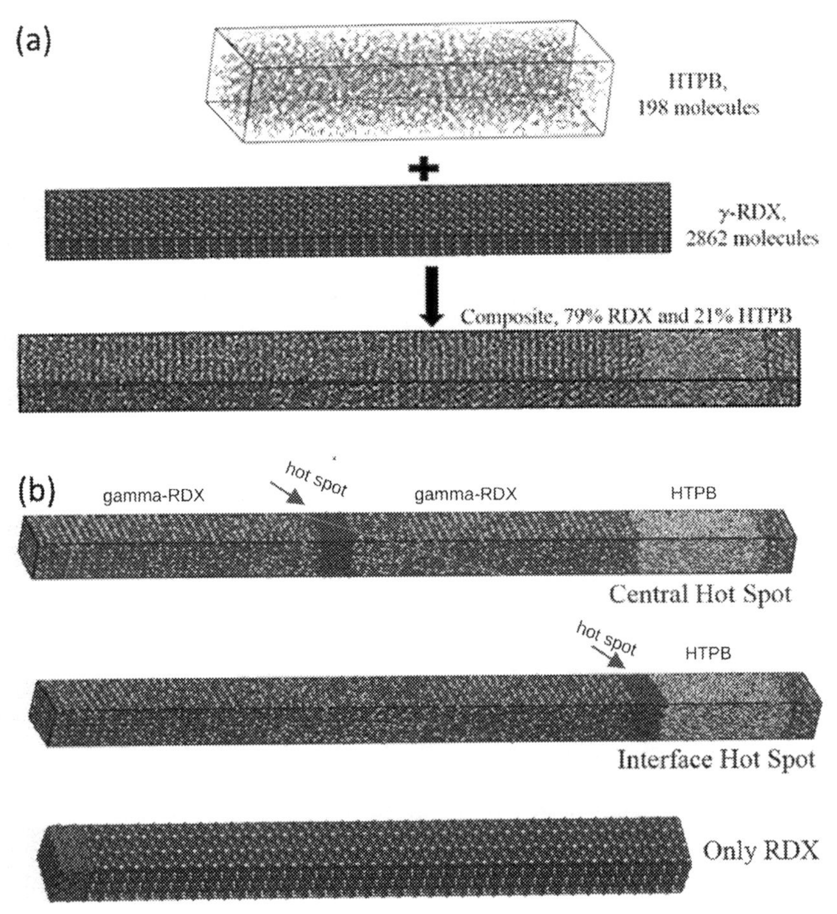

FIG. 12.8 (A) Snapshots of the polymer-bonded HTPB-RDX consisting of γ-RDX and HTPB with chain length of 32 backbone carbon atoms ($C_{32}H_{50}O_2$). The PBX system is 58.2 nm long in the *z*-direction and has a cross-sectional area of 3.768 nm × 2.844 nm. (B) Snapshots of starting structures of a central hot spot, interface hot spot, and RDX-only configurations. For each configuration, the hot spot is shown by *larger red-colored spheres*(*gray* in print version). The size of the hot spot is the same for all configurations. Color scheme: *cyan*(*dark gray* in print version)—C, *blue*(*light gray* in print version)—N, *red*(*gray* in print version)—O, and *white*—H. *Adapted from K. Joshi, S. Chaudhuri, Hot spot interaction with hydroxyl-terminated polybutadiene binder in energetic composites, J. Phys. Chem. C 122 (26) (2018) 1443414446, https://doi.org/10.1021/acs.jpcc.7b11155.*

7 Conclusion

Since the insults to which an explosive may be subjected may differ significantly in their nature or magnitude, and given the range of chemical composition, molecular structure and defects encountered in EMs, initiation can occur in many different ways. A single model is unlikely to describe this wide variety of loads and response mechanisms.

The sensitivities of energetic materials to the various stimuli to which they may be subjected are a broad and complex issue that cannot be fully understood based on experiments and simulations alone. This is due to several reasons:

- the wide variety of stimuli to be considered, which may differ in nature or intensity;
- the wide variety of defects within the material that may facilitate initiation;
- the diversity of mechanisms by which hot spot formation and development of self-sustaining decomposition can occur;
- the possible role of external factors like the initial temperature of the sample, the ambient pressure, and atmosphere composition.

Even if we limit ourselves to a well-defined solicitation in the context of a standardized test carried out under controlled conditions and for samples with a well-characterized structure, a detailed understanding of the phenomenon would require a multiphysics and multiscale approach, including of course the molecular scale [95].

As initiation is a fundamentally dynamic and nonequilibrium process, molecular dynamics (MD) is the most obvious simulation method at the microscopic scale. Until now, its coupling with the higher scales has been realized implicitly via the boundary conditions so as to impose relevant constraints, for instance by ensuring that the final state of the simulated system is consistent with the Hugoniot relationships. The focus has been on systems whose description does not require too many atoms in the simulation cell, such as simple liquids (nitromethane, hydrazine) and monocrystals, either perfect or with small localized defects (especially nanopores) although in recent years, more extended defects like shear bands and dislocations are increasingly considered. MD has been primarily used to get atomistic-level insight into hot spot formation and the onset of chemistry in both homogeneous and defective systems. While the early steps in hot spot formation may be studied with the help of classical force fields, reactive potentials are required to describe the subsequent steps as they involve chemical reactions. Although current progress in computer hardware and models (e.g., the emergence of machine learned interatomic potentials) promises significant advances in the coming years, MD has already made a significant contribution to the current understanding of the initiation of EMs:

- First, simulations of toy models (those deliberately made simple so as to explain the physical mechanism under study) provide a very useful pedagogical material for newcomers to approach the microscopic mechanisms involved in initiation. This is well illustrated by the series of articles by Shi and Brenner [6, 46, 87] which are recommended reading, although the extrapolation of the results to upper scales and more chemically realistic systems requires caution.
- Second, the growing number of simulations of more chemically realistic compounds are providing unprecedented detail on physical initiation mechanisms, as is well illustrated by the comprehensive study of shocked HMX crystals with nanovoids by Zhou et al. [44].
- Third, regarding chemical mechanisms, reactive MD simulations support simple intuitive pictures. For nitro compounds in condensed phases, NO_2 departure, which is the predominant decomposition pathway in gas phase, is hampered by surrounding molecules. As a result, although the role of this mechanism is often important in heat-induced decomposition, the corresponding rate decreases with pressure so that other mechanisms typically involving hydrogen abstraction often play a dominant role in shock-induced chemistry.
- Fourth, in addition to providing qualitative insight into the initiation of EMs and in spite of the complexity of the processes involved, the quantitative correlation linking observed

impact sensitivities to simulated critical decomposition suggests that MD simulations might be sufficient to obtain quantitative estimates of impact sensitivities, provided these values are measured according to a suitable experimental protocol that minimizes the role of factors beyond the reach of current simulation techniques [96].
- Finally, the insight gained from current MD simulation results should be very helpful to design theoretical or semiempirical predictive models accounting for initiation mechanisms and observed sensitivity data. These simulations clearly highlight, among other things, that breaking a trigger bond is not sufficient to initiate an explosion, and that shock-induced reactions do not start within a crystal lattice.

This last point is especially important from a practical viewpoint. To date, most efforts toward the quantitative prediction of sensitivities introduce models either empirical or based on postulated mechanisms. Empirical models increasingly resort to large pools of descriptors and rely on statistical techniques to pick up the relevant ones. However, this is hardly possible with the very limited databases available for sensitivity which show many chance correlations whose purely coincidental nature is not easy to establish.

Regarding theoretical approaches, their underlying assumptions are primarily motivated by the fact that they lend themselves to a theoretical description rather than by experimental evidence. For instance, while the interaction of phonons with molecular vibrational modes is a natural starting point to the study of the shock response of monocrystals, the use of phonons to rationalize impact sensitivities for materials of practical value may seem paradoxical insofar as they are specific to perfect crystals and any minor defect gives rise to very different vibrational modes. Nevertheless, after two decades of efforts, these approaches have begun to show very encouraging results, as discussed in Chapter 10. This might suggest that the vibrational structure of isolated molecules and their ability to mix with external vibrations and provide doorway modes is much more important than the details of the phonon bath.

Notwithstanding phonon up-pumping processes, the excited electronic states have also been suggested to play an important role in the initiation. As discussed in Chapter 11, several papers have studied the change in band gap of perfect and defective crystals upon compression. However, the reduction observed was too small to justify a predominant role of electronic excitations in the initiation process. Nevertheless, in shocked materials, sustained reactions may initiate in regions of high vorticity where high-velocity molecular collisions take place, as observed in MD simulations [42] and extensively discussed on the basis of continuum simulations [43]. Such high-compression regions clearly offer favorable conditions for electronic transitions to excited states, as evidenced by the especially bright flashes experimentally observed at these locations upon void closure [36]. Such results should therefore stimulate a reappraisal of the possible role of electronic excitations.

Finally, discussions of MD simulations with regard to mechanical initiation assume that higher reaction rates imply higher sensitivity [56], as made explicit in the correlation between impact sensitivity and critical temperature for decomposition [96]. The thermokinetic model linking sensitivity to the rate of the self-sustained decomposition [97, 98] may actually be viewed as an extremely simplified version of this MD-based correlation. It is a strength of this model that it is fully consistent with the MD viewpoint.

References

[1] C.B. Storm, J.R. Stine, J.F. Kramer, Sensitivity relationships in energetic materials, in: Chemistry and Physics of Energetic Materials, NATO Science Series, vol. 309, Kluwer Academic Publishers, Dordrecht, Dordrecht, 1990, pp. 605–639.

[2] N.V. Muravyev, D.B. Meerov, K.A. Monogarov, I.N. Melnikov, E.K. Kosareva, L.L. Fershtat, A.B. Sheremetev, I.L. Dalinger, I.V. Fomenkov, A.N. Pivkina, Sensitivity of energetic materials: evidence of thermodynamic factor on a large array of CHNOFCl compounds, Chem. Eng. J. 421 (2021) 129804, https://doi.org/10.1016/j.cej.2021.129804.

[3] Q. An, W.A. Goddard, S.V. Zybin, S.-N. Luo, Inhibition of hotspot formation in polymer bonded explosives using an interface matching low density polymer coating at the polymer-explosive interface, J. Phys. Chem. C 118 (34) (2014) 19918–19928, https://doi.org/10.1021/jp506501r.

[4] R.H.B. Bouma, A.E.D.M. van der Heijden, T.D. Sewell, D.L. Thompson, Simulations of deformation processes in energetic materials, in: Numerical Simulations of Physical and Engineering Processe, 2011, IntechOpen, pp. 29–58, https://doi.org/10.5772/24888. URL https://www.intechopen.com/books/numerical-simulations-of-physical-and-engineering-processes/simulations-of-deformation-processes-in-energetic-materials.

[5] B.C. Barnes, K.W. Leiter, J.P. Larentzos, J.K. Brennan, Forging of hierarchical multiscale capabilities for simulation of energetic materials, Propellants Explos. Pyrotech. 45 (2) (2020) 177–195, https://doi.org/10.1002/prep.201900187.

[6] Y. Shi, D.W. Brenner, Simulated thermal decomposition and detonation of nitrogen cubane by molecular dynamics, J. Chem. Phys. 127 (13) (2007) 134503, https://doi.org/10.1063/1.2779877.

[7] C.T. White, D.R. Swanson, D.H. Robertson, Molecular dynamics simulations of detonations, in: Chemical Dynamics in Extreme Environments, Advanced Series in Physical Chemistry, vol. 11, World Scientific, 2001, pp. 547–592, ISBN: 978-981-02-4177-3, https://doi.org/10.1142/9789812811882_0011.

[8] B.M. Rice, T.D. Sewell, Equilibrium molecular dynamics simulations, in: S.M. Peiris, G.J. Piermarini (Eds.), Static Compression of Energetic Materials, Springer, Berlin, Heidelberg, 2008, pp. 255–290, ISBN: 978-3-540-68151-9,-https://doi.org/10.1007/978-3-540-68151-9_7.

[9] L. Zhang, A.C.T. van Duin, S.V. Zybin, W.A. Goddard III, Thermal decomposition of hydrazines from reactive dynamics using the ReaxFF reactive force field, J. Phys. Chem. B 113 (31) (2009) 10770–10778, https://doi.org/10.1021/jp900194d.

[10] S.V. Bondarchuk, Quantification of impact sensitivity based on solid-state derived criteria, J. Phys. Chem. A 122 (24) (2018) 5455–5463.

[11] M.R. Manaa, L.E. Fried, C.F. Melius, M. Elstner, T. Frauenheim, Decomposition of HMX at extreme conditions: a molecular dynamics simulation, J. Phys. Chem. A 106 (39) (2002) 9024–9029, https://doi.org/10.1021/jp025668+.

[12] V.V. Zhakhovskiı˘, S.V. Zybin, K. Nishihara, S.I. Anisimov, Shock wave structure in Lennard-Jones crystal via molecular dynamics, Phys. Rev. Lett. 83 (6) (1999) 1175–1178, https://doi.org/10.1103/PhysRevLett.83.1175.

[13] A.V. Bolesta, L. Zheng, D.L. Thompson, T.D. Sewell, Molecular dynamics simulations of shock waves using the absorbing boundary condition: a case study of methane, Phys. Rev. B 76 (22) (2007) 224108, https://doi.org/10.1103/PhysRevB.76.224108.

[14] M.J. Cawkwell, T.D. Sewell, L. Zheng, D.L. Thompson, Shock-induced shear bands in an energetic molecular crystal: application of shock-front absorbing boundary conditions to molecular dynamics simulations, Phys. Rev. B 78 (1) (2008) 014107, https://doi.org/10.1103/PhysRevB.78.014107.

[15] T.-R. Shan, R.R. Wixom, A.P. Thompson, Extended asymmetric hot region formation due to shockwave interactions following void collapse in shocked high explosive, Phys. Rev. B 94 (5) (2016) 054308, https://doi.org/10.1103/PhysRevB.94.054308.

[16] S.V. Zybin, W.A. Goddard, P. Xu, A.C.T. van Duin, A.P. Thompson, Physical mechanism of anisotropic sensitivity in pentaerythritol tetranitrate from compressive-shear reaction dynamics simulations, Appl. Phys. Lett. 96 (8) (2010) 081918, https://doi.org/10.1063/1.3323103.

[17] M.M. Kuklja, S.N. Rashkeev, Shear-strain-induced chemical reactivity of layered molecular crystals, Appl. Phys. Lett. 90 (15) (2007) 151913, https://doi.org/10.1063/1.2719031.

[18] V. Guilbaud, J.-P. Dognon, D. Mathieu, C. Morell, A. Grand, P. Maldivi, Chemistry is everywhere, CLEFS CEA 60 (2011) 17–20.

[19] Q. An, Y. Liu, S.V. Zybin, H. Kim, W.A. Goddard III, Anisotropic shock sensitivity of cyclotrimethylene trinitramine (RDX) from compress-and-shear reactive dynamics, 116 (18) (2012) 10198–10206, https://doi.org/10.1021/jp300711m.

[20] T. Zhou, S.V. Zybin, Y. Liu, F. Huang, W.A. Goddard, Anisotropic shock sensitivity for β-octahydro-1,3,5,7-tetranitro-1,3,5,7-tetrazocine energetic material under compressive shear loading from ReaxFF-lg reactive dynamics simulations, J. Appl. Phys. 111 (12) (2012) 124904, https://doi.org/10.1063/1.4729114.

[21] D. Guo, Q. An, W.A. Goddard, S.V. Zybin, F. Huang, Compressive shear reactive molecular dynamics studies indicating that cocrystals of TNT/CL-20 decrease sensitivity, J. Phys. Chem. C 118 (51) (2014) 30202–30208, https://doi.org/10.1021/jp5093527.

[22] M. Riad Manaa, E.J. Reed, L.E. Fried, G. Galli, F. Gygi, Early chemistry in hot and dense nitromethane: molecular dynamics simulations, J. Chem. Phys. 120 (21) (2004) 10146–10153, https://doi.org/10.1063/1.1724820.

[23] Y. Long, J. Chen, Theoretical study of the reaction kinetics and the detonation wave profile for 1,3,5-triamino-2,4,6-trinitrobenzene, J. Appl. Phys. 120 (18) (2016) 185902, https://doi.org/10.1063/1.4967395.

[24] J.-B. Maillet, M. Mareschal, L. Soulard, R. Ravelo, P.S. Lomdahl, T.C. Germann, B.L. Holian, Uniaxial Hugoniostat: a method for atomistic simulations of shocked materials, Phys. Rev. E 63 (1) (2000) 016121, https://doi.org/10.1103/PhysRevE.63.016121.

[25] R. Ravelo, B.L. Holian, T.C. Germann, P.S. Lomdahl, Constant-stress Hugoniostat method for following the dynamical evolution of shocked matter, Phys. Rev. B 70 (1) (2004) 014103, https://doi.org/10.1103/PhysRevB.70.014103.

[26] Q. An, T. Cheng, W.A. Goddard, S.V. Zybin, Anisotropic impact sensitivity and shock induced plasticity of TKX-50 (Dihydroxylammonium 5,5′-bis(tetrazole)-1,1′-diolate) single crystals: from large-scale molecular dynamics simulations, J. Phys. Chem. C 119 (4) (2015) 2196–2207.

[27] M.M. Islam, A. Strachan, Decomposition and reaction of polyvinyl nitrate under shock and thermal loading: a ReaxFF reactive molecular dynamics study, J. Phys. Chem. C 121 (40) (2017) 22452–22464, https://doi.org/10.1021/acs.jpcc.7b06154.

[28] E.J. Reed, L.E. Fried, J.D. Joannopoulos, A method for tractable dynamical studies of single and double shock compression, Phys. Rev. Lett. 90 (23) (2003) 235503, https://doi.org/10.1103/PhysRevLett.90.235503.

[29] Z.-H. He, J. Chen, G.-F. Ji, L.-M. Liu, W.-J. Zhu, Q. Wu, Dynamic responses and initial decomposition under shock loading: a DFTB calculation combined with MSST method for β-HMX with molecular vacancy, J. Phys. Chem. B 119 (33) (2015) 10673–10681, https://doi.org/10.1021/acs.jpcb.5b05081.

[30] T.-R. Shan, R.R. Wixom, A.E. Mattsson, A.P. Thompson, Atomistic simulation of orientation dependence in shock-induced initiation of pentaerythritol tetranitrate, J. Phys. Chem. B 117 (3) (2013) 928–936, https://doi.org/10.1021/jp310473h.

[31] L. Zhang, Y. Yu, M. Xiang, A study of the shock sensitivity of energetic single crystals by large-scale ab initio molecular dynamics simulations, Nanomaterials 9 (9) (2019) 1251, https://doi.org/10.3390/nano9091251. URL https://www.mdpi.com/2079-4991/9/9/1251.

[32] J.E. Field, Hot spot ignition mechanisms for explosives, Acc. Chem. Res. 25 (11) (1992) 489–496, https://doi.org/10.1021/ar00023a002.

[33] M.A. Wood, A. Strachan, Nonequilibrium reaction kinetics in molecular solids, J. Phys. Chem. C 120 (1) (2016) 542–552, https://doi.org/10.1021/acs.jpcc.5b09820.

[34] N.K. Bourne, J.E. Field, Shock-induced collapse of single cavities in liquids, J. Fluid Mech. 244 (1992) 225–240, https://doi.org/10.1017/S0022112092003045. URL https://www.cambridge.org/core/journals/journal-of-fluid-mechanics/article/abs/shockinduced-collapse-of-single-cavities-in-liquids/4E120D1189D623CED808BE81F325641A.

[35] J.P. Dear, J.E. Field, A.J. Walton, Gas compression and jet formation in cavities collapsed by a shock wave, Nature 332 (6164) (1988), https://doi.org/10.1038/332505a0. URL https://www.nature.com/articles/332505a0.

[36] N. Bourne, A. Milne, The temperature of a shock-collapsed cavity, Proc. R. Soc. Lond. A Math. Phys. Eng. Sci. 459 (2036) (2003) 1851–1861, https://doi.org/10.1098/rspa.2002.1101.

[37] A.B. Swantek, J.M. Austin, Collapse of void arrays under stress wave loading, J. Fluid Mechanics 649 (2010) 399–427, https://doi.org/10.1017/S0022112009993545. URL https://www.cambridge.org/core/journals/journal-of-fluid-mechanics/article/abs/collapse-of-void-arrays-under-stress-wave-loading/B93BE2A3058887D160716B1BB0901C91.

[38] B.L. Holian, T.C. Germann, J.-B. Maillet, C.T. White, Atomistic mechanism for hot spot initiation, Phys. Rev. Lett. 89 (28) (2002) 285501, https://doi.org/10.1103/PhysRevLett.89.285501.

[39] T. Hatano, Spatiotemporal behavior of void collapse in shocked solids, Phys. Rev. Lett. 92 (1) (2004) 015503, https://doi.org/10.1103/PhysRevLett.92.015503.

[40] K.-i. Nomura, R.K. Kalia, A. Nakano, P. Vashishta, Reactive nanojets: nanostructure-enhanced chemical reactions in a defected energetic crystal, Appl. Phys. Lett. 91 (18) (2007) 183109, https://doi.org/10.1063/1.2804557.

[41] F. Wang, L. Chen, D. Geng, J. Lu, J. Wu, Molecular dynamics simulations of an initial chemical reaction mechanism of shocked CL-20 crystals containing nanovoids, J. Phys. Chem. C 123 (39) (2019) 23845–23852, https://doi.org/10.1021/acs.jpcc.9b06137.

[42] R.M. Eason, T.D. Sewell, Molecular dynamics simulations of the collapse of a cylindrical pore in the energetic material α-RDX, J. Dyn. Behav. Mater. 1 (4) (2015) 423–438, https://doi.org/10.1007/s40870-015-0037-z.

[43] N.K. Rai, M.J. Schmidt, H.S. Udaykumar, High-resolution simulations of cylindrical void collapse in energetic materials: effect of primary and secondary collapse on initiation thresholds, Phys. Rev. Fluids 2 (4) (2017) 043202, https://doi.org/10.1103/PhysRevFluids.2.043202.

[44] T. Zhou, J. Lou, Y. Zhang, H. Song, F. Huang, Hot spot formation and chemical reaction initiation in shocked HMX crystals with nanovoids: a large-scale reactive molecular dynamics study, Phys. Chem. Chem. Phys. 18 (26) (2016) 17627–17645, https://doi.org/10.1039/C6CP02015A. URL https://pubs.rsc.org/en/content/articlelanding/2016/cp/c6cp02015a.

[45] T.C. Germann, B.L. Holian, P.S. Lomdahl, A.J. Heim, N. Grønbech-Jensen, J.B. Maillet, Molecular dynamics simulation of detonation in defective explosive crystals, tech. rep, 2002. www.intdetsymp.org. San Diego.

[46] Shi, D.W. Brenner, Jetting and detonation initiation in shock induced collapse of nanometer-scale voids, J. Phys. Chem. C 112 (16) (2008) 6263–6270, https://doi.org/10.1021/jp7119735.

[47] M.A. Wood, M.J. Cherukara, E.M. Kober, A. Strachan, Ultrafast chemistry under nonequilibrium conditions and the shock to deflagration transition at the nanoscale, J. Phys. Chem. C 119 (38) (2015) 22008–22015, https://doi.org/10.1021/acs.jpcc.5b05362.

[48] M.P. Kroonblawd, T.D. Sewell, Anisotropic relaxation of idealized hot spots in crystalline 1,3,5-triamino-2,4,6-trinitrobenzene (TATB), J. Phys. Chem. C 120 (31) (2016) 17214–17223, https://doi.org/10.1021/acs.jpcc.6b04749.

[49] J.J. Dick, Effect of crystal orientation on shock initiation sensitivity of pentaerythritol tetranitrate explosive, Appl. Phys. Lett. 44 (9) (1984) 859–861, https://doi.org/10.1063/1.94951.

[50] J.J. Dick, J.P. Ritchie, Molecular mechanics modeling of shear and the crystal orientation dependence of the elastic precursor shock strength in pentaerythritol tetranitrate, J. Appl. Phys. 76 (5) (1994) 2726–2737, https://doi.org/10.1063/1.357576.

[51] R.W. Armstrong, Dislocation mechanisms for shock-induced hot spots, J. Phys. IV France 5 (C4) (1995), https://doi.org/10.1051/jp4:1995407. C4–C4-102.

[52] M.J. Cawkwell, K.J. Ramos, D.E. Hooks, T.D. Sewell, Homogeneous dislocation nucleation in cyclotrimethylene trinitramine under shock loading, J. Appl. Phys. 107 (6) (2010) 063512, https://doi.org/10.1063/1.3305630.

[53] C. Zhang, X. Wang, H. Huang, π-Stacked interactions in explosive crystals: buffers against external mechanical stimuli, J. Am. Chem. Soc. 130 (26) (2008) 8359–8365, https://doi.org/10.1021/ja800712e.

[54] D. Mathieu, Sensitivity of energetic materials: theoretical relationships to detonation performance and molecular structure, Ind. Eng. Chem. Res. 56 (29) (2017) 8191–8201, https://doi.org/10.1021/acs.iecr.7b02021.

[55] X. Xue, Y. Wen, X. Long, J. Li, C. Zhang, Influence of dislocations on the shock sensitivity of RDX: molecular dynamics simulations by reactive force field, J. Phys. Chem. C 119 (24) (2015) 13735–13742, https://doi.org/10.1021/acs.jpcc.5b03298.

[56] C. Deng, J. Liu, X. Xue, X. Long, C. Zhang, Coupling effect of shock, heat, and defect on the decay of energetic materials: a case of reactive molecular dynamics simulations on 1,3,5-trinitro-1,3,5-triazinane, J. Phys. Chem. C 122 (49) (2018) 27875–27884, https://doi.org/10.1021/acs.jpcc.8b09170.

[57] R.E. Winter, J.E. Field, D. Tabor, The role of localized plastic flow in the impact initiation of explosives, Proc. R. Soc. Lond. A Math. Phys. Sci. 343 (1634) (1975) 399–413, https://doi.org/10.1098/rspa.1975.0074.

[58] E. Jaramillo, T.D. Sewell, A. Strachan, Atomic-level view of inelastic deformation in a shock loaded molecular crystal, Phys. Rev. B 76 (6) (2007) 064112, https://doi.org/10.1103/PhysRevB.76.064112.

[59] M.P. Kroonblawd, L.E. Fried, High explosive ignition through chemically activated nanoscale shear bands, Phys. Rev. Lett. 124 (20) (2020) 206002, https://doi.org/10.1103/PhysRevLett.124.206002.

[60] E.E. Fileti, V.V. Chaban, O.V. Prezhdo, Exploding nitromethane in silico, in real time, J. Phys. Chem. Lett. 5 (19) (2014) 3415–3420, https://doi.org/10.1021/jz501848e.

[61] J. Chang, P. Lian, D.-Q. Wei, X.-R. Chen, Q.-M. Zhang, Z.-Z. Gong, Thermal decomposition of the solid phase of nitromethane: ab initio molecular dynamics simulations, Phys. Rev. Lett. 105 (18) (2010) 188302, https://doi.org/10.1103/PhysRevLett.105.188302.

[62] S.-p. Han, A.C.T. van Duin, W.A. Goddard, A. Strachan, Thermal decomposition of condensed-phase nitromethane from molecular dynamics from ReaxFF reactive dynamics, J. Phys. Chem. B 115 (20) (2011) 6534–6540, https://doi.org/10.1021/jp1104054.

[63] N. Rom, S.V. Zybin, A.C.T. van Duin, W.A. Goddard, Y. Zeiri, G. Katz, R. Kosloff, Density-dependent liquid nitromethane decomposition: molecular dynamics simulations based on ReaxFF, J. Phys. Chem. A 115 (36) (2011) 10181–10202, https://doi.org/10.1021/jp202059v.

[64] R. Perriot, M.J. Cawkwell, E. Martinez, S.D. McGrane, Reaction rates in nitromethane under high pressure from density functional tight binding molecular dynamics simulations, J. Phys. Chem. A 124 (17) (2020) 3314–3328, https://doi.org/10.1021/acs.jpca.9b11897.

[65] M.M. Islam, A. Strachan, Reactive molecular dynamics simulations to investigate the shock response of liquid nitromethane, J. Phys. Chem. C 123 (4) (2019) 2613–2626, https://doi.org/10.1021/acs.jpcc.8b11324.

[66] A. Strachan, A.C.T. van Duin, D. Chakraborty, S. Dasgupta, W.A. Goddard, Shock waves in high-energy materials: the initial chemical events in nitramine RDX, Phys. Rev. Lett. 91 (9) (2003) 098301, https://doi.org/10.1103/PhysRevLett.91.098301.

[67] A. Strachan, E.M. Kober, A.C.T. van Duin, J. Oxgaard, W.A. Goddard, Thermal decomposition of RDX from reactive molecular dynamics, J. Chem. Phys. 122 (5) (2005) 054502, https://doi.org/10.1063/1.1831277.

[68] D. Xiang, W. Zhu, Adiabatic and constant volume decomposition process of condensed phase δ-1,-3,5,7-tetranitro-1,3,5,7-tetrazocane at high temperatures: quantum molecular dynamics simulations, J. Mol. Graph. Model. 85 (2018) 68–74, https://doi.org/10.1016/j.jmgm.2018.08.003.

[69] C.-C. Ye, Q. An, W.A. Goddard, T. Cheng, S. Zybin, X.-h. Ju, Initial decomposition reactions of bicyclo-HMX [BCHMX or cis-1,3,4,6-Tetranitrooctahydroimidazo-[4,5-d]imidazole] from quantum molecular dynamics simulations, J. Phys. Chem. C 119 (5) (2015) 2290–2296, https://doi.org/10.1021/jp510328d.

[70] X. Huang, Z. Qiao, X. Dai, K. Zhang, M. Li, G. Pei, Y. Wen, Effects of different types of defects on ignition mechanisms in shocked β-cyclotetramethylene tetranitramine crystals: a molecular dynamics study based on ReaxFF-lg force field, J. Appl. Phys. 125 (19) (2019) 195101, https://doi.org/10.1063/1.5086916.

[71] J.J. Dick, Shock initiation sensitivity of PETN: a steric hindrance model, in: Workshop on Desensitization of explosives and Propellants, TNO Prins Maurits Laboratory, Delft, The Netherlands, 1991. https://www.osti.gov/biblio/6143167.

[72] S.D. McGrane, D.S. Moore, D.J. Funk, Shock induced reaction observed via ultrafast infrared absorption in poly(vinyl nitrate) films, J. Phys. Chem. A 108 (43) (2004) 9342–9347, https://doi.org/10.1021/jp048464x.

[73] L. Zhang, S.V. Zybin, A.C.T. van Duin, S. Dasgupta, W.A. Goddard, E.M. Kober, Carbon cluster formation during thermal decomposition of octahydro-1,3,5,7-tetranitro-1,3,5,7-tetrazocine and 1,3,5-Triamino-2,4,6-trinitrobenzene high explosives from ReaxFF reactive molecular dynamics simulations, J. Phys. Chem. A 113 (40) (2009) 10619–10640, https://doi.org/10.1021/jp901353a.

[74] Q. Wu, H. Chen, G. Xiong, W. Zhu, H. Xiao, Decomposition of a 1,3,5-triamino-2,4,6-trinitrobenzene crystal at decomposition temperature coupled with different pressures: an ab initio molecular dynamics study, J. Phys. Chem. C 119 (29) (2015) 16500–16506, https://doi.org/10.1021/acs.jpcc.5b05041.

[75] B.W. Hamilton, M.P. Kroonblawd, M.M. Islam, A. Strachan, Sensitivity of the shock initiation threshold of 1,3,5-triamino-2,4,6-trinitrobenzene (TATB) to nuclear quantum effects, J. Phys. Chem. C 123 (36) (2019) 21969–21981, https://doi.org/10.1021/acs.jpcc.9b05409.

[76] D. Furman, R. Kosloff, F. Dubnikova, S.V. Zybin, W.A. Goddard, N. Rom, B. Hirshberg, Y. Zeiri, Decomposition of condensed phase energetic materials: interplay between uni- and bimolecular mechanisms, J. Am. Chem. Soc. 136 (11) (2014) 4192–4200, https://doi.org/10.1021/ja410020f.

[77] N. Rom, B. Hirshberg, Y. Zeiri, D. Furman, S.V. Zybin, W.A. Goddard, R. Kosloff, First-principles-based reaction kinetics for decomposition of hot, dense liquid TNT from ReaxFF multiscale reactive dynamics simulations, J. Phys. Chem. C 117 (41) (2013) 21043–21054, https://doi.org/10.1021/jp404907b.

[78] H. Liu, Y. He, J. Li, Z. Zhou, Z. Ma, S. Liu, X. Dong, ReaxFF molecular dynamics simulations of shock induced reaction initiation in TNT, AIP Advances 9 (1) (2019) 015202, https://doi.org/10.1063/1.5047920.

[79] D. Guo, S.V. Zybin, W.A. Goddard, Q. An, Enhancing the detonation properties of liquid nitromethane by adding nitro-rich molecule nitryl cyanide, J. Phys. Chem. C 124 (18) (2020) 9787–9794, https://doi.org/10.1021/acs.jpcc.0c02010.

[80] Y. Liu, S.V. Zybin, J. Guo, A.C.T. van Duin, W.A. Goddard, Reactive dynamics study of hypergolic bipropellants: monomethylhydrazine and dinitrogen tetroxide, J. Phys. Chem. B 116 (48) (2012) 14136–14145, https://doi.org/10.1021/jp308351g.

[81] H. Xie, W. Zhu, Thermal decomposition mechanisms of the energetic benzotrifuroxan:1,3,3-trinitroazetidine cocrystal using ab initio molecular dynamics simulations, J. Chin. Chem. Soc. 67 (2) (2020) 218–226, https://doi.org/10.1002/jccs.201900169.

[82] T.-R. Shan, A.C.T. van Duin, A.P. Thompson, Development of a ReaxFF reactive force field for ammonium nitrate and application to shock compression and thermal decomposition, J. Phys. Chem. A 118 (8) (2014) 1469–1478, https://doi.org/10.1021/jp408397n.

[83] G. Xiong, W. Zhu, Coupling effects of high temperature and pressure on the decomposition mechanisms of 1,1-diamino-2,2-dinitroehethe crystal: ab initio molecular dynamics simulations, J. Chin. Chem. Soc. 67 (9) (2020) 1571–1578, https://doi.org/10.1002/jccs.201900504.

[84] Z.A. Dreger, Y. Tao, Y.M. Gupta, Phase diagram and decomposition of 1,1-diamino-2,2-dinitroethene single crystals at high pressures and temperatures, J. Phys. Chem. C 120 (20) (2016) 11092–11098, https://doi.org/10.1021/acs.jpcc.6b02360.

[85] Q. An, W.-G. Liu, W.A. Goddard, T. Cheng, S.V. Zybin, H. Xiao, Initial steps of thermal decomposition of dihydroxylammonium 5,5′-bistetrazole-1,1′-diolate crystals from quantum mechanics, J. Phys. Chem. C 118 (46) (2014) 27175–27181, https://doi.org/10.1021/jp509582x.

[86] D. Xiang, W. Zhu, H. Xiao, Thermal decomposition mechanisms of energetic ionic crystal dihydrazinium 3,3′-dinitro-5,5′-bis-1,2,4-triazole-1,1-diolate: an ab initio molecular dynamics study, Fuel 202 (2017) 246–259, https://doi.org/10.1016/j.fuel.2017.04.043.

[87] Y. Shi, D.W. Brenner, Molecular simulation of the influence of interface faceting on the shock sensitivity of a model plastic bonded explosive, J. Phys. Chem. B 112 (47) (2008) 14898–14904, https://doi.org/10.1021/jp805690w.

[88] C.D. Yarrington, R.R. Wixom, D.L. Damm, Shock interactions with heterogeneous energetic materials, J. Appl. Phys. 123 (10) (2018) 105901, https://doi.org/10.1063/1.5022042.

[89] X. Bidault, N. Pineau, Granularity impact on hotspot formation and local chemistry in shocked nanostructured RDX, J. Chem. Phys. 149 (22) (2018) 224703, https://doi.org/10.1063/1.5049474.

[90] J. Bakarji, D.M. Tartakovsky, Microstructural heterogeneity drives reaction initiation in granular materials, Appl. Phys. Lett. 114 (25) (2019) 254101, https://doi.org/10.1063/1.5108902.

[91] Q. An, S.V. Zybin, W.A. Goddard, A. Jaramillo-Botero, M. Blanco, S.-N. Luo, Elucidation of the dynamics for hot-spot initiation at nonuniform interfaces of highly shocked materials, Phys. Rev. B 84 (22) (2011) 220101, https://doi.org/10.1103/PhysRevB.84.220101.

[92] Q. An, W.A. Goddard, S.V. Zybin, A. Jaramillo-Botero, T. Zhou, Highly shocked polymer bonded explosives at a nonplanar interface: hot-spot formation leading to detonation, J. Phys. Chem. C 117 (50) (2013) 26551–26561, https://doi.org/10.1021/jp404753v.

[93] K. Joshi, S. Chaudhuri, Hot spot interaction with hydroxyl-terminated polybutadiene binder in energetic composites, J. Phys. Chem. C 122 (26) (2018) 14434–14446, https://doi.org/10.1021/acs.jpcc.7b11155.

[94] Z.-q. Xue, J. He, J. Zhang, X.-l. Zhang, Y.-g. Chen, F.-d. Ren, Theoretical investigation of the safety of nitroguanidine-based PBXs containing the nonpolar desensitizing agent polytetrafluoroethylene, J. Mol. Model. 23 (12) (2017) 346, https://doi.org/10.1007/s00894-017-3519-1.

[95] B.C. Barnes, J.K. Brennan, E.F.C. Byrd, S. Izvekov, J.P. Larentzos, B.M. Rice, Toward a predictive hierarchical multiscale modeling approach for energetic materials, in: N. Goldman (Ed.), Computational Approaches for Chemistry Under Extreme Conditions, Springer International Publishing, Cham, 2019, pp. 229–282, ISBN: 978-3-030-05600-1, https://doi.org/10.1007/978-3-030-05600-1_10.

[96] M.J. Cawkwell, V.W. Manner, Ranking the drop-weight impact sensitivity of common explosives using arrhenius chemical rates computed from quantum molecular dynamics simulations, J. Phys. Chem. A 124 (1) (2020) 74–81, https://doi.org/10.1021/acs.jpca.9b10808.

[97] D. Mathieu, T. Alaime, Predicting impact sensitivities of nitro compounds on the basis of a semi-empirical rate constant, J. Phys. Chem. A 118 (41) (2014) 9720–9726, https://doi.org/10.1021/jp507057r.

[98] D. Mathieu, Modeling sensitivities of energetic materials using the Python language and libraries, Propellants Explos. Pyrotech. 45 (6) (2020) 966–973, https://doi.org/10.1002/prep.201900377.

CHAPTER

13

Quantum chemical investigations of reaction mechanism

*Weihua Zhu**

Institute for Computation in Molecular and Materials Science, School of Chemistry and Chemical Engineering, Nanjing University of Science and Technology, Nanjing, China
*Corresponding author: E-mail: zhuwh@njust.edu.cn

1 Introduction

Energetic materials (EMs) are types of special matter that store high energy and can release high amounts of energy easily, which have been applied widely in the field of the national economy and national defense. EMs mainly include explosives, propellants, and pyrotechnics. Ideal EMs has not only high energy but also excellent stability (low sensitivity). Therefore, in the development of EMs, the energy level and sensitivity (initiation reactivity) have been the core of scientists' research. However, these two kinds of performance are mutually conditioned. How to balance this pair of inherently mutual contradictory performances to get new EMs with both high energy and low sensitivity has always been an ultimate goal that researchers strive for. The power of EMs is determined by their energy level, while their sensitivity is mainly dominated by their molecular structure, molecular environment, and external stimulus. The former two factors are the internal cause of determining the stability of EMs, while the last one is the external conditions. Thus, studying the stability of EMs can provide not only a lot of basic information for designing new high-energy insensitivity EMs but also some guidance and help for controlling their risk during usage and storage.

It is known that EMs have a strong exothermic reactivity since they are in a metastable state. When an external stimulus is applied, EMs will decompose to release a lot of energy, as shown in Fig. 13.1. The changes in energy before and after the decomposition of an EM (ΔE) can be used to measure their detonation performance, while the activation energy barrier (E_a) of its decomposition reaction can be employed to weigh up its stability. Therefore, detailed investigations of the reaction mechanisms and kinetics of EMs will help understand their stability and design new EMs.

Theoretical and Computational Chemistry, Volume 22
https://doi.org/10.1016/B978-0-12-822971-2.00004-8

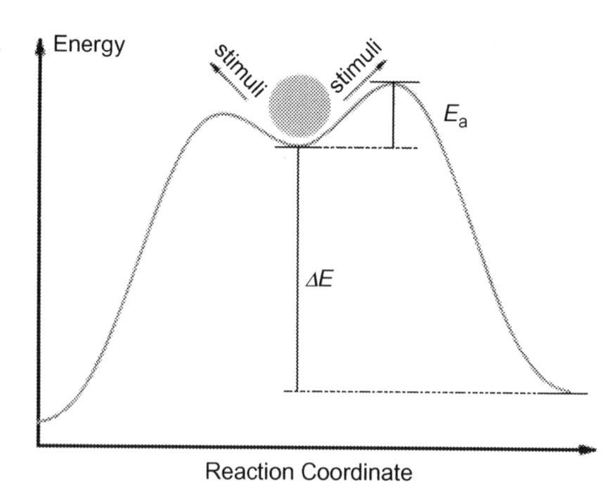

FIG. 13.1 Potential energy profile for the chemical decomposition of EMs.

It is extremely difficult to obtain a clear atomistic picture of the decomposition mechanisms of EMs in experimental measurements due to their complex chemical behaviors and reaction time scales of a picosecond. At present, experimental investigations are mainly confined to the analysis of postreaction residues and thermochemical properties [1], which can only provide indirect insight on global decomposition kinetics and final products. Theoretical simulation is an effective way to study the chemical reactions of EMs at the atomic level as a complement to experimental work [2–8]. At first, researchers used quantum chemical methods to study the thermal decomposition mechanisms of gaseous explosive molecules, such as references [9–15]. However, in most cases, EMs are often in solid-state. Accordingly, detailed descriptions of the chemical reaction mechanisms of condensed EMs at external conditions are very important to understand their stability and performance.

Recently, new simulation methods to study the reactivity of condensed-phase EMs have been developed. There are two different approaches to simulating condensed-phase reactions for EMs: chemical equilibrium method and atomistic simulation [4]. Chemical equilibrium approach is an empirical method that can conveniently predict the detonation and decomposition processes of EMs, but no details of the atomistic reactive mechanisms in the detonation are disclosed. Therefore, this method is often employed to design new EMs, both at the level of synthetic chemistry and formulation design since it depicts the reaction process of EMs using a highly simplified thermodynamic method. Here we will not review these works on this aspect. Atomistic simulations use far fewer empirical parameters than chemical equilibrium calculations. In recent years, with the rapid increment of the computing power of computers, atomistic simulations enter a rapidly emerging era for modeling condensed-phase reactions of EMs.

The atomistic simulation methods modeling the chemical reaction process of condensed-phase EMs mainly include quantum molecular dynamics (MD) and reactive force field MD (RFF-MD) [16]. Quantum MD is a unification of quantum chemistry and classical MD. Its force fields are optimized by quantum chemical calculations at each configuration. Thus, it can depict the rupture and formation of chemical bonds in chemical reactions.

TABLE 13.1 Principles and applications of the three atomistic simulation methods.

Methods	AIMD	DFTB-MD	RFF-MD
Calculation of force field	Evaluated by DFT at every time step	Evaluated by DFTB at every time step	Classical force field
Size of model	Small	Medium	Large
Precision of simulation	High	Medium	Low
Level of particle	Atom and electron	Atom and electron	Atom
Electronic resolution	Including	Including	Lacking
Scope of application	Universal	Limited	Limited

Quantum MD is more accurate than classical MD. It can be divided into ab initio MD (AIMD) [17] and semi-empirical quantum MD. Table 13.1 compares the principles and applications of the three atomistic simulation methods. AIMD can calculate both the atomic and electronic structure consistently and explore how changes in one are correlated with changes in the other. Due to the huge computational cost, AIMD can only be applied to systems with a small number of atoms. Semi-empirical quantum MD such as density functional tight-binding (DFTB) MD (DFTB-MD) [18] is an approximate quantum method due to the extension of the standard tight-binding approach in the context of density functional theory (DFT). Unlike DFT, DFTB contains an empirical repulsive force field based on experimental data and DFT simulations. Thus, DFTB-MD is less costly than AIMD and can treat systems with over 1000 atoms and time with a few hundred picoseconds. RFF-MD is a method that uses reactive empirical force fields to express intermolecular interactions in MD. It is potentially applicable to very large sizes of systems due to its computational requirements scaling almost linearly with the number of atoms. However, the application range of RFF-MD is very narrow because many parameters used in the force field are not unique and are typically refined for particular applications. RFF-MD belongs to classical mechanics, while quantum MD is based on quantum mechanics. Thus, unlike quantum MD, RFF-MD cannot permit us for the first time to study electronic properties at detonation conditions. Here we will review recent progress on the reaction mechanisms of condensed-phase EMs at extreme conditions using atomistic quantum MD. These researches simulated by RFF-MD do not belong to the scope of the chapter.

2 Reactions at high temperatures

In the research field of EMs, it is essential to grasp the knowledge of their decomposition mechanisms at high temperatures. This information could help us to understand the complex behaviors that may occur in the explosion and to control the risk during the usage and storage. Many experimental and theoretical studies have been devoted to investigating the chemical reactions of EMs at different temperatures [19–24].

2.1 HMX

Octahydro-1,3,5,7-tetranitro-1,3,5,7-tetrazocine (HMX) is an important and commonly used energetic ingredient in various high-performance explosives and propellant formulations. HMX is known to exist in four crystalline phases, denoted as α, β, δ, and γ. δ-HMX is often found to exist in the vicinity of damaged regions within HMX-based plastic-bonded explosives. δ-HMX is greatly more sensitive than β-HMX, a thermodynamically stable form under ambient conditions. Therefore, a detailed study on the decomposition mechanisms of condensed-phase δ-HMX offers the possibility of understanding the stability of β-HMX.

Manaa et al. [25] used self-consistent-charge DFTB-MD to simulate the chemical decomposition process of δ-HMX with a density of 1.9 g/cm^3 at a temperature of 3500 K for up to 55 ps, similar to the Chapman-Jouget detonation state. It was found that HMX was in a highly reactive dense supercritical fluid state. The production rates of typical products such as H_2O, N_2, CO_2, and CO were estimated to be 0.48, 0.08, 0.05, and 0.11 ps^{-1}, respectively. The simulated concentrations of dominant species are in agreement with those obtained from thermodynamic calculations.

Xiang and Zhu [26] performed DFTB-MD simulations to investigate the initiation chemistry of condensed-phase δ-HMX at high temperatures by maintaining constant energy and volume to the model adiabatic initiation process. The decomposition of the HMX crystal is primarily triggered by the breaking of the C—N bond in the molecular ring, and the C—H bond cleavage of other HMX molecules at 2400 K. At 2700 K, the decomposition of HMX is triggered only by the C—N bond breaking. When the temperature increases to 3000 K, the decomposition of HMX took place by the breaking of the C—H and N—O bonds only in the branch chains. This indicates that the temperature changes the initial decomposition paths.

Fig. 13.2 displays a schematic diagram of the competing decomposition channels for the HMX crystal. After the initial decomposition stage, further decompositions of the HMX crystal began. At a low temperature of 2400 K, there are seven main decomposition channels found in the simulations. Among them, the five decomposition paths of the HMX molecules are intermolecular reactions, while only two channels are triggered by intramolecular reactions. The N—O bond cleavage is a dominant reaction pathway. The boat configuration of the HMX molecule caused a new reaction Channel C to be happened by forming a new N—N bond. Another new reaction Channel D took place to form a new N—C bond due to intermolecular effects.

Recently, Ye et al. [27] employed AIMD to study the initial decomposition mechanisms of β-HMX and δ-HMX crystals heated from 300 to 3000 K in 20 ps at a heating rate of 135 K/ps. There are two different initial unimolecular decomposition pathways for β-HMX, both taking place simultaneously. One is the HONO elimination reaction, while the other is the N—NO_2 cleavage, as shown in Fig. 13.3. β-HMX first transforms to an intermediate before the HONO elimination reaction. β-HMX first transforms to δ-HMX before the N—NO_2 cleavage. The initial decomposition step of δ-HMX is unimolecular N—NO_2 cleavage and HONO elimination in one δ-HMX molecule at the same time.

Xiang et al. [28] carried out AIMD simulations to study the solid phase β-HMX at different temperatures such as 534 K (initial decomposition temperature), 608 K (thermal ignition temperature), and 873 K (experimentally investigated temperature). The results indicate that the

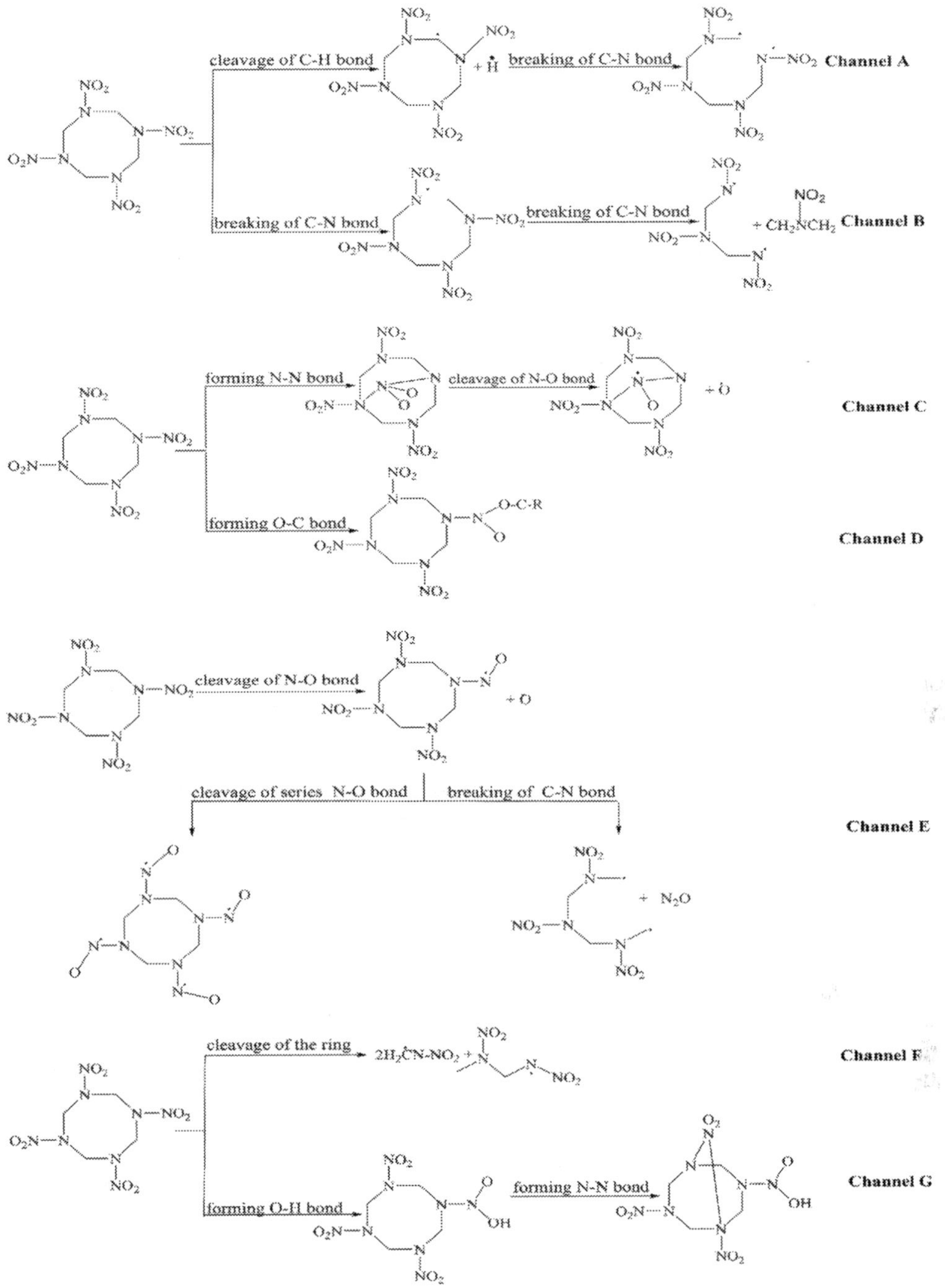

FIG. 13.2 A schematic diagram of the competing decomposition channels for the HMX crystal. *Reproduced from D. Xiang, W. Zhu, Adiabatic and constant volume decomposition process of condensed phase δ-1,3,5,7-tetranitro-1,3,5,7-tetrazocane at high temperatures: Quantum molecular dynamics simulations, J. Mol. Graph. Model. 85 (2018) 68 with permission from Elsevier.*

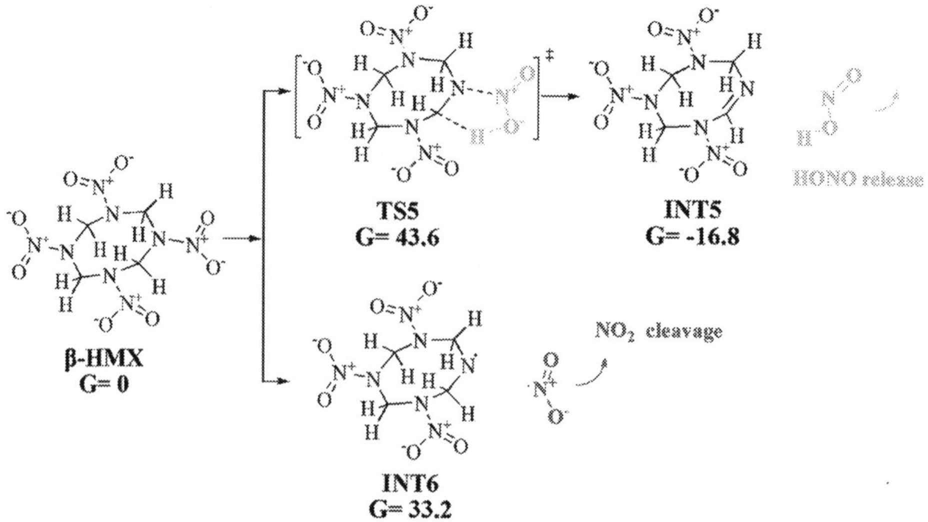

FIG. 13.3 Mechanism of the proposed reaction pathway of HONO release and NO_2 cleavage in β-HMX. The HONO release reaction has a barrier (TS5) of 43.6 kcal/mol, and the NO_2 cleavage reaction energy (barrier) is 33.2 kcal/mol above β-HMX. Units are in kcal/mol. *Reprinted (adapted) with permission from C. Ye, Q. An, W. Zhang, W.A. Goddard, III. Initial decomposition of HMX energetic material from quantum molecular dynamics and the molecular structure transition of β-HMX to δ-HMX, J. Phys. Chem. C 123 (2019) 9231. Copyright (2019) American Chemical Society.*

lattice constants and unit-cell volume of HMX are not obviously changeable under the temperature effects. But the wag angle of HMX suggests slight changes in molecular structure. Most of the vibrational modes display a similar hardening trend under the temperatures, suggesting that the crystal transformation of HMX under heat is through this displacement. Thus, it may be concluded that the low temperatures can not result in the decomposition of the β-HMX crystal without the cooperative effects of the pressures.

Zhu et al. [29] used AIMD to study the crystalline α-HMX at low temperatures of 300, 325, 350, 375, 400, and 425 K. The increasing temperature does not produce any obvious effects on the crystal structure of α-HMX. But the increasing temperature strengthens the free rotation of the nitro groups. The NO_2 asymmetric stretching modes are sensitive to the temperature. The vibrational modes of these side groups (NO_2 and CH_2) couple strongly with the increasing temperature and so could provide an important vibrational channel for H migration reactions and N—NO_2 cleavage. Therefore, the α-HMX crystal also does not decompose at low temperatures without the influence of pressure.

2.2 CL-20

2,4,6,8,10,12-Hexanitro-2,4,6,8,10,12-hexaazaisowurtzitane (CL-20 or HNIW) is a nitramine explosive with high density (about 2.0 g/cm^3) and superior performance (44.8 GPa and 9762 m/s, about 14% higher than HMX). The best-known members of the nitramine explosives are RDX (1,3,4-trinitro-1.3.5-triazocyclohexane), HMX, and the caged nitramine CL-20. CL-20 is regarded as one of the most promising explosives to replace RDX and HMX.

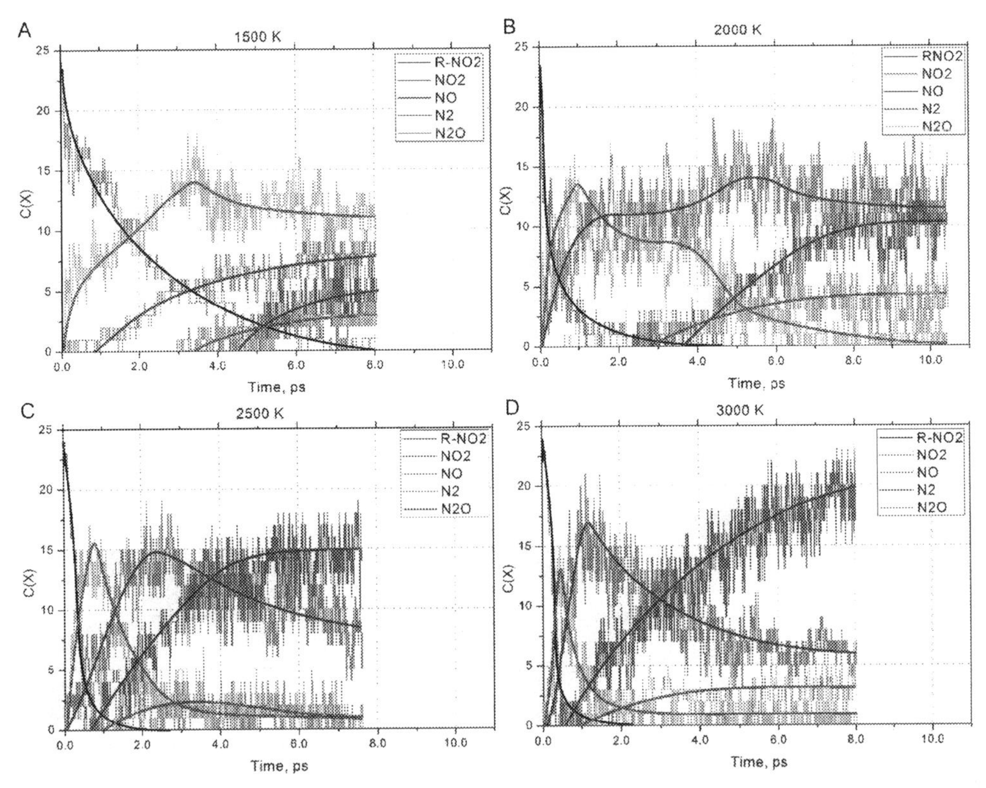

FIG. 13.4 Time evolution of initial CL-20 (R-NO_2), products N_2, N_2O, NO, and intermediate NO_2 for T 1500 K (A), 2000 K (B), 2500 K (C), and 3000 K (D) as calculated from simulations of the small (one unit cell) model. Thick spline trendlines correspond to the actual concentration data of the matching color (*different shades of gray* in the print version) behind. *Reprinted (adapted) with permission from O. Isayev, L. Gorb, M. Qasim, J. Leszczynski, Ab initio molecular dynamics study on the initial chemical events in nitramines: thermal decomposition of CL-20, J. Phys. Chem. B 112 (2008) 11005. Copyright (2008) American Chemical Society.*

Isayev et al. [30] performed AIMD simulations to investigate the chemical events of thermal decomposition of isolated and crystalline CL-20. Fig. 13.4 shows the time evolution of the population of CL-20 molecules, products N_2, N_2O, NO, and intermediate NO_2 at 1500–3000 K. When the temperature increases, the condensed-phase decomposition process gradually accelerates. The primary reactions forming NO_2, NO, N_2O, and N_2 occur at very early stages. The activation barrier for the formation of NO_2 was estimated to be 137 kJ/mol, which essentially determines the overall decomposition kinetics. The effective rate constants for NO_2 and N_2 were also evaluated. The whole reaction mechanism of the condensed-phase CL-20 was not provided.

2.3 RDX

RDX is one of the most important nitramine explosives widely used for both civilian and military applications. To get more insight into the thermal decomposition behavior of RDX, Xu et al. [31] performed AIMD simulations to study its decomposition process at a high temperature of 3000 K. It is found that H_2O, OH, N_2, and NO_2 are the dominant decomposition products, while CO, CO_2, CNO, and N_2O emerge as minor products. With the increase of the simulation time, the numbers of main products H_2O, CO_2, and N_2 generally increase. Charged NO_2 and N_2O radicals may play a promoting role in the decomposition process.

2.4 PETN

Pentaerythritol tetranitrate (PETN), a typical representative of nitrate explosives, has the most stable performance. Wu et al. [32] used DFTB-MD to stimulate the decomposition of crystalline PETN at 4200 K. The results indicate that the decomposition of PETN is triggered by the O—NO_2 bond cleavage of a PETN molecule, which is from its outside (NO_2 and NO_3) groups to inside carbon core by unimolecular mechanism. Some intermediates HONO, NO, HO, and H play an important role in the formation of the products. The reaction rate constants for overall productions of H_2O, CO_2, and N_2 are 1.2, 5.2, and 24.1 ps, respectively.

Afterward, Wu et al. [33] also performed AIMD simulations to study the decomposition reactions of PETN crystal at 3000 and 4200 K. The catalytic mechanism of water in the decomposition of PETN is completely different from previously proposed decomposition mechanisms [32], in which water is just an end product. H and OH can catalytically transport oxygen from NO_2 to CO, thus forming CO_2. H_2O, H, and OH play active roles in detonation chemistry, so the kinetics of water formation may also increase the sensitivity of PETN. Hydrogen atoms retain high mobility at extreme conditions.

Recently, Zhou et al. [34] used AIMD to study the decomposition mechanisms of solid PETN and pentaerythritol tetranitrocarbamate (PETNC) and their sila analogs Si-PETN and Si-PETNC heated continuously from 300 to 3000 K at a heating rate of 135 K ps^{-1}. The first decomposition step of PETN is the O—NO_2 bond breaking. The same is true of Si-PETN. The first decomposition step for Si-PETNC is intermolecular hydrogen transfer between two Si-PETNC molecules, while the initial decomposition of PETNC is the C—CH_2 bond fission.

The sila analogs Si-PETN and Si-PETNC have higher sensitivity than PETN and PETNC due to the formation of the highly exothermic Si—O bond as a predominate initial decomposition step. These Si—O bonds can promote subsequent decomposition reactions to generate various intermediates and products, so accelerating the decomposition process and energy release and increasing the sensitivity of Si-PETN and Si-PETNC.

2.5 NM

Nitromethane (NM) is a model compound of EM. Manaa et al. [35] carried out AIMD simulations to investigate the early chemical events of dense NM ($\rho = 1.97$ g/cm^3) at 3000 and 4000 K. The first step in the decomposition is an intermolecular proton transfer mechanism that produces $CH_3NO_2H^+$ and $H_2CNO_2^-$. The NM molecule can also transform into the aci acid form CH_2NO_2H by an intramolecular hydrogen transfer fashion, accompanying this event.

Chang et al. [36] used AIMD to stimulate the decomposition of solid NM about 2200 K under ambient pressure. The initial decomposition steps include both proton transfer and

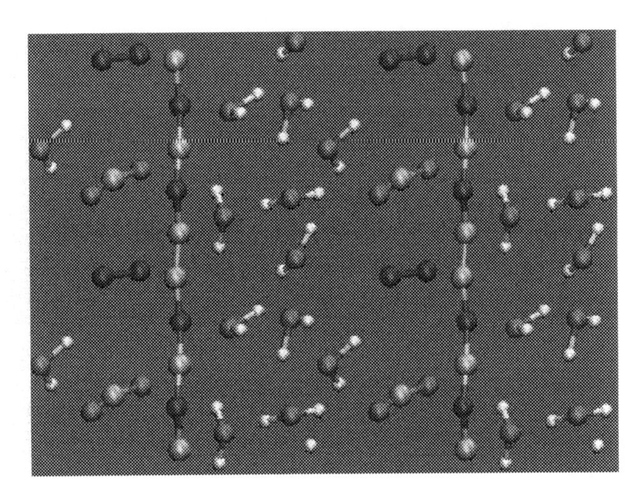

FIG. 13.5 Snapshots of the decomposition of the solid NM by the AIMD simulation at 195 ps. The *red* (*gray* in the print version), *white*, *blue* (*dark gray* in the print version), and *cyan* (*light gray* in the print version) balls refer to oxygen, hydrogen, nitrogen, and carbon atoms, respectively. *Reprinted with permission from J. Chang, P. Lian, D. Wei, X. Chen, Q. Zhang, Z., Gong, Thermal decomposition of the solid phase of nitromethane: Ab initio molecular dynamics simulations, Phys. Rev. Lett. 105 (2010) 188302. Copyright (2010) by the American Physical Society.*

commonly known C—N bond rupture. Fig. 13.5 displays the snapshots of the decomposition of the solid NM by the AIMD simulation at 195 ps. It is found that the final products are H_2O, CO_2, N_2, and CNCNC. There are about 75 species and 100 elementary reactions observed in the decomposition of NM.

2.6 Silver azide

Silver azide is a sensitive and powerful solid explosive that is used extensively as a detonating agent for explosives. Zhu and Xiao [37] performed AIMD simulations to investigate the structure and stability of crystalline silver azide at temperatures 298, 473, 498, 523 (melting point), 548, and 573 K. The azide sublattice structure broke down prior to the silver sublattice. Fig. 13.6 displays the snapshots of (a) N—N rupture, (b) N_2 formation, (c) N radical formation, and (d) Ag cluster formation at 573 K during the thermal equilibration phase. It is found that the initiation decomposition of silver azide is triggered by the N—N bond breaking. This will initiate many decomposition reactions and produce many nitrogen radicals, N_2, and silver clusters.

Fig. 13.7 presents the band gaps of silver azide as a function of temperature. It is seen in Fig. 13.7 that the band gap of silver azide gradually decreases with the increment of temperature. At 573 K, there is a band gap closure in the system. According to the first-principles band gap criterion [38–40], it may be inferred that the sensitivity for silver azide becomes more and more sensitive with the temperature increasing.

2.7 DiAT

3,6-Di(azido)-1,2,4,5-tetrazine (DiAT), a high-nitrogen organic polyazido compound only containing C and N atoms, is an ideal potential candidate for initial explosives because of their high energy and clean and thermodynamically favorable decomposition.

Wu et al. [41] performed AIMD simulations to study the initial chemical processes and decomposition reactions of isolated and crystal DiAT at about 3000 K. The preferential initial

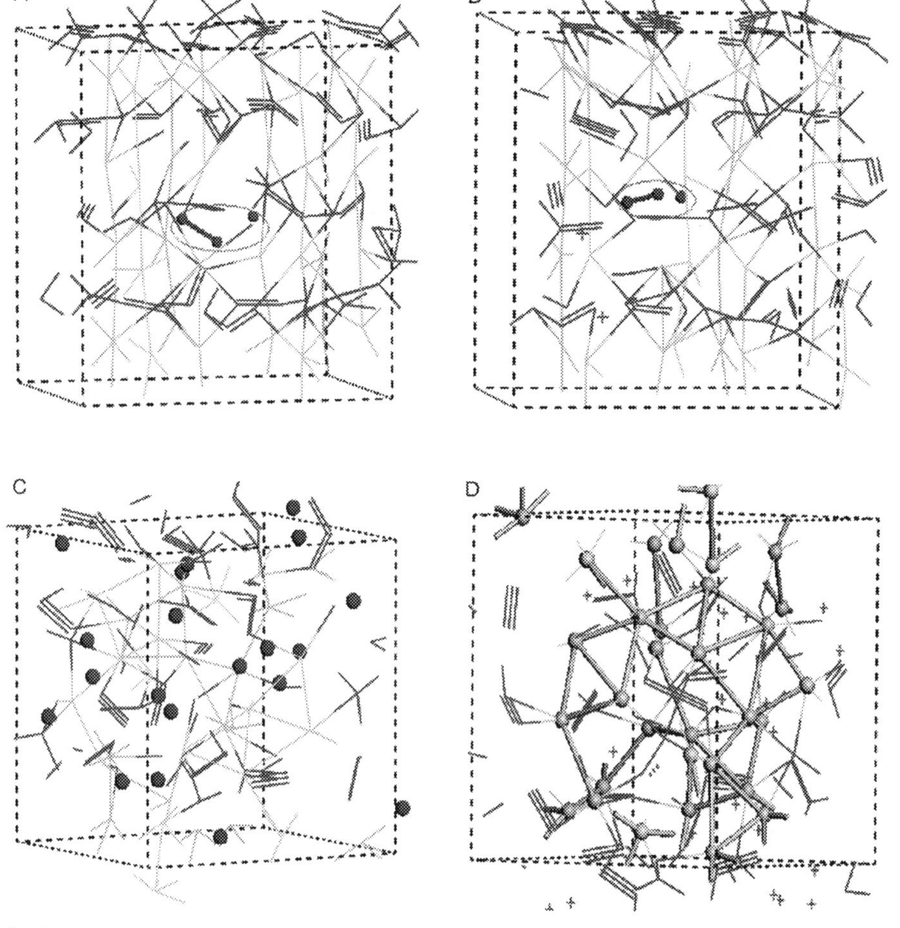

FIG. 13.6 Snapshots of (A) N—N rupture, (B) N_2 formation, (C) N radical formation, and (D) Ag cluster formation at 573 K during the thermal equilibration phase. *Dark cyan* (*dark gray* in the print version) and *blue* (*gray* in the print version) lines or spheres stand for Ag and N atoms, respectively. *Reprinted (adapted) with permission from W. Zhu, H. Xiao, Ab initio molecular dynamics study of temperature effects on the structure and stability of energetic solid silver azide, J. Phys. Chem. C 115 (2011) 20782. Copyright (2011) American Chemical Society.*

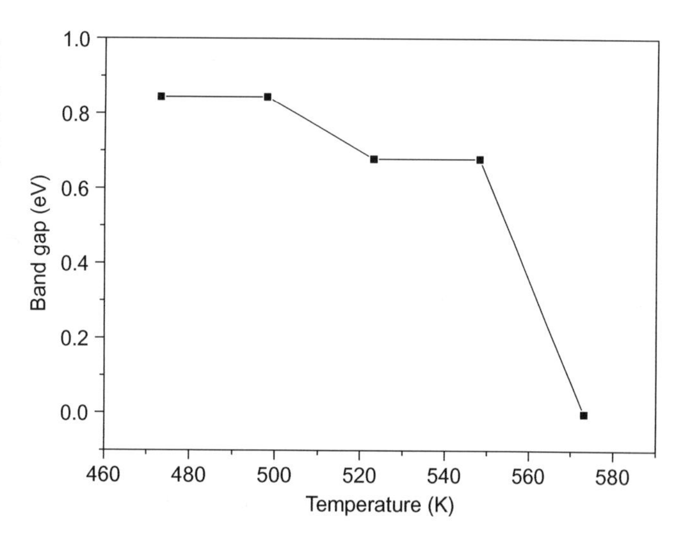

FIG. 13.7 Band gaps of silver azide as a function of temperature. *Reprinted (adapted) with permission from W. Zhu, H. Xiao, Ab initio molecular dynamics study of temperature effects on the structure and stability of energetic solid silver azide, J. Phys. Chem. C 115 (2011) 20782. Copyright (2011) American Chemical Society.*

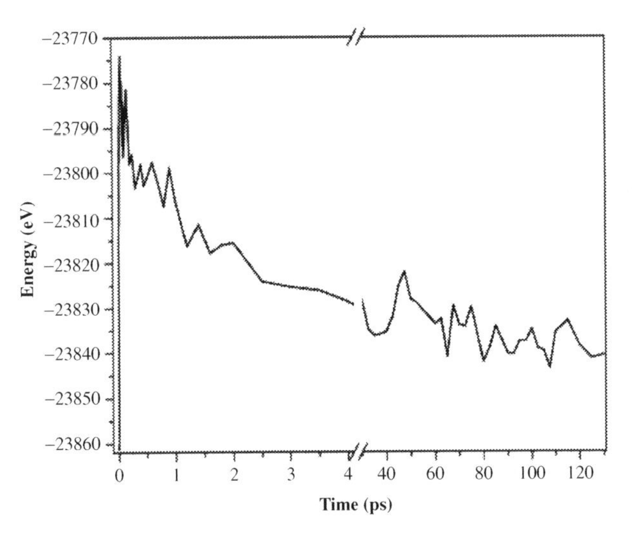

FIG. 13.8 Time dependence of the total energy of DiAT. For reproduction of material from PCCP. *Reproduced from Q. Wu, W. Zhu, H. Xiao, An ab initio molecular dynamics study of thermal decomposition of 3,6-di(azido)-1,2,4,5-tetrazine, Phys. Chem. Chem. Phys. 16 (2014) 21620 with permission from the PCCP Owner Societies.*

decomposition step is the homolysis of the N—N_2 bond in the azido group. In the early decomposition, DiAT decomposes very fast and drastically without forming any stable long chains or heterocyclic clusters, and most of the nitrogen gases are released through rapid rupture of N—N and C—N bonds. But in the later decomposition stage, the release of nitrogen gas is inhibited due to low mobility, a long distance from each other, and strong C—N bonds. To overcome the obstacles, the nitrogen gases are released through the slow formation and disintegration of polycyclic networks.

Fig. 13.8 displays the time dependence of the total energy of DiAT. It is seen that most of the energy is released in the initial 4.0 ps, while the release of energy is not obvious in the following 126.0 ps. This reveals the essence of explosive characteristics of DiAT as an initial explosive that it needs to release most of the energy before the formation of detonation.

2.8 TKX-50

Dihydroxylammonium 5,5,-bistetrazole-1,1′-diolate (TKX-50) is a recently synthesized azole-based ionic explosive with promising performance because of high exothermicity, high density, and insensitivity toward external stimuli.

An et al. [42] employed DFTB-MD to simulate the initial chemical reactions of the TKX-50 crystal with the temperature increasing continuously from 300 to 3000 K at a heating rate of 180 K/ps. Fig. 13.9 displays the fragment analyses during the cook-off simulation of TKX-50 as a function of temperature. The results indicate that the initial decomposition of TKX-50 involves ring breaking to release N_2 rather than N_2O with the release of energy, which in turn drives further decompositions. The proton transfer took place at a much low temperature than the ring-breaking reaction. The proton transfer decreases the reaction barrier for both the releases of N_2 and N_2O by ca. 10 kcal/mol. These simulations suggest that the strategy to design less sensitive azole-based EMs is to prevent the proton transfer between the cation and anion.

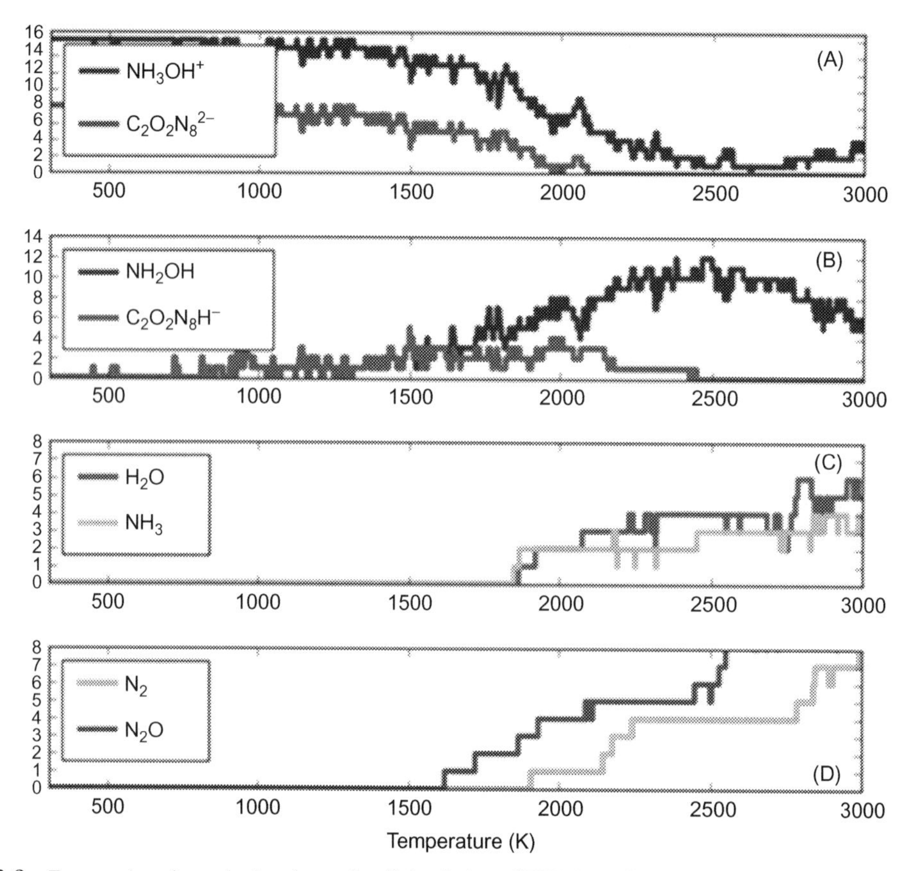

FIG. 13.9 Fragment analyses during the cook-off simulation of TKX-50 as the temperature is ramped at a uniform rate of 180 K/ps from 300 to 3000 K. *Reprinted (adapted) with permission from Q. An, W. Liu, W.A. Goddard III, T. Cheng, S.V. Zybin, et al., Initial steps of thermal decomposition of dihydroxylammonium 5,5′-bistetrazole-1,1′-diolate crystals from quantum mechanics, J. Phys. Chem. C 118 (2014) 27175. Copyright (2014) American Chemical Society.*

Later, Meng et al. [43] performed an AIMD simulation to investigate the chemical decomposition of TKX-50 crystal at the temperature from 300 to 3000 K. It was concluded that the intense hydrogen bonding in TKX-50 acts as a potential source to promote H-transfer, which accelerates the ring breaking of tetrazole, and thus enhances the subsequent thermal decomposition of TKX-50. This effect of the strong hydrogen bonding damages the thermal stability of TKX-50 relative to its notably low impact sensitivity.

Lu et al. [44] used AIMD to simulate the decomposition reactions of noncompressed and compressed TKX-50 crystals at 1800 and 2500 K. The results indicate that the intermolecular H-transfer initiating the TKX-50 decomposition is inhibited by the increasing compression. The compression also decelerates subsequent decompositions.

2.9 Furoxan

Furoxan stores significant quantities of chemical energy and is thought of as an energetic nitrogen-containing compound. It can be used as an important precursor to synthesize new

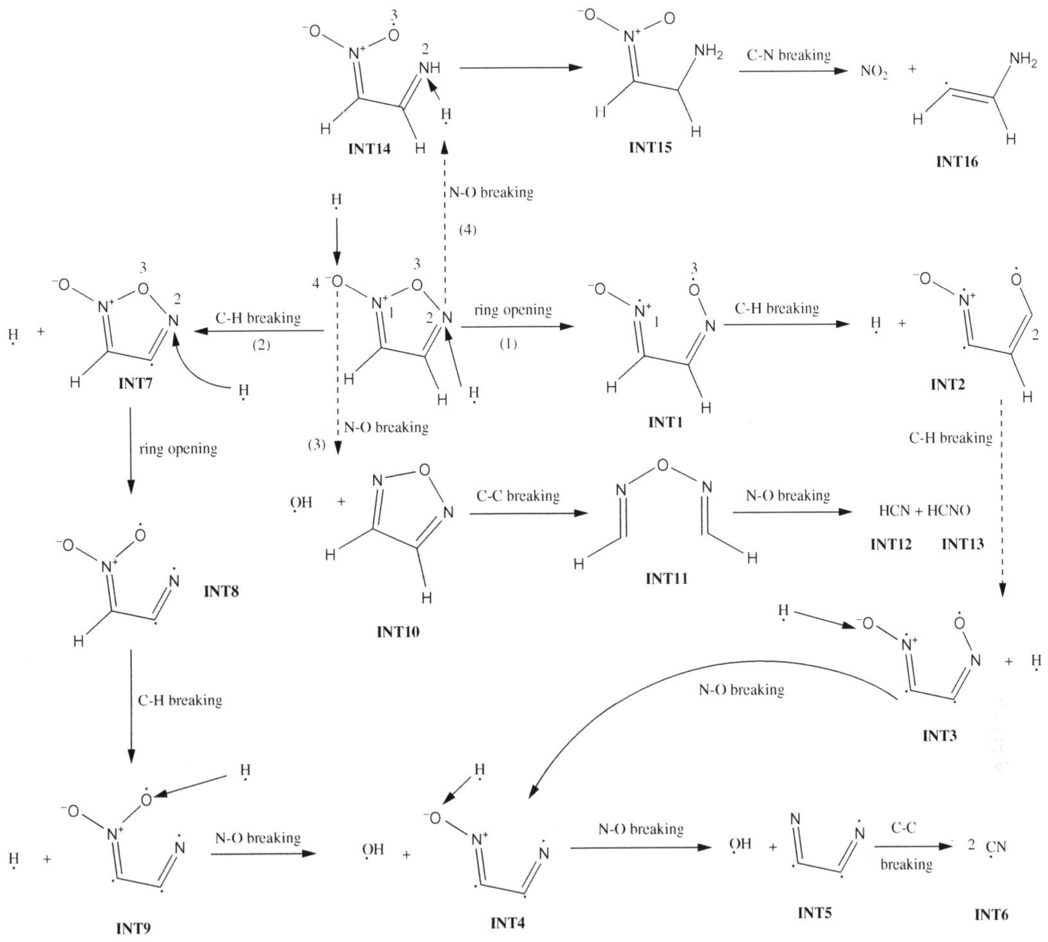

FIG. 13.10 A schematic diagram of the four decomposition processes of furoxan. *For reproduction of material from RSC Adv.: Reproduced from Q. Wu, W. Zhu, H. Xiao, Catalytic behavior of hydrogen radicals in the thermal decomposition of crystalline furoxan: DFT-based molecular dynamics simulations, RSC Adv. 4 (2014) 34454 with permission from The Royal Society of Chemistry.*

powerful energetic nitrogen-rich compounds. Wu et al. [45] selected parent furoxan as a model compound and performed AIMD simulations to investigate its initial chemical processes and decomposition mechanism under 2400 K.

Fig. 13.10 displays a schematic diagram of the four decomposition processes of furoxan. The initial decomposition step in path 1 is the direct opening of the furoxan ring through the homolysis of the N1—O3 bond. In path 2, furoxan can directly decompose into hydrogen radical and INT7 through the breaking of the C—H bond. In path 3, hydrogen radical would attack and capture the oxygen atom (O4 atom) out of the ring to form a furazanmolecule (INT10) and hydroxyl radical. In path 4, hydrogen radical can attack the N2 atom directly to form INT14 by the N2—O3 bond cleavage.

The hydrogen radicals play a catalytic role in the following decomposition of furoxan and can promote the decomposition to a great degree. They not only catalyze the opening of furoxan rings and breaking of N—O bonds, but also capture and transport the oxygen atoms from nitrogen atoms to carbon atoms and promote the release of some small products. The catalytic ability may come from their high mobility that enables the hydrogen to transport fast between different species.

2.10 DTTO

Di-tetrazine-tetroxide (DTTO) was regarded as the performance limit for C—H—N—O explosives, making it a "holy grail" in high-energy materials. Ye et al. [46] carried out AIMD simulations to study the initial reaction mechanisms of DTTO crystal (including two DTTO crystalline phases: c1-DTTO and c2-DTTO) heated from 300 to 3000 K.

It is found that the initial reaction of c2-DTTO starts at 2670 K, involving the intermolecular reaction between the two c2 molecules, with no gas products released. Then, at 2700 K, one N_2 molecule is released from this bimolecular complex. Afterward, one N_2O molecule was also released. This shows that by 2700 K, the primary reactions involve both N_2 and N_2O releasing processes. This is quite different from the c1-DTTO reaction mechanisms, in which the initial reactions involve unimolecular decomposition or oxygen transfer.

2.11 BCHMX

cis-1,3,4,6-Tetranitrooctahydroimidazo-[4,5-d]imidazole (BCHMX or Biocycl-HMX) is a theoretically designed energetic compound with superior detonation performance. To elucidate the initial reaction mechanism of the BCHMX crystal, Ye et al. [47] performed AIMD simulations to investigate its decomposition process of BCHMX.

For noncompressed BCHMX, its initial decomposition path is nitro-aci isomerization reaction, followed by unimolecular NO_2 release. For compressed BCHMX, its initial step is intermolecular hydrogen transfer, followed by bimolecular NO_2 release. The intermolecular hydrogen transfer in compressed BCHMX decreases the reaction barrier of releasing NO_2 by about 7 kcal/mol. The HONO elimination reaction in compressed BCHMX is more favorable, whose reaction barrier is lower than that of unimolecular NO_2 releasing reaction by 10 kcal/mol and that of bimolecular NO_2 releasing reaction by about 3 kcal/mol.

2.12 TEX

4,10-Dinitro-2,6,8,12-tetraoxa-4,10-diazaisowurtzitane (TEX) possesses the same isowurtzitane cage structure as CL-20, which is one of the most powerful high explosives that have been synthesized to date and has a relatively high ambient temperature density of 1.99 g cm^{-3}. The absence of sterically demanding nitramine groups in the TEX structure greatly reduces its sensitivity and makes it an interesting insensitive EM ($h_{50} > 177$ cm). Its heat of formation, detonation velocity, and detonation pressure were calculated to be 541 kJ mol^{-1}, 8665 m s^{-1}, and 41.03 GPa, respectively. Xiang and Zhu [48] performed

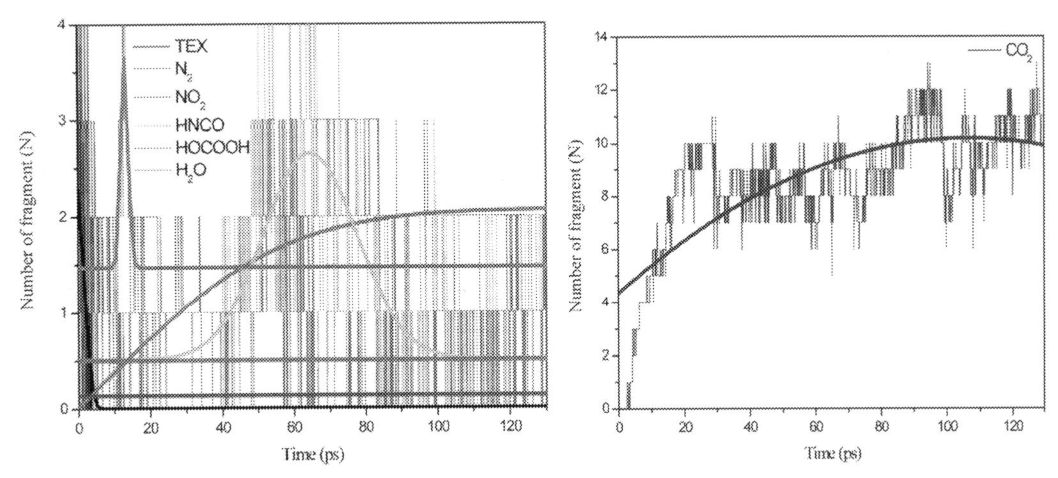

FIG. 13.11 Time evolution of the population of the main fragments during the decomposition of the crystal TEX: N_2, NO_2, NO, HNCO, HOCOOH, H_2O, and CO_2. The *thick trend lines* correspond to the actual concentration data of the corresponding matching color (*different shades of gray* in the print version). *Reproduced from D. Xiang, W, Zhu, Thermal decomposition of isolated and crystal 4,10-dinitro-2,6,8,12-tetraoxa-4,10-diazaisowurtzitane according to ab initio molecular dynamics simulations, RSC Adv. 7 (2017) 8347 with permission from The Royal Society of Chemistry.*

AIMD simulations to study the initiation chemical reaction and subsequent decomposition mechanism of TEX crystal at 2160 K.

The initial decomposition of the TEX crystal is triggered by the unimolecular cleavage of the C—H bond to form hydrogen radical. This is different from those of the isolated TEX molecule: (a) N—NO_2 bond cleaves to release NO_2; (b) N—NO_2 bond cleavage and ring opening in the concerted step; and (c) ring opening through the C—C bond cleavage. The generated H radicals are very active and can prompt other unreacted TEX molecules to decompose. Fig. 13.11 presents the time evolution of the population of the main fragments during the decomposition of the crystal TEX. The decomposition of TEX was completed in 20 ps. CO_2 was mostly produced during the latter decomposition stage. The primary decomposition products are nitrogen and various carboxyl derivatives.

2.13 HMX nanoparticles

Explosive nanoparticles (NPs), as an intermediate state between bulk material and its constituent molecule, possess unusual physical and chemical properties. To understand the evolution of various physical and chemical properties from the constituent molecule to bulk material, the size effects of the explosive NPs on their behaviors have attracted extensive attention. Liu et al. [49] report a DFTB-MD simulation of the initiation chemistry of α-HMX NPs with the diameter range 1.4–2.8 nm (Fig. 13.12) at high temperatures (2400–3000 K) by maintaining constant energy and volume (NVE ensemble) to model adiabatic initiation process.

The initial decomposition process of the HMX NPs includes two sequential stages: (i) competition between rapid expansion and unimolecular decompositions at surfaces; (ii) subsequent complex uni- and bimolecular decompositions. Fig. 13.13 displays a schematic

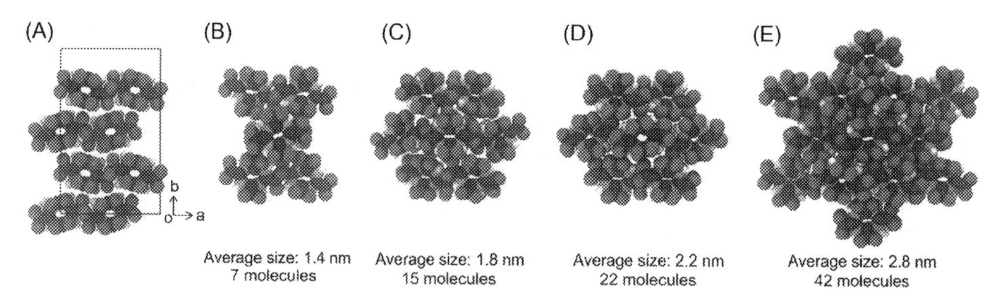

FIG. 13.12 Configurations of relaxed (A) α-HMX bulk and NPs with average diameters of (B) 1.4 nm, (C) 1.8 nm, (D) 2.2 nm, and (E) 2.8 nm. The C, H, O, and N are represented by *gray, white, red* (*dark gray* in the print version), *and blue* (*dark gray* in the print version) balls, respectively. *Reprinted (adapted) with permission from Z. Liu, W. Zhu, G. Ji, K. Song, H. Xiao, Decomposition mechanisms of α-octahydro-1,3,5,7-tetranitro-1,3,5,7-tetrazocine nanoparticles at high temperatures, J. Phys. Chem. C 121 (2017) 7728. Copyright (2017) American Chemical Society.*

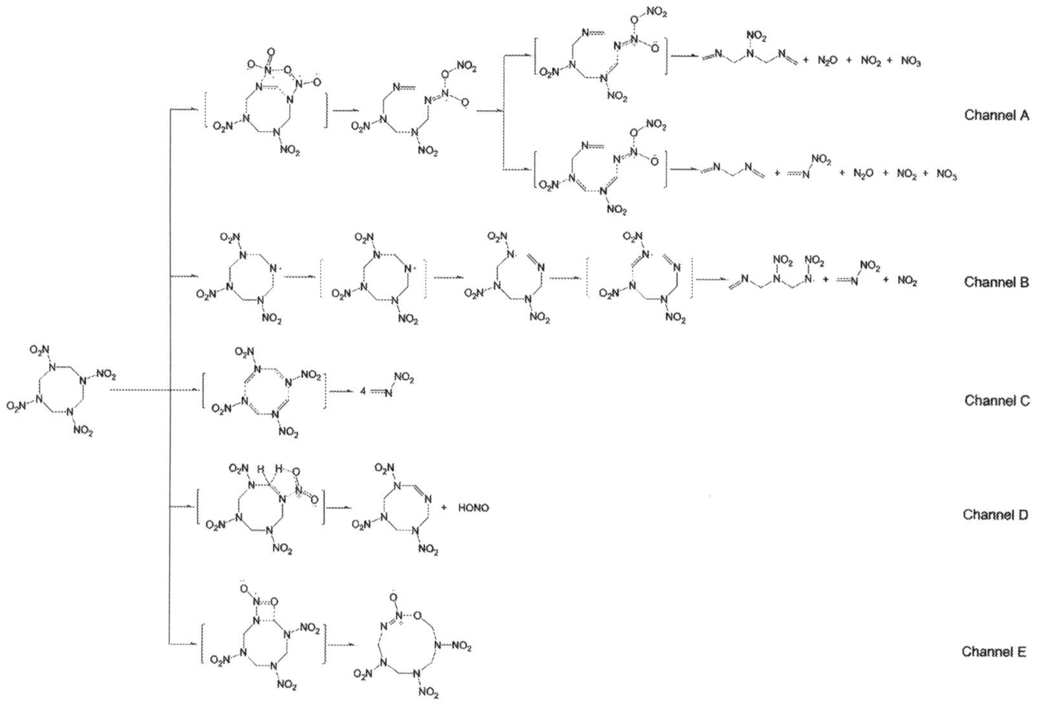

FIG. 13.13 A schematic diagram of the competing unimolecular decomposition channels for the HMX NPs. Two representative decomposition channels are listed in Channel A and one representative decomposition channel is listed in Channel B. *Reprinted (adapted) with permission from Z. Liu, W. Zhu, G. Ji, K. Song, H. Xiao, Decomposition mechanisms of α-octahydro-1,3,5,7-tetranitro-1,3,5,7-tetrazocine nanoparticles at high temperatures, J. Phys. Chem. C 121 (2017) 7728. Copyright (2017) American Chemical Society.*

diagram of the competing unimolecular decomposition channels for the HMX NPs. The decomposition mechanisms of the HMX NPs are governed by five unimolecular competition reaction channels: isomerization into linear isomer (Channel A); N—NO_2 homolysis with ring opening (Channel B); concerted ring opening (Channel C); HONO elimination (Channel D); isomerization into the 10-membered ring (channel E). All the rate-determining steps in predominated decomposition mechanisms are multistep processes. Thus, it may be inferred that the decomposition kinetics of the HMX NPs is different from the solid phase decomposition in the studied temperature range.

Fig. 13.14 presents the time evolution of the proportion of the dominant decomposition channels of the 1.4 nm HMX NPs at different temperatures. At 2400 K, the proportions of Channel A, Channel B, and Channel C are 80.2%, 11.4%, and 8.39%, respectively. This indicates that Channel A is the main decomposition pathway. When the temperature increases, their proportions have large changes. At 3000 K, the proportions of Channel A, Channel B, and Channel C are 31.1%, 65.4%, and 1.31%, respectively, suggesting that Channel B becomes a predominate decomposition pathway. Therefore, the main decomposition pathway of the

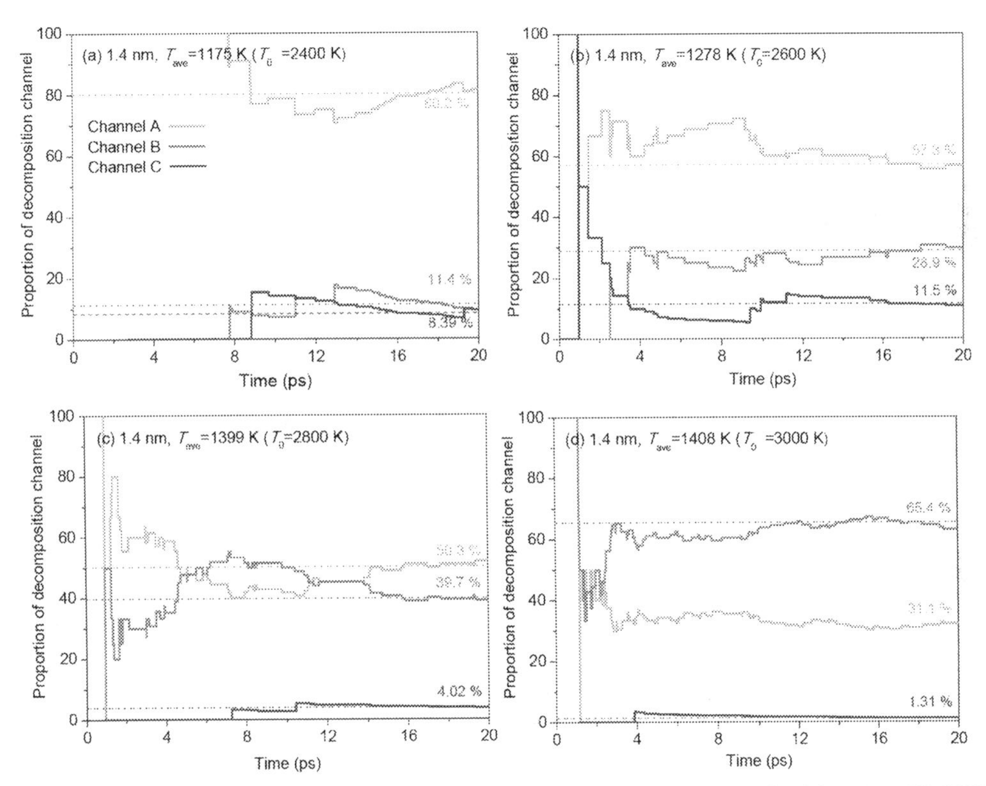

FIG. 13.14 Time evolution of the proportion of the dominant decomposition channels of the 1.4 nm HMX NPs at different temperatures. *Reprinted (adapted) with permission from Z. Liu, W. Zhu, G. Ji, K. Song, H. Xiao, Decomposition mechanisms of α-octahydro-1,3,5,7-tetranitro-1,3,5,7-tetrazocine nanoparticles at high temperatures, J. Phys. Chem. C 121 (2017) 7728. Copyright (2017) American Chemical Society.*

TABLE 13.2 Activation barriers (E_a, kcal mol^{-1}) and pre-exponential factors (lnA, $[N]^{-1}$ ps^{-1}) of the global decomposition and rate-determining steps of the dominant decomposition channels in the decompositions of the HMX NPs predicted by DFTB-MD simulations.

	1.4 nm		1.8 nm		2.8 nm	
Channel	E_a	ln$A$	E_a	ln$A$	E_a	ln$A$
Global	22.9	12.2	32.6	15.0	26.5	14.9
A	9.57	6.24	19.7	9.17	31.7	15.8
B	47.6	20.6	47.2	19.8	22.6	12.9

HMX nanoparticles at low temperature is the isomerization reaction of the HMX molecule, while that at high-temperature changes to the N—NO_2 homolysis with ring opening.

Table 13.2 lists the obtained reaction rates and activation energies of different channels for different sizes of the HMX NPs under different channels (inset) for the HMX NPs as a function of 1000/T. It is found that the global reaction rates for different sizes of HMX NPs are both size- and temperature-dependent. For a particular NP, as the starting temperature increases from 2400 to 3000 K, the global reaction rates increase by 5.1–6.7 times. For different sizes of HMX NPs, the size effect is also significant. The 2.8 nm NP has a faster reaction rate than smaller NPs at the same starting temperature.

2.14 Nanothermite

Nanothermites or metastable intermolecular composites (MICs), a class of energetic material containing both fuel (Aluminum) and oxidizer (Fe_2O_3, CuO, MoO_3, WO_3, Bi_2O_3, etc.), can undergo a rapid redox reaction. They have wide applications in explosives and propellants. Xiong et al. [50] performed AIMD simulations to investigate the reactive properties and reaction processes of the Al/CuO nanothermite at high temperatures (830, 1000, 2000, and 3000 K).

The redox reactions that produced copper metal and aluminum oxide initiate the reactions of the Al/CuO nanothermite at the interface. The higher the temperature is, the faster the atomic configuration changes. Fig. 13.15 displays the time evolution of the position $<z_c(t)>$ of the combustion fronts for the Al/CuO nanothermite. When the temperature increases from 830 to 1000 K, the time duration of the linear reaction fronts increases greatly. This may be because the higher temperature makes the reaction more complete. The reaction times continually decrease with the increasing temperature. Therefore, the higher temperature promotes the thermite reaction. In all, the temperature plays different roles on the reaction fronts, primarily improving the reaction progresses.

Recently, Xiong et al. [51] used AIMD to simulate the adiabatic reaction processes of the Al/CuO nanothermite at 2000 K. Fig. 13.16 displays the interfacial reaction process of the Al/CuO nanolaminate observed at different times during the AIMD simulations. The initial reaction is the decomposition of CuO at the interface: CuO → Cu + O. These resulting O atoms diffused into the Al layer. Then, the Al atoms reacted with the O atoms to form Al_2O_3.

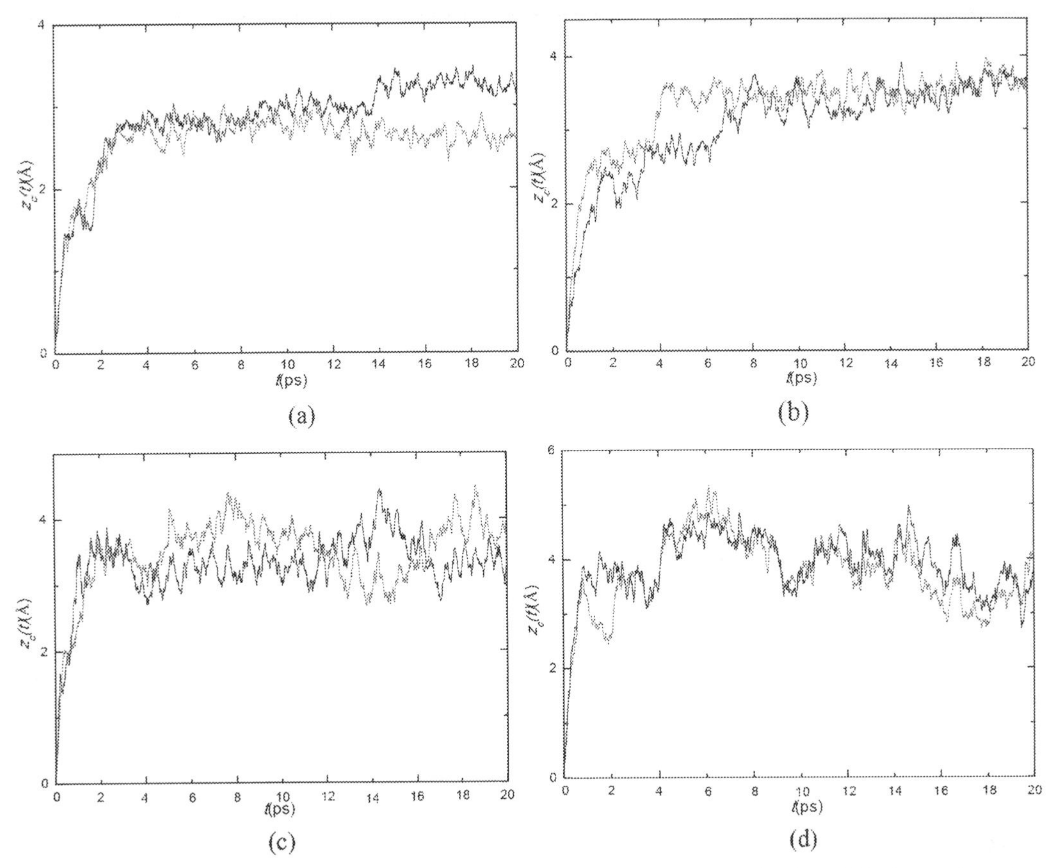

FIG. 13.15 Time evolution of the position $<z_c(t)>$ of the combustion fronts for the Al/CuO nanothermite at (A) 830, (B) 1000, (C) 2000, and (D) 3000 K. *Reproduced from G. Xiong, C. Yang, W. Zhu, Interface reaction processes and reactive properties of Al/CuO nanothermite: an ab initio molecular dynamics simulation, Appl. Surf. Sci. 459 (2018) 835 with permission from Elsevier.*

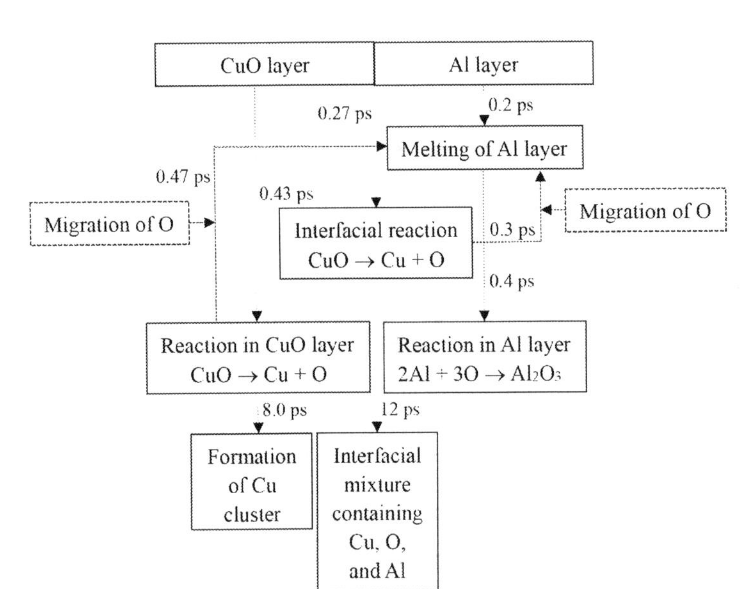

FIG. 13.16 Interfacial reaction process of the Al/CuO nanolaminate observed at different times during the AIMD simulations. *Reproduced from G. Xiong, C. Yang, S. Feng, W. Zhu, Ab initio molecular dynamics studies on the transport mechanisms of oxygen atoms in the adiabatic reaction of Al/CuO nanothermite, Chem. Phys. Lett. 745 (2020) 137278. with permission from Elsevier.*

The decomposition of CuO also took place in the CuO layer. When most of the forming O atoms migrate out, the Cu clusters were formed in the CuO layer. At the end of our AIMD simulations, the interfacial mixture containing Cu, O, and Al was formed. The interfacial reaction of the Al/CuO nanolaminate can be expressed as $3CuO + 2Al \rightarrow 3Cu + Al_2O_3$.

The interfacial reaction gives rise to the movement of inner O atoms. Their successive movement is maintained by the energy from the redox reaction and the Cu vacancies. The forming copper clusters may act as a barrier to weaken its ultimate output performance. Its combustion velocity is related to the consecutive transport of the oxygen atoms that is a rate-limiting process.

2.15 FOX-7

1,1-Diamino-2,2-dinitroethylene (FOX-7) is an EM with high explosive performance comparable to RDX, but with lower sensitivity to impact and friction. Liu et al. [52] used AIMD to study the thermal decomposition mechanisms of solid FOX-7 at 2500, 3000, 3500, and 4000 K. The results indicate that the initial decomposition of FOX-7 has three main reaction paths. The most popular initial decomposition step is the C—NO_2 bond fission reaction, followed by the intramolecular hydrogen transfer, and the least popular one is the intermolecular hydrogen transfer. They also mapped out a complete reaction network for the formations of H_2O, CO_2, and N_2 in the thermal decomposition of FOX-7.

Afterward, Jiang et al. [53] performed DFTB-MD simulations to the reaction mechanism of the temperature-programmed and constant temperature-heated FOX-7. The N—O bond rupture as the initial decomposition step of the FOX-7 was observed in the temperature-programmed heating simulation from 300 to 3000 K. The intra- and intermolecular hydrogen transfer reactions as the initial steps were found in constant temperature heating simulation at 3000 K. Under both heating conditions, the C—NO_2 bond rupture as a dominant step occurs in the decomposition of the FOX-7 crystal.

2.16 LLM-105

2,6-Diamino-3,5-dinitropyrazine-1-oxide (LLM-105) is regarded as a representative of the new generation of low-sensitivity EMs. Wang et al. [54] performed DFTB-MD simulations to investigate the decomposition mechanisms of condensed-phase LLM-105. There are four possible initial pathways in the decomposition of LLM-105, including the intramolecular H transfer (1), the NO_2 partition (2), the acyl O partition (3), and the O partition from a NO_2 group (4). These pathways are strongly temperature-dependent.

Fig. 13.17 displays the potential energy evolution of LLM-105 heated in a programmed manner from 300 to 1700 K, heated constantly at 1700 K for 5 ps and annealed to 300 K. The results indicate that the intramolecular H transfer took place always with a low energy barrier, and the H transferred product can also return to the original molecules with a low barrier too, exhibiting the reversibility of the transfer. They thought that such reversibility can partly be responsible for the low sensitivity of LLM-105, as the reversible H transfer can buffer the external stimuli with an energy transfer and a slight structural variation only.

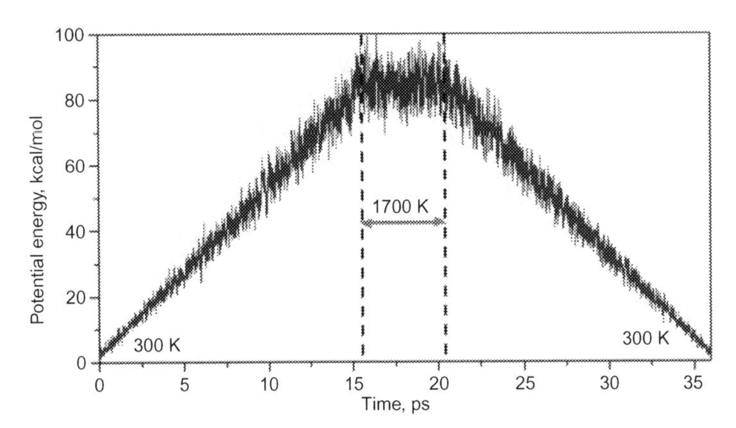

FIG. 13.17 Potential energy evolution of LLM-105 heated in a programmed manner from 300 to 1700 K, heated constantly at 1700 K for 5 ps, and annealed to 300 K. *Reprinted (adapted) with permission from J. Wang, Y. Xiong, H. Li, C. Zhang, Reversible hydrogen transfer as new sensitivity mechanism for energetic materials against external stimuli: a case of the insensitive 2,6-dDiamino-3,5-dinitropyrazine-1-oxide, J. Phys. Chem. C 122 (2018) 1109. Copyright (2018) American Chemical Society.*

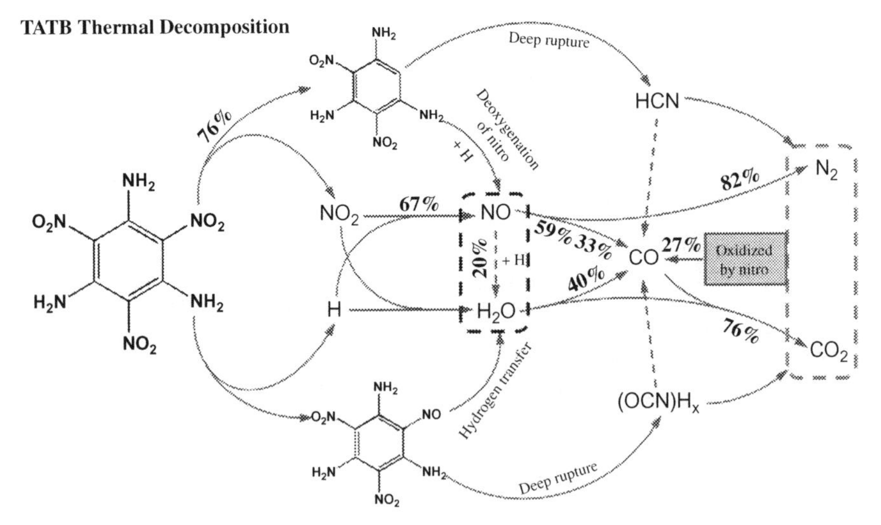

FIG. 13.18 Reaction mechanism for TATB thermal decomposition. *For reproduction of material from NJC. Reproduced from Z. He, Y. Yu, Y. Huang, J. Chen, Q. Wu, Reaction kinetic properties of 1,3,5-triamino-2,4,6-trinitrobenzene: a DFTB study of thermal decomposition, New Journal of Chemistry 43 (2019) 18027 with permission from the Centre National de la Recherche Scientifique (CNRS) and The Royal Society of Chemistry.*

2.17 TATB

1,3,5-Triamino-2,4,6-trinitrobenzene (TATB), a typical nitro explosive, is famous as one of the most stable EMs. TATB has strong hydrogen bonding interactions and special π-stacking of interlayer molecules, resulting in extreme insensitivity toward heat, friction, and shock. He et al. [55] employed DFTB-MD to stimulate the chemical reactions of TATB crystal at 2500, 3000, 3500, and 4000 K.

At the earliest decomposition stage, the cleavage of the C—NO_2 bond shows a considerable reaction activity. Fig. 13.18 presents the reaction mechanisms for TATB thermal decomposition. There are four sequential reaction stages during the thermal dissociation of TATB:

(1) C—NO_2 bond cleavage and amino dehydrogenation; (2) hydrogen transfer to nitro or NO_2, with further conversion to NO and H_2O; (3) carbon oxidation and molecular ring deep fission to generate CO and small —CN fragments; (4) final formation of gaseous N_2 and CO_2. The results also show that the C—NH_2 bond breaking is the rate-controlling step for carbon oxidation, while the N—N bond formation is the rate-controlling step for the generation of N_2.

2.18 Cocrystal explosives

Cocrystallization technology can effectively combine the advantages of two or more explosive components, reduce the sensitivity of the existing explosives, and maintain their energy properties. Therefore, cocrystal explosives have desirable physicochemical properties and detonation properties, such as high chemical stability, low impact sensitivity, and good oxygen balance. Wu et al. [56] used DFTB-MD to stimulate the adiabatic initiation process of various HMX-based cocrystals HMX/DMF (*N,N*-dimethylformamide) and HMX/DNDAP (bis(2,4-dinitro-2,4-diazapentane)) at 2000, 2500, and 3000 K by maintaining a constant energy and volume (NVE ensemble).

The decomposition and reaction mechanisms of the two cocrystals showed great dependence on the temperature. Fig. 13.19 displays the evolution process of the conformation from α-HMX to β-HMX in the HMX/DMF cocrystal at 2000 K. The conformations of the HMX molecules at 0 ps present α crystal forms. Thereafter, the conformation of the α-HMX molecule gradually changed to the β form owing to the effects of the surrounding cocrystal molecules.

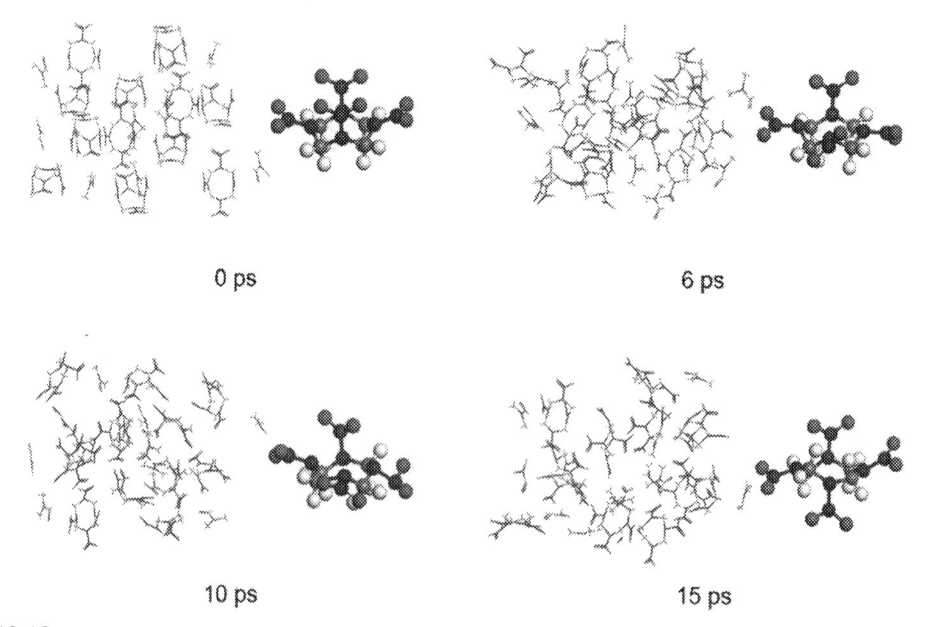

FIG. 13.19 Evolution process of the conformation from α-HMX to β-HMX in the HMX/DMF cocrystal at 2000 K. *Reprinted (adapted) with permission from X. Wu, Z. Liu, W. Zhu, Conformational changes and decomposition mechanisms of HMX-based cocrystal explosives at high temperatures, J. Phys. Chem. C 124 (2020) 25. Copyright (2020) American Chemical Society.*

Finally, the α-HMX molecule completely transformed into the β form. However, the HMX molecules in the HMX/DNDAP cocrystal did not undergo a conformational change. At 2500 K, the initial reaction paths of HMX include N—NO_2 homolysis, HONO elimination, and ring breaking of the HMX molecule (C—N bond cleavage). Meanwhile, the DNDAP molecules decomposed, but the DMF molecules did not. This is the result of the interactions of HMX with different cocrystal molecules at 2500 K.

At 3000 K, the decomposition pathways of HMX in the two cocrystals are isomerization into a linear isomer, N—NO_2 homolysis and ring opening, and HONO formation and isomerization into a 10-membered ring. There are two decomposition channels of the cocrystal DMF molecule: elimination of the aldehyde group and oxidization into the carboxylic acid. However, there is only one decomposition channel of the cocrystal DNDAP molecule: the breaking of the C—N bond.

Fig. 13.20 presents two competing reaction mechanisms between the HMX and DMF molecules (a) and between the HMX and DNDAP molecules (b) at 3000 K. In the HMX/DMF

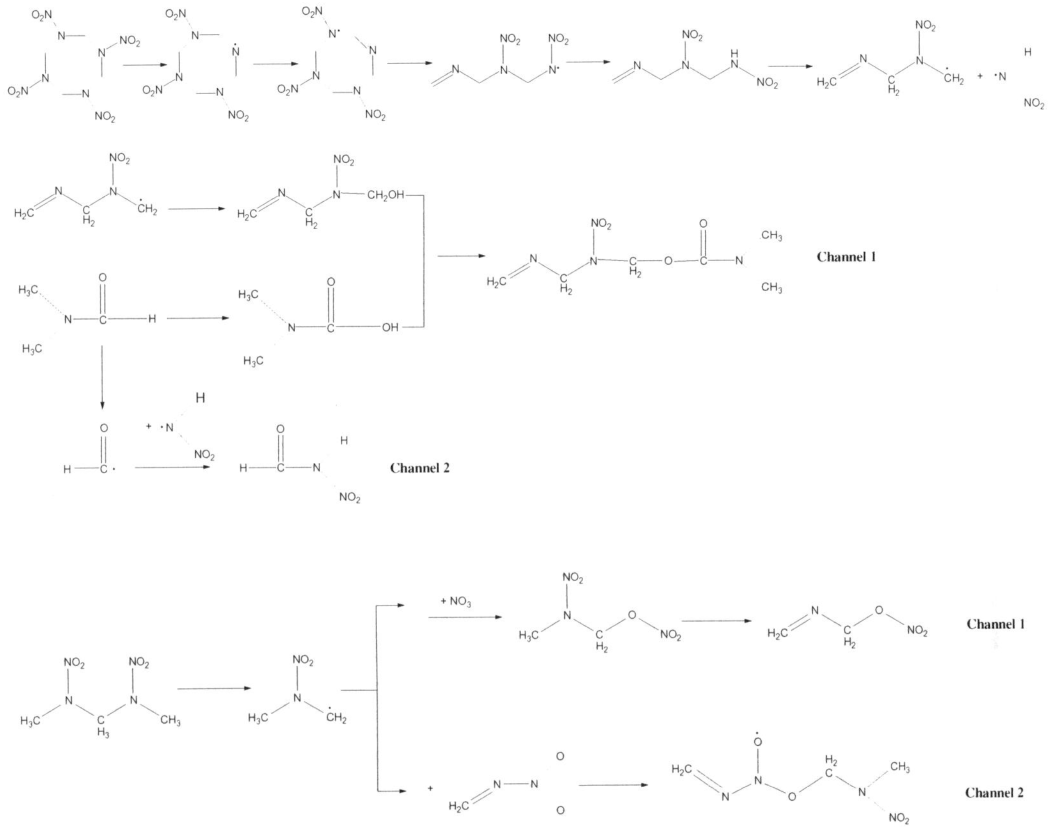

FIG. 13.20 Two competing reaction mechanisms between the HMX and DMF molecules (A) and between the HMX and DNDAP molecules (B) at 3000 K. *Reprinted (adapted) with permission from X. Wu, Z. Liu, W. Zhu, Conformational changes and decomposition mechanisms of HMX-based cocrystal explosives at high temperatures, J. Phys. Chem. C 124 (2020) 25. Copyright (2020) American Chemical Society.*

cocrystal, there are two competing reaction mechanisms: the coupling of the HMX decomposition intermediate radical $NHNO_2$ with the aldehyde group and the esterification reaction between the intermediate reaction product $C_3H_7N_3O_3$ and carboxylic acid. In the HMX/DNDAP cocrystal, there are two competing reaction mechanisms. The competition occurs between the HMX decomposition species $CH_2N_2O_2$ and NO_3. They compete with each other for the reaction with the radical $C_2H_5N_2O_2$ formed by the DNDAP homolysis.

Xie et al. [57] performed AIMD to study the thermal decomposition mechanisms of the energetic benzotrifuroxan/1,3,3-trinitroazetidine (BTF/TNAZ) cocrystal at 3000 K. It is found that there are four different types of initial reactions involved in the decomposition of the cocrystal. The TNAZ molecules are easier to decompose to trigger the decomposition of the cocrystal than the BTF ones. Releasing of the H radicals and oxygen-containing groups plays an important role in the whole decomposition process. After a series of ring opening and recombination reactions, some long chains and complicated carbon-rich heterocyclic rings were formed, and then they broke to form small fragments gradually: CO_2, N_2, NO_2, NO, HNCO, and H_2O. Overall, before the point at which NO begins to reduce, apparently, NO_2 and NO are the main decomposition products of BTF/TNAZ. But after that point, CO_2 and N_2 are the main products.

3 Reactions at low temperatures coupled with high pressures

Except for the high-temperature decompositions, many experimental studies [58–63] have been devoted to investigating the reactions of the EMs at low temperatures coupled with different pressures. Therefore, the initiation mechanisms and chemical decomposition processes of the solid EMs over a range of temperatures and pressures are momentous and useful for their practical storage, usage and disposal, and for developing new EMs.

3.1 HMX

Wu et al. [64] performed AIMD simulations to study coupling effects of temperature (534–873 K) and pressure (1–20 GPa) on the initiation mechanisms and subsequent chemical decompositions of nitramine explosive β-HMX. It is found that the initial decomposition step of HMX is triggered by the unimolecular C—H bond breaking in the coupling of temperatures and pressures investigated, except for the case in which 534 K was coupled with 2 GPa and 3 GPa. After the C—H bond breaking, the homolysis of the N—NO_2 bond of the adjacent equatorial nitro group in the same molecule follows, as shown in Fig. 13.21. The generated isolated hydrogen is very active and can promote other unreacted HMX molecules to decompose as displayed in Fig. 13.22.

After the initial decomposition reactions, some big intermediates (INTs), including INT1, INT2, INT3, and INT4 (Figs. 13.21 and 13.22), were generated; these INTs further decomposed to form other smaller INTs or gas products through the breaking of N—N, N—O, C—H, and C—N bonds. However, compared to the former three bonds, the C—N bond breaking may be much more difficult, especially the first CN bond cleavage in the ring, which would lead to the ring opening of HMX, as shown in Fig. 13.23.

C-H breaking
equatorial
homolysis of N-NO2
HMX
INT1
INT2

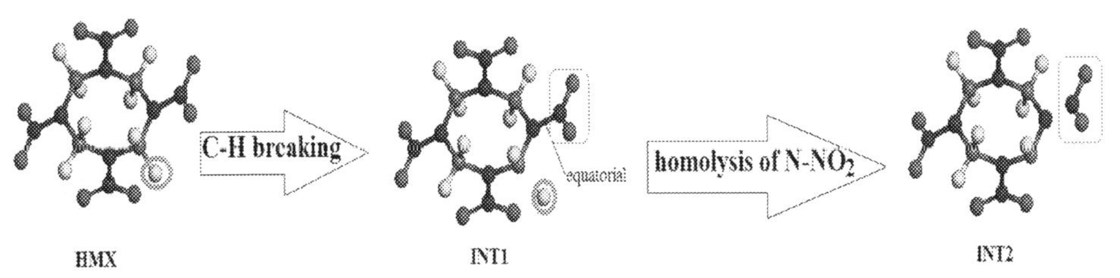

FIG. 13.21 Initial decomposition mechanisms of HMX under coupling effects of different temperatures (534–873 K) and pressures (1–20 GPa). *Gray, white, red* (*dark gray* in the print version), *and blue* (*dark gray* in the print version) *spheres* stand for carbon, hydrogen, oxygen, and nitrogen atoms, respectively. For reproduction of material from PCCP. *Reproduced from Q. Wu, G. Xiong, W. Zhu, H. Xiao, How does low temperature coupled with different pressures affect initiation mechanisms and subsequent decompositions in nitramine explosive HMX? Phys. Chem. Chem. Phys. 17 (2015) 22823 with permission from the PCCP Owner Societies.*

homolysis of N-NO2
+ NO2 (1)
INT3
breaking of N-O
axial
+ OH (2)
INT4

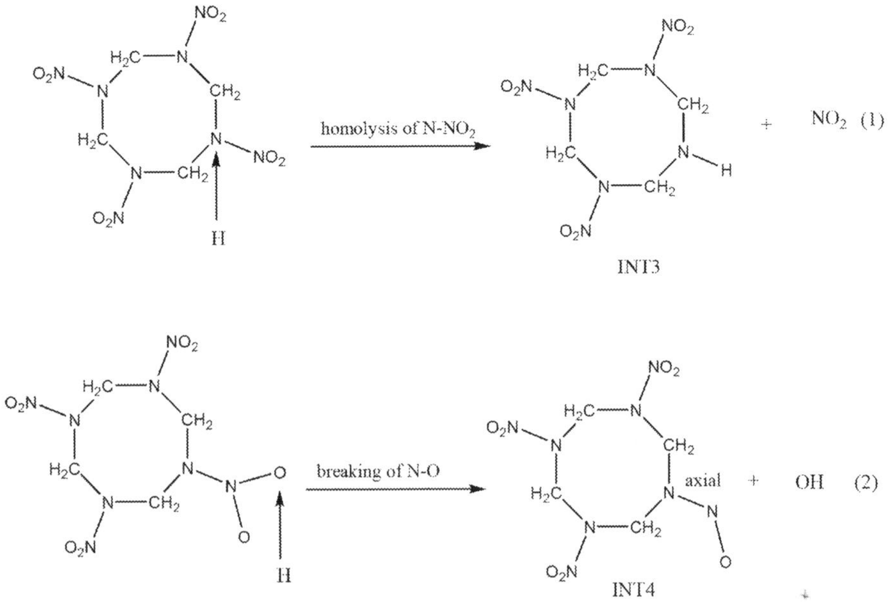

FIG. 13.22 Hydrogen induced two initial decomposition paths of HMX under coupling effects of different temperatures (534–873 K) and pressures (1–20 GPa). *For reproduction of material from PCCP. Reproduced from Q. Wu, G. Xiong, W. Zhu, H. Xiao, How does low temperature coupled with different pressures affect initiation mechanisms and subsequent decompositions in nitramine explosive HMX? Phys. Chem. Chem. Phys. 17 (2015) 22823 with permission from the PCCP Owner Societies.*

At the initial decomposition temperature (534 K) and thermal ignition temperature (608 K), the pressure could decelerate the decompositions of HMX with the pressure increasing from 1 to 10 GPa, while the pressure increases from 10 to 20 GPa produced a positive and negative pressure effect on the decompositions at 534 and 608 K, respectively. At a high temperature (873 K), the temperature had a great influence on the decomposition, while the pressure effect was not obvious.

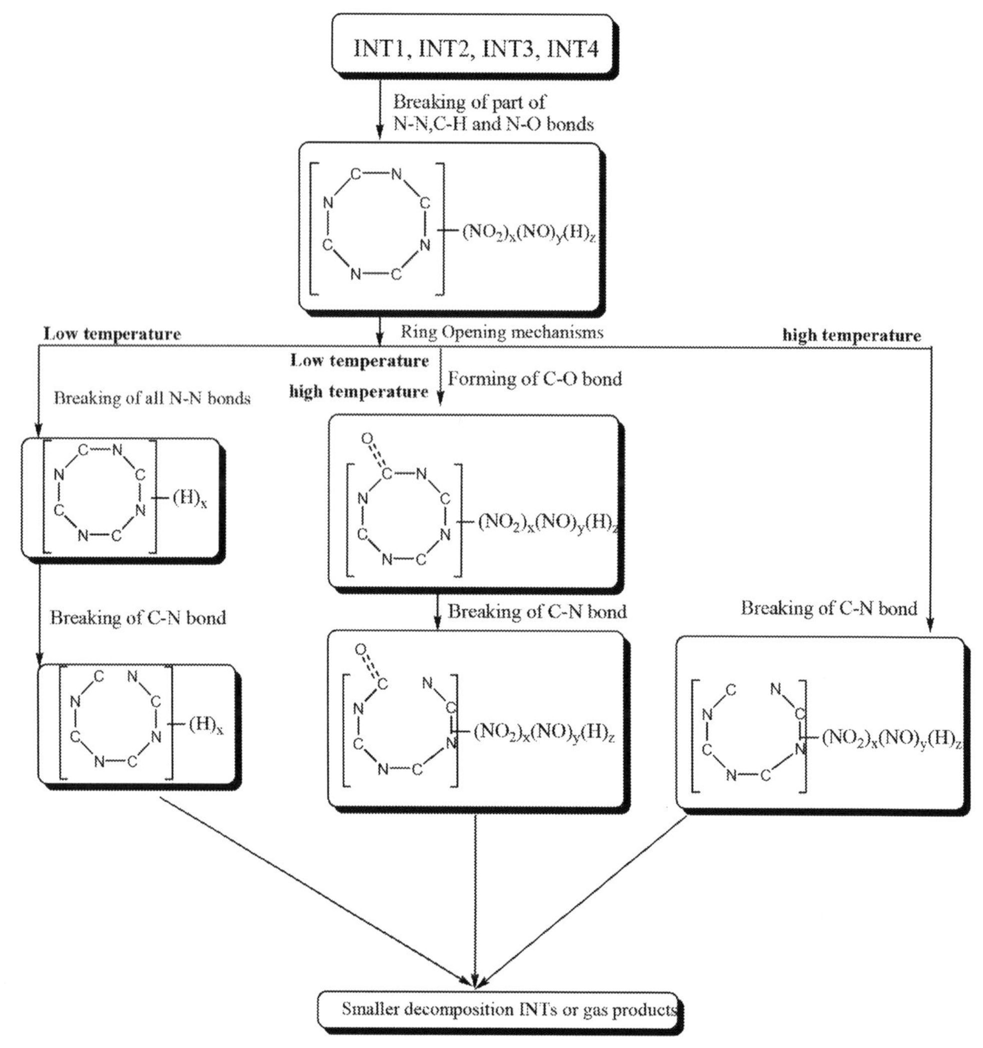

FIG. 13.23 Three different ring opening mechanisms of HMX under coupling effects of different temperatures (534–873 K) and pressures (1–20 GPa). *For reproduction of material from PCCP. Reproduced from Q. Wu, G. Xiong, W. Zhu, H. Xiao, How does low temperature coupled with different pressures affect initiation mechanisms and subsequent decompositions in nitramine explosive HMX? Phys. Chem. Chem. Phys. 17 (2015) 22823 with permission from the PCCP Owner Societies.*

3.2 TATB

Wu et al. [65] used AIMD to study the initiation mechanisms and subsequent decompositions of the crystalline TATB at 623 K (decomposition temperature) coupled with a pressure of 1–20 GPa. The results indicate that the initial decomposition step of TATB is the unimolecular intramolecular hydrogen transfer; moreover, this initiation mechanism is independent of the variation of the pressure. After the initial decomposition reaction, INT1 was

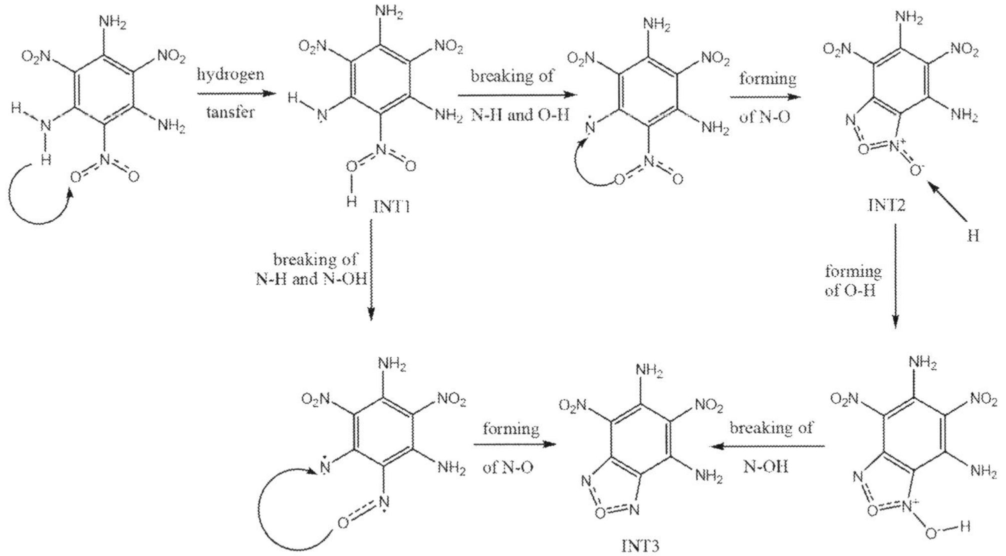

FIG. 13.24 Forming mechanisms of the first furoxan and fuxazan rings in the decomposition of TATB. *Reprinted (adapted) with permission from Q. Wu, H. Chen, G. Xiong, W. Zhu, H. Xiao, Decomposition of a 1,3,5-triamino-2,4,6-trinitrobenzene crystal at decomposition temperature coupled with different pressures: an ab initio molecular dynamics study, J. Phys. Chem. C 119 (2015) 16500. Copyright (2015) American Chemical Society.*

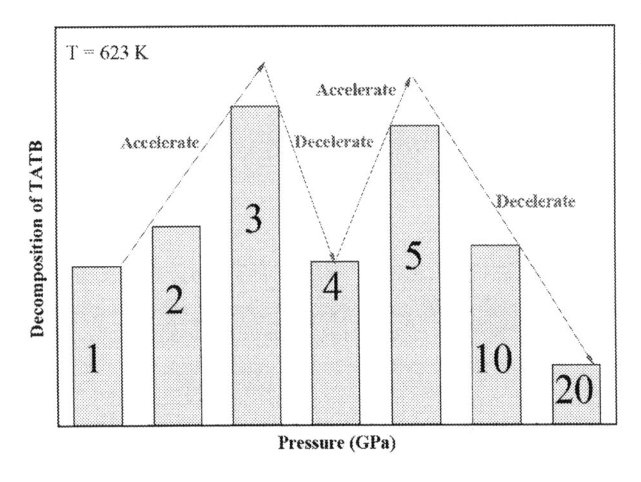

FIG. 13.25 Pressure effects on the decomposition of TATB at decomposition temperature. *Reprinted (adapted) with permission from Q. Wu, H. Chen, G. Xiong, W. Zhu, H. Xiao, Decomposition of a 1,3,5-triamino-2,4,6-trinitrobenzene crystal at decomposition temperature coupled with different pressures: an ab initio molecular dynamics study, J. Phys. Chem. C 119 (2015) 16500. Copyright (2015) American Chemical Society.*

formed. INT1 will further decompose to form other INTs like the benzofuroxan and benzofurazan compounds, as shown in Fig. 13.24, which was supported by experimental studies [66] on the thermal decomposition of TATB. Major products are the same at different pressures, which are NO_2, H_2O, and HNO_2.

Subsequent decomposition of TATB is very sensitive to the applied pressure at decomposition temperature. Fig. 13.25 displays the pressure effects on the decomposition of TATB at decomposition temperature. The pressure accelerates the subsequent decomposition of TATB with the increase of pressure from 1 to 3 GPa and from 4 to 5 GPa, while it would decelerate the decomposition when the pressure increases from 3 to 4 GPa and from 5 to 20 GPa. The deceleration of the pressure on the decomposition of TATB may be caused by inhibiting initiation decomposition.

Also, Wu et al. [67] performed AIMD to study pressure effects in crystalline TATB at room temperature in the pressure range of 1–100 GPa. The results indicate that TATB is chemically stable in the entire investigated pressure range, in agreement with the experiment report [68] that TATB is chemically stable to 150 GPa. Both the intra- and intermolecular hydrogen bonding are strengthened with the increasing pressure from 5 to 50 GPa, consistent with the experimental results up to 40 GPa [68]. Thus, it may be concluded that the external pressures can not destroy the chemical stability of the TATB crystal without the coupling effects of the temperatures.

3.3 DNAAF

3,30-Dinitroamino-4,40-azoxyfurazan (DNAAF) is a good representative of the furazan explosives, which is a new powerful high explosive with good overall performance. Wu et al. [69] employed AIMD to study the initiation mechanisms and subsequent chemical reactions of DNAAF crystal at low temperatures (363–963 K) coupled with different pressures (1–5 GPa). The crystal structure of DNAAF is sensitive to the applied pressure and DNAAF is more compressible at higher temperatures.

At the decomposition temperature (363 K), the initial decomposition step of DNAAF is the intermolecular hydrogen transfer (Fig. 13.26) at 1 and 2 GPa, while DNAAF was unreactive at 3, 4, and 5 GPa. At 663 K, the initial decomposition step is the unimolecular N—NO_2 bond cleavage (Fig. 13.27) at 1 GPa, while at 2–5 GPa, DNAAF decomposes through the intermolecular hydrogen transfer. Finally, at 963 K, the initial decomposition step is the N—NO_2 bond breaking at 1 and 2 GPa, while it is intermolecular hydrogen transfer at 3–5 GPa.

Fig. 13.28 displays the pressure effects on the decomposition of DNAAF at different temperatures. At 363 K, the decomposition of DNAAF decelerates with the increment of pressure from 2 GPa to 3–5 GPa. At 663 K, the pressure accelerates the decomposition with the pressure increasing from 1 to 3 GPa, while it would decelerate the decomposition when from 3 to 5 GPa. At 963 K, the pressure decelerates the decomposition with the increment of pressure from 1 to 3 and 4 to 5 GPa, while it would accelerate the decomposition when from 3 to 4 GPa. The temperature can change the effect of pressure on the decomposition of DNAAF. C_2N_2, NO_2, and NO are three common decomposition products.

3.4 PETN

Wu et al. [70] performed AIMD simulations to study the initiation of decomposition and formation of first products of PETN at its thermal decomposition temperature 475 K coupled with different pressures (1–5 GPa). At thermal decomposition temperature, PETN

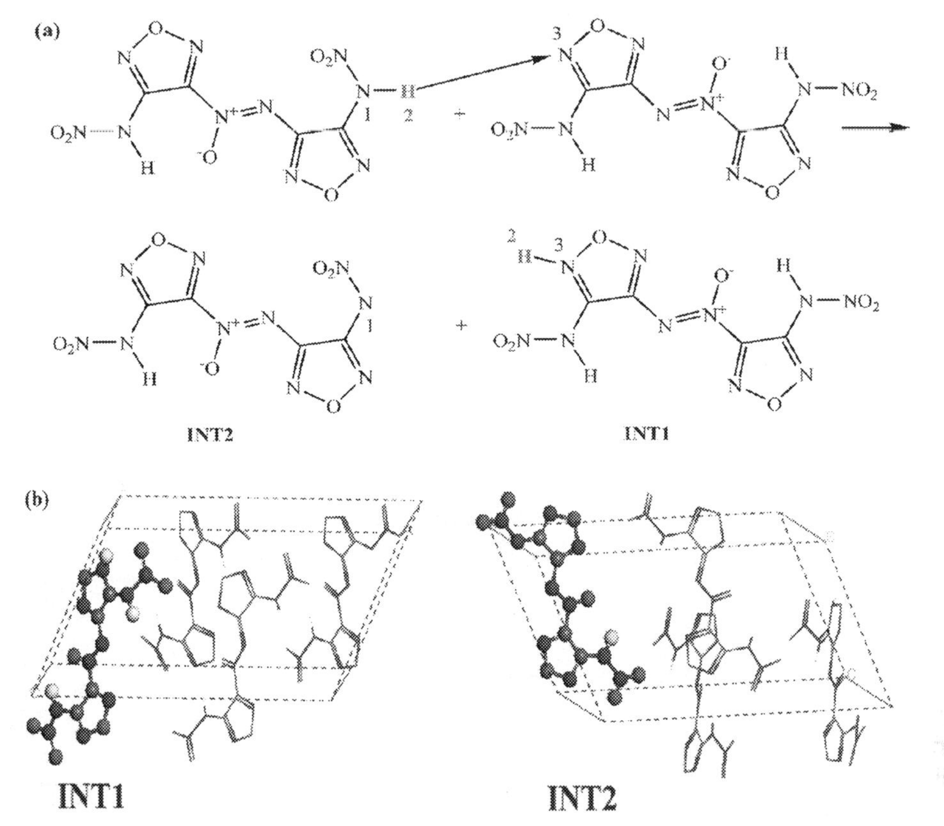

FIG. 13.26 The process (A) and snapshots (B) of the bimolecular intermolecular hydrogen transfer during the decomposition of DNAFF. *For reproduction of material from PCCP: Reproduced from Q. Wu, W. Zhu, H. Xiao, Cooperative effects of different temperatures and pressures on the initial and subsequent decomposition reactions of the nitrogen-rich energetic crystal 3,3′-dinitroamino-4,4′-azoxyfurazan, Phys. Chem. Chem. Phys. 18 (2016) 7093 with permission from the PCCP Owner Societies.*

melted without decomposition at 1 GPa, but it melted and decomposed at 2, 3, 4, and 5 GPa. Fig. 13.29 displays the snapshots before, during, and after the initiation decomposition of PETN. The decomposition of PETN was triggered by intermolecular hydrogen transfer (C—H⋯O). The initial decomposition mechanism was independent of the pressure.

3.5 NTO

5-Nitro-2,4-dihydro-1,2,4-triazole-3-one (NTO), high energy but very insensitive explosive, has been widely used in explosive formulations, PBXs (polymer-bonded explosives), and generators. Its thermal decomposition temperature is about 531 K, but few experimental and theoretical studies have been devoted to investigate its structure and decomposition around this temperature coupled with high pressure.

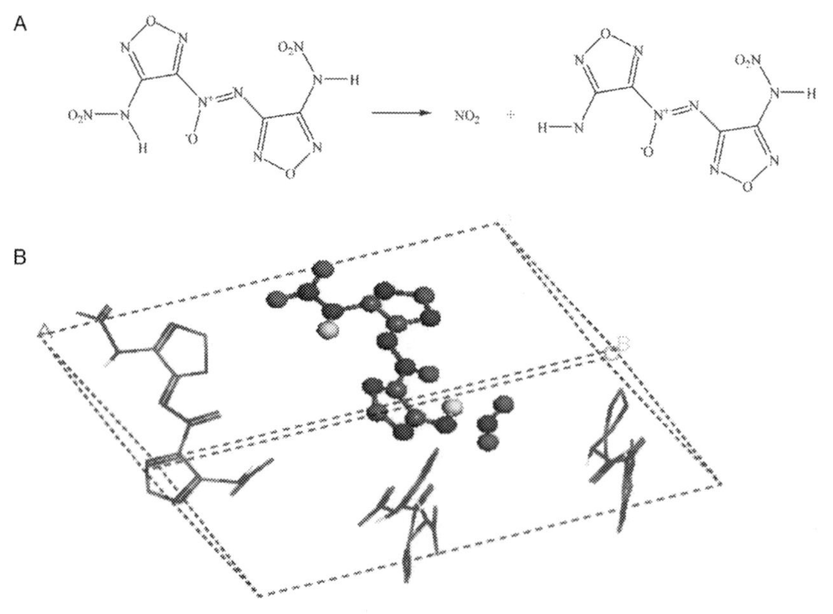

FIG. 13.27 The process (A) and snapshots (B) of the unimolecular N—NO_2 bond breaking during the decomposition of DNAFF. *For reproduction of material from PCCP: Reproduced from Q. Wu, W. Zhu, H. Xiao, Cooperative effects of different temperatures and pressures on the initial and subsequent decomposition reactions of the nitrogen-rich energetic crystal 3,3′-dinitroamino-4,4′-azoxyfurazan, Phys. Chem. Chem. Phys. 18 (2016) 7093 with permission from the PCCP Owner Societies.*

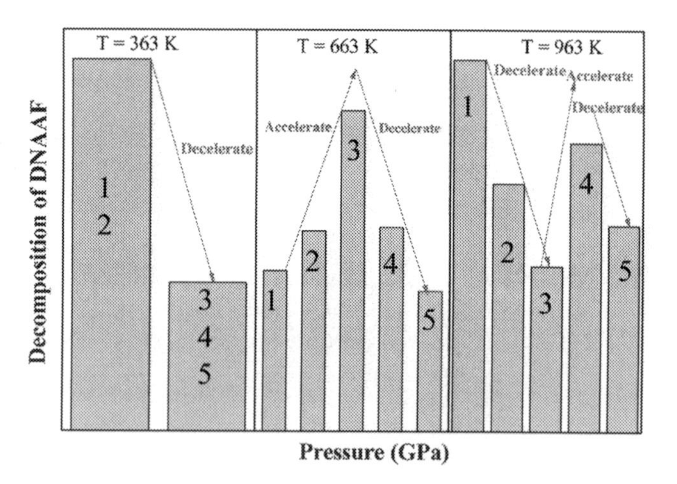

FIG. 13.28 Pressure effects on the decomposition of DNAAF at different temperatures. *For reproduction of material from PCCP: Reproduced from Q. Wu, W. Zhu, H. Xiao, Cooperative effects of different temperatures and pressures on the initial and subsequent decomposition reactions of the nitrogen-rich energetic crystal 3,3′-dinitroamino-4,4′-azoxyfurazan, Phys. Chem. Chem. Phys. 18 (2016) 7093 with permission from the PCCP Owner Societies.*

Wu et al. [70] used AIMD to investigate the chemical reactions of NTO at its thermal decomposition temperature 531 K coupled with different pressures (1–5 GPa). At thermal decomposition temperature, NTO decomposed before melting at 1, 2, 3, 4, and 5 GPa. This is different from that of PETN. Fig. 13.30 displays the snapshots before, during, and after

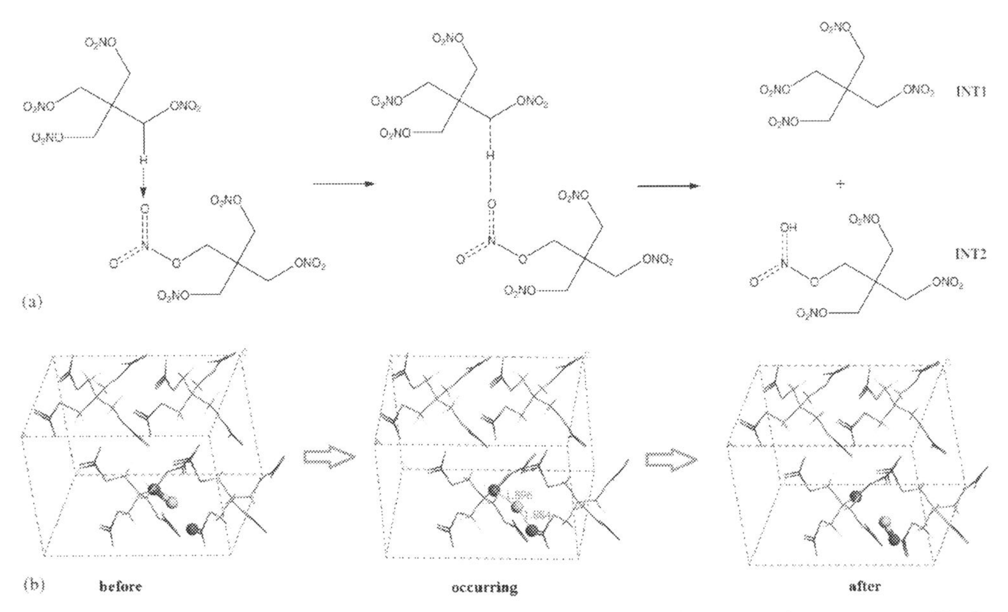

FIG. 13.29 Snapshots before, during, and after initiation decomposition of PETN. Carbon, oxygen, and hydrogen atoms are represented by *gray*, *red* (*dark gray* in the print version), and *white spheres*, respectively. *Reproduced from Q. Wu, D. Xiang, G. Xiong, W. Zhu, H. Xiao, Coupling of temperature with pressure induced initial decomposition mechanisms of two molecular crystals: an ab initio molecular dynamics study, J. Chem. Sci. 128 (2016) 695 with permission from Springer Nature.*

the initiation of the decomposition of NTO through N—H···O and N···H···N intermolecular hydrogen transfers. The initial decomposition mechanism was dependent on the pressure. At 1, 2, and 3 GPa, the initial decomposition mechanism is the N—H···O intermolecular hydrogen transfer, while at 4 and 5 GPa, this is the N—H···N intermolecular hydrogen transfer.

3.6 FOX-7

Xiong and Zhu [71] performed AIMD simulations to study the initiation mechanisms and subsequent decomposition of FOX-7 crystal at a temperature of 504 K (initial decomposition temperature) coupled with a pressure of 1–5 GPa, 604 K at 5GPa, and 704 K at 5 GPa. The initial decomposition step is the intermolecular hydrogen transfer. The initial decomposition mechanism was dependent on the temperature but independent of the pressure. This is supported by previous experimental studies [63] that similar decomposition pathways occur in FOX-7, irrespective of the pressure.

The subsequent decomposition of FOX-7 is sensitive to both temperature and pressure. At 504 K, the decomposition of FOX-7 was accelerated from 1 to 2 GPa and from 3 to 5 GPa but decelerated from 2 to 3 GPa. The temperature exhibits a positive effect on the decomposition. Fig. 13.31 presents the forming mechanism and snapshots of furoxan rings during the

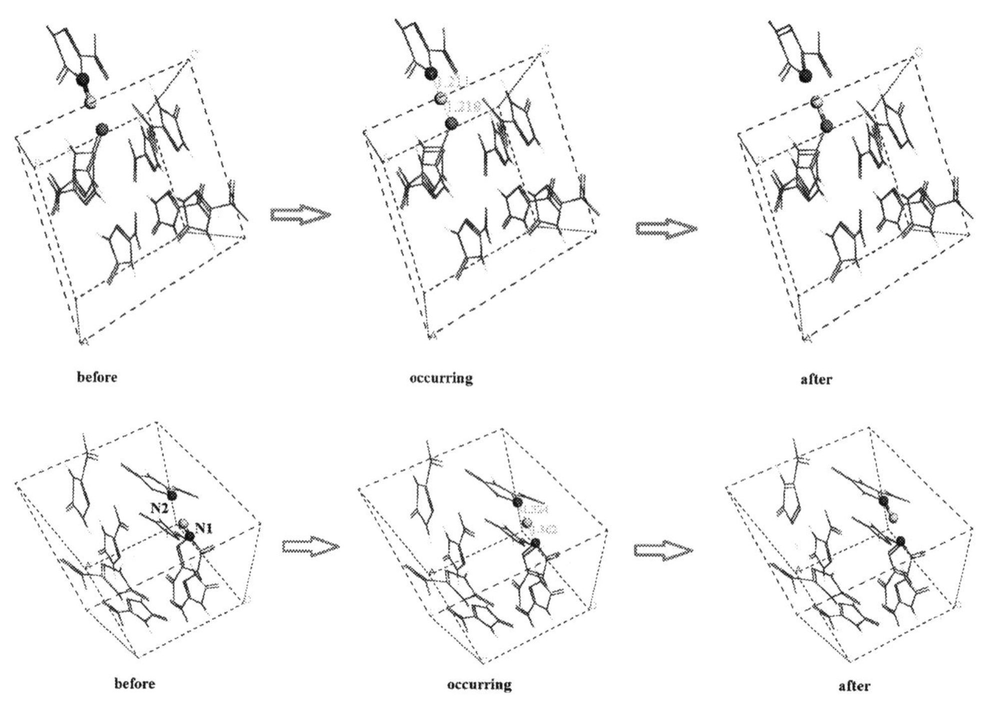

FIG. 13.30 Snapshots before, during, and after the initiation of the decomposition of NTO through N—H···O (A and B) and N···H···N (C and D) intermolecular hydrogen transfer. Nitrogen, oxygen, and hydrogen atoms are represented by *blue* (*dark gray* in the print version), *red* (*gray* in the print version), and *white* spheres, respectively. *Reproduced from Q. Wu, D. Xiang, G. Xiong, W. Zhu, H. Xiao, Coupling of temperature with pressure induced initial decomposition mechanisms of two molecular crystals: an ab initio molecular dynamics study, J. Chem. Sci. 128 (2016) 695 with permission from Springer Nature.*

decomposition of FOX-7. It is found that the intermediates INT1 and INT2 form furoxan derivative INT3 through inter- and intramolecular hydrogen transfer. Also, the higher temperature would promote the decomposition of FOX-7 and reveal the competitive relation between the reaction temperature and pressure.

Xiang and Zhu [72] employed DFTB-MD to simulate the chemical decomposition of the multi-molecular FOX-7 system at the flame temperature of 2799 K with constant energy and volume conditions (NVE ensemble). It is found that the initial decomposition is triggered by the unimolecular cleavage of the N—O bond to form oxygen radical. There are three different decomposition channels during the pyrolysis of the multi-molecular FOX-7 systems. The first initial reaction involves the N—O bond cleavage to form a biradical intermediate, which then transfers oxygen radicals to form more stable intermediates. The second reaction path happened by the N—H bond breaking to form a biradical intermediate, which then transfers a hydrogen atom to accelerate the following reactions. The third reaction path is as

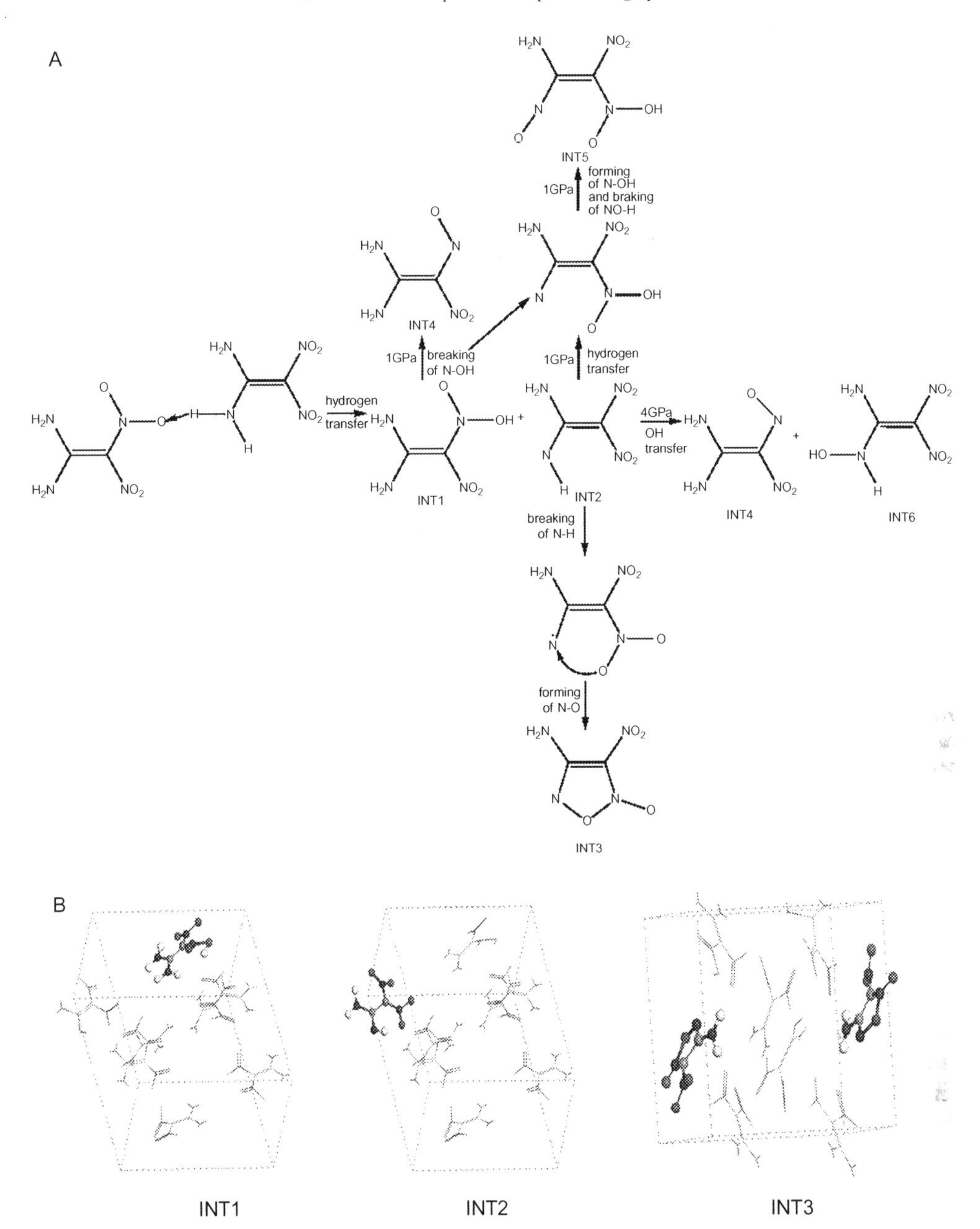

FIG. 13.31 Forming mechanism (A) and snapshots (B) of furoxan rings during the decomposition of FOX-7. *Reproduced from G. Xiong, W. Zhu, Coupling effects of high temperature and pressure on the decomposition mechanisms of 1,1-diamino-2,2-dinitroehethe crystal: Ab initio molecular dynamics simulations. J. Chin. Chem. Soc. 67 (2020) 1571–1578 with permission from John Wiley and Sons.*

follows: the three FOX-7 molecules are triggered by the breaking of C—C bond. The last path has a relatively lower possibility than the former two reactions. The biradical intermediate is very active to accelerate the following complex reactions. In all, the N—O bond cleavage is the main reaction channel during the multi-molecular pyrolysis of FOX-7.

4 Reactions at high temperatures coupled with high pressures

Cooperative effects of high pressure and temperature on the EMs will initiate complex chemical reactions, in turn releasing large amounts of energy to sustain the detonation process [59,61]. Therefore, a detailed description of the chemical processes in condensed EMs under high-temperature coupled with high pressure is very important to understand events that occur at the reactive front of these materials under combustion or detonation conditions.

4.1 DBTD

Recently synthesized dihydroxylammonium 5,50-bistetrazole-1,10-diolate (DBTD), an azole-based ionic high energy density compound by introducing the N-oxides, has an excellent performance due to high exothermicity, high density, and insensitivity toward external stimuli. Xiang et al. [73] investigated the initiation mechanisms and subsequent decompositions of energetic ionic crystal DBTD at pure high temperatures (3842 and 2000 K) and at high temperatures coupled with detonation pressure (34.2 GPa) using AIMD. The decomposition of DBTD is triggered by the N—H bond breaking to release H radical, as shown in Fig. 13.32. The initiation mechanisms are independent of both the temperature and pressure.

Fig. 13.33 displays the subsequent decomposition process of DBTD at (a) 3842 K coupled with 34.2 GPa, (b) 3842 K, (c) 2000 K coupled with 34.2 GPa, (d) 2000 K. Subsequent decompositions were very sensitive to high temperatures, and moreover, the temperature becomes the foremost factor affecting the decomposition. However, the product formation mechanisms indicate that the pressure could decelerate the decompositions. The hydrogen radicals could promote subsequent decompositions.

4.2 CL-20

Xiang et al. [74] performed AIMD simulations to study initial decomposition mechanisms and subsequent decomposition process of CL-20 crystal at extreme conditions (3000 K and

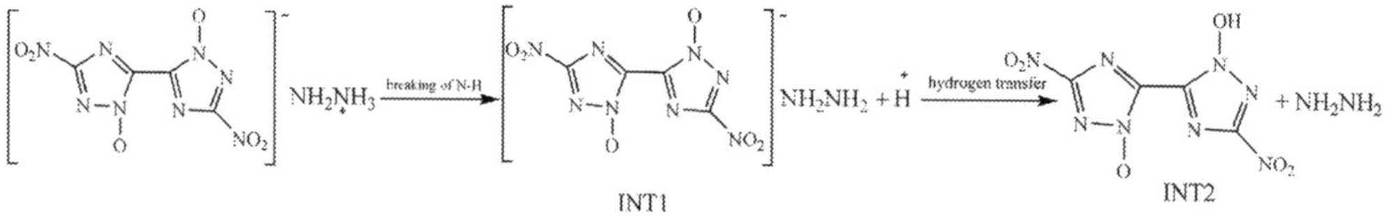

FIG. 13.32 Initial decomposition mechanism of DBTD at high temperatures (2000 and 3842 K) and at detonation pressure 34.2 GPa coupled with different temperatures (298.15, 2000, and 3842 K). *Reproduced from D. Xiang, W. Zhu, H. Xiao, Thermal decomposition mechanisms of energetic ionic crystal dihydrazinium 3,30-dinitro-5,50-bis-1,2,4-triazole-1,1-diolate: an ab initio molecular dynamics study, Fuel 202 (2017) 246 with permission from Elsevier.*

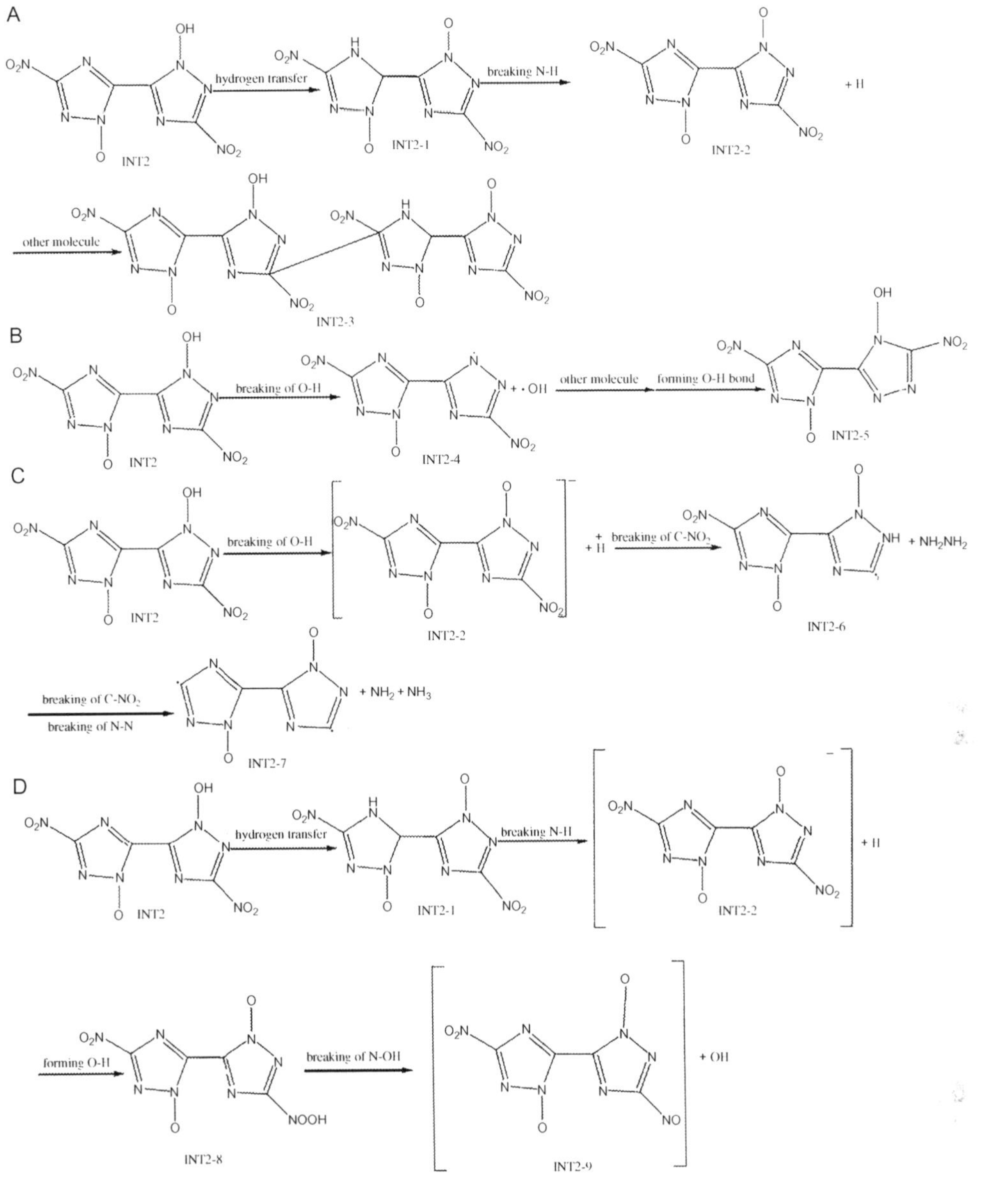

FIG. 13.33 Subsequent decomposition process of DBTD at (A) 3842 K coupled with 34.2 GPa, (B) 3842 K, (C) 2000 K coupled with 34.2 GPa, (D) 2000 K. *Reproduced from D. Xiang, W. Zhu, H. Xiao, Thermal decomposition mechanisms of energetic ionic crystal dihydrazinium 3,30-dinitro-5,50-bis-1,2,4-triazole-1,1-diolate: an ab initio molecular dynamics study, Fuel 202 (2017) 246 with permission from Elsevier.*

3000 K coupled with 44.5 GPa). It is found that the initial decomposition steps of the CL-20 molecules in the crystal at 3000 K have two different paths: the C—C cleavage and the C—H bond breaking. The initial decomposition of the CL-20 crystal was triggered by the C—H cleavage at 3000 K. The initial decomposition steps of the CL-20 molecules in the crystal include the above-mentioned C—C bond breaking, the C—N bond rupture, and the N—NO_2 bond cleavage, as shown in Fig. 13.34. The initial decomposition of the CL-20 crystal was triggered by the C—C cleavage at 3000 K with 44.5 GPa.

Fig. 13.35 presents the subsequence decomposition mechanisms of CL-20 at 3000 K coupled with 44.5 GPa. There is only one major distinctive channel for secondary decomposition of the intermediates. The appearance of special intermediates R-C_xO_y ($x > 2$, $y > 5$) indicates that the high pressure makes the decomposition much more complex. Among these intermediates, C_3O_6 is proved to be a high energy density compound. The decomposition reactions of the CL-20 crystal are sensitive to both high temperature and high pressure. The numbers of corresponding major products at 3000 K coupled with 44.5 GPa are smaller than those at 3000 K. This demonstrates that the high-pressure decelerates the decomposition of CL-20.

4.3 RDX

The initiation and subsequent reaction mechanisms of α-RDX crystal under high temperature (3000 K) coupled with detonation pressure (34.5 GPa) were studied by AIMD simulations [75]. The results indicate that the initial decomposition step of α-RDX is the unimolecular C—H bond breaking to form hydrogen radical, as shown in Fig. 13.36. The H radicals then induced the cleavage of two C—N bonds in other α-RDX molecules.

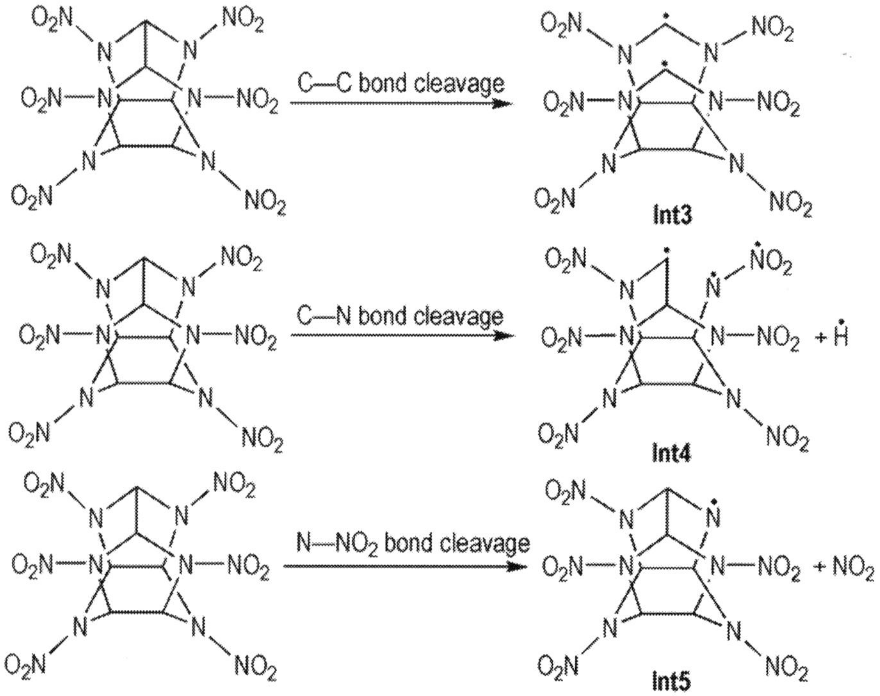

FIG. 13.34 Initial decomposition mechanisms of the CL-20 molecules in the crystal at 3000 K coupled with 44.5 GPa. *Reproduced from D. Xiang, Q. Wu, W. Zhu, Ab initio molecular dynamics studies on the decomposition mechanisms of CL-20 crystal under extreme conditions, Chin. J. Energ. Mater. 26 (2018) 59.*

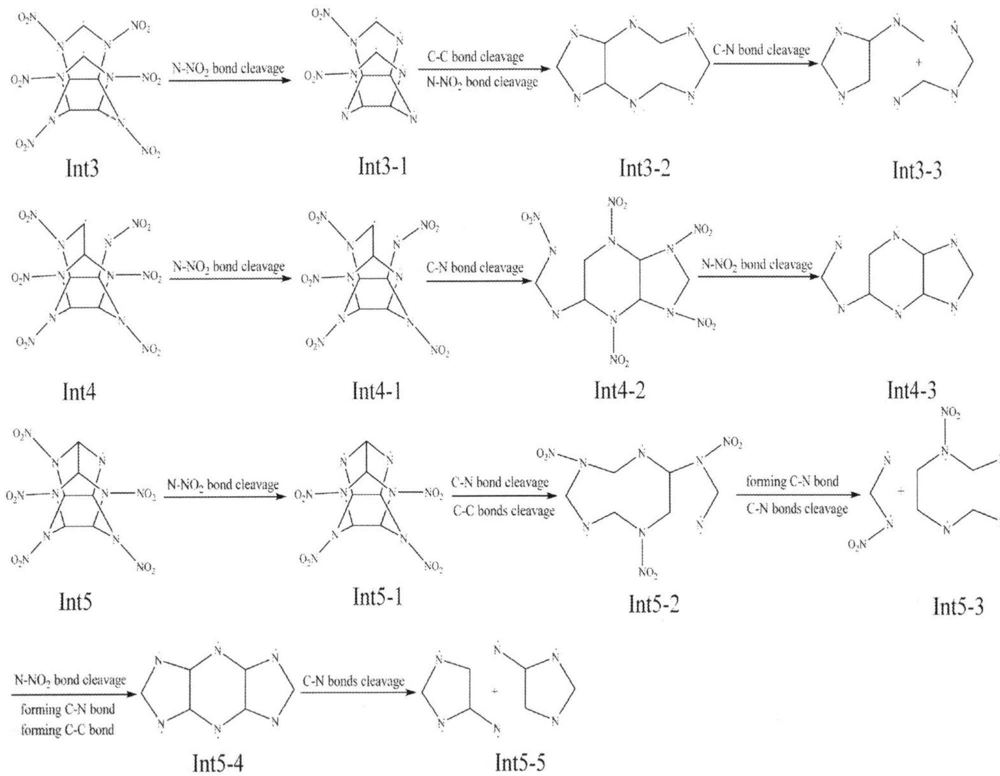

N-NO_2 bond cleavage
C-C bond cleavage
N-NO_2 bond cleavage
C-N bond cleavage
Int3 Int3-1 Int3-2 Int3-3

N-NO_2 bond cleavage
C-N bond cleavage
N-NO_2 bond cleavage
Int4 Int4-1 Int4-2 Int4-3

N-NO_2 bond cleavage
C-N bond cleavage
C-C bonds cleavage
forming C-N bond
C-N bonds cleavage
Int5 Int5-1 Int5-2 Int5-3

N-NO_2 bond cleavage
forming C-N bond
forming C-C bond
C-N bonds cleavage
Int5-4 Int5-5

FIG. 13.35 Subsequence decomposition mechanisms of CL-20 at 3000 K coupled with 44.5 GPa. *Reproduced from D. Xiang, Q. Wu, W. Zhu, Ab initio molecular dynamics studies on the decomposition mechanisms of CL-20 crystal under extreme conditions, Chin. J. Energ. Mater. 26 (2018) 59.*

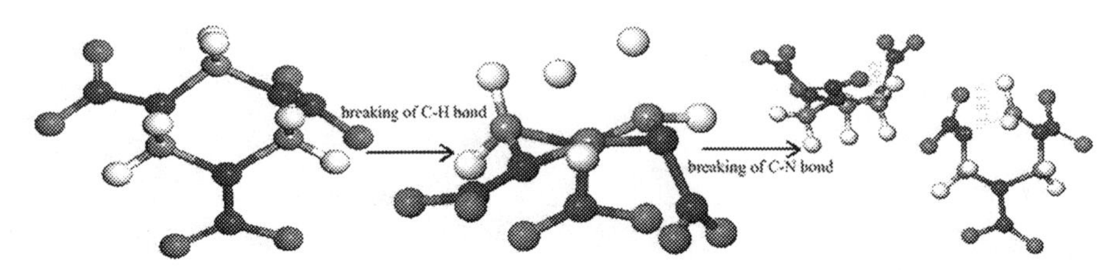

FIG. 13.36 Initial decomposition mechanisms of α-RDX crystal at high temperature 3000 K coupled with detonation pressure 34.2 GPa. *Reproduced from D. Xiang, Q. Wu, W. Zhu, Decomposition mechanisms of crystalline α-RDX under high temperature coupled with detonation pressure from ab initio molecular dynamics simulations, Chin. J. Energ. Mater. 26 (2018) 477.*

The bond lengths of the two C—N bonds increase to 1.857 and 1.762 Å, respectively. This suggests that the H radicals play a catalytic role in promoting subsequent decompositions.

Fig. 13.37 displays the three subsequent decomposition paths after the initiation of α-RDX crystal. After the initial decomposition, three are three subsequent decomposition processes took place. They include: (1) the C—N bond hemolysis triggers other C—N bonds of this ring to break; (2) the dissociation of N—NO_2 bond releases NO_2 gas; (3) the H radical attacks the O

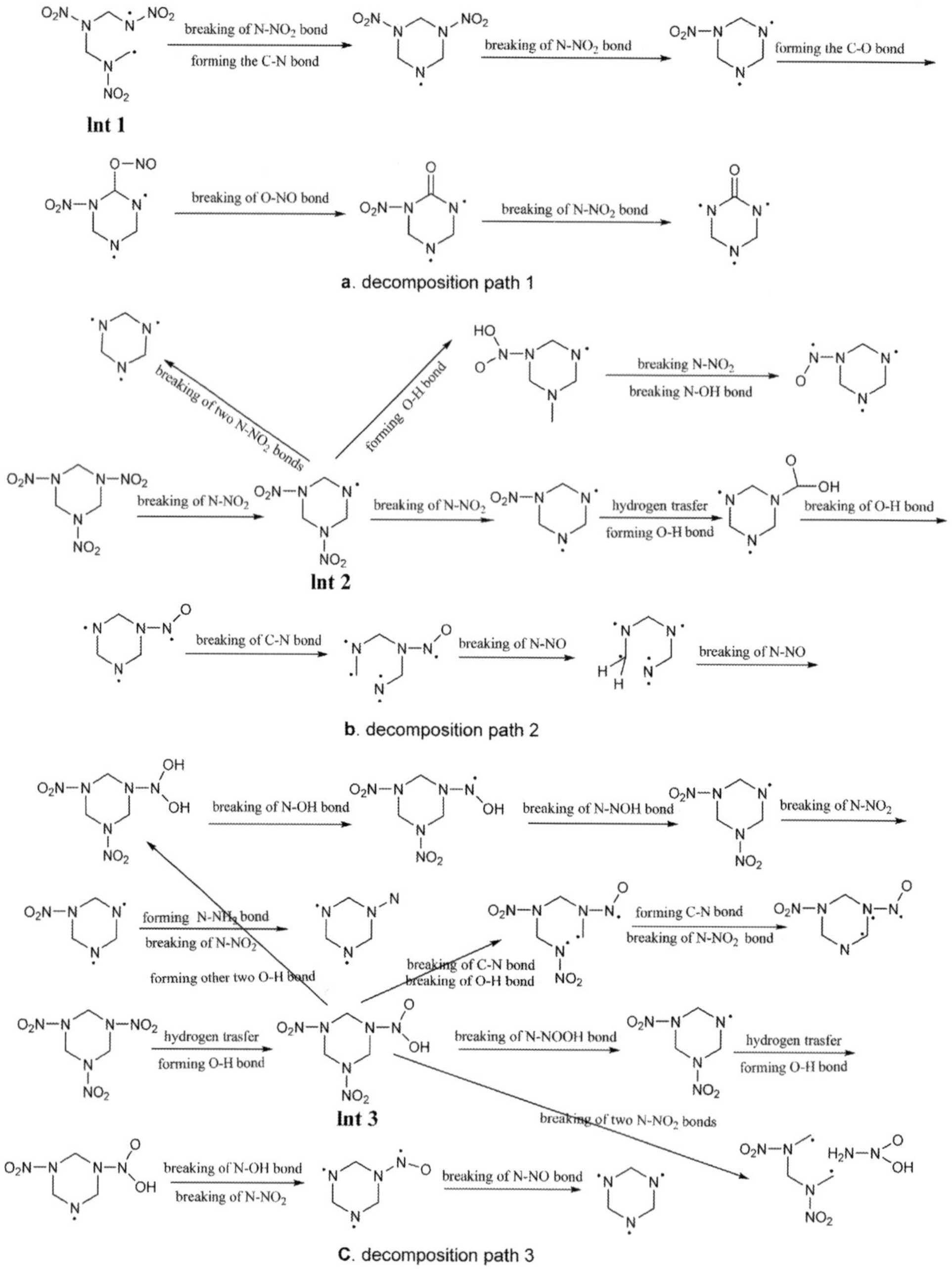

a. decomposition path 1

b. decomposition path 2

c. decomposition path 3

FIG. 13.37 Three subsequent decomposition paths after the initiation of α-RDX crystal. *Reproduced from D. Xiang, Q. Wu, W. Zhu, Decomposition mechanisms of crystalline α-RDX under high temperature coupled with detonation pressure from ab initio molecular dynamics simulations, Chin. J. Energ. Mater. 26 (2018) 477.*

atom to release O—H radical by forming O—H bond. There are a large number of subsequently generated intermediates in the unimolecular decomposition of α-RDX.

4.4 Energetic MOFs

Recently, metal–organic frameworks (MOFs) compounds were found to become potential EMs because of their high densities and high heats of detonation. Nickel hydrazine nitrate ($NiN_8H_{12}O_6$, NHN) is an energetic coordination compound whose explosive performance between primary and secondary explosives, which can replace lead azide as the primary explosive and possess the output of secondary explosive RDX. Xiang and Zhu [76] performed AIMD simulations to study the initiation and subsequent reactions of energetic MOFs NHN crystal under high temperatures (518 and 4000 K) and high temperatures (518 and 4000 K) coupled with high pressures (1, 2, 3, and 20.2 GPa).

Fig. 13.38 displays the initial decomposition mechanisms of the NHN crystal at 518 K, 518 K coupled with 1, 2, and 3 GPa. At the four different external conditions, the collapse of the cage structure triggers the initial decomposition of NHN, but the decomposition sites are different. The initial decomposition of NHN proceeds via the Ni—N bond cleavage at 518 K. At 518 K coupled with 1 GPa, the initial step of NHN is the cleavage of the N—N bond in the first cage. The other reaction site of the initial decomposition step is the N—N bond in

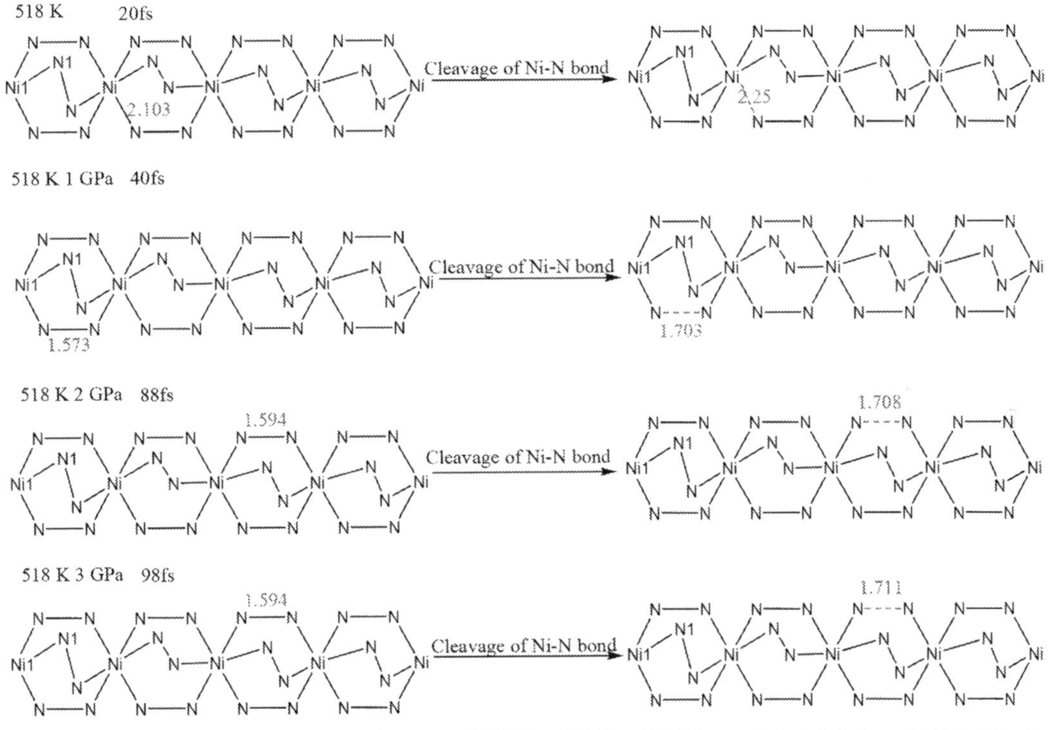

FIG. 13.38 Initial decomposition mechanisms of NHN at 518 K and 518 K coupled with 1, 2, and 3 GPa. The distance is in Å. *Reproduced from D. Xiang, W. Zhu, Thermal decomposition of energetic MOFs nickel hydrazine nitrate crystals from an ab initio molecular dynamics simulation, Comput. Mater. Sci. 143 (2018) 170 with permission from Elsevier.*

the third cage. The initial reaction times increase as the pressure increases. This indicates that the pressure decelerates the initial decomposition of the NHN crystal. Subsequent reactions are only the cleavage of a series of Ni—N bonds at 518 K. Under compression, their subsequent reactions are the breaking of a series of Ni—N and N—N bonds. Therefore, the following decomposition pathway after the initiation of NHN is further dissociation of cage structure to produce small chains. The decomposition mechanisms of the NHN crystal at 518 K coupled with 20.2 GPa are much different from those at 518 K coupled with low pressures.

The initial decomposition steps of the NHN crystal at 4000 K and at 4000 K coupled with 20.2 GPa are the breaking of the electrovalent bonds between Ni and N. After that, there are two subsequent decomposition paths observed at 4000 K and at 4000 K coupled with 20.2 GPa, as shown in Fig. 13.39. These secondary reactions are mainly through two interesting paths at 4000 K: (1) the homolysis of N—H bonds produced the H radicals; (2) the NH_2 radials formed by the N—N bond breaking. There is only one secondary reaction (1) observed at 4000 K coupled with 20.2 GPa. The decomposition reactions are much more complicated at 4000 K than those at 4000 K coupled with 20.2 GPa. This shows that the pressure decelerates the decomposition of NHN. In short, the high temperature plays a dominant role in affecting the decomposition of the NHN crystal.

In the early decomposition stage, NHN decomposed very fast and drastically. But the later decomposition was slow.

4.5 β-HMX

The decomposition reactions of the condensed-phase β-HMX under extreme conditions were extensively investigated [2–4,6,27,64,77–79]. However, initiation mechanisms and subsequent chemical reactions of crystalline β-HMX at flame temperature (3500 K) coupled with different pressures are not clear. Xiang [80] performed AIMD simulations to study the initial chemical processes of HMX at 3500 K coupled with different pressures (1–16 GPa).

Fig. 13.40 displays the initial decomposition paths of the crystal HMX at 3500 K and 3500 K coupled with 1, 5, 10, and 16 GPa. It is found that the initial decomposition steps are the C—H and N—NO_2 bond cleavage at 3500 K. When the pressure (1–10 GPa) is applied, the first reaction steps are primarily the C—N and C—H bond fission. The C—H bond cleavage is a triggering decomposition step of the HMX crystals at 3500 K coupled with 16 GPa.

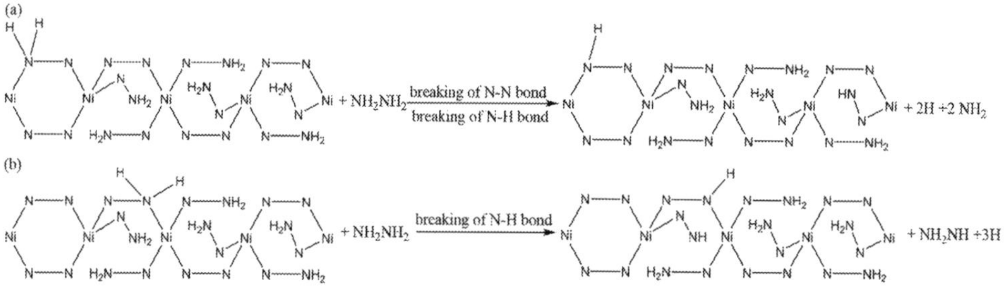

FIG. 13.39 Reaction of (A) N—H and N—N bond cleavage during the decomposition of NHN at 4000 K and (B) N—H bond cleavage during the decomposition of NHN at 4000 K coupled with 20.2 GPa. *Reproduced from D. Xiang, W. Zhu, Thermal decomposition of energetic MOFs nickel hydrazine nitrate crystals from an ab initio molecular dynamics simulation, Comput. Mater. Sci. 143 (2018) 170 with permission from Elsevier.*

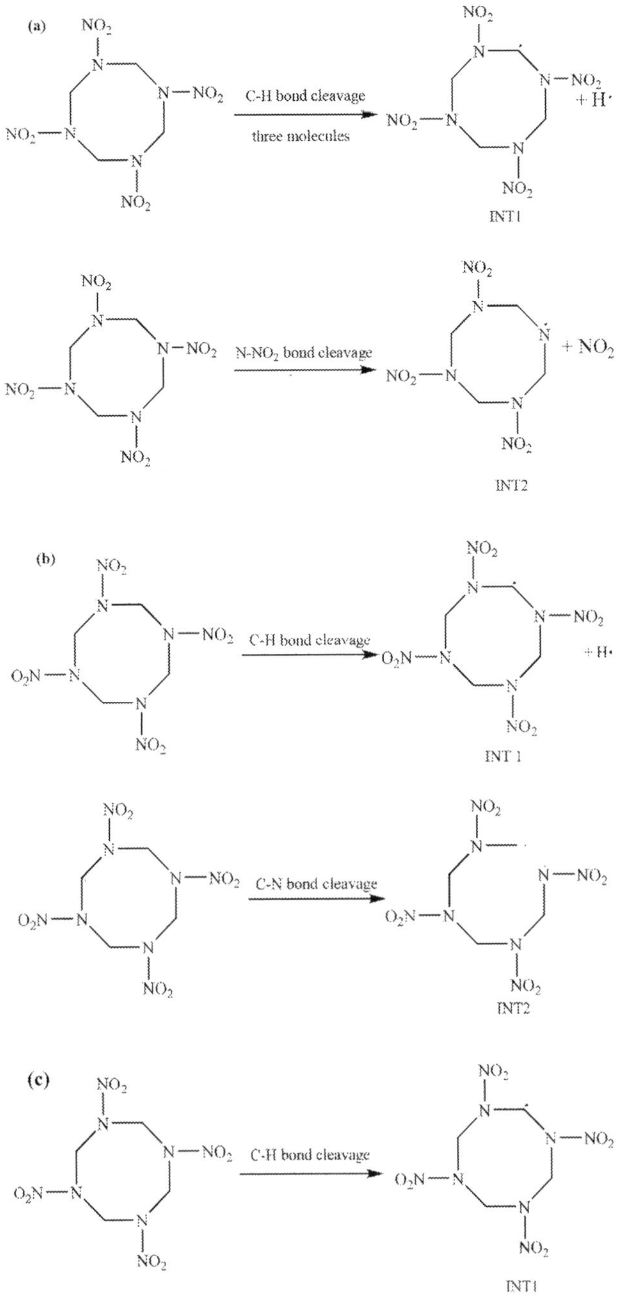

FIG. 13.40 Initial decomposition paths of the crystal HMX at (A) 3500 K, (B) 3500 K coupled with 1, 5, and 10 GPa, and (C) 3500 K coupled with 16 GPa. *Reproduced from D. Xiang, G. Ji, W. Zhu, Ab initio molecular dynamics simulations study on initial decompositions of β-HMX at high temperature coupled with high pressures, J. Chin. Chem. Soc. 66 (2019) 1429 with permission from John Wiley and Sons.*

The decomposition pressure (16 GPa) restrains the N—NO_2 bond breaking, making the hydrogen atoms much more active and the C—H bonds become much weak.

Fig. 13.41 exhibits the subsequent decomposition paths of (a) INT1 and (b) INT2 during the decomposition of the HMX crystals at 3500 K coupled with 1–16 GPa. The secondary

(a)
INT1
C-N bond cleavage
N-O bond cleavage
C-N bond cleavage
N-NO2 bonds cleavage
Smaller fragments

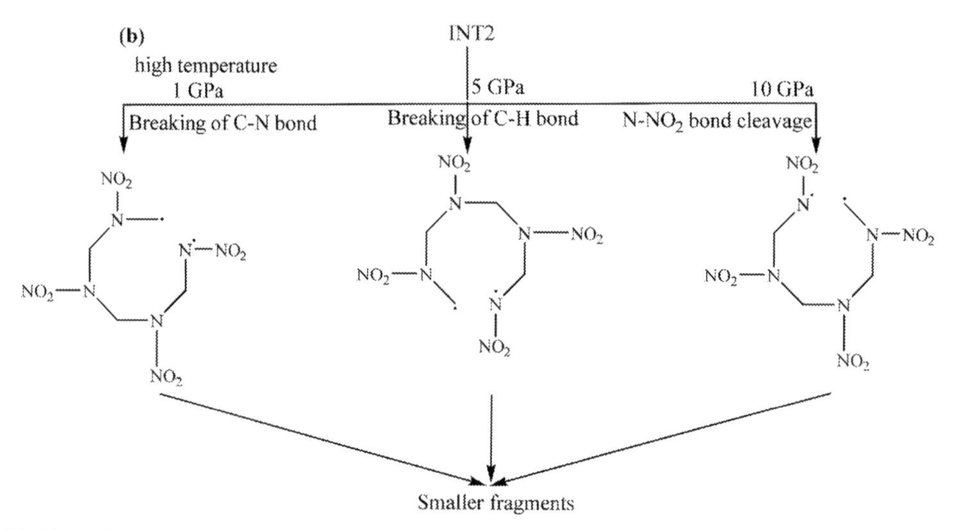

FIG. 13.41 Subsequent decomposition paths of (A) INT1 and (B) INT2 during the decomposition of the HMX crystals at 3500 K coupled with 1–16 GPa. *Reproduced from D. Xiang, G. Ji, W. Zhu, Ab initio molecular dynamics simulations study on initial decompositions of β-HMX at high temperature coupled with high pressures, J. Chin. Chem. Soc. 66 (2019) 1429 with permission from John Wiley and Sons.*

decomposition of INT1 has three pathways: (a) the C—N bond cleavage, (b) the $N—NO_2$ bond breaking, and (c) the N—O bond of fission. Subsequent decompositions of INT2 also have three paths. The C—H and $N—NO_2$ bonds become weaker than the C—N bond in the ring as the pressure increases. The number of main chemical species at the initial decomposition stage is strongly dependent on both the temperature and pressure.

4.6 Explosion products

During the last decade, thermophysical and thermochemical properties of the high temperature and high-pressure state of explosives have been a topic of growing interest. The decomposition reactions of the explosives proceed rapidly under extreme conditions (high-temperature coupled with high pressure). The detonation products of typical explosives containing C, H, N, and O atoms are mixtures of N_2, CO_2, H_2O, and other minor species. These explosion products will proceed with secondary chemical reactions at high temperatures (1000–5000 K) and pressures (1–100 GPa). However, their interaction mechanisms that are very crucial to construct the equation of state (EOS) of the explosives could not be easily observed and measured because of fast detonation reactions and extreme conditions. Xiong et al. [81] performed AIMD simulations to investigate the initiation mechanisms of these binary mixture systems including $CO_2—N_2$, $H_2O—N_2$, and $H_2O—CO_2$ at 2000 and 4000 K coupled with the pressures of 30, 40, and 50 GPa.

It is found that there is no reaction in the $CO_2—N_2$ system at extreme conditions. Only at 4000 K and 50 GPa, the mixture of H_2O and N_2 began to decompose and produce some products such as NH_3, HNO_2, etc. The mixture of H_2O and CO_2 can readily react to each other and ultimately form different reaction products even at a relatively low temperature and pressure in contrast with the above two mixtures. The temperature plays a negative effect on the interactions in the systems.

Fig. 13.42 displays the presence ratio of different clusters containing different amounts of carbon atoms in the $H_2O—CO_2$ system at 2000 K (a) and 4000 K (b) coupled with different pressures, where C_n is the cluster constituted of n carbon atoms and n is greater than 2. Generally, the smaller the ratio of C_n is, the more the fragments remain, and the weaker the system interacts. It demonstrates that the pressure is favorable to form a larger cluster. But the large clusters occur frequently at lower temperatures. This indicates that the temperature is a disadvantage factor for forming clusters.

5 Reactions at shock wave loading

Shock wave loading not only dynamically compresses EMs in a uniaxial direction but also simultaneously heats the materials. The coupling between the mechanical loading and detonation initiation in the EMs will lead to extremely complicated chemical events [82–84]. Shock wave experiments can achieve nanosecond temporal resolution and are not thus appropriate to characterize such heterogeneous and nonequilibrium systems with length scales of nanometers, time scales of picoseconds, and involving a multitude of complex many-body

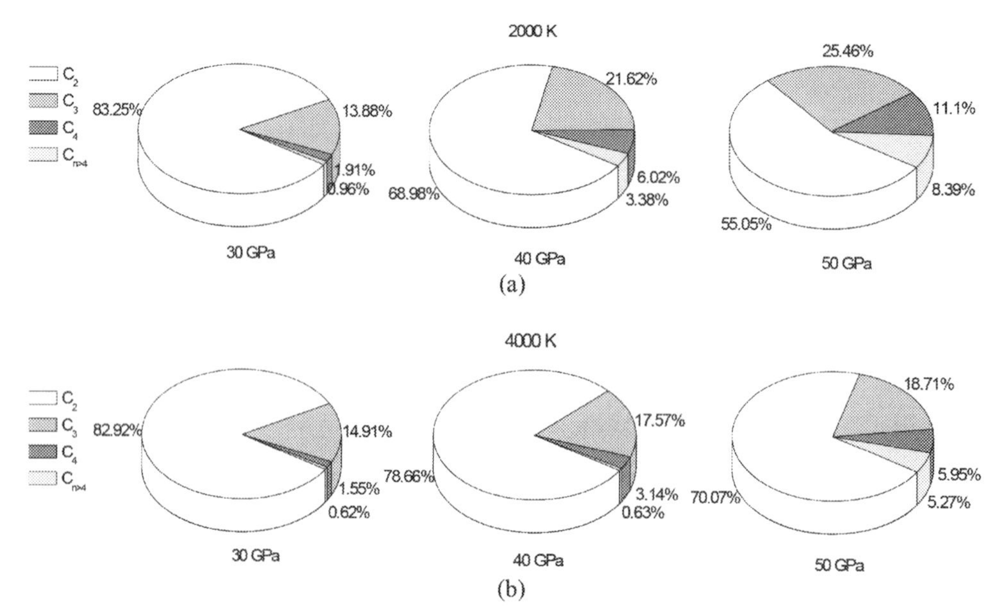

FIG. 13.42 Presence ratio of different clusters containing different amounts of carbon atoms in the $H_2O—CO_2$ system at 2000 K (A) and 4000 K (B) coupled with different pressures. *Reproduced from G. Xiong, S. Zhu, C. Yang, W. Zhu, Insight into interaction mechanisms of binary mixture systems of explosion products (H2O, CO2, and N2) at extreme high pressures and temperatures, Chem. Phys. Lett. 714 (2019) 103 with permission from Elsevier.*

chemical processes and intermediates. Therefore, it has been extremely difficult to obtain a clear molecular-level picture of the chemistry behind shock fronts experimentally.

5.1 NM

Reed et al. [85] employed DFTB-MD in conjunction with multiscale shock technique (MSST) to simulate shock-induced chemistry in NM (CH_3NO_2), one of the simplest detonable molecular materials, although not typically used in practical explosives. The results indicate that the wide-bandgap insulator NM undergoes chemical decomposition and a transformation into a semimetallic state for a limited distance behind the detonation front. This transformation is associated with the production of charged decomposition species. However, some bandgap reduction occurs on initial compression, and moreover, it is most pronounced after initiation. This means that the bandgap lowering or electronically excited states are not required to sustain chemical reactions under the conditions of detonation considered.

5.2 TATB

Manna et al. [86] applied DFTB-MD with MSST to study the chemical transformations of TATB experiencing overdriven shock speeds of 9 and 10 km/s. The simulations indicate that water is the earliest decomposition product, resulting from both inter- and intra-hydrogen transfers. After that, some TATB remaining fragments began to polymerize to form

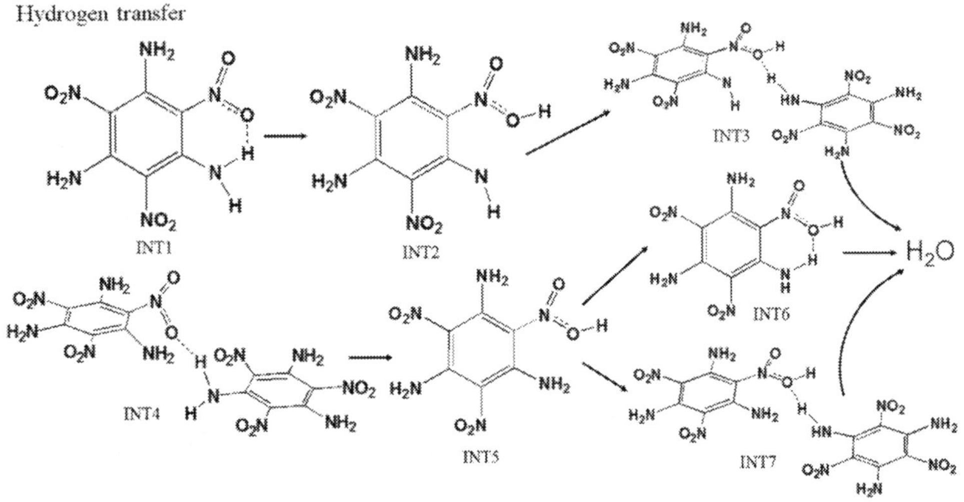

Hydrogen transfer

FIG. 13.43 Hydrogen transfer to form water molecule. *Reprinted (adapted) with permission from Z. He, J. Chen, Q. Wu, G. Ji, Initial decomposition of condensed-phase 1,3,5-triamino-2,4,6-trinitrobenzene under shock loading, J. Phys. Chem. C 121 (2017) 8227. Copyright (2017) American Chemical Society.*

high-nitrogen clusters. This is a slow diffusion-limited process. Finally, these clusters decomposed to produce N_2, CO (or CO_2), and particulate carbon, etc.

Recently, He et al. [87] employed the same method to study the reaction mechanisms of TATB crystal under shock compression with two different shock speeds of 9 and 10 km s^{-1}. The decomposition of TATB is always initiated by intra- and intermolecular hydrogen transfers, as shown in Fig. 13.43. More intriguingly, 28 kinds of heterocyclic structures involved in the TATB decomposition process were observed. Most of the newly formed C—N heterocycles possess high stability, which obviously inhibits further decomposition to release carbon oxides and N_2.

5.3 HMX

Zhu et al. [88] performed AIMD simulations in conjunction with MSST to study the initial chemical processes of HMX crystal under shock wave loading of 6.5 km s^{-1}. The initial decomposition of shocked HMX is triggered by the N—O bond cleavage and the ring opening. This will initiate many decomposition reactions and lead to the production of many small radicals at a moment. As the shock compression continues, these small radicals recombine to produce many large radicals and further form ring-shaped radicals. Then, these radicals begin to further decompose. It is also found that the system transiently produces a large number of metallic states under the shock compression, as shown in Fig. 13.44.

Also, they employed AIMD in conjunction with MSST to stimulate the initial chemical processes of PETN and TATB crystals under shock wave loading (6.5 km s^{-1}) [89]. The

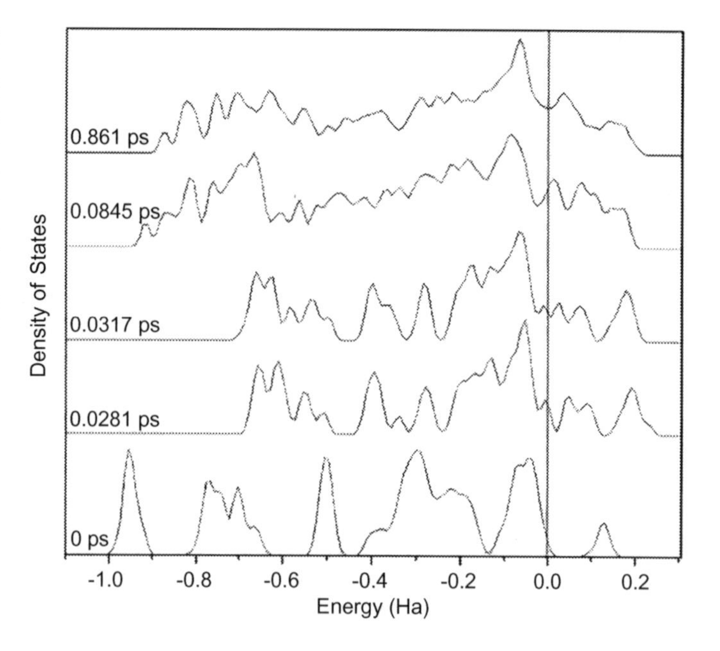

FIG. 13.44 Electronic density of states (DOS) of the HMX crystal under shock compression for a sequence of times. Each *curve* is averaged over one hundred configurations. The Fermi energy is shown as a *vertical line*. *Reproduced from W. Zhu, H. Huang, H.J. Huang, H. Xiao, Initial chemical events in shocked octahydro-1,3,5,7-tetranitro-1,3,5,7-tetrazocine: a new initiation decomposition mechanism, J. Chem. Phys. 136 (2012) 044516 with permission from AIP Publishing.*

results indicate that the decomposition of the PETN and TATB crystals under shock compression is triggered by the C—O bond and N—O bond breaking. As the simulation continues, the electronic delocalization in their systems increases, and the bandgap between conduction and valence bands decreases gradually. Then, metallic states in the systems begin to appear and increase. Under shock wave loading, the decreasing order of the time for the trigger bond cleavage for TATB, HMX, and PETN are in agreement with the increasing sequence of their experimental shock sensitivity [87]. This suggests that there is a relationship between structure or initiation decomposition mechanisms and sensitivity to the shock wave.

Ge et al. [90] performed DFTB-MD simulations with MSST to study the initial chemical processes of condensed-phase β-HMX under shock wave loading. It is found that the initial decomposition of shocked HMX is triggered by the N-NO_2 bond breaking under the low-velocity impact (8 km s^{-1}). As the shock velocity increases (11 km s^{-1}), the C—H bond dissociation becomes the primary pathway for HMX decomposition in its early stages, while the homolytic cleavage of the N—NO_2 bond is suppressed under high pressure. Further decomposition is accompanied by a five-membered ring formation and hydrogen transfer from the CH_2 group to the NO_2 group.

Afterward, they also employed the same method to investigate the initial chemical processes and the anisotropy of shock sensitivity of the condensed-phase β-HMX under shock loadings (10 and 11 km s^{-1}) applied along with the **a**, **b**, and **c** lattice vectors [77]. The response along lattice vectors **a** and **c** are similar to each other, whose reaction temperature is up to 7000 K, but quite different along lattice vector **b**, whose reaction temperature is only up to 4000 K. The small relative sliding between adjacent slipping planes as a buffer is the main reason leading to less sensitivity under shock wave compression along the lattice

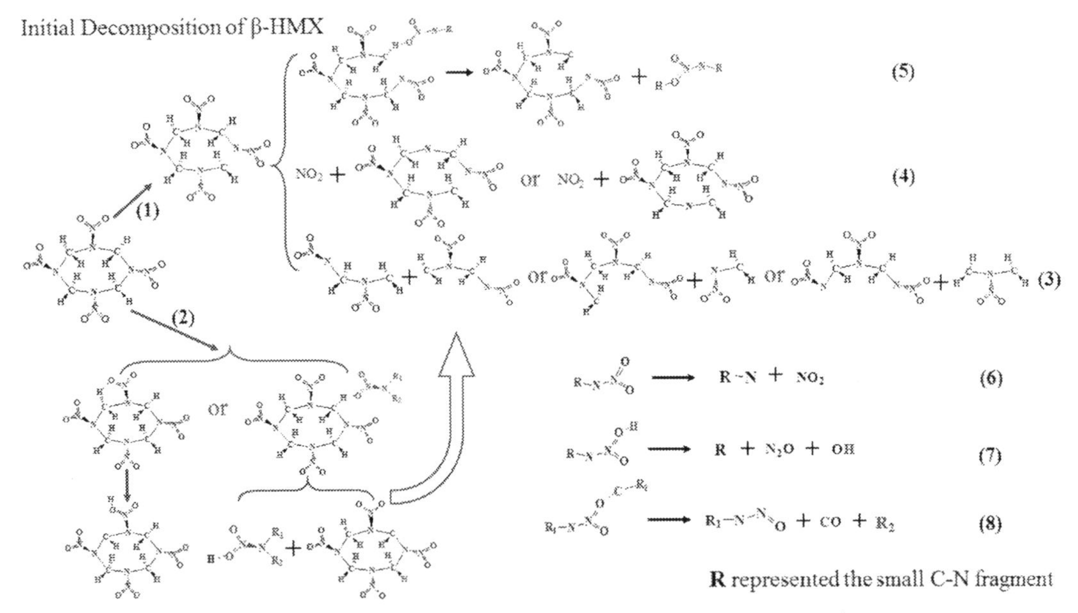

FIG. 13.45 Initial decomposition of β-HMX with two-molecule vacancy under 9 km s^{-1} shock loading. *Reprinted (adapted) with permission from Z. He, J. Chen, G. Ji, L. Liu, W. Zhu, Q. Wu, Dynamic responses and initial decomposition under shock loading: A DFTB calculation combined with MSST method for β-HMX with molecular vacancy, J. Phys. Chem. B 119 (2015) 10673. Copyright (2015) American Chemical Society.*

vector **b**. In addition, the C—H bond dissociation is the primary pathway for the HMX decomposition in early stages under high shock loading from various directions. The homolytic cleavage of the N—NO_2 bond was obviously suppressed with the increasing pressure.

He et al. [78] explored the microscopic dynamic response and initial decomposition of β-HMX with (100) surface and molecular vacancy under shock compression (9 and 10 km s^{-1}) by means of DFTB-MD in conjunction with MSST. Fig. 13.45 presents the initial decomposition of β-HMX with two-molecule vacancy under 9 km s^{-1} shock loading. The first reaction primarily began with the breaking of the C—N bond in the molecular ring. The C—N bonds close to the major axis have an apparent inclination to break for triggering the initial decomposition. Subsequent decomposition of the molecular ring enhanced intermolecular interactions and promoted the cleavage of C—H and N—NO_2 bonds. More significantly, the dynamic response behavior clearly depends on the angle between the chemical bond and shock direction. A bigger angle (close to 90°) enables the chemical bond to avoid the compression efficiently and gets the fast activation with the increasing temperature.

Later, He et al. [91] carried out DFTB-MD simulations with MSST to reveal the rapid initiation mechanism of β-HMX crystal under the shock wave loading of 9 km s^{-1}. During the early decomposition of HMX, water and its derivative (OH) acting as oxidizers dominated the carbon oxidation reaction and promoted the oxygen transport from nitrogen to carbon. Water monomer and its small polymers can efficiently transfer proton and hydroxyl moieties between reaction centers, accelerating carbon oxidation reaction. The carbon oxidation significantly reduced the dissociation energy barrier of the C—N bonds, causing fast and deep decomposition of HMX.

5.4 CL-20

Xue et al. [92] performed DFTB-MD simulations with MSST to disclose the early decay events of shocked ε-CL-20 under four specified shock velocities of 8–11 km s^{-1}. The ring opening was observed to trigger molecular decay at all four shock conditions, while sufficient N—NO_2 fission was observed at $U_s = 8$ and 9 km s^{-1}, and strongly inhibited at $U_s = 10$ and 11 km s^{-1}. Main chemical species in the initial reaction stage are strongly dependent on the shock strength. CO_2, N_2, and H_2O are the main products during the decomposition of the shocked CL-20.

5.5 RDX

Yuan et al. [93] applied DFTB-MD with MSST to stimulate the shock response of condensed-phase RDX under the shock velocity of 10 km s^{-1}. The initial decomposition of shocked condensed-phase RDX is triggered by the C—N bond scission of the ring. The C—N bonds are more sensitive to shock loading than other bonds. Generally speaking, the C—N bonds will be broken preferentially, and the rings are easier to be dissociated in the shocked RDX. The products from the ring opening are produced earlier than other products from the N—N bond scission and unimolecular HONO elimination.

Ge et al. [94] performed DFTB-MD simulations with MSST to study the initial chemical processes of condensed-phase RDX under various shock velocities (8, 10, and 11 km s^{-1}). The N—NO_2 bond dissociation is the primary decomposition pathway of RDX with the NO_2 groups (group 1) facing the shock input, whereas the C—N bond scission is the dominant initiation channel for RDX with the NO_2 groups (group 2) opposite to the shock. The NO_2 groups facing away from the shock are rather inert than those opposite to shock loading. As the shock velocity increases, the decomposition reactions become faster and faster. The reaction pathways of a single RDX molecule under the shock velocity of 11 km s^{-1} have been mapped out in detail, as shown in Fig. 13.46.

5.6 CL-20/TNT cocrystal

Zhang et al. [95] used DFTB-MD with MSST to study the initial chemical reactions of CL-20/TNT cocrystal under shock wave loading with the velocities of 8, 9, 10, and 11 km s^{-1} along the a-axis. The initial decomposition step of the CL-20 molecule in the cocrystal is the N—NO_2 bond cleavage at low velocity, which is inhibited at high velocity, while the H transfer and C—NO_2 bond breaking are the main initial decay step of the TNT molecules. CL-20 decomposes faster than TNT, and moreover, the releasing heat of the CL-20 decomposition is transferred to TNT to accelerate its decomposition. Additionally, the evolution of the main stable products strongly depends on the shock strength.

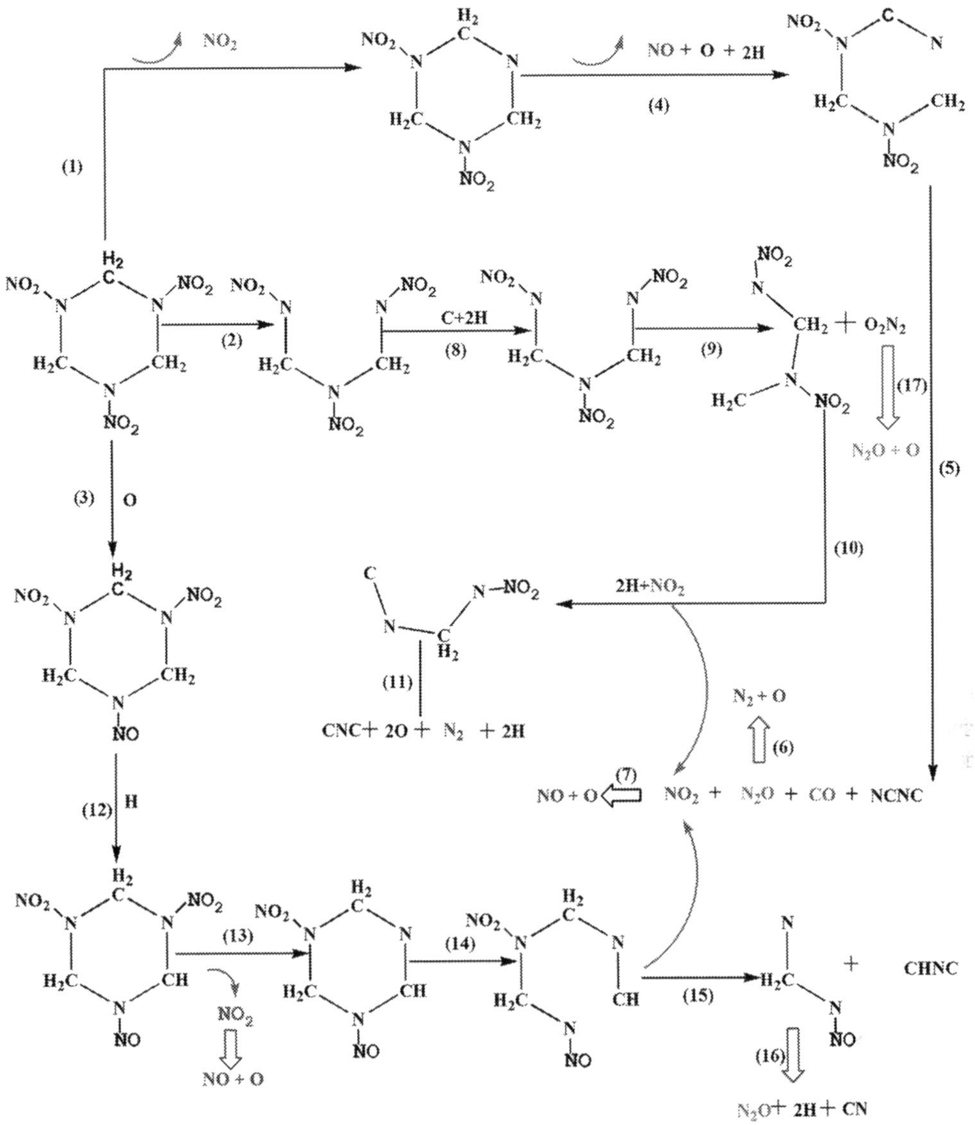

FIG. 13.46 The decomposition scheme for RDX. *Reproduced from N. Ge, S. Bai, J. Chang, G. Ji, Shock response of condensed-phase RDX: molecular dynamics simulations in conjunction with the MSST method, RSC Adv. 8 (2018) 17312 with permission from The Royal Society of Chemistry.*

6 Summary

Recent quantum MD studies on the chemical reaction mechanisms of some important condensed-phase EMs under high temperatures, low temperatures coupled with high pressures, high temperatures coupled with high pressures, and shock wave loading have been reviewed. Quantum MD methods involved here include AIMD and DFTB-MD. Although quantum MD stimulations are computationally intensive and currently limited in the realm of detonations to picosecond time scales, these methods give us a chance to get the first reliable insights into the condensed-phase chemical processes of the EMs responsible for high explosive detonation and electronic properties. In addition, to understand the chemical processes at the reactive front of the EMs is necessary to extend the time scales involved in atomistic simulations to the picosecond scale, which is exactly what the quantum MD can reach. Some of these significant results that have been outlined are as follows: (1) The initial reaction steps of the solid EMs are often the cleavage of the bonds connected with the energetic groups, the ring opening, the inter- or intramolecular hydrogen transfer, the HONO elimination, the breaking of the N—N_2, C—O, N—O N—H, C—C, or C—H bonds, the isomerization, and the cage cracking, dependent on different external conditions. (2) Some large clusters were observed during further decompositions, which may retard the subsequent reaction process. (3) Water and hydrogen radicals formed during the decomposition of the solid EMs play a catalytic role in the subsequent reaction process. (4) Initial conformational changes or molecular rearrangements can act as a buffer to delay the initiation reactions of the EMs. (5) Some small gas molecules were released through slow formation and disintegration of polycyclic networks. (6) The main decomposition pathway of the HMX nanoparticles at low temperature is the isomerization reaction of the HMX molecule, while that at high-temperature changes to the N—NO_2 homolysis with ring opening. This indicates that different external conditions significantly change the proportion of the dominant decomposition channels of the solid EMs. (7) Main explosion products will proceed with secondary chemical reactions at high temperatures (1000–5000 K) coupled with pressures (1–100 GPa). (8) The chemical kinetics of the dominant decomposition channels of the solid EMs can be studied.

The study on the primary chemical processes of the initiation of EMs is indispensable for understanding the sensitivity (initiation reactivity). There are many kinds of pathways to initiate the chemical reactions of solid EMs. Therefore, we can manipulate the chemical composition of new EMs and their morphology to decrease their sensitivity to external stimuli by many different combinations. This is expected to help in the design of new EMs with higher energy and lower sensitivity.

Quantum MD stimulations at the atomic level have been emerging as new tools to realize the sensitivity of solid EMs. But their initiation mechanisms are strongly dependent upon their molecular environment, namely different types of defects [96–105]. Unfortunately, experimental identification of the defects is typically difficult and indirect, usually requiring an ingenious combination of different techniques. Accordingly, the development and application of new methods to model chemical reactions of the solid EMs with types of defects under extreme conditions are imperative to realize the sensitivity of practical EMs.

Acknowledgments

This work is supported by the National Natural Science Foundation of China (Grant No. 21773119).

References

[1] G.B. Manelis, G.M. Nazin, Y.I. Rubtsov, V.A. Strunin, Thermal Decomposition and Combustion of Explosives and Powders, Izdat Nauka, Moscow, 1996.

[2] M.R. Manaa, Initiation and decomposition mechanisms of energetic materials, in: P.A. Politzer, J.S. Murray (Eds.), Energetic Materials, Part 2. Detonation, Combustion, Theoretical and Computational Chemistry, vol. 13, Elsevier B.V., 2003, pp. 71–100.

[3] L.E. Fried, M.R. Manaa, J.P. Lewis, Modeling the reactions of energetic materials in the condensed phase, in: R.W. Shaw, T.B. Brill, D.L. Thompson (Eds.), Overviews of recent research on energetic materials, Advanced Series in Physical Chemistry, vol. 16, World Scientific, 2005, pp. 275–301.

[4] L.E. Fried, The reactivity of energetic materials at extreme conditions, in: K.B. Lipkowitz, T.R. Cundari (Eds.), Reviews in Computational Chemistry, vol. 25, John Wiley & Sons, Inc., Wiley-VCH, 2007, pp. 159–189.

[5] W. Zhu, H. Xiao, Quantum Chemistry of Energetic Crystals, Science Press, Beijing, 2012.

[6] M.R. Manaa, L.E. Fried, The reactivity of energetic materials under high pressure and temperature, in: J.R. Sabin (Ed.), Energetic Materials, Advances in Quantum Chemistry, vol. 69, Elsevier Inc., 2014, pp. 221–252.

[7] J. Xiao, W.H. Zhu, W. Zhu, H. Xiao, Molecular Dynamics of Energetic Materials, Science Press, Beijing, 2014.

[8] M.M. Kuklja, Quantum-Chemical Modeling of Energetic Materials: Chemical Reactions Triggered by Defects, Deformations, and Electronic Excitations, Advances in Quantum Chemistry, vol. 69, Elsevier Inc., 2014, pp. 71–145.

[9] F.J. Owens, Calculation of energy barriers for bond rupture in some energetic molecules, J. Mol. Struct. (THEOCHEM) 370 (1996) 11.

[10] J. Fan, H. Xiao, Theoretical study on pyrolysis and sensitivity of energetic compounds. (2) Nitro derivatives of benzene, J. Mol. Struct. (THEOCHEM) 365 (1996) 225.

[11] H. Xiao, J. Fan, X. Gong, Theoretical study on pyrolysis and sensitivity of energetic compounds. Part I: simple model molecules containing NO_2 group, Propellants Explos. Pyrotech. 22 (1997) 360.

[12] J. Fan, Z. Gu, H. Xiao, H. Dong, Theoretical study on pyrolysis and sensitivity of energetic compounds. Part 4. Nitro derivatives of phenols, J. Phys. Org. Chem. 11 (1998) 177.

[13] Z. Chen, H. Xiao, S. Yang, Theoretical investigation on the impact sensitivity of tetrazole derivatives and their metal salts, Chem. Phys. 250 (1999) 243.

[14] Z. Chen, H. Xiao, Impact sensitivity and activation energy of pyrolysis for tetrazole compounds, Int. J. Quantum Chem. 79 (2000) 350.

[15] B.M. Rice, S. Sahu, F.J. Owens, Density functional calculations of bond dissociation energies for NO_2 scission in some nitroaromatic molecules, J. Mol. Struct. (THEOCHEM) 583 (2002) 69.

[16] A.C.T. van Duin, S. Dasgupta, F. Lorant, W.A. Goddard, ReaxFF: a reactive force field for hydrocarbons, J. Phys. Chem. A 105 (2001) 9396.

[17] R. Car, M. Parrinello, Unified approach for molecular dynamics and density-functional theory, Phys. Rev. Lett. 55 (1985) 2471.

[18] M. Elstner, D. Porezag, G. Jungnickel, J. Elsner, M. Hauk, T. Frauenheim, et al., Self-consistent-charge density-functional tight-binding method for simulations of complex materials properties, Phys. Rev. B 58 (1998) 7260.

[19] J. Owens, J. Sharma, X-ray photoelectron spectroscopy and paramagnetic resonance evidence for shock-induced intramolecular bond breaking in some energetic solids, J. Appl. Phys. 51 (1980) 1494.

[20] D.V. Shalashilin, D.L. Thompson, Monte Carlo variational transition-state theory study of the unimolecular dissociation of RDX, J. Phys. Chem. A 101 (1997) 961.

[21] O. Sharia, M.M. Kuklja, Modeling thermal decomposition mechanisms in gaseous and crystalline molecular materials: application to β-HMX, J. Phys. Chem. B 115 (2011) 12677.

[22] C.M. Tarver, T.D. Tran, Thermal decomposition models for HMX-based plastic bonded explosives, Combust. Flame 137 (2004) 50.

[23] T.B. Brill, P.E. Gongwer, G.K. Williams, Thermal decomposition of energetic materials. 66. Kinetic compensation effects in HMX, RDX, and NTO, J. Phys. Chem. 98 (1994) 12242.

[24] G.T. Long, S. Vyazovkin, B.A. Brems, C.A. Wight, Competitive vaporization and decomposition of liquid RDX, J. Phys. Chem. B 104 (2000) 2570.

[25] M.R. Manaa, L.E. Fried, C.F. Melius, M. Elstner, T. Frauenheim, Decomposition of HMX at extreme conditions: a molecular dynamics simulations, J. Phys. Chem. A 106 (2002) 9024.

[26] D. Xiang, W. Zhu, Adiabatic and constant volume decomposition process of condensed phase δ-1,3,5,7-tetranitro-1,3,5,7-tetrazocane at high temperatures: quantum molecular dynamics simulations, J. Mol. Graph. Model. 85 (2018) 68.

[27] C. Ye, Q. An, W. Zhang, W.A. Goddard III, Initial decomposition of HMX energetic material from quantum molecular dynamics and the molecular structure transition of β-HMX to δ-HMX, J. Phys. Chem. C 123 (2019) 9231.

[28] D. Xiang, G. Ji, W. Zhu, Structural and vibrational properties of crystalline β-octahydro-1,3,5,7-tetranitro-1,3,5,7-tetrazocine at high temperatures: ab initio molecular dynamics studies, ChemistrySelect 4 (2019) 4244.

[29] S. Zhu, W. Zhu, An ab initio molecular dynamics study of low temperature effects in crystalline α-HMX, Phys. Status Solidi B 256 (2019) 1900057.

[30] O. Isayev, L. Gorb, M. Qasim, J. Leszczynski, Ab initio molecular dynamics study on the initial chemical events in nitramines: thermal decomposition of CL-20, J. Phys. Chem. B 112 (2008) 11005.

[31] J. Xu, J. Zhao, L. Sun, Thermal decomposition behaviour of RDX by first-principles molecular dynamics simulation, Mol. Simul. 34 (2008) 961.

[32] C.J. Wu, M.R. Manaa, L.E. Fried, Tight binding molecular dynamic simulation of PETN decomposition at an extreme condition, Mater. Res. Soc. 987 (2007) 139.

[33] C.J. Wu, L.E. Fried, L.H. Yang, N. Goldman, S. Bastea, Catalytic behaviour of dense hot water, Nat. Chem. 1 (2009) 56.

[34] T. Zhou, T. Cheng, S.V. Zybin, W.A. Goddard III, F. Huang, Reaction mechanisms and sensitivity of silicon nitrocarbamate and related systems from quantum mechanics reaction dynamics, J. Mater. Chem. A 6 (2018) 5082.

[35] M.R. Manaa, E.J. Reed, Early chemistry in hot and dense nitromethane: molecular dynamics simulations, J. Chem. Phys. 120 (2004) 10146.

[36] J. Chang, P. Lian, D. Wei, X. Chen, Q. Zhang, Z. Gong, Thermal decomposition of the solid phase of nitromethane: ab initio molecular dynamics simulations, Phys. Rev. Lett. 105 (2010) 188302.

[37] W. Zhu, H. Xiao, Ab initio molecular dynamics study of temperature effects on the structure and stability of energetic solid silver azide, J. Phys. Chem. C 115 (2011) 20782.

[38] W. Zhu, H. Xiao, Ab initio study of energetic solids: cupric azide, mercuric azide, and lead azide, J. Phys. Chem. B 110 (2006) 18196.

[39] W. Zhu, H. Xiao, Ab initio study of electronic structure and optical properties of heavy-metal azides: TlN_3, AgN_3, and CuN_3, J. Comput. Chem. 29 (2008) 176.

[40] W. Zhu, H. Xiao, First-principles band gap criterion for impact sensitivity of energetic crystals: a review, Struct. Chem. 21 (2010) 657.

[41] Q. Wu, W. Zhu, H. Xiao, An ab initio molecular dynamics study of thermal decomposition of 3,6-di(azido)-1,2,4,5-tetrazine, Phys. Chem. Chem. Phys. 16 (2014) 21620.

[42] Q. An, W. Liu, W.A. Goddard III, T. Cheng, S.V. Zybin, et al., Initial steps of thermal decomposition of dihydroxylammonium 5,5′-bistetrazole-1,1′-diolate crystals from quantum mechanics, J. Phys. Chem. C 118 (2014) 27175.

[43] L. Meng, Z. Lu, X. Wei, X. Xue, Y. Ma, Q. Zeng, et al., Two-sided effects of strong hydrogen bonding on the stability of dihydroxylammonium 5,5′-bistetrazole-1,1′-diolate (TKX-50), CrstEngComm 18 (2016) 2258.

[44] Z. Lu, Q. Zeng, X. Xue, Z. Zhang, F. Nie, C. Zhang, Does increasing pressure always accelerate the condensed material decay initiated through bimolecular reactions? A case of the thermal decomposition of TKX-50 at high pressures, Phys. Chem. Chem. Phys. 19 (2017) 23309.

[45] Q. Wu, W. Zhu, H. Xiao, Catalytic behavior of hydrogen radicals in the thermal decomposition of crystalline furoxan: DFT-based molecular dynamics simulations, RSC Adv. 4 (2014) 34454.

[46] C. Ye, Q. An, W.A. Goddard III, T. Cheng, W. Liu, S.V. Zybina, et al., Initial decomposition reaction of di-tetrazine-tetroxide (DTTO) from quantum molecular dynamics: implications for a promising energetic material, J. Mater. Chem. A 3 (2015) 1972.

[47] C. Ye, Q. An, W.A. Goddard III, T. Cheng, S.V. Zybina, X. Ju, Initial decomposition reactions of bicyclo-HMX [BCHMX or cis-1,3,4,6-tetranitrooctahydroimidazo-[4,5-d]imidazole] from quantum molecular dynamics simulations, J. Phys. Chem. C 119 (2015) 2290.

[48] D. Xiang, W. Zhu, Thermal decomposition of isolated and crystal 4,10-dinitro-2,6,8,12-tetraoxa-4,10-diazaisowurtzitane according to ab initio molecular dynamics simulations, RSC Adv. 7 (2017) 8347.

[49] Z. Liu, W. Zhu, G. Ji, K. Song, H. Xiao, Decomposition mechanisms of α-octahydro-1,3,5,7-tetranitro-1,3,5,7-tetrazocine nanoparticles at high temperatures, J. Phys. Chem. C 121 (2017) 7728.
[50] G. Xiong, C. Yang, W. Zhu, Interface reaction processes and reactive properties of Al/CuO nanothermite: an ab initio molecular dynamics simulation, Appl. Surf. Sci. 459 (2018) 835.
[51] G. Xiong, C. Yang, S. Feng, W. Zhu, Ab initio molecular dynamics studies on the transport mechanisms of oxygen atoms in the adiabatic reaction of Al/CuO nanothermite, Chem. Phys. Lett. 745 (2020) 137278.
[52] Y. Liu, F. Li, H. Sun, Thermal decomposition of FOX-7 studied by ab initio molecular dynamics simulations, Theor. Chem. Acc. 133 (2014) 1567.
[53] H. Jiang, Q. Jiao, C. Zhang, Early events when heating 1,1-diamino-2,2-dinitroethylene: self-consistent charge density-functional tight-binding molecular dynamics simulations, J. Phys. Chem. C 122 (2018) 15125.
[54] J. Wang, Y. Xiong, H. Li, C. Zhang, Reversible hydrogen transfer as new sensitivity mechanism for energetic materials against external stimuli: a case of the insensitive 2,6-dDiamino-3,5-dinitropyrazine-1-oxide, J. Phys. Chem. C 122 (2018) 1109.
[55] Z. He, Y. Yu, Y. Huang, J. Chen, Q. Wu, Reaction kinetic properties of 1,3,5-triamino-2,4,6-trinitrobenzene: a DFTB study of thermal decomposition, New J. Chem. 43 (2019) 18027.
[56] X. Wu, Z. Liu, W. Zhu, Conformational changes and decomposition mechanisms of HMX-based cocrystal explosives at high temperatures, J. Phys. Chem. C 124 (2020) 25.
[57] H. Xie, W. Zhu, Thermal decomposition mechanisms of the energetic benzotrifuroxan:1,3,3-trinitroazetidine cocrystal using ab initio molecular dynamics simulations, J. Chin. Chem. Soc. 67 (2020) 218.
[58] A.K. Burnham, R.K. Weese, J.F. Wardell, T.D. Tran, A.P. Wemhoff, J.G. Koerner, et al., Proceedings of the 13th International Detonation Symposium, Office of Naval Research, Norfolk, VA, 2006.
[59] G.J. Piermarini, S. Block, P.J. Miller, Effects of pressure and temperature on the thermal decomposition rate and reaction mechanism of beta-octahydro-1,3,5,7-tetranitro-1,3,5,7-tetrazocine, J. Phys. Chem. 91 (1987) 3872.
[60] P.J. Miller, S. Block, G.J. Piermarini, Effects of pressure on the thermal decomposition kinetics, chemical reactivity and phase behavior of RDX, Combust. Flame 83 (1991) 174.
[61] E.A. Glascoe, J.M. Zaug, A.K. Burnham, Pressure-dependent decomposition kinetics of the energetic material HMX up to 3.6 GPa, J. Phys. Chem. A 113 (2009) 13548.
[62] Z.A. Dreger, M.D. McCluskey, Y.M. Gupta, High pressure-high temperature decomposition of γ-cyclotrimethylene trinitramine, J. Phys. Chem. A 116 (2012) 9680.
[63] Z.A. Dreger, Y. Tao, Y.M. Gupta, Phase diagram and decomposition of 1,1-diamino-2,2-dinitroethene single crystals at high pressures and temperatures, J. Phys. Chem. C 120 (2016) 11092.
[64] Q. Wu, G. Xiong, W. Zhu, H. Xiao, How does low temperature coupled with different pressures affect initiation mechanisms and subsequent decompositions in nitramine explosive HMX? Phys. Chem. Chem. Phys. 17 (2015) 22823.
[65] Q. Wu, H. Chen, G. Xiong, W. Zhu, H. Xiao, Decomposition of a 1,3,5-triamino-2,4,6-trinitrobenzene crystal at decomposition temperature coupled with different pressures: an ab initio molecular dynamics study, J. Phys. Chem. C 119 (2015) 16500.
[66] J. Sharma, J.W. Forbes, C.S. Coffey, T.P. Liddiard, The physical and chemical nature of sensitization centers left from hot spots caused in triaminotrinitrobenzene by shock or impact, J. Phys. Chem. 91 (1987) 5139.
[67] Q. Wu, W. Zhu, H. Xiao, Comparative DFT- and DFT-D-based molecular dynamics studies of pressure effects in crystalline 1,3,5-triamino-2,4,6-trinitrobenzene at room temperature, RSC Adv. 4 (2014) 53149.
[68] A.J. Davidson, R.P. Dias, D.M. Dattelbaum, C. Yoo, "Stubborn" triaminotrinitrobenzene: unusually high chemical stability of a molecular solid to 150 GPa, J. Chem. Phys. 135 (2011) 174507.
[69] Q. Wu, W. Zhu, H. Xiao, Cooperative effects of different temperatures and pressures on the initial and subsequent decomposition reactions of the nitrogen-rich energetic crystal 3,3′-dinitroamino-4,4′-azoxyfurazan, Phys. Chem. Chem. Phys. 18 (2016) 7093.
[70] Q. Wu, D. Xiang, G. Xiong, W. Zhu, H. Xiao, Coupling of temperature with pressure induced initial decomposition mechanisms of two molecular crystals: an ab initio molecular dynamics study, J. Chem. Sci. 128 (2016) 695.
[71] G. Xiong, W. Zhu, Coupling effects of high temperature and pressure on the decomposition mechanisms of 1,1-diamino-2,2-dinitroehethe crystal: ab initio molecular dynamics simulations, J. Chin. Chem. Soc. 54 (2020). https://doi.org/10.1002/jccs.201900504.
[72] D. Xiang, W. Zhu, Mechanisms and kinetics of initial pyrolysis and combustion reactions of 1,1-diamino-2,2-dinitroethylene from density functional tight-binding molecular dynamics simulations, Can. J. Chem. 97 (2019) 795.

[73] D. Xiang, W. Zhu, H. Xiao, Thermal decomposition mechanisms of energetic ionic crystal dihydrazinium 3,30-dinitro-5,50-bis-1,2,4-triazole-1,1-diolate: an ab initio molecular dynamics study, Fuel 202 (2017) 246.
[74] D. Xiang, Q. Wu, W. Zhu, Ab initio molecular dynamics studies on the decomposition mechanisms of CL-20 crystal under extreme conditions, Chin. J. Energ. Mater. 26 (2018) 59.
[75] D. Xiang, Q. Wu, W. Zhu, Decomposition mechanisms of crystalline α-RDX under high temperature coupled with detonation pressure from ab initio molecular dynamics simulations, Chin. J. Energ. Mater. 26 (2018) 477.
[76] D. Xiang, W. Zhu, Thermal decomposition of energetic MOFs nickel hydrazine nitrate crystals from an ab initio molecular dynamics simulation, Comput. Mater. Sci. 143 (2018) 170.
[77] N. Ge, Y. Wei, Z. Song, X. Chen, G. Ji, F. Zhao, D. Wei, Anisotropic responses and initial decomposition of condensed-phase β-HMX under shock loadings via molecular dynamics simulations in conjunction with multiscale shock technique, J. Phys. Chem. B 118 (2014) 8691.
[78] Z. He, J. Chen, G. Ji, L. Liu, W. Zhu, Q. Wu, Dynamic responses and initial decomposition under shock loading: a DFTB calculation combined with MSST method for β-HMX with molecular vacancy, J. Phys. Chem. B 119 (2015) 10673.
[79] Y. Long, J. Chen, Systematic study of the reaction kinetics for HMX, J. Phys. Chem. A 119 (2015) 4073.
[80] D. Xiang, G. Ji, W. Zhu, Ab initio molecular dynamics simulations study on initial decompositions of β-HMX at high temperature coupled with high pressures, J. Chin. Chem. Soc. 66 (2019) 1429.
[81] G. Xiong, S. Zhu, C. Yang, W. Zhu, Insight into interaction mechanisms of binary mixture systems of explosion products (H_2O, CO_2, and N_2) at extreme high pressures and temperatures, Chem. Phys. Lett. 714 (2019) 103.
[82] J.J. Gilman, Chemical-reactions at detonation fronts in solids, Philos. Mag. B 71 (1995) 1057.
[83] J.J. Gilman, Mechanochemistry, Science 274 (1996) 65.
[84] F. Williams, Electronic states of solid explosives and their probable role in detonations, Adv. Chem. Phys. 21 (1971) 289.
[85] E.J. Reed, M.R. Manaa, L.E. Fried, K.R. Glaesemann, J.D. Joannopoulos, A transient semimetallic layer in detonating nitromethane, Nat. Phys. 4 (2008) 72.
[86] M.R. Manaa, E.J. Reed, L.E. Fried, N. Goldman, Nitrogen-rich heterocycles as reactivity retardants in shocked insensitive explosives, J. Am. Chem. Soc. 131 (2009) 5483.
[87] Z. He, J. Chen, Q. Wu, G. Ji, Initial decomposition of condensed-phase 1,3,5-triamino-2,4,6-trinitrobenzene under shock loading, J. Phys. Chem. C 121 (2017) 8227.
[88] W. Zhu, H. Huang, H.J. Huang, H. Xiao, Initial chemical events in shocked octahydro-1,3,5,7-tetranitro-1,3,5,7-tetrazocine: a new initiation decomposition mechanism, J. Chem. Phys. 136 (2012) 044516.
[89] W. Zhu, H. Huang, H.J. Huang, H. Xiao, Initial decomposition mechanisms of three explosive crystals under shock loading by ab initial molecular dynamics, Chin. J. Energ. Mater. 21 (2013) 557.
[90] N. Ge, Y. Wei, G. Ji, X. Chen, F. Zhao, D. Wei, Initial decomposition of the condensed-phase β-HMX under shock waves: molecular dynamics simulations, J. Phys. Chem. B 116 (2012) 13696.
[91] Z. He, J. Chen, Q. Wu, G. Ji, Special catalytic effects of intermediate-water for rapid shock initiation of β-HMX, RSC Adv. 6 (2016) 93103.
[92] X. Xue, Y. Wen, C. Zhang, Early decay mechanism of shocked ε-CL-20: a molecular dynamics simulation study, J. Phys. Chem. C 120 (2016) 21169.
[93] J. Yuan, Y. Wei, X. Zhang, X. Chen, G. Ji, M.K. Kotni, D. Wei, Shock response of 1,3,5-trinitroperhydro-1,3,5-triazine (RDX): the C-N bond scission studied by molecular dynamics simulations, J. Appl. Phys. 122 (2017) 135901.
[94] N. Ge, S. Bai, J. Chang, G. Ji, Shock response of condensed-phase RDX: molecular dynamics simulations in conjunction with the MSST method, RSC Adv. 8 (2018) 17312.
[95] X. Zhang, X. Chen, S. Kaliamurthi, G. Selvaraj, G. Ji, D. Wei, Initial decomposition of the co-crystal of CL-20/TNT: sensitivity decrease under shock loading, J. Phys. Chem. C 122 (2018) 24270.
[96] M.M. Kuklja, Quantum-chemical modeling of energetic materials: chemical reactions triggered by defects, deformations, and electronic excitations, in: J.R. Sabin (Ed.), Energetic Materials, Advances in Quantum Chemistry, vol. 69, Elsevier Inc., 2014, pp. 71–145.
[97] M.M. Kuklja, A.B. Kunz, Simulation of defects in energetic materials. 3. The structure and properties of RDX crystals with vacancy complexes, J. Phys. Chem. B 103 (1999) 8427.
[98] S. Boyd, J.S. Murray, P. Politzer, Molecular dynamics characterization of void defects in crystalline (1,3,5-trinitro-1,3,5-triazacyclohexane), J. Chem. Phys. 131 (2009) 204903.

[99] A.V. Kimmel, P.V. Sushko, M.M. Kuklja, The structure and decomposition chemistry of isomer defects in a crystalline DADNE, J. Energ. Mater. 28 (2010) 128.
[100] M.M. Kuklja, S.N. Rashkeev, Self-accelerated mechanochemistry in nitroarenes, J. Phys. Chem. Lett. 1 (2010) 363.
[101] R.W. Armstrong, Dislocation mechanics aspects of energetic material composites, Rev. Adv. Mater. Sci. 19 (2009) 13.
[102] M.M. Kuklja, A.B. Kunz, Electronic structure of molecular crystals containing edge dislocations, J. Appl. Phys. 89 (2001) 4962.
[103] Z. Liu, Q. Wu, W. Zhu, H. Xiao, Vacancy-induced initial decomposition of condensed phase NTO via bimolecular hydrogen transfer mechanisms at high pressure: a DFT-D study, Phys. Chem. Chem. Phys. 17 (2015) 10568.
[104] W. Zhu, X. Zhang, T. Wei, H. Xiao, DFT study of effects of potassium doping on band structure of crystalline cuprous azide, Chin. J. Chem. 26 (2008) 2145.
[105] Z. Liu, Q. Wu, W. Zhu, H. Xiao, Formation and growth mechanisms of natural metastable twin boundary in crystalline 1,3,5,7-tetranitro-1,3,5,7-tetrazocane: a computational study, RSC Adv. 5 (2015) 86041.

CHAPTER

14

Ranking explosive sensitivity with chemical kinetics derived from molecular dynamics simulations

M.J. Cawkwell, S.R. Ferreira, N. Lease, and V.W. Manner*

Los Alamos National Laboratory, Los Alamos, NM, United States

*Corresponding author: E-mail: cawkwell@lanl.gov

1 Introduction

The handling safety and sensitivity of explosives are of critical importance. The propensity of explosives for accidental initiation spans primary explosives, which may detonate under innocuous stimuli, to insensitive secondary explosives, which are difficult to initiate and hence can be handled relatively safely. Suites of empirical tests have been developed over decades to assess the handling safety of explosives [1,2]. These tests insult an explosive by impact, shock, friction, elevated temperature, or spark discharge of various strengths and the response of the material is measured. The various safety tests are sensitive to different characteristics of the explosive, ranging from its underlying chemistry to mesoscale structure and porosity.

The drop weight impact test, which is the focus of this chapter, is used widely to characterize the handling safety of explosives relative to one another [3,4]. Its popularity is tied to the fact that only small quantities of energetic materials are needed to obtain a fairly rapid turnaround of statistically significant results. Despite the apparent simplicity of the test, where a small quantity of explosive is placed between a metal anvil and a striker, upon which a mass is dropped; it imparts a complex insult to the sample and it is not entirely clear what causes the sample to react. Indeed, different apparatus, indicators of reactivity, and even operators can give significantly different drop height sensitivities for the same material [5,6]. The key idea behind the drop weight test is that the kinetic energy of the drop weight is converted into heat in the sample that drives chemical reactions toward a detectable deflagration. Those materials

Theoretical and Computational Chemistry, Volume 22
https://doi.org/10.1016/B978-0-12-822971-2.00007-3

that require higher temperatures to start chemical reactions are considered to be less sensitive. The mechanisms through which heat is generated in the samples are thought to include friction and fluid flow, plastic flow, fracture, and adiabatic cavity collapse [7–11].

We and others are designing new energetic molecules with the aim and hope that they may exceed the performance and handling safety of current energetic materials [12–18]. However, synthesizing and fully characterizing new molecules is a costly and time-consuming process, in part due to safety concerns during novel synthesis. The capability to reliably predict certain aspects of the handling safety of explosives based only on their molecular structure could guide explosive synthesis by identifying in silico those molecules with useful sensitivity. An analogous capability exists for predicting explosive performance metrics where the results of electronic structure-based calculations that estimate density and the heat of formation are entered into thermochemical codes such as Cheetah or machine-learning methods to estimate detonation velocity [19–21].

To date, it has not been possible to predict the handling safety of a new explosive without resorting to empirical interpolation over the properties of similar molecules [22]. Numerous authors have discovered correlations between various properties of related explosive molecules and their handling safety. These properties include oxygen balance, bond charge densities, hydrogen bonding, detonation velocity, and steric hindrance within the crystal structure [12,23–31]. Interpolation schemes based on artificial neural networks or other machine-learning methods that use nonlinear combinations of multiple descriptors have been developed over the last few decades and show impressive performance [32–34]. Unfortunately, such formalisms are rather opaque and provide little physical insight into the origin of explosive sensitivity.

Wenograd, in the early 1960s, published a somewhat overlooked paper on the connection between the kinetics of time to explosion under elevated temperatures and the drop weight sensitivity, H_{50}, of a diverse set of explosives [35]. Wenograd found an approximate linear correlation between the logarithm of H_{50} and the critical temperature required for the explosive to react within the 250 μs duration of the drop weight test. Those molecules with a high critical temperature, that is, those which require high temperatures to transition to a deflagration, have a relatively large value of H_{50}. The critical temperature for each explosive was computed from Arrhenius kinetics parameterized to time to explosion experiments. Wenograd's study implies that the drop height test initiates reactions through a thermal mechanism and that the intrinsic chemical dynamics of the explosives have a significant influence on handling sensitivity [36–38].

The correlation between chemical kinetics and drop weight impact sensitivity that was first revealed by Wenograd suggests that other physical properties of explosives, such as phase, crystal size, porosity, or crystal structure, play secondary roles in determining handling sensitivity. Our recent experimental work, in addition to existing literature data, generally supports this view. For instance, drop weight impact tests on a set of RDX lots that were engineered for reduced shock sensitivity as part of the reduced sensitivity RDX round robin showed no statistical difference in H_{50} values with respect to regular RDX varieties [5]. Hence, the properties of explosive crystals that promote shock insensitivity such as the minimization of porosity or the rounding of corners, do not affect drop weight sensitivity. Similarly, Manner et al. developed a series of TNT lots by recrystallization that had a range of well-defined particle sizes and shapes [39]. Drop weight impact tests on these lots revealed H_{50} values that were essentially independent of particle size and shape. Manner et al. also synthesized a series of explosive molecules based on substituted pentaerythritol tetranitrate (PETN) [12] and erythritol tetranitrate (ETN) [13]. The PETN derivatives had one nitrate ester (ONO_2) group substituted with a non-energetic moiety, including -NH_2, -H, -CH_3, -PO, or halide salts whereas the ETN derivates had some or all of nitrate ester groups substituted by other energetic moieties such as azides, N_3, or nitramines, $NHNO_2$. With the exception of the halide salt derivative, the impact sensitivities of the substituted PETN derivatives did not change markedly from those of PETN despite the fact that several of the derivates were liquids. Hence, although the crystal structures, densities, and phases of the molecules differed significantly from those of PETN, their H_{50} values remained PETN-like, likely due to the fact that all contained the nitrate ester as the energetic functional group. In contrast, the substitution of energetic moieties in ETN resulted in significant changes to the measured H_{50} values. The substitution of nitrate esters for azides led to very sensitive, azide-like responses and the substitution of nitramine groups gave rise to impact sensitivities in the vicinity of those of the cyclic nitramines cyclotrimethylene trinitramine (RDX) and cyclotetramethylene tetranitramine (HMX). Taken together, these systematic studies of particle shape, size, and chemistry support the view that the energetic functional groups in explosive molecules have a dominant effect on drop weight impact sensitivity, that is, molecules with similar underlying chemistry have similar drop weight sensitivities. It is important to emphasize that drop weight impact tests are just one measure of explosive sensitivity. For instance, the shock sensitivity of secondary explosives is highly dependent on porosity and even crystal orientation, while drop weight tests are evidently not as sensitive to these properties.

The correlation between chemical kinetics and drop weight sensitivity observed experimentally provides a path toward a noninterpolative, parameter-free method for predicting the handling safety of any energetic molecule, that is, if we can reliably predict the Arrhenius kinetics of the time to explosion for explosives, we can infer their relative impact sensitivity. We demonstrated in a recent publication that condensed-phase quantum molecular dynamics simulations are now capable of predicting the kinetics of time to explosion for a variety of secondary explosives containing only carbon, hydrogen, nitrogen, and oxygen [40] and that the critical temperatures obtained from these rates exhibit the same dependence on $log(H_{50})$ as seen by Wenograd [37]. Our theoretical studies, which provide the first independent confirmation of Wenograd's results, are reviewed in Section 4.

Condensed-phase quantum-based molecular dynamics simulations are computationally expensive, even when parameterized, semi-empirical models for the interatomic forces are used as in Refs. [40–44]. The computational expense of the simulations is compounded by the need to perform multiple independent simulations at each initial temperature to obtain a statistically meaningful distribution of the times to explosion. Hence, months of computational time are typically required to evaluate the Arrhenius kinetics of each explosive. We have sought to develop a more computationally efficient scheme for ranking explosive sensitivity that is still based on the concept of the critical temperatures required for the prompt reaction. Careful analyses of our large, condensed-phase quantum molecular dynamics simulations showed that the runaway exothermic reactions that characterize a thermal explosion are a consequence of the very first bond-breaking event in the system, which can be thought of as a trigger linkage [31,38]. The initial chemical reactions seed a cascade of reactions by generating highly reactive intermediate species that induce the decomposition of neighboring molecules, which in turn yield additional reactive intermediates. The first reactions are preceded by relatively long incubation periods and the distribution of the times to the first reactions is consistent with Arrhenius kinetics. Since the onset of thermal explosion in molecular dynamics simulations follows naturally from the first bond-breaking event there is potentially no need to run our simulations out to the thermal explosion stage. More importantly, because the first bond-breaking event occurs on a single molecule within the larger periodic supercell, we can instead compute the rates of the first bond-breaking events for molecules in the gas phase. Simulations of individual molecules in the gas phase, albeit with many independent trajectories to obtain adequate statistics, is orders of magnitude faster than the corresponding simulations in periodic supercells because of the $O(N^3)$ scaling of the computational time with respect the number of atoms, N. Despite the vast simplifications that we have made to real explosive systems, we show that the critical temperatures derived from reaction rates fitted to the results of molecular dynamics simulation of single molecules in the gas phase are positively correlated with drop weight impact sensitivity.

2 Arrhenius kinetics of time to explosion

Numerous authors [35,45–48] have shown that the time, τ, required for an explosive to undergo a violent deflagration for a given initial temperature, T, is consistent with Arrhenius kinetics,

$$\kappa = 1/\tau = Aexp(-E_a/k_B T) \tag{14.1}$$

where κ is the reaction rate, A the pre-exponential factor, E_a the activation energy, and k_B the Boltzmann constant. Owing to a large number of individual reactions and short-lived intermediates that precede and give rise to a thermal explosion, the activation energy, E_a, is an effective barrier that captures all of the underlying chemistry. For explosive chemistry, it is more accurate to use a pressure-dependent activation enthalpy, H_a, rather than pressure-independent activation energy because reaction rates are known to depend sensitively on pressure [40]. However, because drop weight testing does not generate the large volumetric compression or pressures that are typically seen in shock experiments, it is a reasonable approximation to neglect any pressure dependences [30].

Wenograd measured the time to explosion for samples of trinitrotoluene (TNT), trinitrobenzene (TNB), dinitropropyl trinitrobutyrate (DNPTB), *N*-methyl trinitrophenylnitramine (tetryl), trinitroethyl trinitrobutyrate (TNETB), nitroglycerin (NG), dinitroxyethyl nitramine (DINA), bistrinitroethyl nitramine (BTNEN), and PETN confined within resistively heated hypodermic needles as a function of temperature. From the pre-exponential factors and activation energies fitted to these rate data, the critical temperatures,

$$T = \frac{-E_a}{k_B \ln(1/A\tau)} \tag{14.2}$$

were obtained for a thermal explosion to occur within the $\tau = 250\mu s$ duration of the impact event in a drop weight test. As depicted in Fig. 14.1, there is an approximate linear correlation between $log(H_{50})$ and T, which implies a strong connection between chemical rates and drop

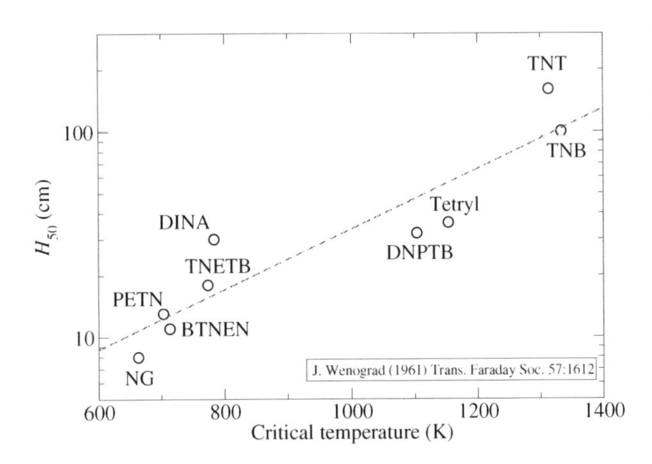

FIG. 14.1 The dependence of impact sensitivity H_{50} for the explosives NG, PETN, DINA, TNETB, BTNEN, DNPTB, tetryl, TNB, and TNT vs the critical temperatures required for explosions to occur within 250 μs obtained from Arrhenius rates. The sensitivity and critical temperatures were obtained from Table 1 of Ref. [35]. The broken line depicts a linear regression to the data and highlights the trend that higher critical temperatures correlated with lower impact sensitivity.

weight impact sensitivity. Wenograd used values for H_{50} obtained from the Type 12 Explosive Research Laboratory apparatus with a 2.5 kg drop weight and grit paper on the anvil. As noted earlier, different versions of the drop weight test can give significantly different results so it is important that the drop weight sensitivities are measured consistently.

3 Quantum molecular dynamics formalism

Reactive molecular dynamics simulations are becoming a popular tool for the study of explosive chemistry. Simulations are used routinely to develop insights into the reaction mechanisms in explosives as they transition from reactants to products through a sequence of short-lived intermediate species [49–51]. Perhaps the primary reason that molecular dynamics simulations play such a prominent role in explosive research is that it has been extremely difficult to interrogate explosives with sufficient chemical and temporal resolution during deflagration or detonation. Time-resolved ultrafast spectroscopic measurements have provided a probe into chemical changes on sub-ns time scales, but supporting atomistic simulations are still needed to interpret the measured spectra [44,52,53].

Reactive molecular dynamics simulations of explosives require a model of the interatomic forces that can accurately describe the making and breaking of covalent bonds. For secondary explosives containing C, H, N, and O, this means that the interatomic potential should be able to accurately describe most of organic chemistry and the associated reaction barriers. While *ab initio* electronic structure methods, such as density functional theory, provide a suitably accurate and reliable framework for describing interatomic forces and reactions barriers in organic materials at high temperatures, their computational cost prohibits most applications to explosive chemistry. Condensed-phase molecular dynamics simulations of energetic materials require periodic systems containing hundreds to thousands of atoms and time scales above 100s of picoseconds. If simulations are limited to smaller systems or shorter time scales then the results may lose accuracy because of self-interaction with periodic images, temperature fluctuations that are inconsistent with canonical dynamics [40], or that reactions occur before the systems are properly thermalized. Hence, although density functional theory is widely considered to provide the gold standard in terms of accuracy for condensed phases, it is not yet a practical tool for computing chemical rates based on time to explosion. As a result, empirical interatomic potentials and semi-empirical electronic structure methods that approach density functional theory in terms of accuracy but are computationally faster are used widely. We have combined semi-empirical density functional tight-binding theory [54,55] with Niklasson's extended Lagrangian Born–Oppenheimer molecular dynamics [56–58] to perform quantum-based molecular dynamics simulation in both the gas and condensed phases.

Molecular dynamics simulations naturally provide an atomic-scale view of reactions in organic materials. Detailed analyses of the trajectories are not required to identify when runaway exothermic chemistry has occurred—we can obtain this information by simply monitoring the temperature provided that the simulations are performed in the microcanonical (constant number of particles, volume, and energy) ensemble. Similarly, our analyses of the gas-phase molecular stability requires knowing only that a covalent bond has broken and not the identity of the bond or the resulting fragments. The simplicity of the analyses that are required to obtain rates from the otherwise complex quantum molecular dynamics simulations allow these computational tools to be applied to essentially any organic explosive without significant investments in post-processing.

3.1 Density functional tight-binding theory

The gas- and condensed-phase quantum molecular dynamics simulations reported here use density functional tight-binding (DFTB) theory with the *lanl*31 parameterization to describe interatomic forces and binding energies in explosives containing C, H, N, and O [59]. Density functional tight-binding theory can be derived from an expansion of the total energy in Kohn-Sham density functional theory to second order in fluctuations in the charge about neutral-atom charge densities [54,60,61]. It provides a physically intuitive picture of the formation of covalent bonds between atom-centered atomic orbitals and charge transfer between species with different electronegativity.

The potential energy in DFTB theory is.

$$u = 2Tr((P - P_0)H_0) + \sum_{i=1, j\neq i}^{N} \frac{1}{2} q_i q_j \gamma_{ij} + E_{pair} \tag{14.3}$$

where $Tr(X)$ denotes the trace of matrix X, P is the self-consistent density matrix, P_0 the density matrix for neutral noninteracting atoms, H_0 the charge-independent, two center Slater–Koster tight-binding Hamiltonian, q_i the Mulliken partial charge on atom i, γ_{ij} a screened Coulomb potential, N the number of atoms, and E_{pair} a sum of atom-centered pair potentials that provide strong repulsion at short interatomic distances. The self-consistent density matrix, P, is computed from the charge-dependent DFTB Hamiltonian,

$$H = H_0 + H_1 \tag{14.4}$$

where,

$$(H_1)_{i\alpha,j\beta} = \frac{1}{2} S_{i\alpha,j\beta}(V_i + V_j). \tag{14.5}$$

Here, i and j label atoms, α and β label orbitals, $S_{i\alpha,\, j\beta}$ are elements of the overlap matrix, and $\{V\}$ are the electrostatic potentials and Hubbard U terms arising from the set of Mulliken partial charges, $\{x\}$,

$$V_i = Uq_i + \sum\nolimits_{k \neq i=1}^{N} q_k \gamma_{ik}. \tag{14.6}$$

The Mulliken charges are computed from the density matrix, P,

$$q_i = \sum\nolimits_{\alpha \in i} \left(P_{i\alpha,j\beta} S_{j\beta,i\alpha} + S_{i\alpha,j\beta} P_{j\beta,i\alpha}\right) - n_i^e, \tag{14.7}$$

where n_i^e is the number of valence electrons on neutral atom i. The total energy and forces are solved self-consistently because the DFTB Hamiltonian depends on the Mulliken charges obtained from the density matrix. Short-range repulsion is provided by a sum of pair potentials, Φ,

$$E_{pair} = \frac{1}{2} \sum\nolimits_{i=1}^{N} \sum\nolimits_{j \neq i=1}^{N} \Phi(R_{ij}) \tag{14.8}$$

where R_{ij} is the distance between atoms i and j.

Density functional tight-binding theory is semi-empirical, meaning that although it is derived from *ab initio* electronic structure theory, it contains terms that are approximated and parameterized to speed up the calculations with respect to parameter-free first-principles methods. We have used the *lanl*31 DFTB parameterization of the bond and overlap integrals, on-site energies, Hubbard *U*s, and repulsive potentials, Φ, that was developed via multiparameter optimization to high-quality density functional theory calculations on a set of small organic molecules. The *lanl*31 DFTB parameterization was shown to exhibit outstanding transferability to the binding energies and geometries of gas-phase organic molecules. It also yields very good predictions for the Hugoniot loci of organic materials and explosives despite using no condensed-phase properties in its construction.

3.2 Extended Lagrangian Born–Oppenheimer molecular dynamics

In Born–Oppenheimer molecular dynamics the interatomic forces that are used to propagate the atomic positions are evaluated at the self-consistent electronic ground state [62]. This requirement can be computationally expensive because multiple self-consistent field (SCF) iterations might be required to obtain the self-consistent density matrix, P, at each time step. The time required for the calculation of the density matrix using standard algorithms, such as matrix diagonalization, scale with the cube of the number of atoms, $O(N^3)$, which makes it imperative that the number of SCF iterations per time step is minimized. To improve the rate of SCF convergence it has been customary to extrapolate the electronic degrees of freedom, such as the charge density or Mulliken charges, from previous time steps. While such schemes help SCF convergence, it was discovered that they also lead to an unphysical systematic drift in the total (kinetic plus potential), energy during microcanonical dynamics [63,64].

The drift of the total energy causes a drift in the temperature that is often addressed by the addition of a thermostat or by tightening the tolerance on the SCF procedure. Neither of these solutions are desirable since they affect the dynamics or incur an additional computational expense. The extended Lagrangian Born–Oppenheimer molecular dynamics formalism, which has been developed and advanced by Niklasson and coworkers [56–58], allows the number of SCF cycles per time step to be minimized while simultaneously removing the systematic drift in the total energy during microcanonical dynamics. The invention of extended Lagrangian Born–Oppenheimer molecular dynamics has been pivotal to our ability to simulate thermal explosions using accurate, semi-empirical electronic structure methods [12,37,65].

The systematic drift in the total energy in regular Born–Oppenheimer molecular dynamics is caused by the ad hoc propagation of electronic degrees of freedom from one-time step to the next to reduce the number of SCF cycles per time step. These propagation schemes, combined with the highly nonlinear SCF solvers, lead to a broken time-reversal symmetry in the dynamics which appears as a systematic energy drift. The extended Lagrangian Born–Oppenheimer molecular dynamics formalism removes the systematic energy drift by restoring time-reversal symmetry to the propagation of the electronic degrees of freedom. The formalism restores time-reversal symmetry by introducing a set of auxiliary electronic degrees of freedom that are constrained to evolve in harmonic potentials centered on the self-consistent electronic degrees ground state. In the case of DFTB molecular dynamics, auxiliary degrees of freedom are introduced for the Mulliken partial charges and chemical potential [66,67].

The Lagrangian that governs the equations of motion for Born–Oppenheimer dynamics is

$$L(R,\dot{R}) = \frac{1}{2}\sum_{i=1}^{N} m_i \dot{R}_i^2 - u(R,P) \tag{14.9}$$

where $R = \{R_i\}$ are the atomic positions, m_i the mass of atom i, and $u(R,P)$ the potential energy evaluated at the self-consistent electronic ground state, P. The dots denote time derivatives. The extended Lagrangian, L^X, which includes the auxiliary degrees of freedom, n, with mass μ and frequency parameter ω is,

$$L^X(R,\dot{R},n,\dot{n}) = L + \frac{\mu}{2}\sum_{i=1}^{N} \dot{n}_i^2 - \frac{\mu\omega^2}{2}\sum_{i=1}^{N} (q_i - n_i)^2 \tag{14.10}$$

where q_i is the self-consistent Mulliken charge on atom i. We can derive the equations of motion for the atomic positions and auxiliary degrees of freedom from the extended Lagrangian, where in the limit $\mu \to 0$,

$$m_i \ddot{R}_i = \frac{-\partial u(R,P)}{\partial R_i} \tag{14.11}$$

and

$$\ddot{n}_i = \omega^2 (q_i - n_i). \tag{14.12}$$

The first equation of motion is that of standard Born–Oppenheimer molecular dynamics where using forces evaluated at the self-consistent electronic ground state. The second gives the equation of motion for the auxiliary degrees of freedom, which can be integrated using a time-reversible integrator like the atomic degrees of freedom. Because the auxiliary degrees of freedom can be propagated using a time-reversible integrator and because they are constrained to remain close to the self-consistent ground state, they make excellent guesses for the SCF procedure at each time step and restore time reversibility to the propagation of the electronic degrees of freedom. Hence, the extended Lagrangian Born–Oppenheimer molecular dynamics formalism allows us to greatly reduce the number of SCF cycles per time step while simultaneously removing the systematic drift in the total energy seen in regular Born–Oppenheimer molecular dynamics.

The extended Lagrangian Born–Oppenheimer molecular dynamics formalism enables computationally cheap, stable, and precise microcanonical trajectories. The ability to perform accurate microcanonical dynamics is crucial for simulating exothermic chemistry because the use of a thermostat would suppress the temperature changes that are the very signature of endo- or exothermic chemistry that we seek. It was also shown that the accurate underlying microcanonical dynamics that this scheme provides are required to give the correct fluctuations when performing simulations in the canonical or isothermal-isobaric ensembles [68]. The extended Lagrangian Born–Oppenheimer molecular dynamics scheme was applied to generate the microcanonical trajectories used to generate time to explosion data in Section 4.1 and in the canonical trajectories used to study the rate of the first bond-breaking events for gas-phase explosive molecules in Section 4.2.

4 Simulations of explosive chemistry and Arrhenius kinetics

4.1 Time to explosion for condensed-phase explosives

Quantum molecular dynamics simulations of explosive chemistry were performed in the microcanonical ensemble using DFTB theory with the *lanl*31 parameterization and the extended Lagrangian Born–Oppenheimer molecular dynamics scheme. Eight explosives, which are depicted in Fig. 14.2, from four distinct families were studied (Fig. 14.2): 2,4,6-trinitrotoluene (TNT), 3,3′-diamino-4,4′-axoxyfurazan (DAAF), 1,3,5-trinitro-1,3,5-triazinane (RDX), 1,3,5,7-tetranitro-1,3,5,7-tetrazoctane (HMX), erythritol tetranitrate (ETN) and its stereoisomer L-ETN [69], pentaerythritol tetranitrate (PETN), and erythritol tetranitramine (ETNA). These explosives include representatives from four distinct families (nitrobenzene, axoxyfurazan, nitramine, and nitrate ester) and impact sensitivities from $H_{50}=5.4$ to 293 cm using the ERL apparatus with the Type 12 tool (Table 14.1).

With the exception of L-ETN, which is a liquid at room temperature, all of the simulations were initialized from the corresponding published crystal structures under ambient conditions. Periodic supercells were created for most systems in order to minimize interactions between periodic images. Owing to the speed and $O(N^3)$ scaling of DFTB calculations and the requirement to perform multiple simulations over 100s picoseconds, we limited the system sizes to a few hundred atoms. Along each principal axis, the supercells measured $1\times2\times1$ for TNT and ETN, $2\times2\times1$ for ETNA, $1\times1\times2$ for PETN, $2\times1\times2$ for HMX, and $2\times2\times2$ for DAAF. Short isothermal-isobaric simulations were performed on each structure to obtain structures consistent with the target temperatures and pressures for each set of reactive molecular dynamics trajectories. Following the estimates of the pressures accessible by drop

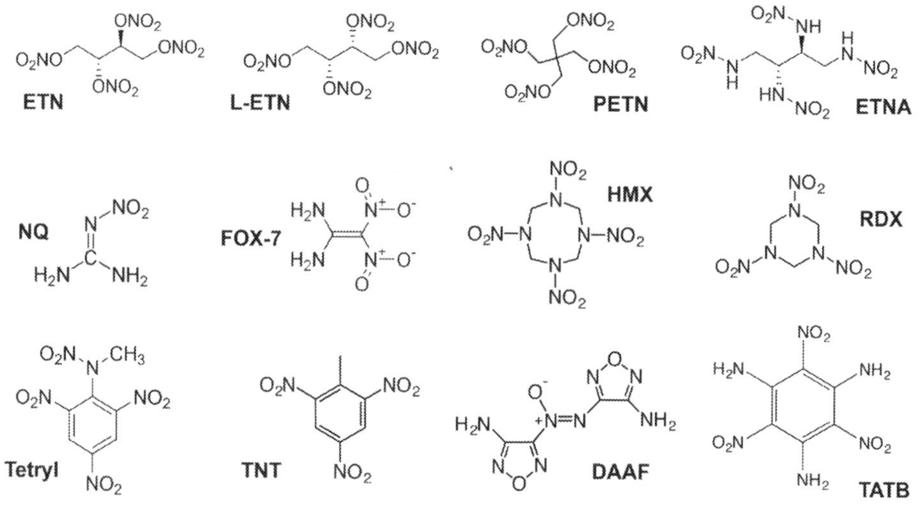

FIG. 14.2 Structures of ETN, L-ETN, PETN, ETNA, NQ, FOX-7, HMX, RDX, Tetryl, TNT, DAAF, and TATB.

TABLE 14.1 Drop weight impact sensitivities, H_{50}, in cm for ETN, L-ETN, PETN, RDX, HMX, ETNA, DAAF, and TNT measured using the ERL apparatus with Type 12 and Type 12 B tools.

	H_{50}(cm)	
	Type 12 (grit)	**Type 12 B (no grit)**
ETN	5.4	14.7
L-ETN	—	3.0
PETN	10.2	20.8
RDX	19.1	17.4
HMX	32.1	19.8
ETNA	26.7	52.5
DAAF	292.6	314.5
TNT	228.1	123.5

weight impact tests suggested by Storm, we thermalized all of the systems to pressures of 0.5 or 1 GPa [30]. We created a liquid-like state for L-ETN by placing eight molecules at random positions in a large, cubic supercell such that no molecules overlapped, followed by an isothermal-isobaric simulation with a target temperature 1000 K and pressure of 1.0 GPa. The combination of high temperature and random starting positions for the L-ETN molecules gave a structure with no long-range order.

The simulations used a time step of either 0.125 or 0.25 fs for the integration of the equations of motion. The density matrix was calculated using a finite electronic temperature corresponding to 0.25 eV to smear the occupancies in the vicinity of the chemical potential to greatly improve numerical stability during the reactive stage of the simulations. We used both a linear SCF mixing algorithm with a small mixing coefficient with the 0.125 fs time step or the DIIS algorithm of Pulay [70,71] with the larger 0.25 fs time step. Both algorithms yielded stable trajectories but the DIIS algorithm proved to be slightly faster overall.

The time to explosion for small systems containing only a handful of molecules is inherently stochastic [40,72–74]. For the effective single-step model considered here, where all of the underlying chemistry is captured by a single activation energy, E_a, and pre-exponential factor, A, the system spends a significant amount of time exploring its initial (reactants) state until it finds an escape pathway to the thermal explosion stage. The probability distribution for the waiting time, $p(t)$, until the system finds a transition is,

$$p(t) = \kappa exp(-\kappa t), \tag{14.13}$$

where κ is the Arrhenius rate. Hence, if the rate, κ, is small because the thermal energy is small with respect to the energy barrier, the distribution of escape times can be broad [73]. As a result, sufficient independent trajectories are required so the distribution of escape times (time to explosion in our case) is sampled adequately. For the condensed-phase simulations, we found that 10 trajectories per initial temperature give a distribution of times to explosion that are clearly consistent with Arrhenius kinetics. Our gas-phase simulations, on account of the smaller system sizes, required considerably more independent trajectories at each temperature. Independent trajectories were created by reassigning the velocities of the atoms in each trajectory using a different random number seed. The trajectories were then allowed to decorrelate while being re-thermalized by computing a short trajectory with a simple velocity rescaling thermostat. Following thermalization, the thermostat was removed and the systems were allowed to evolve in the microcanonical ensemble. We used a velocity rescaling thermostat rather than a stochastic thermostat such as the Langevin thermostat because the former maintains much better control over the mean temperature. Stochastic thermostats give rise to large temperature fluctuations in small systems, which can result in the microcanonical trajectories being launched at temperatures that differ significantly from the target values.

Our reactive molecular dynamics simulations of explosive chemistry exhibit at least four stages, (i) incubation, (ii) endothermic or thermal neutral chemistry, (iii) thermal runaway, and (iv) thermal neutral product chemistry, which is illustrated in Fig. 14.3 using an L-ETN trajectory that was initialized at $T = 1000$ K and a pressure of 1 GPa. Fig. 14.3A depicts the evolution of the kinetic temperature over the nearly 250 ps duration of the simulation. The temperature is constant for the first 105 ps during the incubation stage until the first bond-breaking event occurs, as denoted by the closure of the highest occupied molecular orbital—lowest unoccupied molecular orbital (HOMO-LUMO) gap (Fig. 14.3C). A small increase in the temperature occurs at about 112 ps, but the temperature remains essentially constant until about 180 ps while a series of reactions occur. The evidence for continued reactions despite no change in the global temperature can be seen in the evolution of the HOMO-LUMO gap. The time between the first bond-breaking event and thermal runaway corresponds to stage ii. At 180 ps the temperature and pressure, as depicted in Fig. 14.3B, begin to increase precipitously. This thermal runaway arises from the cascade of strongly exothermic

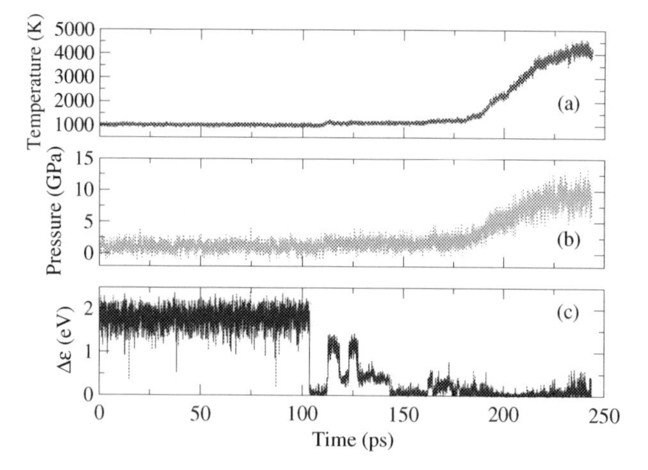

FIG. 14.3 Evolution of the (A) kinetic temperature, (B) pressure, and (C) the HOMO-LUMO gap, $\Delta\epsilon$, during a DFTB molecular dynamics simulation of liquid L-ETN. The initial temperature and pressure were 1000 K and 1 GPa, respectively.

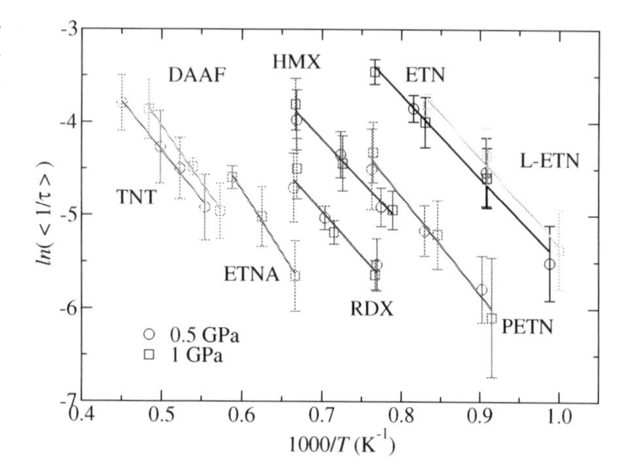

FIG. 14.4 Arrhenius plot derived from the time to explosion from condensed-phase DFTB molecular dynamics simulations for TNT, DAAF, ETNA, HMX, RDX, PETN, ETN, and liquid L-ETN.

chemistry that is characteristic of an explosion. After the completion of the exothermic chemistry at about 235 ps, the system enters stage iv, which comprises hot product species that continue to react but do not lead to further increases in the temperature or pressure. All of the explosives that we have studied using DFTB-based molecular dynamics exhibit similar behavior and evolve through the same four stages. In this work, we define the time to explosion, τ, of each trajectory as the time at which the temperature is mid-way between its initial value and the final value once the system has entered stage iv [37].

The rates derived from the time to explosion data, $\kappa = <1/\tau>$, are presented in an Arrhenius plot in Fig. 14.4. Here, $ln(\kappa)$ is plotted against the inverse temperature, which gives a straight line with gradient $-E_a/k_B$ and intercept $ln(A)$, provided that the rates are consistent with Arrhenius kinetics. The error bars on each data point are equal to $\sigma(1/\tau)/<1/\tau>$, where $\sigma(1/\tau)$ denotes the standard deviation of the inverse of the times to explosion [75]. Linear regressions to the rate data for each explosive are plotted as solid lines. Fig. 14.4 shows that the

TABLE 14.2 Activation energy, E_a, pre-exponential factor, A, for the Arrhenius rates derived from DFTB molecular dynamics simulations.

	E_a(eV)	$A(ps^{-1})$	T(K)
ETN	0.77	30.8	1059
L-ETN	0.81	59.3	1033
PETN	0.91	37.6	1222
HMX	0.78	8.51	1266
RDX	0.83	5.76	1424
ETNA	1.17	30.7	1610
TNT	0.91	2.69	1760
DAAF	1.05	7.65	1729

The critical temperatures, T, required for explosion within 150 ps are calculated from Arrhenius rates using Eq. (14.2).

dependence of the time to explosion on temperature is consistent with Arrhenius kinetics. The activation energies and pre-exponential factors obtained from linear regressions to these data are presented in Table 14.2.

The activation energies, E_a, presented in Table 14.2 for the eight explosives do not differ significantly from one another. Fig. 14.4 shows that the slopes of the rate data on the Arrhenius plot are similar, which denotes similar activation energies, and the off-sets of the rates from low to high-temperature result from the changes in the pre-exponential factors (the intercepts). Interestingly, there is apparently little correlation between the activation energies and sensitivity. For instance, PETN ($H_{50}=10.2$ cm) and TNT ($H_{50}=228$ cm) both have $E_a=0.91$ eV, and ETNA, with $H_{50}=26.7$ cm has larger activation energy than DAAF with $H_{50}=293$ cm. In contrast, the pre-exponential factors, A, vary significantly between the explosives and show a good correlation with impact sensitivity. The two explosives with the lowest impact sensitivity, TNT and DAAF, have the smallest pre-exponential factors, with $A=2.69$ and 7.65 ps^{-1}, respectively, while the most sensitive explosives, L-ETN, ETN, and PETN have the highest values. The correlations between activation energies, pre-exponential factors, and impact sensitivities can be seen clearly in Fig. 14.4—the explosives with low impact sensitivity lie on the left of the Arrhenius plot, meaning they require high temperatures to transition to a thermal explosion, while the more sensitive explosives are found on the right of the plot because they will undergo thermal explosions on the time scales accessed by our simulations at relatively low temperatures. The trends revealed by the data presented in Fig. 14.4 and Table 14.2 are consistent with Wenograd's results, that is, the explosives studied in Ref. [35] had similar activation energies but significantly different pre-exponential factors.

Following Wenograd, we have used the Arrhenius rates presented in Table 14.2 to estimate the critical temperatures, T, required for thermal explosions. As in Ref. [35], we have set the time for the explosion $\tau=150$ps rather than the 250 μs used by Wenograd because the former lies within the time scale of our molecular dynamics simulations and avoids extrapolation over orders of magnitude in time. We present in Fig. 14.5 the dependence between experimentally measured drop heights (Table 14.1) on the critical temperatures estimated from

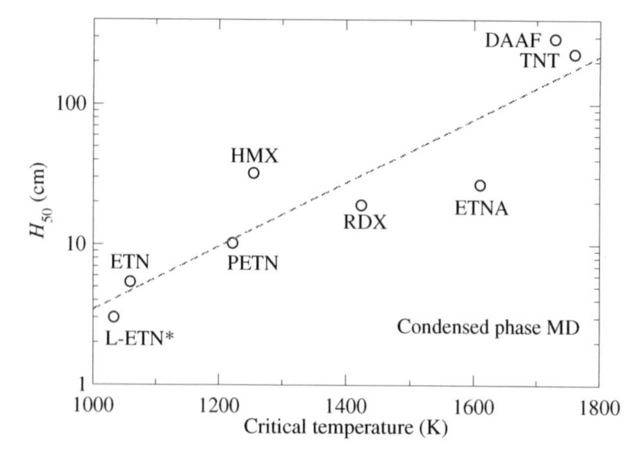

FIG. 14.5 Dependence of the measured drop height impact sensitivity of L-ETN, ETN, PETN, RDX, HMX, ETNA, DAAF, and TNT on the critical temperature required for explosion in 150 ps. All H_{50} values were measured using the Type 12 tool, except for liquid L-ETN which used a bare anvil. The broken line is a linear regression to the data.

the Arrhenius kinetics. Fig. 14.5 uses drop heights measured using the Type 12 tool with grit paper for the seven solid explosives but uses the value of H_{50} measured using the Type 12B tool without grit paper for liquid L-ETN. Fig. 14.5 reveals that $log(H_{50})$ depends linearly on T, in excellent accord with Wenograd's data (Fig. 14.1) and our earlier publication. Our DFTB-based molecular dynamics simulations predict the ETN derivative ETNA to be less reactive and less sensitive than indicated by the measured H_{50}, which suggests that ETNA is of similar sensitivity to the nitramines RDX and HMX. Nevertheless, the predicted rates show that L-ETN is slightly more sensitive than ETN, though it is not yet possible to attribute this result to intrinsic differences in the properties of the stereoisomers or to the disordered, liquid-like state used to simulate L-ETN chemistry. Experimentally, molten ETN exhibits similar sensitivity to L-ETN, though this is likely due to the initiation mechanisms involved in impact tests with liquids [69,76].

4.2 Time to first bond-breaking event in gas-phase explosive molecules

The cascade of reactions that give rise to a thermal explosion starts with a single bond-breaking event. The breaking of the first bond in our condensed-phase simulations signals the onset of stage ii, that is, a period of endothermic or thermal neutral chemistry. This is illustrated in Fig. 14.3C by the closure of the HOMO-LUMO gap at about 105 ps. While we need multiple molecules and periodic boundary conditions to capture the chemistry that leads up the exothermic runaway (stage iii), we propose that the first bond-breaking event is mainly a characteristic property of the individual molecules that depends on the various energetic barriers to bond scission. Furthermore, those molecules with higher barriers to bond scission can be expected to be less sensitive than those with smaller barriers because in the former greater thermal activation will be required to start stage ii.

The energy barriers for bond-breaking in gas-phase explosive molecules have been studied in detail by *ab initio* electronic structure methods [77–79]. These calculations, while very accurate, are expensive and tend to be limited to a few of the most likely or sensible reaction mechanisms. Reactive molecular dynamics simulations instead probe the entire potential energy surface without bias and sample all possible escape paths (bond-breaking events) with the appropriate rates. We have performed gas-phase DFTB-based molecular dynamics simulations of explosive molecules to evaluate the kinetics of the first bond-breaking events. The gas-phase simulations were performed in the canonical ensemble using the implementation of the Langevin thermostat of Martinez et al. [68] and with an electronic temperature of 0.2 eV but with otherwise the same methods as were used for the condensed-phase simulations. Bond-breaking events were identified in the trajectories using an algorithm that counts molecules by bond connectivity. We performed 50 independent trajectories on each molecule at each temperature in order to obtain a statistically meaningful distribution of the time required for the first bond-breaking events by using different random number seeds as input to the Langevin thermostat.

The temperature dependence of the rate of the first bond-breaking events is presented in an Arrhenius plot in Fig. 14.6 for the explosive molecules ETN, PETN, RDX, HMX, ETNA, TNT, DAAF, TATB (triamino trinitrobenzene), FOX-7 (1,1-diamino-2,2-dinitroethene), Tetryl (2,4,6-trinitrophenylmethylnitramine), and NQ (nitroguanidine). We have normalized the rates by the number of energetic groups per molecule, N_x. The rates were normalized to take into account that the first bond-breaking event should occur N_x times sooner in a large molecule that contains N_x energetic groups than in a small molecule that contains one otherwise identical energetic group. The number of energetic groups for each molecule was set equal to the number of nitros, nitramines, nitrates, or azoxyfurazans. We used $N_x = 1$ for tetryl because analysis of the trajectories showed that the bond-breaking steps occurred at the nitramine rather than the three nitro groups. The activation energies and pre-exponential factors obtained from the Arrhenius plot are presented in Table 14.3.

The Arrhenius plot of the normalized rates for the first bond-breaking events, Fig. 14.6, for the 12 explosive molecules confirms that the results of the series of reactive molecular dynamics simulations are consistent with Arrhenius kinetics. As with the condensed-phase simulations of time to explosion, the Arrhenius plot shows relatively small differences between the activation energies across the set of molecules, but larger differences between the values of the pre-exponential factors. The data presented in Table 14.3 confirms this view. TNT and NQ are found to be outliers with unusually large activation energies (3.17 and 2.49 eV, respectively),

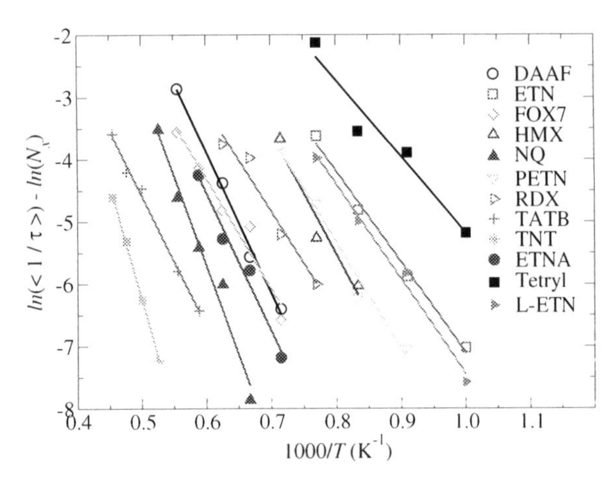

FIG. 14.6 Arrhenius plot for the rates at which the first bond-breaking events occur in DFTB-based molecular dynamics simulations for the explosive molecules ETN, L-ETN, PETN, RDX, HMX, ETNA, FOX-7, TNT, TATB, DAAF, NQ, and Tetryl. The rates are normalized by the number of energetic groups per molecule. The uncertainties on the rate data have been omitted for clarity.

TABLE 14.3 Activation energy, E_a, and pre-exponential factor, A, for the Arrhenius kinetics of the first bond-breaking event explosive molecules from DFTB-based molecular dynamics.

	E_a(eV)	A(ps^{-1})	T(K)
DAAF	1.96	17,489	1364
ETN	1.26	1774	1016
L-ETN	1.31	2395	1035
FOX-7	1.53	572	1339
HMX	1.69	26,160	1148
NQ	2.49	112,196	1559
PETN	1.48	4577	1120
RDX	1.44	928	1216
TATB	1.81	342	1648
TNT	3.17	188,905	1930
ETNA	1.91	6568	1412
Tetryl	1.06	1210	878

The critical temperatures, T, *for a bond to be broken in 1 ns are computed from the Arrhenius rates using Eq. (14.2).*

and pre-exponential factors (188,905 and 112,196 ps^{-1}, respectively) in comparison to the other molecules. The activation energies for the other nine molecules lie between 1 and 2 eV. Furthermore, the more sensitive explosives such as ETN, Tetryl, and RDX tend to have smaller activation energies than the less sensitive materials such as TATB, TNT, NQ, and DAAF.

The critical temperatures, T, required to break bonds in the explosive molecules in $\tau = 1$ ns have been computed using Eq. (14.2) and are presented in Table 14.3. The dependence of the drop height sensitivities on the critical temperatures, presented in Fig. 14.7, is consistent with the experimental results of Wenograd (Fig. 14.1) and our condensed-phase molecular dynamics simulations (Fig. 14.5) because higher critical temperatures are correlated with larger values of H_{50}, but with significantly greater scatter in the data. The scatter in the gas-phase results indicate that while the sensitivity of explosives depends in part on the thermal stability of the molecules, bimolecular reactions and/or the full spectrum of chemistry leading up a thermal explosion, which is captured in condensed-phase simulations, yields a more reliable predictive capability.

5 Discussion and conclusions

The handling safety of explosives is exceedingly complex owing to the variety of stimuli that can be applied to a material, both singly and in combination, in addition to the challenges associated with understanding which properties of the material control its response. Drop height sensitivity has been shown to exhibit correlations with stoichiometry, the strength

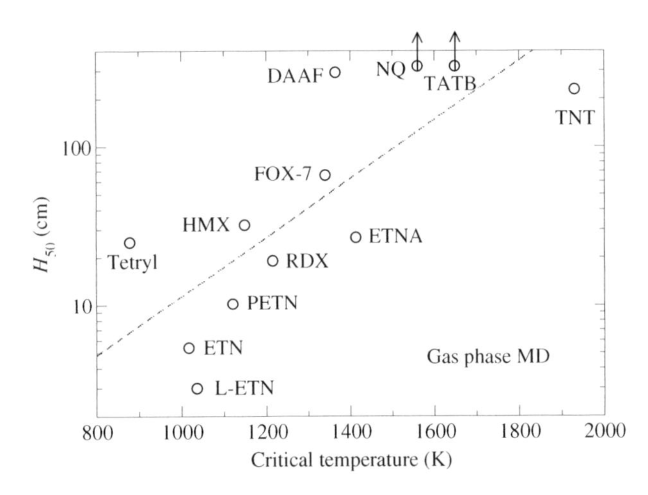

FIG. 14.7 Dependence of the measured drop height impact sensitivity of Tetryl, ETN, PETN, RDX, HMX, FOX-7, NQ, ETNA, DAAF, TATB, and TNT on the critical temperatures required for the first bond-breaking event to occur in $\tau = 1$ ns in the gas phase. The drop weight sensitivities were measured using the Type 12 tool except for FOX-7, which used the Type 12B tool. The arrows denote that the $H_{50}$for NQ and TATB are both in excess of 320 cm, the maximum possible drop height using the ERL apparatus. The broken line is linear regression to the data.

and characteristics of individual covalent bonds, intermolecular bonding, crystal structure symmetry, phase, elastic constants, and impurities, along with physical effects such as material strength, voids [12]. In this work, we have examined correlations between impact sensitivity and the underlying chemical dynamics of the energetic materials and molecules. Using the concept of the critical temperatures required for reactions proposed originally by Wenograd, the Arrhenius kinetics for reactions can provide a parameter-free, physics-based scheme for estimating the impact sensitivity of any organic secondary explosive. The critical temperatures obtained from the rates derived from time to explosion simulations exhibit a positive correlation with $log(H_{50})$. The scatter in these data (Fig. 14.5) is of a similar magnitude to that seen using experimentally-derived rates. On the other hand, while computationally cheaper, gas-phase molecular dynamics simulations of the time to the first bond-breaking event yield Arrhenius kinetics and critical temperatures that also correlate positively with $log(H_{50})$, the scatter in the data (Fig. 14.7) is notably larger.

Our results show that the handling sensitivity of explosives depends, in part, on the underlying chemical dynamics of the systems and that quantum molecular dynamics simulations can predict those dynamics with good accuracy. A physical interpretation of the observed trends is that less sensitive explosives have to be subjected to higher temperatures before runaway reactions can initiate and that those high temperatures arise from the kinetic energy of the drop weights. Nevertheless, despite these positive results, drop weight impact sensitivity also depends on numerous other factors which reactive molecular dynamics simulations cannot address directly. It is currently uncertain whether the Arrhenius rates that can be obtained from quantum molecular dynamics simulations have the fidelity to reliably distinguish the relatively small changes in drop height seen within families of molecules with similar underlying chemistry. For instance, the changes in drop height sensitivity seen in the series of PETN derivatives developed by Manner et al. may depend on the differences in crystal structure and compressibility rather than any change in the reaction rates. Systematic studies that combine synthesis, sensitivity testing, and simulations will be required to answer these open questions.

Acknowledgments

M.J.C. thanks Matt Holmes, Enrique Martinez, Anders Niklasson, Romain Perriot, and Kyle Ramos for numerous helpful discussions and Ed Sanville use of his algorithm for identifying molecular fragments. V.W.M. thanks Geoffrey W. Brown and Lisa M. Kay for impact test results and helpful discussions. This work was supported by the Laboratory Directed Research and Development program of Los Alamos National Laboratory under project number 20200234ER, Campaign 2 at Los Alamos National Laboratory, and used resources provided by the Los Alamos National Laboratory Institutional Computing Program, which are supported by the U.S. Department of Energy through the Los Alamos National Laboratory. Los Alamos National Laboratory is operated by Triad National Security, LLC, for the National Nuclear Security Administration of the U.S. Department of Energy (Contract No. 89233218CNA000001).

References

[1] US Department of Energy Standard on Explosive Safety, 2012 (DOE-STD-1212-2012).
[2] V.W. Manner, B.C. Tappan, B.L. Scott, D.N. Preston, G.W. Brown, Crystal structure, packing analysis, and structural-sensitivity correlations of Erythritol Tetranitrate, Crystal Growth & Design 14 (2014) 6154–6160.
[3] C.S. Coffey, V.F. Devost, Impact testing of explosives and propellants, Propellants Explosives Pyrotechnics 20 (1995) 105–115.
[4] S.M. Walley, J.E. Field, R.A. Biers, W.G. Proud, D.M. Williamson, A.P. Jardine, The use of glass anvils in drop-weight studies of energetic materials, Propellants Explosives Pyrotechnics 40 (2015) 351–365.
[5] R.M. Doherty, D.S. Watt, Relationship between RDX properties and sensitivity, Propellants Explosives Pyrotechnics 33 (2008) 4–13.
[6] G.W. Brown, M.M. Sandstrom, D.N. Preston, C.J. Pollard, K.F. Warner, D.N. Sorensen, D.L. Remmers, J.J. Philliips, T.J. Shelley, J.A. Reyes, P.C. Hsu, J.G. Reynolds, Statistical analysis of an inter-laboratory comparison of small-scale safety and thermal testing of RDX, Propellants, Explosives, Pyrotechnics 40 (2015) 221–232.
[7] F.P. Bowden, M.F.R. Mulcahy, R.G. Vines, A. Yoffe, The detonation of liquid explosives by gentle impact - the effect of minute gas spaces, Proceedings of the Royal Society of London Series A-Mathematical and Physical Sciences 188 (1947) 291–311.
[8] F.P. Bowden, H.T. Williams, Initiation and propagation of explosion in Azides and fulminates, Proceedings of the Royal Society of London Series A-Mathematical and Physical Sciences 208 (1951) 176.
[9] F.P. Bowden, A.D. Yoffe, Initiation and Growth of Explosion in Liquids and Solids, Cambridge University Press, 1952.
[10] J.E. Field, N.K. Bourne, S.J.P. Palmer, S.M. Walley, Hot-spot ignition mechanisms for explosives and propellants, Philosophical Transactions of the Royal Society A-Mathematical Physical and Engineering Sciences 339 (1992) 269.
[11] Y. MeBar, On the collapse of vapour cavities in a thin liquid layer under high pressure transient, Proceedings of the Royal Society A-Mathematical Physical and Engineering Sciences 452 (1996) 757–768.
[12] V.W. Manner, M.J. Cawkwell, E.M. Kober, T.W. Myers, G.W. Brown, H.Z. Tian, C.J. Snyder, R. Perriot, D.N. Preston, Examining the chemical and structural properties that influence the sensitivity of energetic nitrate esters, Chemical Science 9 (2018) 3649–3663.
[13] N. Lease, L.M. Kay, G.W. Brown, D.E. Chavez, D. Robbins, E.F.C. Byrd, G.H. Imler, D.A. Parrish, V.W. Manner, Synthesis of erythritol tetranitrate derivatives: Functional group tuning of explosive sensitivity, Journal of Organic Chemistry 85 (2020) 4619–4626.
[14] T.M. Klapotke, Energetic Materials Encyclopedia, De Gruyter, Berlin/Boston, 2018.
[15] E.C. Johnson, J.J. Sabatini, D.E. Chavez, L.A. Wells, J.E. Banning, R.C. Sausa, E.F.C. Byrd, J.A. Orlicki, Bis(Nitroxymethylisoxazolyl) Furoxan: A promising standalone melt-castable explosive, ChemPlusChem 85 (2020) 237–239.
[16] C.J. Snyder, T.W. Myers, G.H. Imler, D.E. Chavez, PArrish, D. A., Veauthier, J. M. and Scharff, R. J., Tetrazolyl Triazolotriazine: A new insensitive high explosive, Propellants, Explosives, Pyrotechnics 42 (2017) 238–242.
[17] L.M. Barton, J.T. Edwards, E.C. Johnson, E.J. Bukowski, R.C. Sausa, E.F.C. Byrd, J.A. Orlicki, J.J. Sabatini, P.S. Baran, Impact of stereo- and regiochemistry on energetic materials, Journal of the American Chemical Society 141 (2019) 12531–12535.

[18] T.W. Myers, J.A. Bjorgaard, K.E. Brown, D.E. Chavez, S.K. Hanson, R.J. Scharff, S. Tretiak, J.M. Veauthier, Energetic chromophores: Low-energy laser initiation in explosive Fe(ii) tetrazine complexes, Journal of the American Chemical Society 138 (2016) 4685–4692.
[19] K.O. Christe, D.A. Dixon, M. Vasiliu, R. Haiges, B. Hu, How energetic are Cyclo-Pentazolates? Propellants Explosives Pyrotechnics 44 (2019) 263–266.
[20] E.F.C. Byrd, B.M. Rice, Improved prediction of heats of formation of energetic materials using quantum mechanical calculations, Journal of Physical Chemistry A 110 (2006) 1005–1013.
[21] D.C. Elton, Z. Boukouvalas, M.S. Butrico, M.D. Fuge, P.W. Chung, Applying machine learning techniques to predict the properties of energetic materials, Scientific Reports 8 (2018) 9059.
[22] M.J. Kamlet, The relationship of impact sensitivity Wih structure of organic high explosives. I. Polynitroaliphatic explosives, in: Proceedings of the Sixth Symposium (International) on Detonation, 1976.
[23] P. Politzer, J.S. Murray, C-NO_2 dissociation-energies and surface electrostatic potential maxima in relation to the impact sensitivities of some nitroheterocyclic molecules, Molecular Physics 86 (1995) 251–255.
[24] P. Politzer, J.S. Murray, Relationships between dissociation energies and electrostatic potentials of C-NO_2 bonds: Applications to impact sensitivities, Journal of Molecular Structure 376 (1996) 419–424.
[25] P. Politzer, J.S. Murray, Impact sensitivity and crystal lattice compressibility/free space, Journal of Molecular Modeling 20 (2014) (2223).
[26] P. Politzer, J.S. Murray, Some molecular/crystalline factors that affect the sensitivities of energetic materials: Molecular surface electrostatic potentials, lattice free space and maximum heat of detonation per unit volume, Journal of Molecular Modeling 21 (2015) 2525.
[27] P. Politzer, J.S. Murray, Impact sensitivity and the maximum heat of detonation, Journal of Molecular Modeling 21 (2015) 262.
[28] D. Mathieu, Sensitivity of energetic materials: Theoretical relationships to detonation performance and molecular structure, Industrial & Engineering Chemistry Research 56 (2017) 8191–8201.
[29] D. Mathieu, T. Alaime, Impact sensitivities of energetic materials: Exploring the limitations of a model based only on structural formulas, Journal of Molecular Graphics & Modelling 62 (2015) 81–86.
[30] C.B. Storm, J.R. Stine, J.F. Kramer, Sensitivity relationships in energetic materials, Los Alamos National Laboratory LA-UR 89-2936 (1989).
[31] M.J. Kamlet, H.G. Adolph, Relationship of impact sensitivity with structure of organic high explosives. 2. Polynitroaromatic explosives, Propellants and Explosives 4 (1979) 30–34.
[32] H. Nefati, J.M. Cense, J.J. Legendre, Prediction of the impact sensitivity by neural networks, Journal of Chemical Information and Computer Sciences 36 (1996) 804–810.
[33] M.H. Keshavarz, M. Jaafari, Investigation of the various structure parameters for predicting impact sensitivity of energetic molecules via artificial neural network, Propellants Explosives Pyrotechnics 31 (2006) 216.
[34] R. Wang, J.C. Jiang, Y. Pan, Prediction of impact sensitivity of nonheterocyclic nitroenergetic compounds using genetic algorithm and artificial neural network, Journal of Energetic Materials 30 (2012) 135–155.
[35] J. Wenograd, The behaviour of explosives at very high temperatures, Transactions of the Faraday Society 57 (1961) 1612.
[36] D. Mathieu, T. Alaime, Predicting impact sensitivities of nitro compounds on the basis of a semi-empirical rate constant, Journal of Physical Chemistry A 118 (2014) 9720–9726.
[37] M.J. Cawkwell, V.W. Manner, Ranking the drop-weight impact sensitivity of common explosives using Arrhenius chemical rates computed from quantum molecular dynamics simulations, Journal of Physical Chemistry A 124 (2020) 74–81.
[38] D. Mathieu, Toward a physically based quantitative modeling of impact sensitivities, Journal of Physical Chemistry A 117 (2013) 2253–2259.
[39] V.W. Manner, C. Tiemann, L.M. Kay, N. Lease, M.J. Cawkwell, G.W. Brown, J.D. Yeager, S.P. Anthony, D. Montanari, Examining explosives handling sensitivity of trinitrotoluene (TNT) with different particle sizes, in: Proceedings of the 21st biennial confernce of the APS topic group on shock compression of condensed matter, 2019.
[40] R. Perriot, M.J. Cawkwell, E. Martinez, S.D. McGrane, Reaction rates in nitromethane under high pressure from density functional tight binding molecular dynamics simulations, Journal of Physical Chemistry A 124 (2020) 3314–3328.
[41] M.R. Manaa, E.J. Reed, L.E. Fried, N. Goldman, Nitrogen-rich heterocycles as reactivity retardants in shocked insensitive explosives, Journal of the American Chemical Society 131 (2009) 5483–5487.

[42] E.J. Reed, M.R. Manaa, L.E. Fried, K.R. Glaesemann, J.D. Joannopoulos, A transient semimetallic layer in detonating nitromethane, Nature Physics 4 (2008) 72–76.
[43] M.J. Cawkwell, A.M.N. Niklasson, D.M. Dattelbaum, Extended Lagrangian Born-Oppenheimer molecular dynamics simulations of the shock-induced chemistry of phenylacetylene, Journal of Chemical Physics 142 (2015) (064512064512).
[44] P. Bowlan, M. Powell, R. Perriot, E. Martinez, E.M. Kober, M.J. Cawkwell, S. McGrane, Probing ultrafast shock-induced chemistry in liquids using broad-band mid-infrared absorption spectroscopy, Journal of Chemical Physics 150 (2019) 204503.
[45] R.N. Rogers, Thermochemistry of explosives, Thermochimica Acta 11 (1975) 131.
[46] T.B. Brill, K.J. James, Thermal decomposition of energetic materials. 62. Reconciliation of the kinetics and mechanisms of TNT on the time scale from microseconds to hours, Journal of Physical Chemistry 97 (1993) 8759.
[47] B.F. Henson, B.W. Asay, L.B. Smilowitz, P.M. Dickson, Ignition chemistry in HMX from thermal explosion to detonation, in: Shock Compression of Condensed Matter-2001, Pts 1 and 2, Proceedings, 2002.
[48] C.M. Tarver, S.K. Chidester, A.L. Nichols, Critical conditions for impact- and shock-induced hot spots in solid explosives, Journal of Physical Chemistry 100 (1996) 5794–5799.
[49] A. Strachan, A.C.T. van Duin, D. Chakraborty, S. Dasgupta, W.A. Goddard, Shock waves in high-energy materials: The initial chemical events in nitramine RDX, Physical Review Letters 91 (2003) 098301.
[50] M.R. Manaa, L.E. Fried, C.F. Melius, M. Elstner, T. Frauenheim, Decomposition of HMX at extreme conditions: A molecular dynamics simulation, Journal of Physical Chemistry A 106 (2002) 9024–9029.
[51] C.J. Wu, L.E. Fried, L.H. Yang, N. Goldman, S. Bastea, Catalytic behaviour of dense hot water, Nature Chemistry 1 (2009) 57–62.
[52] E. Martinez, R. Perriot, E.M. Kober, P. Bowlan, M. Powell, S. McGrane, M.J. Cawkwell, Parallel replica dynamics simulations of reactions in shock compressed liquid benzene, Journal of Chemical Physics 150 (2019) (244108244108).
[53] M.S. Powell, M.N. Sakano, M.J. Cawkwell, P.R. Bowlan, K.E. Brown, C.A. Bolme, D.S. Moore, S.F. Son, A. Strachan, S.D. McGrane, Insight into the chemistry of PETN under shock compression through ultrafast broadband mid-infrared absorption spectroscopy, The Journal of Physical Chemistry A 124 (35) (2020) 7031–7046.
[54] M. Elstner, D. Porezag, G. Jungnickel, J. Elsner, M. Haugk, T. Frauenheim, S. Suhai, G. Seifert, Self-consistent-charge density-functional tight-binding method for simulations of complex materials properties, Physical Review B 58 (1998) 7260–7268.
[55] T. Frauenheim, G. Seifert, M. Elstner, Z. Hajnal, G. Jungnickel, D. Porezag, S. Suhai, R. Scholz, A self-consistent charge density-functional based tight-binding method for predictive materials simulations in physics, chemistry and biology, Physica Status Solidi B-Basic Solid State Physics 217 (2000) 41–62.
[56] A.M.N. Niklasson, C.J. Tymczak, M. Challacombe, Time-reversible Born-Oppenheimer molecular dynamics, Physical Review Letters 97 (2006) 123001.
[57] A.M.N. Niklasson, C.J. Tymczak, M. Challacombe, Time-reversible ab initio molecular dynamics, Journal of Chemical Physics 126 (2007) 144103.
[58] A.M.N. Niklasson, Extended Born-Oppenheimer molecular dynamics, Physical Review Letters 100 (2008) (123004123004).
[59] M.J. Cawkwell, R. Perriot, Transferable density functional tight binding for carbon, hydrogen, nitrogen, and oxygen: Application to shock compression, Journal of Chemical Physics 150 (2019) 024107.
[60] M.W. Finnis, A.T. Paxton, M. Methfessel, M. van Schilfgaarde, Crystal structures of zirconia from first principles and self-consistent tight binding, Physical Review Letters 81 (1998) 5149–5152.
[61] M.W. Finnis, Interatomic Forces in Condensed Matter, Oxford University Press, Oxford, 2003.
[62] P. Pulay, Ab initio calculation of force constants and equilibrium geometries in polyatomic molecules. I. Theory, Molecular Physics 17 (1969) 197.
[63] P. Pulay, G. Fogarasi, Fock matrix dynamics, Chemical Physics Letters 386 (2004) 272–278.
[64] J.M. Herbert, M. Head-Gordon, Accelerated, energy-conserving Born-Oppenheimer molecular dynamics via Fock matrix extrapolation, Physical Chemistry Chemical Physics 7 (2005) 3269–3275.
[65] M.J. Cawkwell, E.M. Kober, R. Perriot, T.W. Myers, V.W. Manner, Molecular dynamics simulations of the first reactions in nitrate ester-based explosives, in: R. Chau, T.C. Germann, J.M.D. Lane, E.N. Brown, J.H. Eggert, M.D. Knudson (Eds.), Shock Compression of Condensed Matter - 2017, AIP Conference Proceedings, American Institute of Physics, Melville, NY, 2018.

[66] G. Zheng, A.M.N. Niklasson, M. Karplus, Lagrangian formulation with dissipation of Born-Oppenheimer molecular dynamics using the density-functional tight-binding method, Journal of Chemical Physics 135 (2011) 044122.
[67] M.J. Cawkwell, A.M.N. Niklasson, Energy conserving, linear scaling Born-Oppenheimer molecular dynamics, Journal of Chemical Physics 137 (2012) 134105.
[68] E. Martinez, M.J. Cawkwell, A.F. Voter, A.M.N. Niklasson, Thermostating extended Lagrangian Born-Oppenheimer molecular dynamics, Journal of Chemical Physics 142 (2015) 154120.
[69] N. Lease, L.M. Kay, G.W. Brown, D.E. Chavez, P.W. Leonard, D. Robbins, V.W. Manner, Modifying nitrate ester sensitivity properties using explosive isomers, Crystal Growth & Design 19 (2019) 6708–6714.
[70] P. Pulay, Convergence acceleration of iterative sequences: The case of SCF iteration, Chemical Physics Letters 73 (1980) 393–398.
[71] P. Pulay, Improved SCF convergence acceleration, Journal of Computational Chemistry 3 (1982) 556–560.
[72] D.T. Gillespie, General method for numerically simulating stochastic time evolution of coupled chemical-reactions, Journal of Computational Physics 22 (1976) 403–434.
[73] A.F. Voter, Parallel replica method for dynamics of infrequent events, Physical Review B 57 (1998) 13985–13988.
[74] A.F. Voter, in: K.E. Sickafus, E.A. Kotomin, B.P. Uberuaga (Eds.), Introduction to the Kinetic Monte Carlo Method. Radiation Effectts in Solids, 2007, pp. 1–23.
[75] P.R. Bevington, Data Reduction and Error Analysis for the Physical Sciences, McGraw-Hill, New York, 1969.
[76] N. Lease, L. Kay, D.E. Chavez, D. Robbins, V.W. Manner, Increased handling sensitivity of molten erythritol Tetranitrate (ETN), Journal of Hazardous Materials 367 (2019) 546–549.
[77] E.R. Bernstein, On the release of stored energy from energetic materials, Advances in Quantum Chemistry 69 (2014) 31–69. Edited by Sabin, J. R.
[78] B.M. Rice, S. Sahu, F.J. Owens, Density functional calculations of bond dissociation energies for NO_2 scission in some nitroaromatic molecules, Journal of Molecular Structure: THEOCHEM 583 (2002) 69–72.
[79] R.V. Tsyshevsky, O. Sharia, M.M. Kuklja, Molecular theory of detonation initiation: Insight from first principles modeling of the decomposition mechanisms of organic nitro energetic materials, Molecules 21 (2016) 236.

CHAPTER

15

Chemical kinetics and the decomposition of secondary explosives

B.F. Henson and L. Smilowitz*

Chemistry Division, Los Alamos National Laboratory, Los Alamos, NM, United States

***Corresponding author: E-mail: henson@lanl.gov**

1 Introduction

Understanding the physical and chemical mechanisms of energy release in solid explosives has been an active area of research in physical chemistry for over a century. While considerable progress has been made, particularly concerning processes of steady combustion and detonation, significant uncertainties have remained in our understanding of the way by which stored energy is released from the crystalline explosive due to at least two fundamental difficulties, the relatively poorly understood field of solid chemical decomposition and the exponential dependence of the rate of chemical reaction on temperature.

Historically the conceptual model of these processes derives from

$$\rho C_{avg} \frac{\partial T}{\partial t} = \lambda \nabla^2 T + S(T, t) \tag{15.1}$$

a simple energy balance equation. If an internal source of energy $S(T,t) = 0$, a solid of density ρ, heat capacity C_{avg} and thermal conductivity λ may be heated from a boundary, and the resulting spatial thermal profile is simply calculated. The defining attribute of a solid explosive is that $S(T,t)$ does not equal zero, but rather takes the form

$$S(T, t) = \Delta H_r f([c_i]) A_i \exp(-E_i/RT) \tag{15.2}$$

due to exothermic reactions liberating energy, ΔH_r, and increasing temperature, T, as a function of time, t, via chemical decomposition.

The two fundamental difficulties are simply that models of how reagents react in crystalline solids, shown here as a generic function of concentrations $f[c_i]$), are relatively immature

Theoretical and Computational Chemistry, Volume 22
https://doi.org/10.1016/B978-0-12-822971-2.00002-4

compared to the chemical kinetics of well-mixed gases or liquids, and that the accuracy of a calculated rate of energy generation is exponentially sensitive to the accuracy of the activation energy, E_i, in $\exp(-E_i/RT)$.

In the absence of a proper mathematical form for the kinetic order, $f[c_i]$), or an ability to measure and experimentally constrain the E_i over a sufficiently broad range of temperature, models describing thermal ignition or the detonation reaction zone based on simple concentration or progress variable approximations, and poorly constrained E_i, have been continually reparameterized for half a century in an attempt to calibrate empirical models with sufficient extrapolative reliability (prediction). These efforts have until now been the only available simulation strategy.

In this work, we advance hypotheses that address both of these difficulties. We first show a compilation of thermal ignition data for a number of secondary explosives that span the entire 200–3000°C range of energetic responses. From these data E_i emerges with sufficient precision to associate them with known thermodynamic enthalpies in these crystalline systems. We construct a simple model of the kinetic order, $f[c_i]$), describing the decomposition of the solid that reproduces this observed coupling of kinetics and thermodynamics. We couple this solid-state mechanism with subsequent reactivity in intermediate and product phases to produce complete models of thermal decomposition to ignition or initiation for HMX, TATB, and PETN from initial crystalline solid to final products. We show that without any engineering modification or optimization to account for the specifics of a real application the model can already reproduce the known temporal, thermal, and pressure profiles of thermal ignition, deflagration, and detonation in these three explosives with considerable accuracy.

1.1 Thermal response phenomenology

We divide the thermal response of a secondary explosive according to the observed components of the total response, from the initial heating of the explosive to the possibility of a final transition to detonation. In Fig. 15.1 we illustrate this as a chronology of successive phenomena, labeled according to the dominant phenomena in each regime and separated by three bifurcations, indicating possible terminations of the response dependent on initial conditions, or a shift to a new dominant mechanism of propagation or consumption. Relatively steady (and well understood) regimes of Heating, Conduction, Deflagration, and Detonation are separated by abrupt transitions at these bifurcations. These transitions include *Ignition*, the abrupt pressure, and temperature increase that accompanies the involvement of gas-phase chemistry separated spatially from the solid phase, *Convection*, the abrupt acceleration of solid-phase consumption by deflagration when sufficient pressures enable gaseous convective heating into the solid and, easily the least understood transition, *Acceleration*, the change in the mechanism that accompanies the significant application of pressure to the solid and the possible association of chemical energy release with shock-wave processes in the sample. This is the Deflagration to Detonation Transition (DDT).

Each transition involves acceleration in the time scale of roughly three orders of magnitude, from seconds in the direct heating regime to milliseconds in the conductive burning regime, microseconds during convection and nanoseconds at detonation. These transition mechanisms are extremely important both in the modeling and simulation of response and in the technology of thermal response experimentation for at least three reasons;

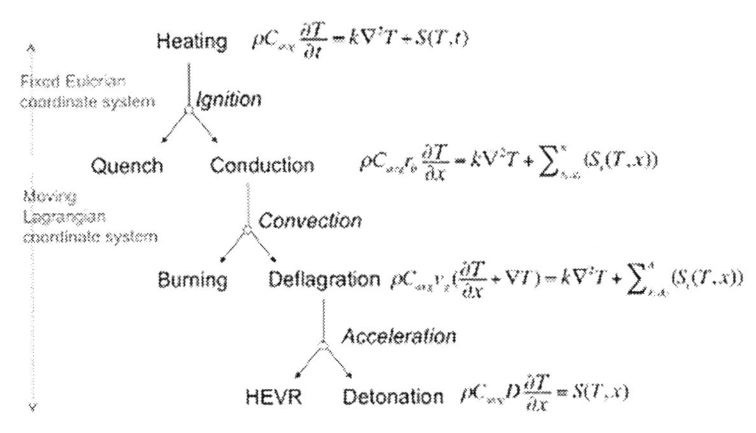

FIG. 15.1 Decision tree illustrating the chronology of thermal response.

(1) In direct simulations of these coupled regimes, a sudden temporal acceleration at a transition necessitates a change in the simulation time step of the integration. This has implications that have been long appreciated and are the reason for applications of implicit and explicit integration methods.

(2) The possibility of establishing firm causal connections between these regimes could be extremely useful. If the likelihood of a response proceeding through to each steady regime could be constructed as a product of dependent probabilities of the progress through each successive bifurcation, then real probabilities could be considered for thermal response outcome as a function of initial conditions. This is a very exciting quantitative possibility that is an objective of this work.

(3) Entirely different techniques of measurement are needed to acquire data for each regime. Available technology limits the observables that are available for measurement, e.g., spatially resolved temperatures are crucial to the heating regime, pressure in deflagration and particle velocity and wall motion in Shock to Detonation Transition (SDT) and detonation. Radiography is invaluable to all regimes. A growing understanding of the mechanisms of transition has enabled the development of trigger and synchronization techniques that are essential to enable the acquisition of data spanning each bifurcation and in the steady regime on either side.

1.2 Regimes and phenomena

The regime of Heating denotes the application of an external thermal boundary condition on the solid explosive in its initial state. This can involve complex thermal boundary conditions, including isothermal or an arbitrary rate of heating. In an early and important result of this work, we illustrated the invariance of HMX thermal response to the mechanism of heating, including direct, laser, frictional, electron beam, shear, flow, and strong shock heating [1].

The boundary condition induces energy transport into the solid. The phenomena involve solid morphology changes such as defect propagation by ratchet growth in TATB and crystal fracture during the β-δ phase transition in HMX induced by thermal expansion and molar density change. Important changes as a function of explosive/binder formulation may occur, most importantly related to the absence or restriction of gas-phase permeability. Gas-phase energy and mass transport are critical early in a problem where rates of heating from the boundary are comparable to chemical rates of energy generation. This is likely due to sublimation/condensation processes involving the explosive, pressure-dependent gas-phase reaction rates and the transport of sensible heat. We have shown many of these complex spatial phenomena with the first pRad [2] and then the LARS experiments [3], as well as establishing correlations with these phenomena and subsequent rates of deflagration after ignition [2,4].

Ultimately heating of sufficient rate and magnitude leads to successive chemical reactions participating in the generation/consumption of energy, the rates of which depend exponentially on temperature. We will show the results of isothermal, adiabatic calculations to illustrate this progression, from solid to two gas-phase systems in HMX, one of the two gas-phase systems in TATB, and the other one in PETN. In this way, we will isolate the contribution of each chemistry to the generation of energy (temperature increase) and pressure during the progress to ignition. We will neglect the complexities of solid conduction and gas-phase energy and mass transport. While these processes are crucial, significant aspects of the model may be illustrated without their inclusion. Models of these processes may be included with a larger scale full simulation to compare to the experiment. With simple adiabatic calculations, we will reproduce the range of direct thermal response for each system, from approximately 140°C in PETN to 200°C for HMX and 300°C for TATB, and begin to map the entire fully confined response from the beginning of heating through to complete chemical consumption, final temperature, and adiabatic thermal explosion pressure. The temporal structure in the evolution of P and T as a function of ρ, t is already apparent in such simple calculations and it will be shown that inherent and defining differences even among these three representative secondary explosives are reproduced.

1.3 The scope of the models presented here

The purpose of these models is to capture the relevant and defining features of the complex, coupled thermal response of secondary explosives with a single global model of intermediate complexity. This is accomplished as discussed above using a unique representation of the solid-state chemical decomposition. A highly reduced gas-phase chemistry, drawn from reactions in the literature, is then utilized to complete the decomposition and provide a final product distribution. Intermediate chemistries that link the solid phase with the gas-phase referred to as dark zone chemistry in the literature, are used for HMX and PETN but are unnecessary for TATB. Both HMX and TATB require the bright zone chemistry of HCN oxidation. Coupling the final product distribution of chemical composition with product equations of state, taken from the literature, enable the calculation of gaseous and hydrodynamic pressures at different stages in the response.

These are fully constrained, adiabatic models. The solid decomposition rates are determined independently from several different decomposition and ignition experiments and roughly reproduce times to ignition in the slow heating regime near the critical temperature for HMX, PETN, and TATB. Combined with final temperatures that result from the coupled gas-phase and final product chemistries it has also been shown to reproduce high-pressure velocities in the deflagration regime. With hypotheses regarding the thermal boundary condition in the detonation regime this model also reproduces a temporal profile consistent with the detonation reaction zone of all three explosives, and rates of pressurization from ignition near the critical temperature consistent with the very different shock sensitivities of these three molecules.

The rest of this chapter is structured as follows. Section 2 contains the presentation of an extensive set of ignition data on these materials that are used to model initial steps in the kinetic mechanism, as well as the underlying theoretical justification for the mathematical form of the models. The three kinetic models for HMX, TATB, and PETN are then described in detail in Section 3. Example calculations illustrating the use of these kinetics in the various regimes of response for these materials are shown in Section 4. Summary and conclusions are presented in Section 5 and future directions are briefly described in Section 6.

2 Rate-limiting mechanisms: Solid-state chemistry and secondary explosives

2.1 Thermal ignition data

We present a compiled suite of ignition data for PETN, HMX, and TATB in Fig. 15.2. These data have been collected from experiments for which a well defined and directly measured thermal boundary condition have been applied, either as an external boundary in larger thermal explosion experiments or as a direct measure of temperature in the volume of the sample where ignition occurs subsequent to heating, as in a number of impact experiments. The experiments also represent conditions of gas-phase confinement, either imposed by the boundary or inertially confined under conditions of high heating rate relative to rates of material or thermal diffusion. These experiments also represent conditions of the low pressure or pressurization, either by the lack of initial pressurization or under conditions where stress release by material failure into internal free volume is possible prior to full compaction. The data in Fig. 15.2 are plotted as the ignition time as a function of the inverse temperature. The time is plotted linearly on a logarithmic scale and the inverse temperature is plotted with the origin on the right. Ignition time, therefore, descends to the right with increasing temperature, to an infinite temperature at $1/T=0$.

HMX: Thermal explosion data, plotted as the ignition time as a function of the inverse boundary temperature are shown as the open [5] and filled [6] circles for data from the spherical LLNL ODTX experiment and as open diamonds for data from the cylindrical LANL LARS experiment [7]. Laser-induced temperature jump [8], circles with an included x and laser ignition data [9], circles with an included dot, are plotted as the ignition time as a function of the inverse of the directly measured surface temperature. Ignition induced by fast friction, and shear, square with an included dot [10], are plotted as the ignition time as a function of the inverse of the directly measured surface temperature. Ignition induced by fast compression and flow under isothermal heating conditions, open squares with an included cross [11], are plotted as the ignition time as a function of the inverse of the directly measured surface temperature. Ignition induced by fast compression and flow under linear heating conditions, open squares with an included x [11], are plotted as the inverse of the linear heating rate as a

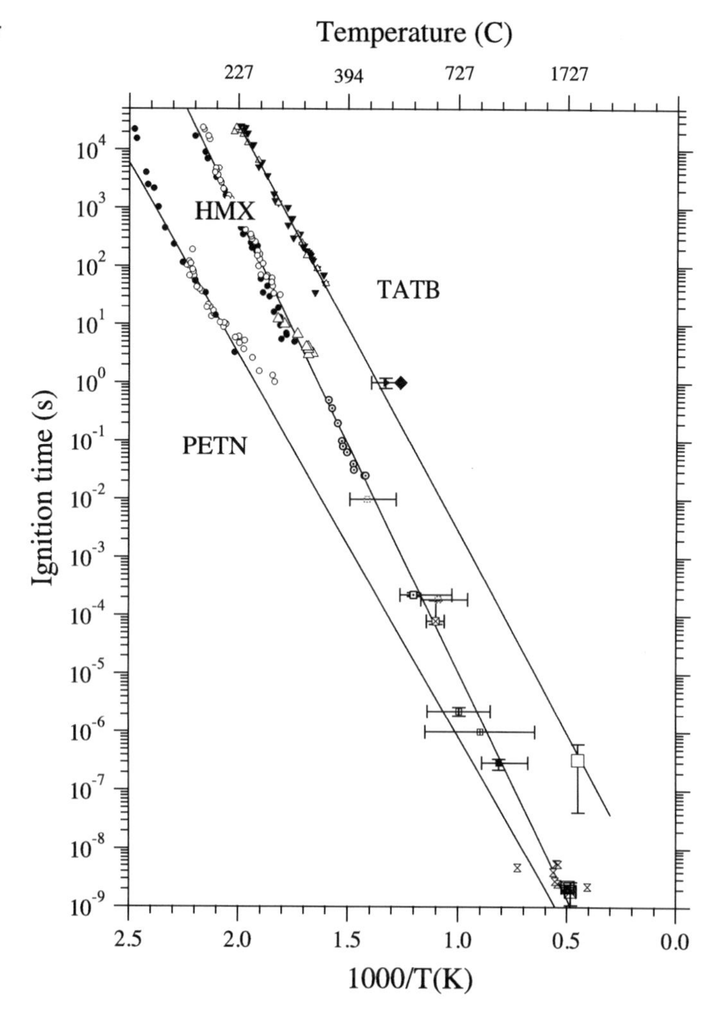

FIG. 15.2 Thermal ignition data for HMX, TATB, and PETN.

function of the peak temperature at ignition. Laser-assisted ignition temperatures [12] presented here are plotted as the ignition time as a function of the isothermal temperature, and as the inverse heating rate as a function of the maximum observed temperature at ignition, hexagon with an included dot. Data from run to detonation experiments are also plotted as the inverse heating rate as a function of the maximum observed temperature at ignition, open double triangle [13]. TATB: Thermal explosion data plotted as the ignition time as a function of the inverse boundary temperature are shown as the open [5] and inverted filled [6] triangles. Combustion data are shown as the filled diamonds [8,14,15]. The reaction zone temperature and time are approximated from detonation curvature data and plotted as the large open square [16]. PETN: Thermal explosion data, plotted as the ignition time as a function of the inverse boundary temperature are shown [17,18]. The calculation of the solid lines will be discussed in Section 3.2.1.

The data in Fig. 15.2 are the principle data from which the model of solid decomposition will be constructed. As discussed in detail previously [19], of these data do not represent the entirety of the historical data on these molecules, but rather represent data obtained under particularly defined boundary conditions. Times to ignition can be modified at any particular temperature in a number of ways that must ultimately be addressed, particularly the role of ignition under conditions of unconfined gas-phase flow, but we assert that these data best simplify and reflect the mechanisms pertinent to the role of the solid state in thermal decomposition.

2.2 Analysis and determination of rate constants

In this section we provide an analysis of the data of Fig. 15.2. We base this analysis on the thermally activated Arrhenius rate constant of the form

$$k_i = A_i \exp\left(\frac{-(\Delta E_i + P\Delta V_i)}{RT}\right) \tag{15.3}$$

or, equivalently

$$k_i = v_i \exp\left(\frac{\Delta S_i}{R}\right) \exp\left(\frac{-(\Delta E_i + P\Delta V_i)}{RT}\right) \tag{15.4}$$

where

$$A_i = v_i \exp\left(\frac{\Delta S_i}{R}\right). \tag{15.5}$$

Eq. (15.3) is the traditional form of the Arrhenius rate constant, expressing the exponential dependence of rate on temperature through the activation energy ΔE_i. This is the rate constant associated with the source term in Eq. (15.1). In our earliest reporting on the compilation of ignition data in this way, we reiterated the consequences of using Eq. (15.3) to represent rate processes over such a large range of rate and temperature, including the implications for detonation, internal molecular thermal equilibrium and the invariance of the chemical kinetic mechanism to the mechanism of heating [1].

The frequency factor A_i in Eq. (15.3) has been the object of considerable interest in fundamental chemical kinetics. In the earliest treatments of Transition State Theory (TST) the activation energy was associated with an activation free energy, $\Delta G = \Delta H\text{-}T\ \Delta S$, as in Eq. (15.4) with $\Delta H = \Delta E + P\ \Delta V$, and A_i derived from the assumptions of an activated spatial degree of freedom and statistical equilibrium of the concentration of an activated state constructed along this coordinate. The value of ν in Eqs. (15.4) and (15.5) could then be related to the activated entropy through some formula reflecting the equilibrated transition state and calculated from the first principles [19]. While we will present a preliminary mechanism for these kinetics we do not yet have a sufficiently detailed molecular mechanism to determine ν, and therefore the relation between the activated entropy and the rate, from first principles.

The activation energy, however, is directly measurable as the slope in Fig. 15.2. Assuming the ignition time to represent a characteristic time associated with the rate as $t = 1/k$ a plot of the line $\ln(t) = \ln(1/A) + (E/R)(1/T)$ for the linear ignition data immediately yields the single activation energy E. The activation energies derived from the kinetic analysis of the three lines of Fig. 15.2 are quantitatively consistent with the known enthalpies of sublimation and vaporization of the constituent molecules. It is a cornerstone of this work and these models that the observed slopes of the ignition data in Fig. 15.2 for all three explosives lead to composite activation energy that is equal (to within uncertainty) to the average of the energies of sublimation and vaporization

$$E = \frac{E_{sub} + E_{vap}}{2} \tag{15.6}$$

The lines of Fig. 15.2 have been calculated using Eq. (15.6). These calculations will be discussed in detail in Section 3.2.

Therefore the rate limiting elements of the solid decomposition mechanism are not related to subsequent covalent bond breaking but rather to features defining the thermodynamic state of the crystalline solid as a function of temperature.

3 Thermal ignition model

3.1 Introduction

We present a model of thermal ignition that is significantly different in the solid-state kinetics and significantly expanded over the version presented in Chapter 3 of Non-Shock Initiation [19]. It remains constrained by global chemistry and parameterized entirely by independent and more elemental measurements of rate. Specific chemical reactions and their stoichiometry determine the enthalpies of reaction, based on measured enthalpies of formation reported for all species in the NIST Chemistry Webbook (Tables 15.1 and 15.2). The heat capacity, C_p, is a mole fraction average of solid and gas-phase heat capacities. Most importantly all rate constants used here have been determined independently of this model, e.g., thermodynamic rates of solid decomposition are based on measurements and compilations from Taylor and Crookes [20] and Lyman et al. [21], (Tables 15.3 and 15.4) gas-phase rates were taken from the measurements and modeling of Melius and Lin [22–25] and from compilations of gas-phase nitrogen chemical kinetics (Table 15.4) [26,27].

The resulting models are comprised of 2 states of the crystalline solid, 1 extensive variable for the solid, 10 (HMX), 11 (TATB), and 6 (PETN) gas-phase species, product carbon for TATB and PETN, all coupled in a reaction system of 16, 16 and 11 partial differential equations, respectively, including an equation for the generation of temperature. The solid decomposition is based on the solid and liquid activity to be discussed below such that the order of the equations used to calculate the rate of solid loss and initial gas-phase generation are highly constrained. The kinetic order of the gas-phase chemistry is either first or second and determined experimentally in measurements of the rate constants for each reaction.

There are no arbitrary modeling degrees of freedom in the calculations shown here. All calculations presented in this report derive from a single, independent, parameterization of the same model construction. These applications encompass the thermal response in the regimes of slow heating to ignition, deflagration, shock to detonation and steady detonation.

Most importantly these will be adiabatic, homogeneous applications of this model and so there is no gas-phase diffusion or mass transport. The "pressure," P, is calculated using the

TABLE 15.1 Gas-phase NIST product thermochemical properties (J/mol)

	$\Delta H_{formation}$	A	B	C	D	E
N_2O	82050	27.67988	51.14898	−30.6454	6.847911	−0.157906
CH_2O	−115900	5.193767	93.23249	−44.85457	7.882279	0.551175
NO_2	33095	16.10857	75.89525	−54.38740	14.30777	0.239423
HCN	135143.2	32.69373	22.5920	−4.369142	−0.407697	−0.282399
HNC	189143.2	0.0	0.0	0.0	0.0	0.0
HNCO	−101671.2	32.73428	63.81479	−38.46142	9.728052	−0.317587
HCO	43514	1.13803	40.43610	−14.71337	0.96901	0.239639
HC	59413	32.94210	−16.71056	24.18595	−7.784709	−0.065198
H	217998	20.78603	0.0	0.0	0.0	0.0
NO	90291	23.83491	12.58878	−1.13901	−1.497458	0.214194
H_2O	−241826.4	30.092	6.832514	6.793435	−2.534480	0.082139
CO	−110527	25.56759	6.09613	4.054657	−2.6713	0.131021
CO_2	−393522	24.99735	55.18696	−33.69137	7.948388	−0.136638
N_2	0.0	19.50583	19.88705	−8.598535	1.369784	0.000117

TABLE 15.2 Solid NIST product thermochemical properties (J/mol)

	$\Delta H_{formation}$	A	B	C	D
δ-HMX	74894	104.5732	714.3017	1.14×10^{-4}	-1.43×10^{-4}
PETN	−538500	316.32	1.06	0.0	0.0
		a_1	a_2	a_3	a_4
TATB	−7470	195.25	583.27	3.5877	419.9

TABLE 15.3 Thermodynamic and kinetic parameters

	ΔH_{sub} (kJ/mol)	ΔS_{sub} (J/mol K)	ΔH_{vap} (kJ/mol)	ΔS_{vap} (J/mol K)	ΔH_{fus} (kJ/mol)	T_m (C)	ν
HMX [21]	174.930	257.942	105.067	131.149	69.900	278	0.085
PETN	150.400 [47]	215.400 [47]	101.062	96.270	49.338 [48]	141 [47]	16.00
TATB	176280 [47]	246.26 [47]	85.640	109.114	90.640 [49]	388 [47]	1.000

All thermodynamic data for HMX was taken from Lyman et al. [21]. For TATB and PETN ΔG_{sub} and T_m were taken from Cundall et al. [47] ΔG_{vap} was then determined using the heat and entropy of fusion by $\Delta H_{vap} = \Delta H_{sub} - \Delta H_{fus}$, $\Delta S_{vap} = \Delta S_{sub} - \Delta S_{fus}$.

TABLE 15.4 Rate constants

	A s^{-1}, or (m^3/mol s)	E(kJ/mol)
δ-HMX k_{2a}	4.957×10^{15}	182.423
k_{2b}	3.516×10^{9}	110.310
TATB k_{2a}	7.30×10^{12}	176.280
k_{2b}	5.01×10^{5}	85.640
PETN k_{2a}	5.35×10^{15}	150.400
k_{2b}	3.20×10^{9}	101.062
HMX, PETN k_3	1.373×10^{6}	71.966
HMX, TATB k_4	3.500×10^{13}	197.458
HMX, TATB k_{4a}	1.100×10^{6}	63.593
HMX, TATB k_{4a}	2.500×10^{6}	108.913

concentration, $[g]$, which is the sum of all gas-phase densities and the temperature in a standard JWL equation of state. The initial conditions required for calculations are an initial temperature, T_o, an initial crystalline density (ambient in all calculations presented here) $\rho_o=5599$ mol/m^3 (PETN), 6300 mol/m^3 (HMX) and 7247 mol/m^3 (TATB), an intensive initial condition for the coupled solid states $x(0)=1$, and $y(0)=0$ and an initial density of zero for all gas-phase species.

3.2 Thermal ignition model

3.2.1 Solid decomposition

The first step in the decomposition of solid is the evolution of two coupled states of the crystalline solid, x and y

$$x \xleftrightarrow{k_{2a},k_{2b}} y \qquad \text{R1}$$

x and y are not meant to indicate crystallographic phases of the solid. It is not known at this time what sort of progress variables are indicated by these states, but there is mounting evidence associating them with the interfacial quasi liquid phase and coupled defect states within the crystalline solid [28]. x and y are coupled via a set of first-order differential equations that, when solved in isolation, generate a sinusoidal oscillatory system with period t,

$$t = \frac{1}{\sqrt{k_{2a}k_{2b}}}\frac{\pi}{4}. \qquad (15.7)$$

where k_{2a} and k_{2b} are rate constants activated by the free energies of sublimation and vaporization, respectively, and where t follows the temperature dependence exhibited in thermal explosion data for confined, unpre-pressurized experiments, Section 2. Eq. (15.3) with $P=0$ and (15.7), with the thermodynamic constants of Table 15.4, were used to calculate the solid lines of Fig. 15.2 for each explosive.

3.2.2 Intermediate gas-phase species production

The solid decomposition to form gas-phase species has been observed for all three explosives discussed here in a number of laboratories. For HMX in particular these species appear as two families, distinguished by a different temperature dependence in the rate of appearance, and are assumed here, as elsewhere, to represent the products of two competing decomposition channels [29]. One channel is characterized by the observation of CH_2O and N_2O, and the other by HCN, NO_2, NO, and H_2O.

In previous versions of these kinetics for HMX, we have based product formation explicitly on these species. With that representation, it was not possible to include the endothermicity indicated by the stoichiometry of branch k_{2a} and achieve ignition in the lower temperature range of response, as discussed at the time [30]. Here we retain the dual branching of the previous authors, with the same thermodynamically derived rates k_{2a} and k_{2b} as we have previously introduced, with a product distribution weighted more heavily toward NO_2 and final product formation.

$$\delta\text{-HMX} \xrightarrow{k_{2a}} 4HCN + 3NO_2 + 2H_2O + (1/2)N_2 \qquad \text{R2a}$$

$$\delta\text{-HMX} \xrightarrow{k_{2b}} 4CH_2O + 2NO_2 + 3N_2 \qquad \text{R2b}$$

Solid transformation and decomposition are governed by the oscillatory step with overall time given by a coupling of both k_{2a} and k_{2b} through Eq. (15.7). R2a and R2b individually represent the rate of generation of the two gas-phase channels. The assumed stoichiometries of R2a and R2b are taken in part from the T-jump experiments and modeling of Brill's group [31], modified as discussed above.

The solid decomposition of TATB to form gas-phase species has been observed in a number of laboratories. Prominent among the condensed phase product species is furazan, although we do not include this species in this model of decomposition. Among the typical gas-phase product species identified are HCN, NO_2, NO and H_2O, and HNCO [31]. These species are prominent in the high-temperature oxidation of HCN following the key isomerization reaction [24,25]. While there is no literature on two intermediate product families in the thermal decomposition of TATB, we evoke this system in analogy with HMX and because one product family-based upon observed intermediates is so endothermic that it is not possible to generate ignition. This endothermic product family is thus generated via branch k_{2a}, again in analogy with HMX as it is the HCN, NO_2 branch that is governed by this rate constant. The second branch, governed by k_{2b}, leads to the prompt formation of final products, with an exothermicity very nearly balancing the endothermicity of k_{2a}. This construction must be regarded as a hypothesis at this point, but it extends the spirit of the mechanism of solid decomposition linking the class of secondary explosives, now involving analogies in product formation as well. The initial product formation is then coupled to the same system of equations governing HCN isomerization, bright zone burning and final product formation to be discussed below.

$$\text{TATB} \xrightarrow{k_{2a}} 3HCN + 3NO_2 + 3H + 3C \qquad \text{R2a}$$

$$\text{TATB} \xrightarrow{k_{2b}} 3H_2O + 3\,N_2 + 3CO + 3C \qquad \text{R2b}$$

Kinetic parameters governing R2a and R2b are taken from rates of vaporization and sublimation of TATB, as described for HMX. The assumed stoichiometry of R2a is taken from the T-jump experiments and modeling of Brill's group [31]. The stoichiometry of R2b is a presumed fast exothermic path to final products.

There is no literature on two intermediate product families in the thermal decomposition of PETN. We again evoke this system in analogy with HMX, and because, as for TATB, one product family-based upon observed intermediates is sufficiently endothermic that it is not possible to generate ignition. This endothermic product family is thus generated via k_{2a}. In contrast to both HMX and TATB molecules that were previously representative of either branch, CH_2O by k_{2b} and NO_2 by k_{2a} are both produced via the 2a branch of the decomposition. This is hypothesized to reflect the relatively simple chemical structure of PETN compared to the ringed structures of HMX and TATB. The second branch, governed by k_{2b}, leads to the prompt formation of final products, with an exothermicity nearly twice that of the endothermicity of k_{2a}. This construction must also be regarded as a hypothesis at this point, as for TATB. The initial product formation is then coupled to the same system of equations governing CH_2O+NO_2 dark zone burning and final product formation as for HMX. This simplified stoichiometry leads to carbon formation, which is clearly incorrect after times long compared to ignition. This is then a preliminary version of this model strategy to investigate its applicability to PETN. The rate constants governing both solid loss and the rates of appearance of both branches are again determined by the PETN energies of sublimation and vaporization, and this investigation further tests these ideas of thermodynamic and kinetic coupling. The assumed stoichiometries of R2a and R2b reflect our best current guess of product branching, in analogy with HMX and TATB, as discussed above.

$$\text{PETN} \xrightarrow{k_{2a}} 4NO_2 + 4CH_2O + C \qquad \text{R2a}$$

$$\text{PETN} \xrightarrow{k_{2b}} 4H_2O + 2N_2 + 3CO_2 + 2CO \qquad \text{R2b}$$

3.2.3 *Dark zone burning mechanism*

The complex reaction of CH_2O with NO_2 provides a source of exothermicity that is relatively low compared to subsequent reactions but important to the overall decomposition of both HMX and PETN [32]. We use a simplified mechanism based on a series of experimental and modeling studies undertaken by Melius and Lin over a temperature range spanning some 1200 K [24]. These results, in combination with the earlier work of Pollard [33,34] and He [35] are represented here as the following overall reaction system.

$$CH_2O + NO_2 \xleftrightarrow{k_3} CO_2 + H_2O + (1/2)N_2 \qquad \text{R3}$$

This system has been shown to be valid from 400 to 1600 K and in particular, is very insensitive at low temperature to a number of radical reactions that presumably proceed concurrently [24].

3.2.4 *Bright zone burning mechanism*

We also include a system of reactions, again determined by Lin and Melius, for HCN oxidation based upon the initial isomerization of HCN [23,25].

The system is extremely exothermic overall, including subsequent final products described in the next section, but is relevant only at temperatures higher than the dark zone chemistry, and is simply limited by the isomerization reaction R4.

$$HCN + M \xrightarrow{k_4} HNC + M \quad \text{R4}$$

$$HNC + NO_2 \xrightarrow{k_{4a}} HNCO + NO \quad \text{R4a}$$

$$HNCO + NO_2 \xrightarrow{k_{4b}} H + N_2O + CO_2 \quad \text{R4b}$$

These reactions are less important to calculations of time to ignition in the low-temperature ignition regime but are critical to calculations of pressure and ignition temperature in the low-temperature regime and necessary to the calculation of time to ignition as well at higher temperatures. The temperature dependence of the rates in dark and bright zone burning is shown in Fig. 15.3.

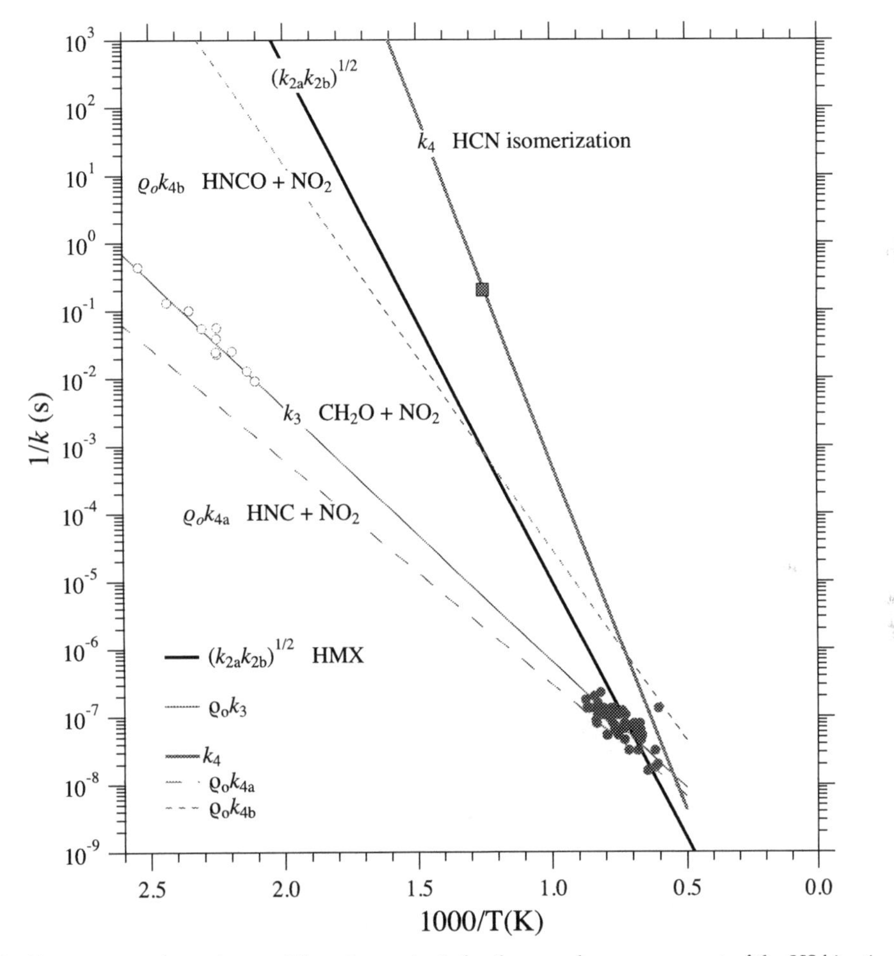

FIG. 15.3 Temperature dependence of the rate constants for the gas-phase component of the HS kinetic system.

3.2.5 Final product mechanism

We have included a number of reactions that serve to take intermediate species to the observed final products N_2, H_2O, and CO or CO_2. These reactions include a composite mechanism of N_2O reduction by free hydrogen, H, R6, and a second-order oxidation of HNCO by NO, R5. R5 is globalization of reactions following the unimolecular decomposition of HNCO.

$$\mathrm{HCNO} + \mathrm{NO} \xrightarrow{k_{4b}} \mathrm{H} + \mathrm{N_2O} + \mathrm{CO} \qquad \text{R5}$$

$$2\mathrm{H} + \mathrm{N_2O} \xrightarrow{k_{4b}} \mathrm{H_2O} + \mathrm{N_2} \qquad \text{R6}$$

Measured rates are slower than those of NO_2 oxidation [36], however for simplicity we approximate the rate somewhat faster in this chemistry. Both of these systems are second order in the gas-phase concentrations and taken to be limited by the slowest rate in the bright zone system, k_{4b}.

3.2.6 Overall stoichiometries

The final composition, and therefore energy release and pressure, vary for each explosive depending on the trajectory of temperature with time and the resulting balance of k_{2a} and k_{2b}. The ultimate yield of CO vs CO_2 is determined by the ratio of k_{2a} and k_{2b} and or k_{4a} and k_{4b} during a given thermal trajectory. The formation of carbon is assumed to yield bulk graphite, and some unreacted chemical potential is "stranded" in intermediates, e.g., HNCO or N_2O, due to the nonintegral stoichiometric relationship between each explosive and a final assumed mixture of only nitrogen, water, and carbon oxides.

3.2.7 The kinetic law

The set of coupled differential equations describing the evolution of the key chemical species begins with

$$\frac{dx}{dt} = -k_{2a}y \qquad \text{(M1a)}$$

$$\frac{dy}{dt} = k_{2b}x \qquad \text{(M1b)}$$

The rate equations governing product formation are driven by a combination of the intensive state x of the coupled defect system x and y, multiplied by the constant crystal density and rate constant as

$$\frac{d[c_i]}{dt} = C_i k_i \rho_o \frac{1-x}{1+x} \qquad \text{(M2ab)}$$

where the c_i are the various concentrations appearing from solid decomposition, C_i are the integer factors indicated by the stoichiometry, and k_i is either k_{2a} or k_{2b} depending on the branch from which c_i derives. The intensive description of the two defect states is reflected in the $(1-x)/(1+x)$ term and is at this point somewhat arbitrary, although based upon considerations of these progress variables and their relation to real condensed phase activity. An extensive description of the amount of material produced is scaled by the constant crystal density ρ_o. The actual loss of solid is captured indirectly through the appearance of products through mass balance as

$$\frac{d[\text{solid}]}{dt} = -\rho_o(k_{2a} + k_{2b})\frac{1-x}{1+x} \quad \text{(M2)}$$

It is important to note that x and y must be coupled without loss to other species in the model, and are therefore unusual as kinetic progress variables. Also, the system is cyclic and negative going in the second half of the period. The concentration of solid must therefore be held to zero after complete solid decomposition using, e.g., a Heaviside function.

The remaining reactions are calculated using the traditional equations of gas-phase kinetics using either

$$\frac{d[c_i]}{dt} = C_i[c_i][c_j]\,k_i \quad \text{(M3)}$$

or

$$\frac{d[c_i]}{dt} = C_i[c_i]k_i \quad \text{(M4)}$$

depending on the order or stoichiometry of the reaction. These formulas are shown for each reaction explicitly in Table 15.5.

The rate constants are parameterized as in Eq. (15.3) with $P=0$ using

$$k_i(T) = A_i e^{-\frac{E}{RT}} \quad (15.8)$$

where $k_i(T)$ is a canonical rate constant with a frequency factor A_i in s^{-1} or $m^3/\text{mol s}$, depending on reaction order, and ΔE_i is the activation energy. The values of the parameters in Eq. (15.8) for all reactions are shown in Table 15.4.

TABLE 15.5 HMX Reaction network: Labeled by reaction R2–R6

Stoichiometry	No.	Energy (kJ/mol)
sublimation / vaporization $x \underset{k_{2b}}{\overset{k_{2a}}{\rightleftarrows}} y$ lattice dynamics	R2	0.0
δ-HMX $\xrightarrow{k_{2a}}$ $4HCN + 3NO_2 + 2H_2O + (1/2)N_2$	R2a	71.511
δ-HMX $\xrightarrow{k_{2b}}$ $4CH_2O + 2NO_2 + 3N_2$	R2b	−482.140
$CH_2O + NO_2 \overset{k_3}{\longleftrightarrow} CO_2 + H_2O + (1/2)N_2$	R3	−552.543
$HCN + M \xrightarrow{k_4} HNC + M$	R4	54.000
$HNC + NO_2 \xrightarrow{k_{4a}} HNCO + NO$	R4a	−233.618
$HNCO + NO_2 \xrightarrow{k_{4b}} H + N_2O + CO_2$	R4b	−24.898
$HNCO + NO \xrightarrow{k_{4b}} H + N_2O + CO$	R5	−122.975
$2H + N_2O \xrightarrow{k_{4b}} H_2O + N_2$	R6	−759.872

3.3 Thermochemistry

The thermo-physical properties of the model are quite simplified. The heat capacity, C_p, is linearly dependent on temperature for HMX and PETN using

$$C_i = A_i + B_i T, \tag{15.9}$$

with T in K and C in units of J/mol K. A and B are again taken from NIST in Tables 15.1 and 15.2 for each species i. The gas-phase heat capacity is approximated as a single phase using the parameterization for NO_2. The heat capacity in the calculation is a mole weighted linear combination of two states, [solid + graphite], and gas products [g], using

$$[c] = [\mathbf{solid} + \mathbf{graphite}] + [g], \tag{15.10}$$

where [g] is the sum over gas-phase species

$$[g] = \sum_{i-1}^{k} [c_i], \tag{15.11}$$

and

$$\mathbf{C_{avg}} = \frac{[\mathbf{solid}]\boldsymbol{C}_{\mathbf{solid}} + [g]\boldsymbol{C}_{\mathbf{NO_2}}}{[c]}. \tag{15.12}$$

The heat capacity of solid TATB is given by [37]

$$C^{\alpha} = a_1 + \frac{(a_2 - a_1)}{\left(1 + \left(\frac{a_3}{T}\right)^{a_4}\right)}. \tag{15.13}$$

3.4 Model summary

The stoichiometries and differential equations for each model are summarized in Tables 15.5–15.10

The equations for chemical progress are coupled to an energy balance equation to calculate the generation of heat and temperature increase using a final differential equation in the system

$$[c]C_{avg}\frac{dT}{dt} = \sum_{i}^{n} \Delta H_r^i f[c_i]k_i \tag{15.14}$$

where the total moles, [c] and average heat capacity C_{avg} are calculated as above, the ΔH_i are calculated for each reaction i from Hess's law and the heats of formation in Table 15.1, and the order $f[c_i]$ and rate constant, k_i, are determined by the reaction.

The pressure is calculated using the total moles of product fluid, [c], suitably normalized as a density in an equation of state (EOS). We use a JWL EOS for a single assumed product fluid in the calculations shown here, although the details of the parameterization are not important to the qualitative results and interpretations. It is, however, critical to use a nonideal EOS,

TABLE 15.6 HMX ODE system: Labeled by species M2–M15

$\frac{dx}{dt} = -k_{2a}y$	**M2a**
$\frac{dy}{dt} = k_{2b}x$	M2b
$\frac{d[HMX]}{dt} = -(k_{2a} + k_{2b})\rho_o \frac{1-x}{1+x}$	M3
$\frac{d[NO_2]}{dt} = (3k_{2a} + 2k_{2b})\rho_o \frac{1-x}{1+x} - k_3[NO_2][CH_2O] - k_{4a}[NO_2][HNC] - k_{4b}[NO_2][HNCO]$	M4
$\frac{d[CH_2O]}{dt} = 4k_{2b}\rho_o \frac{1-x}{1+x} - k_3[NO_2][CH_2O]$	M5
$\frac{d[HCN]}{dt} = 4k_{2a}\rho_o \frac{1-x}{1+x} - k_4[HCN]$	M6
$\frac{d[HNC]}{dt} = k_4[HCN] - k_{4a}[NO_2][HNC]$	M7
$\frac{d[HNCO]}{dt} = k_{4a}[NO_2][HNC] - k_{4b}[NO_2][HNCO] - k_{4b}[NO][HNCO]$	M8
$\frac{d[N_2O]}{dt} = k_{4b}[NO_2][HNCO] + k_{4b}[NO][HNCO] - k_{4b}[H][N_2O]$	M9
$\frac{d[NO]}{dt} = k_{4a}[NO_2][HNC] - k_{4b}[NO][HNCO]$	M10
$\frac{d[H]}{dt} = k_{4b}[NO_2][HNCO] + k_{4b}[NO][HNCO] - 2k_{4b}[H][N_2O]$	M11
$\frac{d[H_2O]}{dt} = 2k_{2a}\rho_o \frac{1-x}{1+x} + k_3[NO_2][CH_2O] + k_{4b}[H][N_2O]$	M12
$\frac{d[N_2]}{dt} = \left(\frac{1}{2}k_{2a} + 3k_{2b}\right)\rho_o \frac{1-x}{1+x} + \frac{1}{2}k_3[NO_2][CH_2O] + k_{4b}[H][N_2O]$	M13
$\frac{d[CO_2]}{dt} = k_3[NO_2][CH_2O] + k_{4b}[NO_2][HNCO]$	M14
$\frac{d[CO]}{dt} = k_{4b}[NO][HNCO]$	M15

TABLE 15.7 TATB Reaction network: Labeled by reaction R2–R6

Stoichiometry	No.	Energy (kJ/mol)
sublimation/vaporization $x \underset{k_{2b}}{\overset{k_{2a}}{\rightleftarrows}} y$ lattice dynamics	R2	0.0
$TATB - \xrightarrow{k_{2a}} 3HCN + 3NO_2 + 3H + 3C$	R2a	1,166.180
$TATB - \xrightarrow{k_{2b}} 3H_2O + 3N_2 + 3CO + 3C$	R2b	−1,049.590
$HCN + M \xrightarrow{k_4} HNC + M$	R4	54.000
$HNC + NO_2 \xrightarrow{k_{4a}} HNCO + NO$	R4a	−233.618
$HNCO + NO_2 \xrightarrow{k_{4b}} H + N_2O + CO_2$	R4b	−24.898
$HNCO + NO \xrightarrow{k_{4b}} H + N_2O + CO$	R5	−122.975
$2H + N_2O \xrightarrow{k_{4b}} H_2O + N_2$	R6	−759.872

TABLE 15.8 TATB ODE system: Labeled by species M2–M15

$\frac{dx}{dt} = -k_{2a}y$	**M2a**
$\frac{dy}{dt} = k_{2b}x$	M2b
$\frac{d[TATB]}{dt} = -(k_{2a}+k_{2b})\rho_o\frac{1-x}{1+x}$	M3
$\frac{d[C]}{dt} = 3(k_{2a}+k_{2b})\rho_o\frac{1-x}{1+x}$	M3a
$\frac{d[NO_2]}{dt} = 3k_{2a}\rho_o\frac{1-x}{1+x} - k_{4a}[NO_2][HNC] - k_{4b}[NO_2][HNCO]$	M4
$\frac{d[HCN]}{dt} = 3k_{2a}\rho_o\frac{1-x}{1+x} - k_4[HCN]$	M6
$\frac{d[HNC]}{dt} = k_4[HCN] - k_{4a}[NO_2][HNC]$	M7
$\frac{d[HNCO]}{dt} = k_{4a}[NO_2][HNC] - k_{4b}[NO_2][HNCO] - k_{4b}[NO][HNCO]$	M8
$\frac{d[N_2O]}{dt} = k_{4b}[NO_2][HNCO] + k_{4b}[NO][HNCO] - k_{4b}[H][N_2O]$	M9
$\frac{d[NO]}{dt} = k_{4a}[NO_2][HNC] - k_{4b}[NO][HNCO]$	M10
$\frac{d[H]}{dt} = 3k_{2a}\rho_o\frac{1-x}{1+x} + k_{4b}[NO_2][HNCO] + k_{4b}[NO][HNCO] - 2k_{4b}[H][N_2O]$	M11
$\frac{d[H_2O]}{dt} = 3k_{2b}\rho_o\frac{1-x}{1+x} + k_{4b}[H][N_2O]$	M12
$\frac{d[N_2]}{dt} = 3k_{2b}\rho_o\frac{1-x}{1+x} + k_{4b}[H][N_2O]$	M13
$\frac{d[CO_2]}{dt} = k_{4b}[NO_2][HNCO]$	M14
$\frac{d[CO]}{dt} = 3k_{2b}\rho_o\frac{1-x}{1+x} + k_{4b}[NO][HNCO]$	M15

TABLE 15.9 PETN Reaction network: Labeled by reaction R2–R6

Stoichiometry	No.	Energy (kJ/mol)
sublimation / vaporization $x \underset{k_{2b}}{\overset{k_{2a}}{\leftrightarrow}} y$ lattice dynamics	R2	0.0
PETN – $\xrightarrow{k_{2a}}$ $4NO_2 + 4CH_2O + C$	R2a	207.28
PETN – $\xrightarrow[k_{2b}]{}$ $4H_2O + 2N_2 + 3CO_2 + 2CO$	R2b	−1,830.43
$CH_2O + NO_2 \overset{k_3}{\longleftrightarrow} CO_2 + H_2O + (1/2)N_2$	R3	−552.543

capable of at least approximating the transition from ideal gas to incompressible fluid. This is particularly critical to understanding phenomena in the Acceleration bifurcation of Fig. 15.1 and the transition to regimes where pressurization due to thermal decomposition can couple to local shock physics.

TABLE 15.10 PETN ODE system: Labeled by species M2–M15

$\frac{dx}{dt} = -k_{2a}y$	M2a
$\frac{dy}{dt} = k_{2b}x$	M2b
$\frac{d[PETN]}{dt} = -(k_{2a} + k_{2b})\rho_o \frac{1-x}{1+x}$	M3
$\frac{d[C]}{dt} = k_{2a}\rho_o \frac{1-x}{1+x}$	M3a
$\frac{d[NO_2]}{dt} = 4k_{2a}\rho_o \frac{1-x}{1+x} - k_3[NO_2][CH_2O]$	M4
$\frac{d[CH_2O]}{dt} = 4k_{2b}\rho_o \frac{1-x}{1+x} - k_3[NO_2][CH_2O]$	M5
$\frac{d[H_2O]}{dt} = 4k_{2b}\rho_o \frac{1-x}{1+x} + k_3[NO_2][CH_2O]$	M12
$\frac{d[N_2]}{dt} = 2k_{2b}\rho_o \frac{1-x}{1+x} + \left(\frac{1}{2}\right)k_3[NO_2][CH_2O]$	M13
$\frac{d[CO_2]}{dt} = 3k_{2b}\rho_o \frac{1-x}{1+x} + k_3[NO_2][CH_2O]$	M14
$\frac{d[CO]}{dt} = 2k_{2b}\rho_o \frac{1-x}{1+x}$	M15

4 Adiabatic applications

We present results in this section of applying the HS kinetic model to each of the major regions of response illustrated in Fig. 15.1. The applications are based on simple adiabatic thermal decomposition in a dimensionless reacting volume without loss. This of course leaves out many important phenomena that are known to contribute to this response in specific instances, but we hope to isolate and validate general elements of the chemical decomposition represented here. We represent the various regimes, and illustrate the mechanisms of transition between them, with approximations in the initial thermal boundary condition and composition.

All calculations presented here are conducted using the models of HMX, TATB and PETN described above with no degrees of freedom or optimization. The models as described above differ only in the thermodynamic energies activating solid decomposition, the linear prefactors for solid decomposition determined from the data of Fig. 15.2, the absence of dark zone reactions in TATB, the absence of bright zone reactions in PETN, and the formation of graphite by TATB and PETN. They also differ in the particular stoichiometries that combine the reactions, and the solid phase heat capacities.

4.1 Heating to adiabatic thermal explosion

In this section, we present calculations utilizing thermal boundary conditions that reflect ignition near the critical temperature. We show the final milliseconds of approximately 300 s time to ignition. A similar time to ignition is achieved with adiabatic boundary conditions approximating the known critical temperatures, 145°C for PETN, 205°C for HMX, and 300°C for TATB. The calculations are taken to completion, and illustrate the progress of temperature and, using an EOS, pressure through to adiabatic thermal explosion. Final pressures calculated in this way are in agreement with the adiabatic thermal explosion pressure derived by Forbes [38].

4.1.1 HMX

The evolution of the solid fraction, several intermediate and product gas-phase concentrations, the temperature, and the pressure are shown for HMX as a function of time subject to the 205 C adiabatic boundary in the three panels of Fig. 15.4. Solid decomposition proceeds for the length of the ~300 s trajectory in the left panel, with intermediate NO_2 and HCN formation and some product formation. The solid fraction is consumed sharply over the millisecond timescale in the middle panel, accompanied by a stabilized HCN, NO_2 concentration. At the end of this intermediate period, the temperature and pressure are again seen to rise but the evolution of concentration is not resolved. The onset of HCN oxidation, which is complete over a microsecond time scale as the final temperatures are approached, is resolved in the right panel. The separation of time scale in solid, intermediate, and final reaction regimes is likely the defining mechanism from which an understanding of the ignition of deflagration and the subsequent transition to detonation may be understood in relatively dense solids, as will be discussed below.

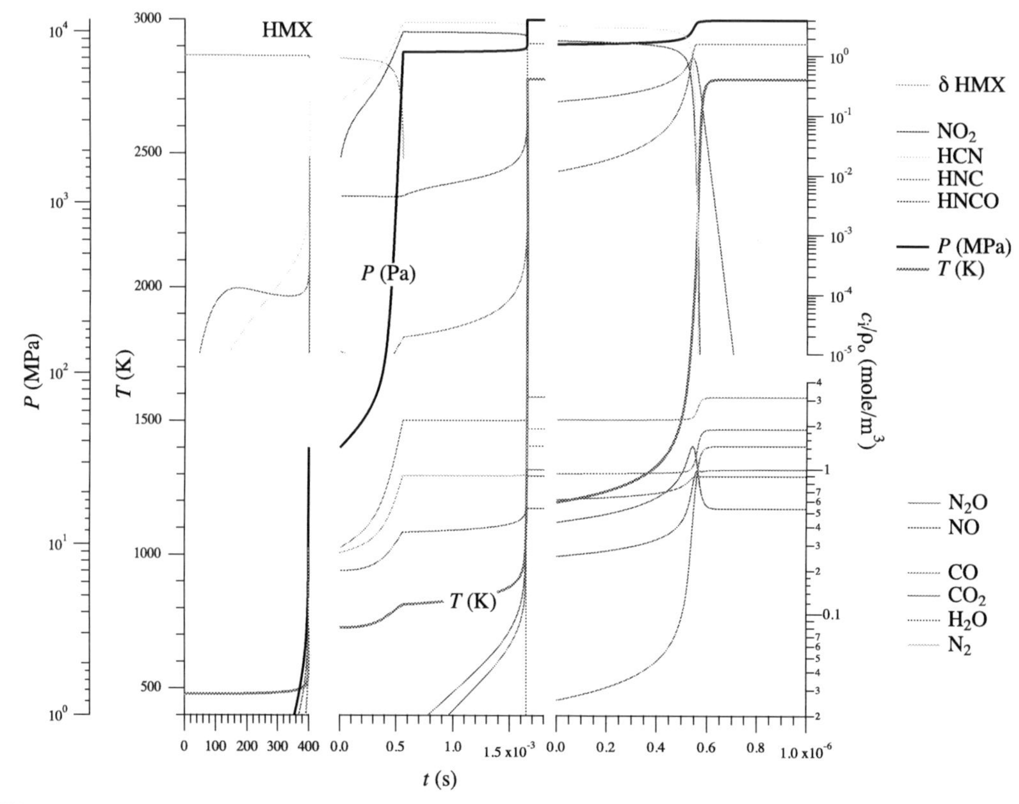

FIG. 15.4 Adiabatic thermal ignition in HMX at 200 C.

4.1.2 TATB

The evolution of the solid fraction and several intermediate and product gas-phase concentrations are shown for TATB as a function of time subject to the 300 C adiabatic boundary in Fig. 15.5. Also shown are the resulting temperature and pressure as a function of time. Solid decomposition again proceeds for the length of the ~300 s trajectory in the left panel, now without the intermediate chemistry available to HMX, descending sharply at the onset of HCN oxidation, which is not resolved. HCN oxidation, which is now complete over a slower, millisecond time scale as the final temperatures are approached at a much slower rate, is resolved in the right panel. The absence of intermediate chemistry, and therefore the lack of temporal separation of solid-state consumption and HCN ignition, and the resulting slower rate of final heating are likely the defining mechanistic differences between the Conventional High Explosive (CHE) HMX and the Insensitive High Explosive (IHE) TATB, particularly in observations of deflagration [39].

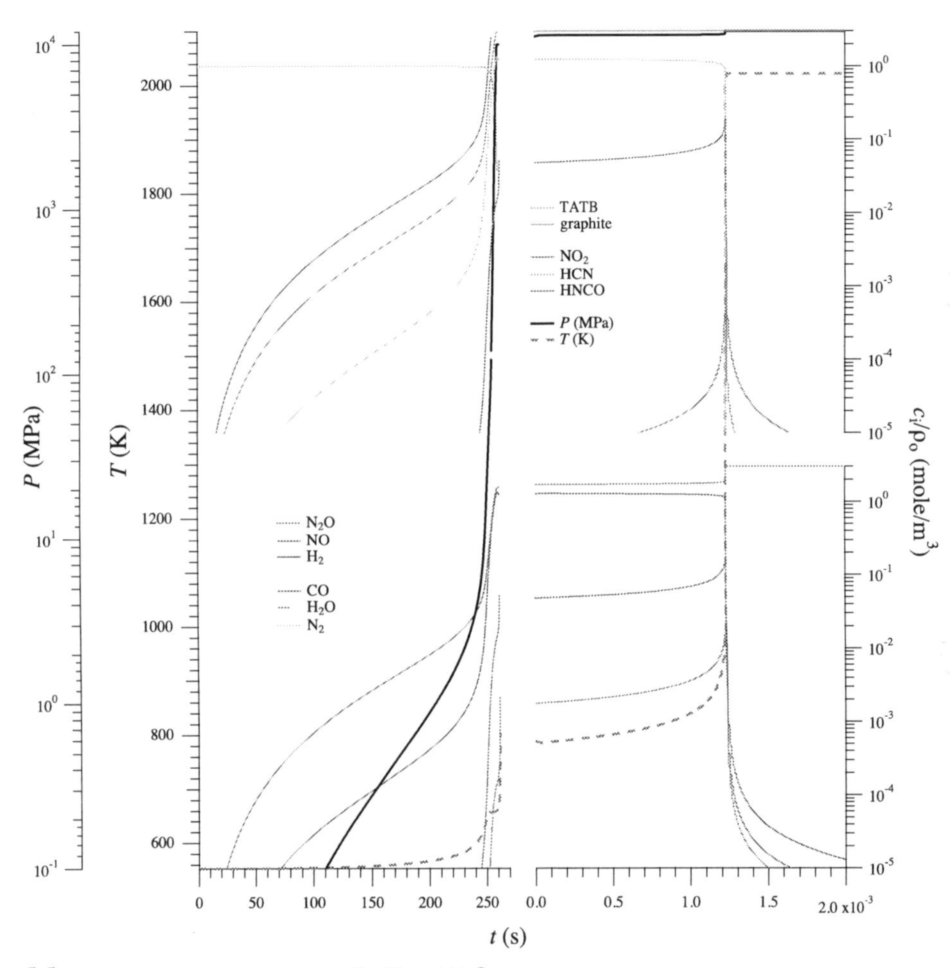

FIG. 15.5 Adiabatic thermal ignition in TATB at 300 C.

4.1.3 PETN

The evolution of the solid fraction and several intermediate and product gas-phase concentrations are shown for PETN as a function of time subject to the 145°C adiabatic boundary in Fig. 15.6. Also shown are the resulting temperature and pressure as a function of time. Solid decomposition again proceeds for the length of the ~300 s trajectory, now descending sharply at the onset of CH_2O oxidation, which is complete at the fast rates of solid decomposition and much faster rates of heating, over a nanosecond time scale. The absence of bright zone chemistry and the very high rates of the intermediate dark zone chemistry at these temperatures, and therefore the lack of temporal separation from solid-state consumption, are likely the defining mechanistic differences between the response of PETN, nearly as sensitive as a primary explosive, and either the CHE HMX or the IHE TATB.

4.2 Deflagration to detonation

The transition to detonation is a difficult and relatively poorly understood phenomenon. Models of the DDT are typically focused on the compaction of material ahead of a fast,

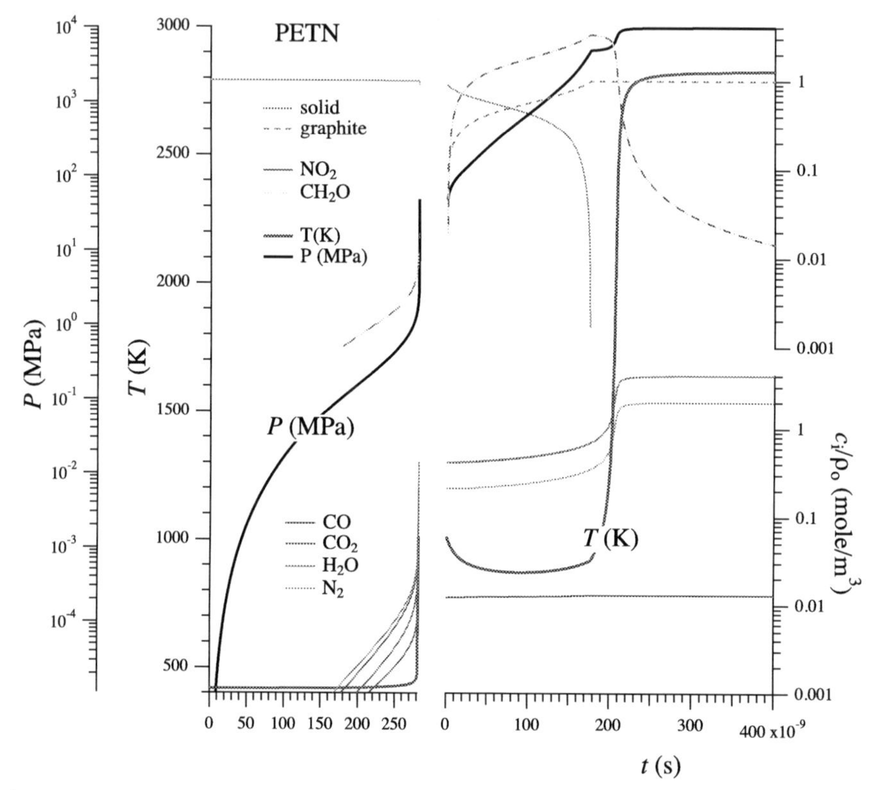

FIG. 15.6 Adiabatic thermal ignition in PETN at 140 C.

pressurizing deflagration front. If material ahead of the front compacts to a higher density than porous original material even further ahead, then the relatively fast speed of sound in the intermediate compressed material can lead to a steepening of the subsonic ramp wave being delivered ahead of the deflagration. If confinement, pressure, deflagration rate, and rates of compaction are such that this steepening ramp wave can become a supersonic shock wave, it is then compared in shock pressure to that necessary in the SDT to determine the length and time to initiation of detonation. We explore two key ideas in the transformation of the product chemistry in the HS kinetics into a high-pressure working fluid to generate detonation for comparison with these ideas.

The first is an assessment of the state of the product fluid as the high pressures implied by the adiabatic thermal explosion calculations above are approached. We show a very high fidelity compilation of pressure, density, and temperature data for N_2 in gray in Fig. 15.7 [40]. This compilation is represented in this plot of pressure as a function of density through

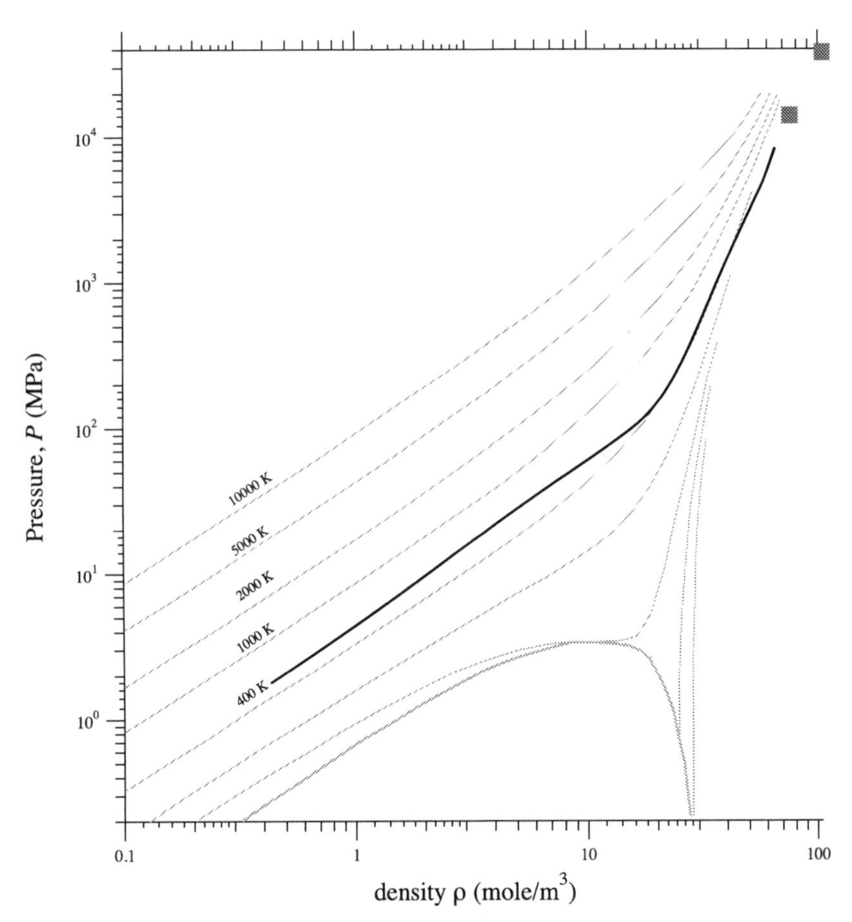

FIG. 15.7 Product equation of state trajectory for HMX from Fig. 15.4. *Curves in gray* are from the EOS for N_2.

a complex Equation of State (EOS) relating P and ρ as curves of constant T. The linear regime of ideal gas behavior is apparent, as is the spinodal region near the critical point. At high ρ the temperature dependence becomes much less pronounced, i.e., the curves begin to overlap, and a very steep dependence of P on ρ is seen.

This is the extraordinarily important regime of transition from a relatively ideal, highly compressible gas amenable to stable combustion to a very incompressible hydrodynamic fluid whose interactions are dominated by the hard sphere core of the molecular potential, whose energy is almost entirely kinetic and for which transitions to shock physics and detonation may be observed. In the representative product species N_2 *shown here, this transition occurs over approximately 100 to 1000 MPa, depending on temperature.*

The pressure in many hydrocode simulations is calculated from a JWL EOS. Both the adiabatic thermal explosion pressure predicted by Forbes [38] at 0.43 CJ ~ 14 GPa and the HMX CJ pressure at 35 GPa are shown as blue squares, where CJ denotes the Chapman-Jouget pressure. We have superimposed in Fig. 15.7. a bold black curve of P vs ρ resulting from the calculations of Fig. 15.9 for adiabatic thermal explosion in HMX. There is a nonlinearity in the ideal gas region of pressure due to the increasing temperature of the gas, and the transition to the incompressible region occurs abruptly at 100 MPa and ends near the adiabatic explosion pressure. The first important point is the clear excursion into the nonideal region of the potential as the pressure rises through to the completion of the reaction. A comparison of the pressure as a function of time in Fig. 15.9 with the pressure in Fig. 15.7 illustrates the stages in the chemistry that drive this excursion. It is clear that pressurization during solid decomposition and the accompanying significant increase in molar density as many product species are formed per mole of solid loss, very quickly drives the pressure into the hydrodynamic regime. In these adiabatic simulations, there is no flow, however, the necessity (and practical difficulty) of sufficiently confining an experiment to yield such high pressures is already apparent.

The second key point is the absence of any time dependence in Fig. 15.7. Indeed the pressure as a function of time for any of PETN, HMX, or TATB rises in an adiabatic calculation to something close to the thermal explosion pressure upon complete reaction, which is very similar for a given explosive from thermodynamic considerations, like those applied by Forbes [38,41]. This is clear from a comparison of peak pressures attained in Fig. 15.9. For that matter detonation velocities are also similar for these three explosives, from 7 to 9 km/s, again from thermodynamic considerations. However, the passage from the initially slow thermal decomposition to the fast final release of energy is kinetically controlled and defines the very different overall behavior of the near primary like PETN, the CHE HMX, and the IHE TATB.

This is shown quantitatively in Fig. 15.8. as simply the derivative of P as a function of time during adiabatic thermal explosion shown in Figs. 15.9. The pressure is shown in blue for each and plotted on the left axis as a function of a time window that captures most of the energy release. The power is plotted in bold black on the right. All three materials exhibit a double pulse in the time derivative, corresponding to peaks in power at the end of solid decomposition and a second at the end of the remaining chemistry when most of the conversion of chemical potential to heat occurs upon final product formation. The peak pressure is very

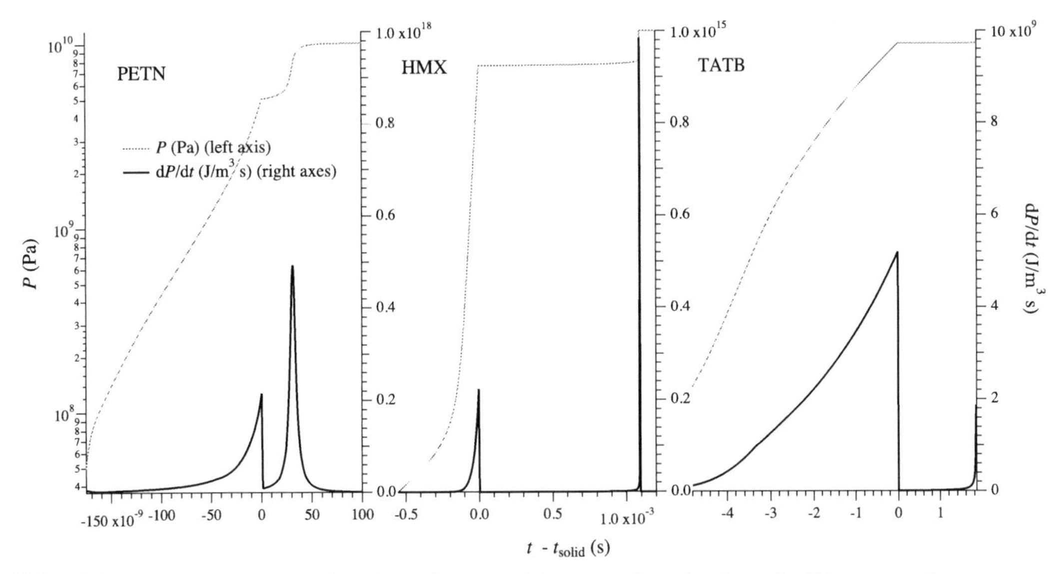

FIG. 15.8 Pressure derivative plotted as a function of the time minus the time of solid consumption.

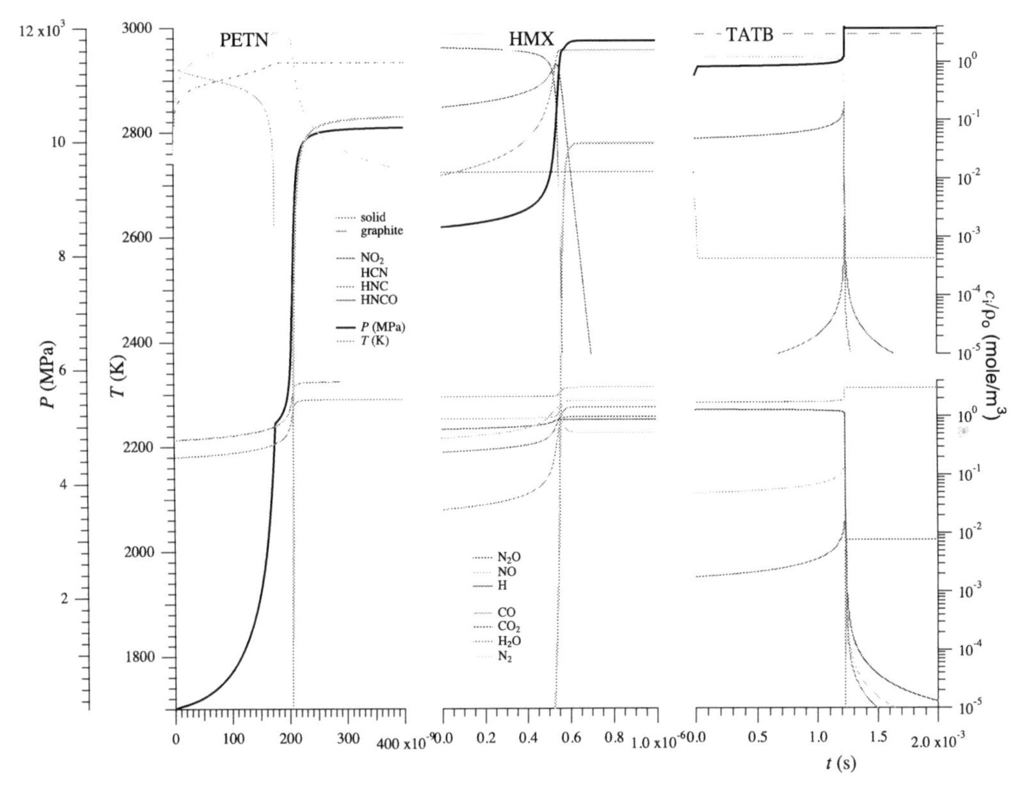

FIG. 15.9 Adiabatic thermal ignition of PETN, HMX, and TATB.

nearly the same for all three materials, and approximately the adiabatic thermal explosion temperature, as expected. The difference in the magnitude of the power is striking, however. PETN, with much of the release of energy concentrated in the last several hundred nanoseconds, achieves peak power on the order of 10^{18} J/m^3 s, HMX 10^{15} over microseconds, and TATB 10^9 over several seconds.

This represents the potential of each material to reach such levels of power generation with sufficient confinement, and to serve as a power source in reactive flow if the experiment (or simulation) is contrived to convert this adiabatic derivative into mass flow along a reacting trajectory, e.g., flow away from the reacting front subsequent to impact in a 1D gas gun experiment in the SDT regime or the same behind an accelerating deflagration front. For reference, in HMX an impact of sufficient momentum and impedance matching to generate a 1 GPa pressure in HMX is sufficient to result in a transition to detonation. The rise time of the impact is no more than approximately 1 ns, limited by instrument resolution, giving an approximate derivative imparted to the sample at an impact of 10 GPa/1 ns $\sim 10^{18}$. PETN easily satisfies this order of magnitude limit from the lowest thermal boundary condition of 145°C and over time, nanoseconds, which will serve to insulate the adiabatic response from many environmental perturbations, e.g., is fast compared to mass or thermal loss, and even perhaps mechanical failure at the boundary. HMX at 10^{15} Pa/s delivers significantly less power at ignition, consistent with its observed behavior and response relative to PETN, but sufficiently close that a transition to detonation is possible along a thermal trajectory, as is often, but not always observed. TATB at 10^9 is clearly significantly less than either. Six orders of magnitude separating TATB from HMX seems a comforting margin. And indeed a TATB based explosive formulation is not believed to have ever unambiguously detonated in a thermal ignition experiment. We have been working deliberately recently toward a rigorous scientific statement that not only has such a detonation never been observed with TATB, it cannot transition to detonation from a thermal ignition [39]. If we can determine rigorously a shock-change power necessary to initiate detonation along a thermal trajectory that lies somewhere on the high side of the six orders of magnitude gulf between TATB and HMX, we will have made the most significant progress yet toward this high consequence goal.

4.3 Detonation

A detailed understanding of the chemistry and energy release within the detonation reaction zone is widely accepted as necessary for understanding detonation curvature, turning and diffraction in complex geometries. Although considerable effort has been dedicated to this class of reactive burn modeling surprising little is known about the conditions that obtain at the leading edge of a detonation wave, particularly as these conditions at the leading edge are the boundary conditions that determine the kinetic response.

In this section, we show the results of adiabatic calculations for PETN, HMX, and TATB simulating this uncertainty in the most general way. We have chosen initial temperatures for each material which lead to an overall consumption time consistent with the known detonation reaction zone time for each. These temperatures are 2000°C for PETN, 2200°C for TATB and 1600°C for HMX. The calculations are then performed as they were for adiabatic thermal explosion, the only difference being the initial thermal boundary temperature.

The results are shown in Fig. 15.10. PETN and TATB show more or less smooth and abrupt passage to the completion of reaction with pressure rising smoothly to the adiabatic explosion pressure. Higher pressures would result from a state of compression in the solid, i.e., if a state ρ_o (compressed) of 1.55ρo (uncompressed) was chosen for HMX then the EOS would yield approximately the detonation CJ pressure at reaction completion. HMX again exhibits a staged energy release, with the discontinuity in reaction rate in dark and bright zone chemistry still affecting reaction structure in this regime. Finally, we show the product species ratio for all three systems under both thermal and detonation conditions in Figs. 15.11 and 15.12.

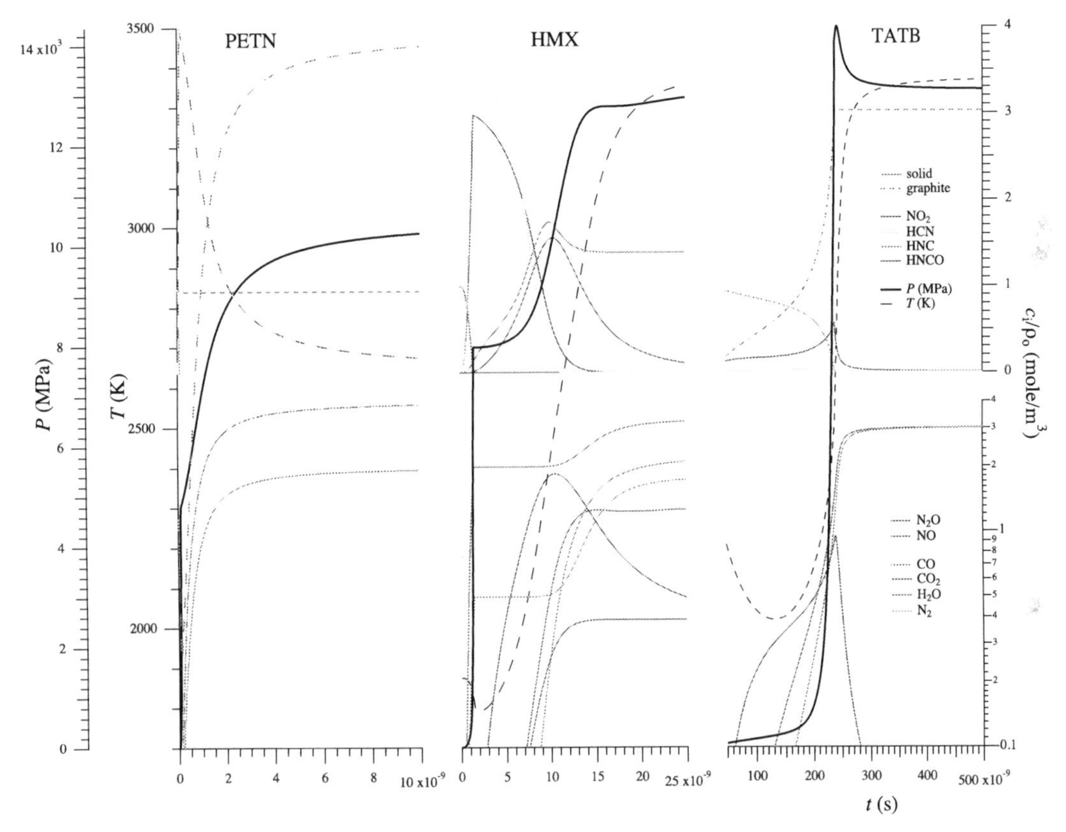

FIG. 15.10 Thermal ignition of secondary explosives in the detonation regime.

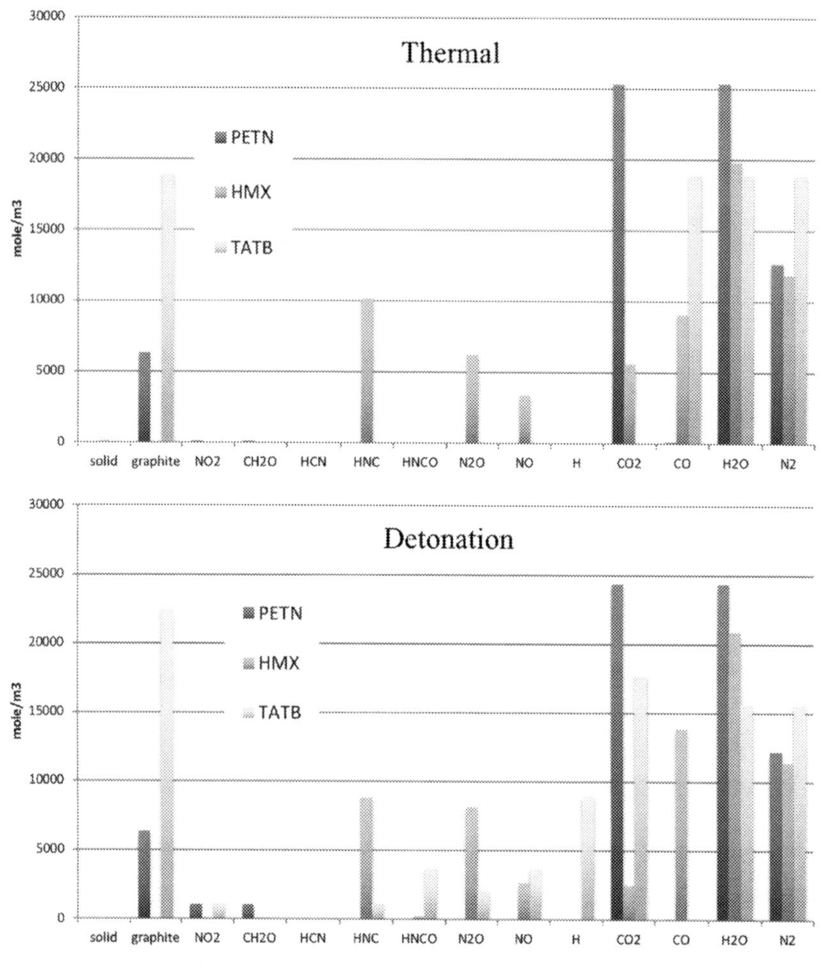

FIG. 15.11 Final product ratio for adiabatic thermal explosion and detonation in HMX.

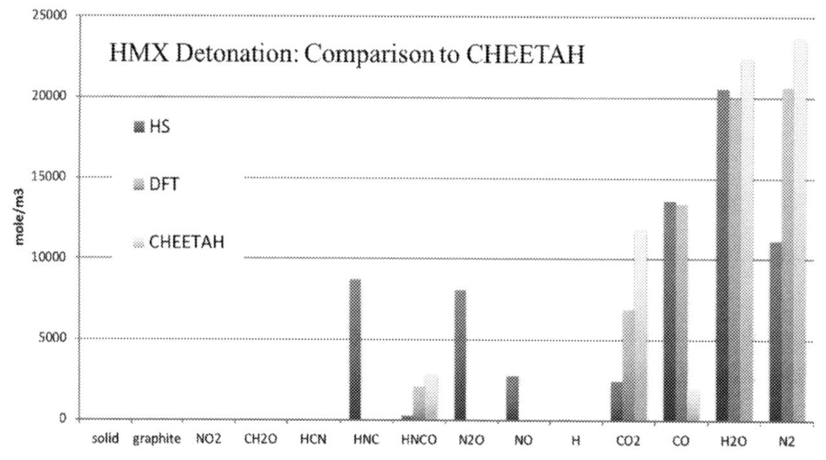

FIG. 15.12 Comparison of final product ratio for HMX ignition under detonation temperatures with CHEETAH and DFTB.

5 Summary and comparison to experiment

The breadth of data woven into this paradigm indicates a far greater accuracy of representation than can be indicated by the precision in fitting any given experiment. Indeed, the kinetic models presented above are parameterized using nearly every defining measurement of the crystalline solid and liquid phases, e.g., density, the heat of formation, enthalpies of sublimation, vaporization and fusion, and heat capacity. These are coupled to global product kinetics that, with the important exception of HCN isomerization, calculated by *ab initio* quantum mechanics, have been measured at temperatures relevant to this application. The synthesis of these experimental components into a comprehensive global kinetic system that accommodates these three very different explosives with no degrees of freedom is unprecedented. This very important qualitative accomplishment, as well as a sample of important quantitative comparisons with the experiment, will be summarized in this section.

5.1 Adiabatic thermal explosion

These adiabatic simulations are not adequate to provide detailed time or location of ignition under any specific heating scenario as spatial transport of mass and energy, particularly that related to sublimation and condensation, must be included in larger, multiphysics simulations as discussed in Section 2. These kinetic are, however, the time and temperature-dependent source terms that will be included in such simulations in the future. The new formula for solid decomposition and the precision with which the exponentially sensitive activation energies in the model have been constrained are expected to yield a significant increase in the confidence and fidelity of such simulations. At this point, the models accurately reproduce the range of temperature over which direct heating leads to a thermal explosion, from first principles and determined by the condensed phase free energies of each explosive.

These simulations are also extremely valuable in illuminating the mechanistic differences between these three secondary explosives that, while unified in their thermodynamic representation of the solid decomposition, respond so dramatically differently when coupled with the two globalized representations of subsequent gas-phase chemistry.

This difference in response has been known empirically for several decades. HMX is considered a CHE of reasonable sensitivity. It finds applications both as a high explosive main charge and in propellant formulations for which stable burning is desired. TATB is an IHE. While deflagration is observed it is not used as a propellant. PETN, while considered a secondary explosive and a CHE, is widely recognized to be much more sensitive than either HMX or TATB and considered to be nearly as sensitive as a primary explosive. It does see some application as a high explosive, typically blended with a large fraction of plastique or other soft components. While deflagration is observed it is never used in propellant formulations. The principal application of PETN is in commercial detonators.

These broad characterizations of behavior clearly have their origins in the chemistry, temperature, and pressure profiles of Figs. 15.4–15.6 and 15.8. In these calculations, typical combustion temperatures approaching 3000 K and pressures approaching the adiabatic (i.e., thermally isolated *and* uncompressed) thermal explosion pressure are obtained. More important are relationships between the presence or absence of dark and bright zone chemistries and solid rates of decomposition and heating. The absence of bright zone chemistry in PETN and the prodigious concentrations of fast-reacting dark zone intermediate species allow the very fast rate of heating to accompany the solid decomposition, even starting at the lowest critical temperature of 145°C. This leads to a final time scale of energy release in the nanosecond regime. The power delivered by this chemistry promptly on decomposition is thus consistent with the propensity of PETN to detonate. The presence of both dark and bright zone chemistries in HMX, and the reduction of immediate chemical potential available on solid decomposition to provide fuel for the slower reacting bright zone species leads to an intermediate temporal regime subsequent to solid decomposition in the millisecond time scale. At this scale, and given the solid and gas-phase thermal conductivities and observed rates of deflagration, the bright zone may spatially separate from the solid given sufficient free volume such that bright zone chemistry effectively establishes a new thermal boundary condition on the moving coordinate system of a stable conductive or convective burn front. Given this and the reduced power delivered by the chemistry due to delayed bright zone reactivity explains the role of HMX as a CHE and stable propellant material. Finally, the absence of dark zone burning in TATB, lower final temperatures, and the slower resulting rates of heating and power delivered by the bright zone chemistry explain the reluctance of this defining IHE to undergo a violent response on heating.

5.2 Detonation to deflagration transition

By following the temporal trajectory of both the fluid (gas phase) density and temperature evolution during the later stages of ignition it is immediately apparent that highly nonideal and incompressible states of the product fluid are potentially achieved during ignition, certainly by PETN and HMX. These incompressible states are furthermore obtained at rates that may enable wave coalescence and shock physics to arise independently of more complex processes of compaction.

This result challenges previous notions that it is solid compaction that is solely responsible for establishing the relative sound speeds along the front that enable the transition from subsonic ramp wave pressurization to supersonic shock velocities, and that the chemistry is responsible primarily as a means of pressurizing and compacting the solid ahead of a deflagration front. It seems clear that, given the highly confined conditions currently believed necessary for DDT, and closely resembling the adiabatic conditions of these calculations, the product fluid itself may be a sufficient medium of compression to enable the shock transition, with the solid ahead of the reactive front serving as the acceptor, rather than the generator, of the shock wave.

These calculations afford some quantitative comparison with an experiment in the measurements of pressure subsequent to thermal ignition and violent response in HMX (PBX 9501) of Urtiew et al. [42] In those experiments pressures subsequent to ignition in the range of 100 to 1000 MPa were observed over times of 10s of microseconds. The detonation was not observed at the level of confinement in those experiments. These measurements of pressure agree well with the results of HMX adiabatic thermal ignition in Fig. 15.4. We have also measured similar pressure, time histories in the LARS experiment at Los Alamos during steady deflagration experiments subsequent to thermal ignition in two HMX formulations PBX 9501 and PBXN-9 [43].

These transitional phenomena are extremely rate dependent, and this aspect of the response is also apparent from these kinetics in the calculation of the pressure derivatives of Fig. 15.8. Again quantitative comparison to experiment is available from the series of Mechanically Coupled Cook-Off experiments of Dickson et al. [44,45] In these experiments thin disks (5 mm) of PBX 9501 were heated to thermal ignition and in one case observed to undergo DDT. Pressures were inferred from hoop strain measurements in the ring of confining steel. Measured pressures are again in good agreement with the results of these calculations for HMX but, more importantly, calculations of dP/dt in these experiments yielded values approaching 10^{14} Pa/s just prior to the DDT transition and the failure of the mechanical boundary. This is in remarkable agreement with the peak values of 10^{14}–10^{15} calculated for HMX and shown in Fig. 15.8.

5.3 Adiabatic detonation

As with thermal ignition, these adiabatic simulations are not adequate to provide the detailed structure of the hydrodynamic coupling of product formation and flow, as conservation of mass and energy must be incorporated through, e.g., the Euler equations for a complete spatial representation. These kinetics are, however, the time and temperature-dependent source terms that will be included in such simulations in the future. The new formula for solid decomposition provided by the defect mechanism and the precision with which the exponentially sensitive activation energies in the model have been constrained are expected to yield a significant increase in the confidence and fidelity of such simulations. At this point, the models accurately reproduce a reasonable first approximation of reaction zone time, final temperature, and evolution of pressure subject to the assumed initial, adiabatic temperature. This is achieved from the first principles and determined by the condensed phase free energies of each explosive.

Current assumptions regarding the continuum temperature at the von Neuman spike are most robust for TATB, 1700 K from modeling using the newest parameterization of the Aslam, Wescott, Scott and Davis (AWSD) model. Temperature for HMX is perhaps 100 K higher and PETN somewhat higher still. The temperatures used for these calculations, 1600 – 2200 K, were chosen to bound the relative order of magnitudes of the three reaction zone structures assumed for TATB, HMX, and PETN. The effect of the uncertainty on the duration of the reaction zone at these temperatures is within reasonable uncertainty in the location of the sonic plane behind the detonation front. The reaction zone is defined here as the end of chemistry and presumed peak of product density when coupled to hydrodynamic flow. Such comparisons are complicated by the fact that many results utilizing current Equations of State are necessarily indirectly parameterized as a function of pressure or reaction progress.

The use of a detailed chemistry model also forces a reconsideration of the dependent variables and boundary conditions of the problem. From the point of view of continuum mechanics, the trajectory of temperature (and ultimately solid stress) at the leading edge of the detonation wave is implied to occur over a finite rise time likely in the first 100 ps (a distance of approximately 1 μm in the coordinate frame moving at 10 mm/μs). Little is known about this process for any energetic material. The use of the HS kinetics is thus another model paradigm through which to reconstruct something we cannot yet measure, the leading edge of a detonation wave.

Finally, a comparison of the final product composition of the HS adiabatic detonation calculations for HMX is made with CHEETAH and a Molecular Dynamics/Density Functional approach in Fig. 15.12 [46]. Again comparison is somewhat qualitative, even among the CHEETAH/DFTB comparisons in the original paper. Overall agreement is good on the formation of H_2O, HS is in better agreement with DFTB than CHEETAH regarding CO and there is general disagreement among all three regarding CO_2. HS is lacking significantly in N_2 formation due to significant nitrogen being stranded in the intermediates HNC, N_2O, and NO. Interestingly the free energy minimization techniques of CHEETAH and DFTB also strand chemical potential in the form of HNCO. R5 in the HS system, HNCO+NO, was specifically introduced to attempt to ameliorate this, although with only modest success.

6 Future directions

The direction of future work may now focus on further elucidating two important elements of the overall problem not considered here, the spatially resolved transport of energy and mass in the slower regimes of Heating, the transition through Ignition to Convection, and the complex thermal and solid stress boundary conditions that govern the temperature and hydrodynamic flow as Deflagration evolves through Acceleration to Detonation.

Both of these directions will require a combination of simple calculations such as those presented here used to illuminate the specific impact of various components of these processes, and larger multiphysics simulation in order to assess the coupling of the HS kinetic model with, e.g., mass and energy conservation in both slow thermal and fast detonative regimes.

Many such directions are possible. The objective of this chapter is to present the HS system as a comprehensive and robust chemical kinetic mechanism for use as the source term, or as a detailed model from which less complex, though still locally accurate, source term models may be created, for application in many problems in secondary explosive response.

References

[1] B.F. Henson, L. Smilowitz, B.W. Asay, P.M. Dickson, P.M. Howe, Twelfth Symposium (International) on Detonation, 987-992 (2002).

[2] L. Smilowitz, B.F. Henson, J.J. Romero, B.W. Asay, A. Saunders, F.E. Merrill, C.L. Morris, K. Kwiatkowski, G. Grim, F. Mariam, C.L. Schwartz, G. Hogan, P. Nedrow, M.M. Murray, T.N. Thompson, C. Espinoza, D. Lewis, J. Bainbridge, W. McNeil, P. Rightley, M. Marr-Lyon, The evolution of solid density within a thermal explosion. I. Proton radiography of pre-ignition expansion, material motion, and chemical decomposition, J. Appl. Phys. 111 (10) (2012) 103515.

[3] L.B. Smilowitz, B.F. Henson, J.J. Romero, D.O. Oschwald, Thermal decomposition of energetic materials viewed via dynamic x-ray radiography, Appl. Phys. Lett. 104 (2014).

[4] L. Smilowitz, B.F. Henson, J.J. Romero, B.W. Asay, C.L. Schwartz, A. Saunders, F.E. Merrill, C.L. Morris, K. Kwiatkowski, G. Hogan, P. Nedrow, M.M. Murray, T.N. Thompson, W. McNeil, P. Rightley,

M. Marr-Lyon, Direct observation of the phenomenology of a solid thermal explosion using time-resolved proton radiography, Phys. Rev. Lett. 100 (22) (2008) 228301–228304.
[5] C.M. Tarver, S.K. Chidester, A.L. Nichols, Critical conditions for impact and shock induced hot spots in solid explosives, J. Phys. Chem. 100 (14) (1996) 5794–5799.
[6] C.M. Tarver, T.D. Tran, Thermal decomposition models for HMX-based plastic bonded explosives, Combust. Flame 137 (1–2) (2004) 50–62.
[7] L. Smilowitz, B. F. Henson, M. M. Sandstrom, B. W. Asay and J. Romero, Thirteenth Symposium (International) on Detonation, 1026-1034 (2006).
[8] T. B. Brill and P. J. Brush, Seventh Symposium (International) on Detonation, 228 (1989).
[9] G. Lengelle, A. Bizot, J. Duterque, J.-C. Amiot, Ignition of solid propellants, in: Rechearche Aerospace, vol. 2, 1991, p. 1.
[10] B. F. Henson, Asay, B. W, Dickson, P. M., Fugard, C. and Funk, D. J., Eleventh Symposium (International) on Detonation, 325-331 (1998).
[11] B. F. Henson, L. Smilowitz, J. Romero, B. W. Asay and P. M. Dickson, AIP Conference Proceedings 845 (1), 1077-1080 (2006).
[12] L. Smilowitz, B. F. Henson, M. M. Sandstrom, B. W. Asay, D. M. Oschwald, J.J. Romero and A. M. Novak, AIP Conference Proceedings 845 (1), 1211-1214 (2006).
[13] W. G. Von Holle and C. M. Tarver, Seventh Symposium (International) on Detonation, 993-1003 (1989).
[14] T.B. Brill, K.J. James, Thermal decomposition of energetic materials. 61. Perfidy in the amino-2,-4,6-trinitrobenzene series of explosives, J. Phys. Chem. 97 (34) (1993) 8752–8758.
[15] S.F. Son, B.W. Asay, B.F. Henson, R.K. Sander, A.N. Ali, P.M. Zielinski, D.S. Phillips, R.B. Schwarz, C.B. Skidmore, Dynamic observation of a thermally activated structure change in 1,3,5-triamino-2,4,6-trinitrobenzene (TATB) by second harmonic generation, J. Phys. Chem. B 103 (26) (1999) 5434–5440.
[16] A. Campbell and R. Engelke, Sixth Symposium (International) on Detonation, 642-652 (1976).
[17] C. M. Tarver, R. R. McGuire, E. L. Lee, E. W. Wrenn and K. R. Brein, Seventeenth International Symposium on Combustion, (1978).
[18] J. Zinn, R.N. Rogers, Thermal initiation of explosives, J. Phys. Chem. 66 (12) (1962) 2646–2653.
[19] B.W. Asay, The chemical kinetics of solid thermal explosions, in: Y. Hori (Ed.), Shock Wave Science and Technology Reference Library, Volume 5, Springer-Verlag, Berlin, 2009, p. 617 (Chapter 3).
[20] J.W. Taylor, R.J. Crookes, Vapor-pressure and Enthalpy of Sublimation of 1,3,5,7-Tetranitro-1,3,5,7-Tetra-Azacyclo-octane (HMX), J. Chem. Soc.-Faraday Trans. I 72 (1976) 723–729.
[21] J.L. Lyman, Y.C. Liau, H.V. Brand, Thermochemical functions for gas-phase, 1,3,5,7-tetranitro-1,3,5,7-tetraazacyclooctane (HMX), its condensed phases, and its larger reaction products, Combust. Flame 130 (3) (2002) 185–203.
[22] Y. He, X.P. Liu, M.C. Lin, C.F. Melius, Thermal reaction of HNCO at moderate temperatures, Int. J. Chem. Kinet. 23 (12) (1991) 1129–1149.
[23] Y. He, X.P. Liu, M.C. Lin, C.F. Melius, Thermal reaction of HNCO with NO_2 at moderate temperatures, Int. J. Chem. Kinet. 25 (10) (1993) 845–863.
[24] C.Y. Lin, H.T. Wang, M.C. Lin, C.F. Melius, A Shock-tube Study of the CH_2O+NO_2 Reaction at High-temperatures, Int. J. Chem. Kinet. 22 (5) (1990) 455–482.
[25] M.C. Lin, Y. He, C.F. Melius, Communication: implications of the HCN YL HNC process to high-temperature nitrogen-containing fuel chemistry, Int. J. Chem. Kinet. 24 (12) (1992) 1103–1107.
[26] W. Tsang, J. Heron, Chemical kinetic data base for propellant combustion I. Reactions involving NO, NO2, HNO, HNO2, HCN and N2O, J. Phys. Chem. Ref. Data 20 (4) (1991) 609–663.
[27] T. Wing, J.T. Herron, Chemical kinetic data base for propellant combustion I. Reactions involving NO, NO_2, HNO, HNO_2, HCN and N_2O, J. Phys. Chem. Ref. Data 20 (4) (1991) 609–616.
[28] B.F. Henson, J.M. Robinson, Dependence of quasiliquid thickness on the liquid activity: a bulk thermodynamic theory of the interface, Phys. Rev. Lett. 92 (24) (2004) 246107.
[29] T.B. Brill, Multiphase chemistry considerations at the surface of burning nitramine monopropellants, J. Propuls. Power 11 (4) (1995) 740–751.
[30] B. F. Henson, L. Smilowitz, J. J. Romero and B. W. Asay, AIP Conference Proceedings 1195, 257-262 (2009).
[31] S.T. Thynell, P.E. Gongwer, T.B. Brill, Condensed-phase kinetics of cyclotrimethylenetrinitramine by modeling the T-jump/infrared spectroscopy experiment, J. Propuls. Power 12 (5) (1996) 933–939.

[32] T.B. Brill, P.J. Brush, S.A. Kinloch, P. Gray, Condensed phase chemistry of explosives and propellents at high-temperature—HMX, RDX and BAMO, Philos. Trans. R. Soc. London, Ser. A 339 (1654) (1992) 377–385.

[33] F.H. Pollard, P. Woodward, Reactions between formaldehyde and nitrogen dioxide: the explosive reaction, Faraday Soc. Trans. 45 (1949) 767–771.

[34] F.H. Pollard, R.M.H. Wyatt, Reactions between formaldehyde and nitrogen dioxide: the kinetics of the slow reaction, Faraday Soc. Trans. 45 (1949) 760–767.

[35] Y. He, E. Kolby, P. Shumaker, M.C. Lin, Thermal reaction of CH_2O with NO_2 in the temperature range of 393-476K: FTIR product measurement and kinetic modeling, Int. J. Chem. Kinet. 21 (11) (1989) 1015–1027.

[36] W. Tsang, Chemical kinetic data base for propellant combustion. II. Reactions involving CN, NCO, and HNCO, J. Phys. Chem. Ref. Data Monogr. 21 (4) (1992) 753–791.

[37] B.M. Dobratz, The insensitive high explosive triaminotrinitrobenzene (TATB): development and characterization, 1888 to 1994, Los Alamos National Laboratory Report Number LA-13014-H, 1995.

[38] J.W. Forbes, Shock Wave Compression of Condensed Matter: A Primer (Chapter 9), Springer Science & Business Media, 2013, p. 282.

[39] B.F. Henson, L. Smilowitz, A.M. Novak, D.M. Oschwald, M.D. Holmes, E.V. Baca, Fifteenth Symposium (International) on Detonation, (2014).

[40] R. Span, E.W. Lemmon, R.T. Jacobsen, W. Wagner, A. Yokozeki, A reference equation of state for the thermodynamic properties of nitrogen for temperatures from 63.151 to 1000 K and pressures to 2200 MPa, J. Phys. Chem. Ref. Data Monogr. 29 (6) (2000) 1361–1433.

[41] W. Fickett, W.C. Davis, Detonation: Theory and Experiment, Courier Corporation, 2012.

[42] P. Urtiew, J. Forbes, C. Tarver, F. Garcia, D. Greenwood, K. Vandersall, Thermal cook-off of an HMX based explosive: pressure gauge experiments and modeling, Russ. J. Phys. Chem. B 1 (1) (2007) 46–51.

[43] L. Smilowitz, B.F. Henson, G. Rodriguez, D. Remelius, E. Baca, D. Oschwald, N. Suvorova, Relationship between pressure and reaction violence in thermal explosions, AIP Conf. Proc. 1793 (2017) 040039.

[44] P. Dickson, B. Asay, B. Henson and C. Fugard, Proceedings, 11th International Symposium on Detonation, Office of Naval Research, Washington, DC, 1998.

[45] P.M. Dickson, B.W. Asay, B.F. Henson, L.B. Smilowitz, Thermal cook-off response of confined PBX 9501, Proc. R. Soc. London, Ser. A 460 (2052) (2004) 3447–3455.

[46] M.R. Manaa, L.E. Fried, C.F. Melius, M. Elstner, T. Frauenheim, Decomposition of HMX at extreme conditions: a molecular dynamics simulation, Chem. A Eur. J. 106 (39) (2002) 9024–9029.

[47] R.B. Cundall, T. Frank Palmer, C. E. C., Wood, Vapour pressure measurements on some organic high explosives, J. Chem. Soc., Faraday Trans. 1 74 (6) (1978) 1339–1345.

[48] R.N. Rogers, R.H. Dinegar, Thermal analysis of some crystal habits of pentaerythritol tetranitrate, Thermochim. Acta 3 (1972) 367–378.

[49] G. Krien, H.H. Licht, J. Zierath, Thermochemische untersuchungen an nitraminen, Thermochim. Acta 6 (5) (1973) 465–472.

PART V

Applications to the design of new materials

CHAPTER

16

Implementation of predictive models: Practical aspects

Didier Mathieu[a,*], *Romain Claveau*[a], *and Julien Glorian*[b]

[a]CEA, DAM, Le Ripault, Monts, France [b]French-German Research Institute of Saint-Louis, Saint-Louis, France

*Corresponding author: E-mail: didier.mathieu@cea.fr

1 Introduction

While it is not that difficult for organic chemists to think of energetic compounds with high energy content and good performances, it is a different matter to imagine compounds that are also safe and reliable [1]. Therefore, there is a significant interest for methods to estimate sensitivities. Ideally, such methods should be available through an user-friendly piece of software, such as the energetic materials designing bench (EMDB), a commercial package providing fast estimates for over 30 physicochemical and detonation properties of energetic materials [2]. Unfortunately, the models implemented in this software lack predictive value regarding sensitivities [3]. This comes as no surprise since the mechanisms by which an energetic material explodes under the effect of complex external loads like friction, electric discharge, or irradiation are far from being fully elucidated.

In fact, there is no consensus within the scientific community regarding the essential physical aspects to be taken into account to explain the differences in sensitivities between materials of different chemical composition. A wide range of unpredictable factors appear to play a role [4], including crystal morphology, particle sizes, number and nature of defects, ambient experimental conditions, and many others. In the last decade, the crystal packing has been given an important role by many researchers [5]. Relationships linking impact sensitivities to the vibrational structure of energetic molecular crystals are extensively studied as well [6, 7]. However, phonons in such materials are tedious to compute and this requires a knowledge of the crystal structure. In this context, empirical or semiempirical models based on analytical relationships provide a reasonable and widely accessible alternative in the early

Theoretical and Computational Chemistry, Volume 22
https://doi.org/10.1016/B978-0-12-822971-2.00016-4

stages of extensive screening of candidate compounds, as the only input data required are structural formulas [3] and possibly molecular geometries [8].

Empirical models extrapolate available experimental sensitivity data to new materials. This can be done on the basis of correlations linking the sensitivities to simpler properties, more amenable to measurement or theoretical evaluation. For instance, impact sensitivities were previously correlated to vibrational frequencies [9], NMR chemical shifts [10], or bond dissociation energies [11]. However, one-to-one correlations between distinct physical properties are necessarily restricted in scope. Furthermore, relationships involving experimental data are not suitable in view of extensive screening of candidate synthesis targets.

Therefore, most empirical correlations for sensitivities are based on so-called molecular descriptors, i.e., arbitrary numerical quantities readily calculable from the molecular structure. Although a particular descriptor has generally no physical meaning, it is assumed that a set of suitably selected descriptors provides a signature of the molecule, thus reflecting its structure, hence the name of descriptor. The selection of a particular subset of descriptors is usually done so as to maximize the robustness of the model, i.e., its ability to yield good predictions for compounds not used to fit the parameters (internal validation). Such a model is commonly referred to as a quantitative structure–property relationship (QSPR). The actual predictive value of a QSPR can only be evaluated empirically through its application to an external test set of unseen (i.e., not use in any way to develop the model) compounds. Unfortunately, complex models require a large training set to fit the parameters and select the hyperparameters (i.e., the constants not modified in the fitting process). As a consequence, it is difficult to save a significant number of entries as would be required to obtain an external test set suitable to draw firm conclusions. In the frequent case where the test set contains only a few entries (e.g., about 10) owing to a lack of available data, conclusions should be viewed cautiously [12]. Finally, since the scope of a QSPR depends on many factors (composition of the training set, nature of the selected descriptors, regression method used for the parameterization, etc.), it requires some expertise from the user or even specific adjustments on a case-by-case basis depending on the type of compound being studied.

In contrast to such purely empirical QSPR models, semiempirical ones rely on physical assumptions. This enables a more general description of materials properties through widely applicable models involving fewer adjustable parameters. For instance, a single semiempirical model for impact sensitivity was shown to apply successfully to a wide range of explosives, including nitroalkanes, nitramines, nitric esters, and nitroarenes [3]. On the other hand, basing a model on clear assumptions makes it possible to define the applicability domain (AD) on physical grounds, instead of resorting to statistical definitions which often prove unreliable in the field of energetic materials due to the limited amount of data at hand [13].

Despite the lack of reliable ready-to-use software to estimate the sensitivities of energetic materials, cutting-edge models can be leveraged and further developed by end users without waiting for their implementation by third parties, as outlined in this chapter. This is made possible by the many tools freely available on the web to handle molecules and examine how material properties depend on their structures [14]. Although models that require knowledge of the crystal structure are discussed, the focus is put on those based on 3D or 2D molecular structures, since they are the only option in the early stages of designing new energetic material components. Finally, friction sensitivity is considered as a worked-out example. We detail how anyone with a basic knowledge of the Python language and associated chemical toolkits can easily implement published empirical models and identify more convincing correlations in a couple of hours.

2 Models based on crystal structure

2.1 Phonon up-pumping models

As far as energetic material sensitivities are concerned, models taking advantage of a knowledge of crystal structure have been mostly developed for the sensitivity to impact, i.e., for the drop weight impact height h_{50}. They assume h_{50} to correlate with a measure of the rate of transfer of the impact energy to intramolecular vibrations. The latter must be sufficiently excited so that the corresponding enhancement of their vibrational amplitudes eventually causes the system to cross energy barriers leading to decomposition. These large amplitudes can also promote transitions to excited electronic states by causing the system to explore configurations where vibronic coupling is significant. For the decomposition process to propagate to the whole material, this energy transfer must be fast enough compared to the rate of dissipation of the chemical energy released upon formation of the final products.

Over the last three decades, much effort has been put in rationalizing impact sensitivities on this basis, as recently reviewed by Michalchuk [7]. These models make a distinction between phonons and internal vibrations, where this last term is used only for crystal vibration modes associated with intramolecular movements of the atoms, whereas phonons denote both acoustic and optical modes associated with intermolecular motions. It is assumed that the primary consequence of an impact on a crystal is to excite the phonons, while the initiation of a chemical decomposition is triggered by the excitation of internal vibrations. Therefore, energy must be transferred from the phonon bath (with frequencies up to Ω_{max}) to internal vibrations, especially to those most likely to trigger decomposition, like the stretching of R–NO_2 trigger bonds in nitro compounds. Since the latter occur at frequencies much higher than Ω_{max}, direct energy transfer from the phonons to these high frequency modes is unlikely. It is thus assumed to occur via so-called doorway modes, i.e., internal vibrations lying in the doorway region ranging roughly from Ω_{max} to $2\Omega_{max}$.

Models developed on the basis of this up-pumping mechanism lead to purely theoretical indicators of the rate of energy transfer from the phonons to internal vibrations that have been put forward as valuable indicators of impact sensitivities [6, 15, 16]. In principle, evaluating these indicators sound challenging as they depend on the cubic anharmonic coupling constants describing the strength of interactions between stationary (harmonic) vibrational states. In particular, the up-pumping efficiency depends on the anharmonic constant for two low-energy phonons coupling to form a higher energy phonon. In practice, broadly similar values of these coupling constants are obtained for a range of energetic materials [17]. Therefore, it might be possible to explain sensitivity differences while ignoring the differences in anharmonic coupling constants between materials. As a potentially more discriminating factor, the density of vibrational states in the doorway region appears as especially significant and a good candidate for a practical indicator of impact sensitivity based only on first-principles calculations.

2.2 Practical aspects

Evaluating this indicator requires that the phonon density of states of the explosive crystal be calculated, typically from plane-wave density functional theory (DFT) calculations with dispersion corrections. The phonons can be obtained either from the forces resulting from atomic displacements away from their equilibrium positions (the so-called frozen phonon method) or from the linear response method, i.e., the application of density functional perturbation theory to the electronic response to atomic motion. The former approach is implemented in the open-source package PHONOPY, which has interfaces to a wide range of periodic electronic structure codes [18]. The latter is more complex and implemented only in few packages, including CASTEP [19], VASP [20], or Quantum Expresso [21]. A especially promising package to process electronic and vibrational data derived from periodic atomistic computations is the Python atomic simulation environment (python-ASE) [22]. Although it does not presently handle phonons calculations, it may prove useful to implement the subsequent postprocessing steps required to quantify the transfer of vibrational energy through the crystal lattice, as detailed in Chapter 10.

In spite of rather promising results, these models have been scarcely applied so far. There are at least two reasons behind this. The first one is that they require the crystal structure of the material, which is usually not available for compounds whose sensitivity is needed. Crystal structure prediction (CSP) is a long-standing problem for which satisfactory solution has emerged only recently [23]. The approach based on a tailor-made force field implemented in the GRACE program proves to be especially successful for molecular crystals. In view of its significant computational requirements and of the advanced methods involved, this approach has been mostly used by drug manufacturers. It should be of great value as well in the final stages of in silico design of energetic materials, although more efficient methods are clearly needed for prescreening of the chemical space.

The second reason why such models to predict sensitivity are not widely used might be due to the fact they involve the arduous (albeit relatively straightforward compared to CSP) task of computing the vibrational structure of complex molecular crystals [7]. However, the number of doorway modes, whose role is especially significant, primarily depends on the structure of a single molecule. This suggests that useful sensitivity estimates might be obtained without considering the actual crystal structure, but simply from the examination of the doorway region for isolated molecules.

3 Models based on molecular geometry

3.1 Generation of 3D molecular models

**** Although the estimation of sensitivity from the vibrational frequencies of isolated molecules has not been studied until now, to the best of our knowledge, many empirical approaches involve 3D descriptors [24–26]. Their application to new compounds thus requires that 3D molecular conformations are first generated. Compounds with flexible geometries exhibit a wealth of distinct conformers, many of which are very similar in energy. This is illustrated in Fig. 16.1 in the case of HMX. With regard to sensitivity estimation, any reasonable conformer may be used as long as the model involves only properties that

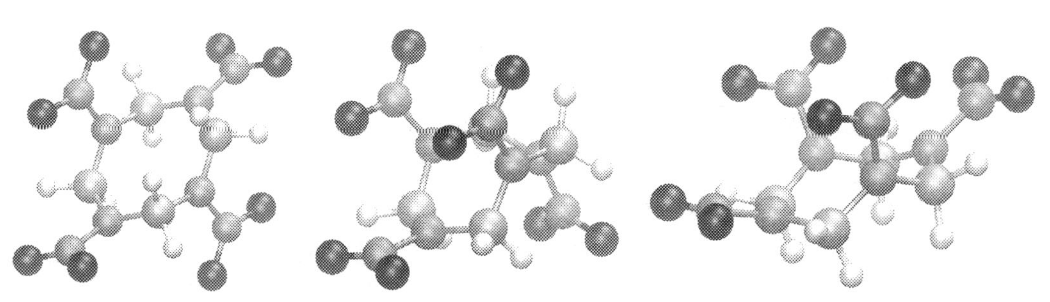

FIG. 16.1 Some conformations obtained for HMX: conformer ranked #0 (i.e., the one with lowest energy) obtained using Open Babel, and conformers ranked #0 and #11 obtained using CREST. The latter is the boat conformer observed in the δ phase of the HMX crystal.

do not critically depend on the specific conformation considered, e.g., bond dissociation energies, energy content, and atomic charges. However, some sensitivity models involve descriptors that may exhibit a significant dependence on molecular conformation. This is especially the case for models derived from the so-called general interaction properties function (GIPF) descriptors [27]. Although the latter were primarily designed to describe properties determined primarily by noncovalent interactions, they might prove valuable descriptors to estimate other properties, including sensitivities of energetic materials [28, 29]. Such models are usually parameterized with the descriptors calculated for conformers taken from X-ray crystal structures. In practice, they are usually applied to unknown compounds whose solid-phase conformation is unknown, which can only be done with caution.

Procedures aimed at building theoretical 3D models of the molecules are implemented in a wide range of conformer generators, either freely available [30] or commercial [31]. The former are usually sufficient for the small molecules considered as high energy materials. They usually aim at searching for the lowest-energy gas-phase conformer. However, some of them incorporate knowledge of preferred torsional angles derived from experimental crystal structures [32]. Algorithms leading to geometries in close agreement with crystal conformations may be especially advantageous to apply empirical models based on fitted relationships between the property of interest and 3D descriptors calculated from X-ray conformations, like the above-mentioned GIPF models. On the other hand, privileging conformations representative of crystals could be a source of ambiguity as crystal structures are usually not available for compounds whose sensitivity is to be estimated. Furthermore, some compounds may change their conformation under the effect of an external load.

Sensitivity models based on 3D molecular geometries usually involve rather costly ab initio or DFT calculations of the electronic structure. In this case, there is no point in focusing on highly efficient conformer generation procedures based on classical molecular mechanics (MM) force fields like MMFF, which were mostly designed with biochemical applications in mind. A procedure based on semiempirical quantum chemical calculations which are likely to provide a more transferable force field may be preferred, like the iMTD-GC workflow based on extensive metadynamic sampling (MTD) with an additional gradient conjugate (GC) optimization at the end, as implemented in the CREST software [33] which calculate energies at the GFN2-xTB tight-binding level [34] through calls to the xtb program [35].

3.2 Evaluation of 3D descriptors

With regard to the prediction of energetic materials sensitivities, 3D molecular models might be used to estimate the number of doorway modes involved in the up-pumping mechanism outlined in Section 2.1, in an attempt to avoid considering the complex vibrational structure or molecular crystals. However, they are rather used so far in the framework of empirical QSPR approaches to determine the values of 3D descriptors, including GIPF descriptors, conceptual DFT descriptors, atomic charges, bond dissociation energies, and bond lengths. These quantities are classified as 3D descriptors since they are derived from quantum chemical (QC) calculations. While most of them require QC calculations only for the equilibrium geometry of the studied compound, this is not always the case. For instance, a bond dissociation energy (BDE) is typically obtained as:

$$\mathrm{BDE} = E(\mathrm{R}_1^{\cdot}) + E(\mathrm{R}_2^{\cdot}) - E(\mathrm{R}_1 - \mathrm{R}_2) \tag{16.1}$$

where R_1–R_2 represents the energetic molecule, $R_1^{\cdot}$ and $R_2^{\cdot}$ are the radicals resulting from the homolytic rupture of the bond between them, and $E(X)$ denotes the energy of any species X. For instance, evaluating the BDE of a trigger bond R–NO_2 in a nitro compound involves two radicals $NO_2^{\cdot}$ and $R^{\cdot}$. For every BDE to be estimated, three QC calculations are needed, namely one for the molecule studied and one for each of the two radicals resulting from the bond scission. To avoid redundant calculations, the different species involved in the calculations of the BDEs of a set of molecules should first be identified. In view of the significant number of species typically involved in such studies, it is desirable that the computations be automated as much as possible so as to lessen the need to restart calculations that have not converged or that have converged to critical points other than minimums. For instance, IRC calculations [36] should be run automatically until the critical point obtained exhibits no imaginary frequency.

QC calculations aimed at deriving descriptors to estimate sensitivities are mostly computed using DFT, e.g., using the popular hybrid B3LYP functional. In addition to programs widely used by the quantum chemistry community like Gaussian or ORCA [37], there are many freely available alternatives, including Psi4 [38] or PySCF [39]. Many descriptors such as atomic charges, bond orders, and bond lengths are directly reported in the output files provided suitable options were provided as input. More advanced descriptors, including those involved in GIPF models, can be computed using a wavefunction analyzer like Multiwfn [40].

The fact that BDEs are mostly calculated at the DFT level is understandable since conventional semiempirical Hamiltonians such as AM1 or PM3 are primarily fitted against experimental data for closed-shell molecules and may lead to unreliable results for radicals. Similarly, GIPF descriptors may require this level of calculation as they critically depend on the electron density. However, faster-than-DFT models are desirable as the design of new materials requires that many potential compounds are considered. Modern semiempirical Hamiltonians tend to be especially fast, relying on a tight-binding (rather than NDDO) approximation. They include the density functional tight-binding (DFTB) approaches as well as the so-called GFN (Geometries–Frequencies—Non covalent interactions) family of models. These recent QC procedures are available in open-source codes like DFTB+ [41], SCINE Sparrow [42], or xtb [35]. In the near future, it should even be possible to estimate some

QC descriptors while avoiding explicit QC calculations altogether. For instance, recent analytical models of electron density [43, 44] should provide better estimates of the GIPF descriptors than previously used point-charge approximations [45, 46].

4 Models based on structural formula

4.1 Handling 2D molecular representations

Structural formulas are particularly meaningful representations in the eyes of the chemist, but do not constitute an efficient means of storing 2D molecular information. In chemo-informatics, a molecule is preferably encoded as a string of characters. The SMILES notation [47] is especially attractive as it is human-readable and lends itself to reduction into canonical forms, thus allowing easy detection of duplicates in large databases. A very good introduction to this notation is provided on Wikipedia [48]. SMILES can be automatically downloaded from online databases like PubChem [49] through interfaces like PubChemPy [50], or generated from MDL. mol files, either downloaded also from databases or prepared using chemical drawing software like Marvin Sketch [51]. Finally, SMILES can be generated directly from chemical names using OPSIN [52].

They are a number of libraries focused on handling molecules, including the Indigo Toolkit [53], Open Babel [54], or RDKit [55]. Indigo provides excellent 2D rendering capabilities leveraged by mol2chemfig, a tool to draw structural formulas [56]. RDkit is a C++/Python library providing a broad range of features, suitable for quick and dirty scripting and large-scale cheminformatics projects as well. Applications of this library to modeling the properties of organic materials from structural formulas typically require only few lines of Python code, as illustrated in Section 5.

4.2 Descriptors derived from 2D formulas

Any quantity that can be unambiguously derived from the molecular structure may be considered as a descriptor, including for instance the wealth of data provided by quantum chemical calculations: atomic charges, polarizability, orbital energies, and geometric parameters. The latter are classified as 3D descriptors because 3D molecular geometries are needed for their calculation. In order to avoid the intricacies and ambiguities inherent to the use of 3D structures, it is advantageous to restrict the pool of descriptors considered to simpler quantities directly calculable from the structural formula. Those quantities, classified as either 0D, 1D, or 2D descriptors, most often make it possible to develop easy to use empirical models, with reliability comparable to more complex models based on 3D conformations.

The 0D descriptors are count descriptors. In other words, a 0D descriptor is simply the number of occurrences in the molecule of a specific substructure: atom, bond or group. For instance, it could be the total number $n(\mathrm{N})$ of nitrogen atoms in the molecule. For satisfactory results, atoms in different chemical environments must usually be considered as belonging to distinct atom types, even if they are the same element. For instance, a nitrogen atom in a primary amine is usually considered as different from a nitrogen atom in a nitro group. This atom-based approach is similar to considering more extended molecular fragments like functional groups. For instance, the number $n(\text{N}-\text{nitro})$ of nitrogen atoms in a nitro group obviously amounts to the number of nitro groups $n(\text{nitro})$. A drawback of groups by comparison with typed atoms is that the decomposition of a molecule in terms of groups can be ambiguous with some popular methods like the original UNIFAC [57]. Fig. 16.2 illustrates such an ambiguous decomposition of etabonic acid (CH_3–CH_2–O–COOH) into groups, assuming that the group contribution model considered does not provide any parameterization for the etabonic group –OCOOH, but exhibits parameters for the following groups: –O–, –COOH, –OC(=O)–, and –OH.

More serious ambiguities may be faced as a result of tautomerism, e.g., with amides which may be viewed as either –NH–C(=O)– or –N=C(OH)–. Such ambiguities cannot be resolved with any additive scheme, unless some constraints are put on the parameters to ensure that both representations are equivalent.

The 1D descriptors are molecular fingerprints, i.e., vectors encoding the structural features of the molecule and allowing fast similarity comparisons. They are used mostly in the field of biochemical sciences for mapping chemical space, virtual screening, and structure–activity relationships based on machine learning techniques.

The 2D descriptors are more general graph invariants known as topological indices [58]. Unlike the 0D descriptors, they do not always allow a straightforward interpretation of their values.

Several computer programs are available to compute molecular descriptors, including RDkit. Besides proprietary codes implementing extremely broad ranges of descriptors, PaDEL-descriptor [59] and Morded [60] are most popular among software specifically dedicated to the calculation of descriptors. Mordred is very easy to use either as a stand-alone program or as a Python library. However, these tools are not required to calculate 0D descriptors. This can easily be done using RDkit, especially for models based on typed atoms or bonds, as illustrated by a recently published Python script for estimating standard sublimation enthalpies [61]. For models involving more specific fragments, this becomes somewhat more complicated, as illustrated by the scripts reported for other properties like Hansen solubility parameters [62] or dielectric constants [63]. An extremely valuable tool to identify and count arbitrary substructures in molecules are SMARTS strings, which are an extension of SMILES allowing to impose conditions on atomic environments [64].

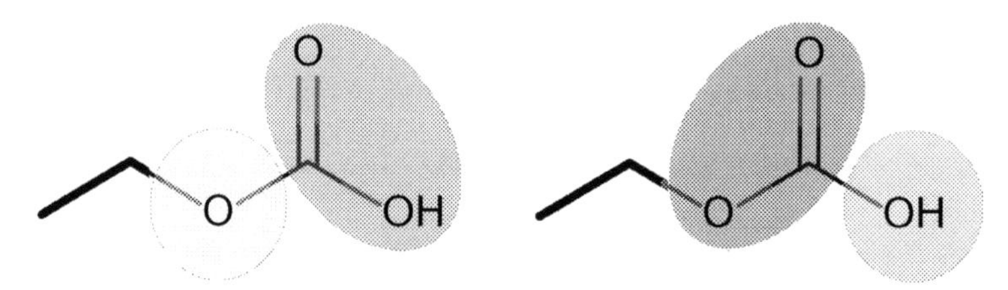

FIG. 16.2 Ambiguity in the decomposition of etabonic acid into groups. In the lack of specific parameters for the etabonic group –OCOOH, the latter can be viewed as a combination of either –O– and –COOH or –O(C=O)– and –OH.

4.3 Additivity schemes and 0D descriptors

The 0D descriptors fall into two main groups depending on the principles guiding the decomposition of molecules into fragments. The first category includes the descriptors resulting from a systematic splitting of molecules into fragments considered on the same footings. Such fragments will be referred here as constitutive descriptors. They are especially useful to estimate extensive properties that scale linearly with respect to molecular size. These properties can be approximately estimated by adding contributions associated with all fragments that make up the molecule (hence called additive fragments). The resulting additive models are named after the nature of the fragments. Those based on atom and bond counts are the simplest but they usually lack in accuracy. Therefore, group contribution (GC) methods [65] are more commonly used, despite their greater complexity. A good introduction to this technique was recently published [66]. Although sensitivities are not extensive properties, their estimation may involve ancillary quantities that lend themselves to decomposition into additive contributions, including for instance heat capacities [67].

Distinctions between different groups are somehow arbitrary. For instance, the two methylene moieties ($-CH_2-$) in propylamine ($H_3C-CH_2-CH_2-NH_2$) can be described either by a single group type i, or by two distinct groups i (for $-CH_2-$ between two sp3 carbon atoms) and j (for $-CH_2-$ between an sp3 carbon and nitrogen atoms) as shown in Fig. 16.3. In any case, the function f linking a property to the occurrences n_i, n_j... of the various groups should return a value in line with experiment, i.e.:

$$f(\ldots.,n_i,n_j\ldots.)=f(\ldots.,n_i+n_j,0\ldots.)=f(\ldots,0,n_i+n_j\ldots.) \tag{16.2}$$

This puts significant constraints on the form of this relationship that are not necessarily obeyed by the analytic expressions commonly used to relate group counts to properties of interest in QSPR modeling. For instance, the so-called group contribution neural network (GCNN) models are not consistent in this respect. As a consequence, a slight change in the definition of the fragments can lead to completely different predictions. Remarkably, this does not prevent models of this kind from providing excellent predictions in practice [68].

The second category of constitutive descriptors includes those aimed at capturing the special role of some substructures (e.g., explosophores) on the property of interest. Such descriptors are used by various models, either semiempirical [4, 13, 69] or purely empirical, to estimate sensitivities of explosives [70] or their decomposition temperature [71]. Semiempirical models are usually more reliable [71, 72]. However, purely empirical models might provide in some cases valuable extrapolations based on experimental trends observed within similar compounds, as shown in Section 5.

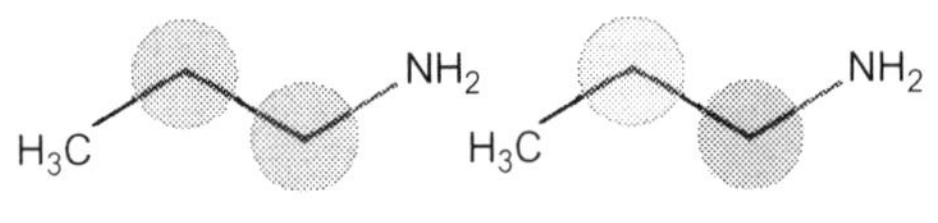

H_3C NH_2 H_3C NH_2

FIG. 16.3 The two methylene moieties ($-CH_2-$) in propylamine can be treated as two instances of the same group (*left*) or as belonging to distinct groups (*right*). This must not have any dramatic effect on the properties to be predicted.

4.4 Models based on general 2D descriptors

QSPR models are especially useful in the field of biochemical sciences, in which this acronym is declined under a number of variants depending on the property "P" to be estimated: the term QSAR refers to QSPRs aimed at evaluating the biological activity of compounds, while QSTR refers to those aimed at evaluating their toxicity. It is well known that such properties depend on the ability of molecules to interact with specific binding sites. Therefore, they are not primarily determined by the functional groups present on the molecule, but depend critically on the mutual positions of these groups. In general, biological properties cannot be studied on the basis of only 0D descriptors. For instance, popular definitions of molecular similarity rely on molecular fingerprints (1D descriptors). Such similarity measures are required in the study of activity cliffs (ACs), i.e., cases where chemically similar compounds exhibit very large differences in biological activity [73]. In addition to molecular fingerprint, a large number of general 2D descriptors have been introduced to overcome the limitations of 0D descriptors in pharmaceutical sciences.

In contrast to biological properties, most physical and chemical properties of organic materials are primarily determined by their composition in functional groups, as clear from the success of GC methods [65]. With regard to sensitivities, the term functional group should be taken here in a rather broad sense, including extended molecular fragments such as tetrazoles, furazans, and furoxans rings, and taking the surroundings of trigger bonds into account, as done to evaluate the decomposition temperatures of organic peroxides [71].

Although GC methods should in principle enable reliable predictions of the properties of energetic materials provided sufficiently specific groups are introduced, this approach is hampered by the large amount of experimental data needed to fix the values of the numerous parameters involved. Alternative 2D descriptors are attractive as some of them could more efficiently capture significant contributions to the property of interest than any group constant. For instance, topological indices are extensively used to make up for missing group parameters in predicting polymer properties [74]. Surprisingly, such empirical models taking advantage of but restricted to the wide range of 2D descriptors are seldom considered for sensitivities, with most models either restricted to constitutive (0D) descriptors [70] or involving 3D descriptors. QSPR models deliberately including general 2D descriptors (like topological indices) while excluding 3D ones have been seldom investigated [75], a prominent example being the neural network of Wang et al. [76]. The fairly good results obtained suggest that there is little point in considering 3D descriptors for impact sensitivities.

5 Worked-out example: Modeling friction sensitivity

This section demonstrates the predictive modeling of friction sensitivity (FS). First, we show how simple models reported in literature can be implemented in no time and examined in some depth to better understand their limitations. Then, we show how to leverage the limited amount of data typically available in the field of energetic materials in order to obtain reasonable sensitivity estimates for similar compounds.

5.1 Implementation of a model from literature

We start with the implementation of one of the few methods currently available to estimate FS. In the lack of a satisfactory understanding of the molecular factors responsible for the different FS values observed between materials, this property is assumed by Keshavarz et al. [77] to depend on the activation energy E_a for thermolysis. While this is reasonable, this allows the authors to compile only 29 nitramine materials, including 6 salts, with both FS and E_a available. Such a limited data set is usually not considered as sufficient to develop a quantitative structure–property relationship using statistical regression methods, let alone machine learning (ML) techniques. In practice, a more exhaustive compilation of experimental data should be done before undertaking seriously the development of empirical predictive schemes. However, in view of its simplicity and the fact it involves highly specific substructures in addition to well-known nitramine explosophores, the Keshavarz et al. model is ideally suited to demonstrate how short Python scripts can be developed in no time to extrapolate experimental sensitivity from a small data set to structurally similar compounds. This model assumes the friction sensitivity FS (in Newton units) to be given as a sum of two components:

$$FS = FS_0 + FS_{cor} \tag{16.3}$$

where FS_0 depends only on the stoichiometry and activation energy for thermolysis of the material:

$$FS_0 = 212 + 32.67n_C - 14.50n_O - 10.21n_N - 85.07E_a/M_w \tag{16.4}$$

FS_{cor} is a correction to FS_0 including positive and negative contributions as follows:

$$FS_{cor} = 81.92FS^+ - 48.19FS^- \tag{16.5}$$

where FS^+ and FS^- are nonnegative quantities defined empirically on the basis of molecular structures. As a first step in the Python implementation of this model on the basis of the RDkit library, the required external libraries have to be imported as follows:

```
from collections import Counter
from rdkit import Chem
from rdkit.Chem import Descriptors
```

The Counter object from the standard collections library is used to obtain the empirical molecular formula of the compound. The Chem module from rdkit provides the main tools to handle a molecule, and Descriptors will be used only to calculate the molecular weight.

Then comes the only nontrivial part, namely the definition of the substructures (i.e., fragments) to be considered in order to apply the rules introduced by the model [77]. Fortunately, this is relatively straightforward using the SMARTS notation discussed above. The lines below initialize a dictionary FRAGS in such a way that FRAGS['Nitramine'] returns the pattern corresponding to a >N–NO_2 substructure, and likewise for the other defined substructures. This dictionary exhibits 10 entries accessed using arbitrary codes for the substructures as keys:

```
NNTRO = '[$([NX3](=O)=O),$([NX3+](=O)[O-])]' # N atom in nitro
RTERMI = '[N;$(N([CX4H3])%s)]' %NNTRO # N(NO2)CH3 terminaison
explolist = (
    ('Picryl',          'c1c(%s)[cH1]c(%s)[cH1]c1(%s)' %tuple(3*[NNTRO])),
    ('Nitramine',       '%s[#7;X3]' %NNTRO),
    ('CyclicNitramine','%s[#7;X3R]' %NNTRO),
    ('Methylene',       '[CX4H2]'),
    ('CH',              '[#6&X3&H1,#6&X4&H1]'),
    ('NNO2Picryl',      '[NX3](%s)c1c(%s)[cH1]c(%s)[cH1]c1(%s)' %tuple
                           (4*[NNTRO])),
    ( 'tetrazole',      'c1nnnn1'),
    ( 'NX3H0',          '[#7;X3;H0]'),
    ( 'ch1',            '%s[CX4H2]%s' %(RTERMI, RTERMI)),
    ( 'ch2',            '%s[CX4H2][CX4H2]%s' %(RTERMI, RTERMI)),
)
# required fragments stored in a global dictionary
FRAGS = dict((c, Chem.MolFromSmarts(s)) for c,s in explolist)
```

The variable NNTRO stores the SMARTS string for any nitrogen atom in a nitro group. Note the use of the logical "or" operator to handle the two possible representations of the nitro group, namely $-N(=O)=O$ or $-N^+(-O^-)=O$. A simpler but less accurate SMARTS for a nitrogen atom in a nitro group could be simply `'[#7;$([NX3](~[OX1])~[OX1])]'`. This matches any trivalent nitrogen atom bonded to two monovalent oxygen atoms.

The model involves empirical rules depending on the number of substructures encountered on the molecule studied. Although these rules are a bit convoluted and sometimes ambiguous, they are quite easy to implement, as done in the function get_correction below which returns FS_{cor} for any molecule mol provided as input:

```
def get_correction(mol):
    if len(Chem.GetMolFrags(mol)) > 1: return 0
    m = dict((c,mol.GetSubstructMatches(fr)) for c, fr in FRAGS.items())
    n = dict((c, len(m[c])) for c in FRAGS)
    cor = 0
    if n['CyclicNitramine'] > 0:
        if n['Nitramine'] < n['Methylene'] or n['Nitramine'] <= n['CH']:
            cor = -0.7
        elif n['Nitramine'] == n['Methylene']:
            cor = +0.8
    else:
        if n['ch1'] > 0 or n['ch2'] > 0:
            cor = -1.4
        elif n['tetrazole'] > 0:
            tetrazoles = [set(t) for t in m['tetrazole']]
```

```
            in_tetrazole_rings = set.union(*tetrazoles)
            inNX3H0 = set.union(*[set(t) for t in m['NX3H0']])
            NX3inTetrazoleRing = in_tetrazole_rings & inNX3H0
            if NX3inTetrazoleRing:
                cor = +1
            else:
                cor = -0.8
        elif n['NNO2Picryl'] > 0:
            cor = +1.4
    return cor*81.92 if cor>0 else cor*48.19
```

The corrections defined in this model apply only to neutral molecules. Therefore, the first line in get_correction returns zero whenever the RDkit molecule provided as mol exhibits more than one fragment, assuming that all compounds provided as input exhibit a single species except salts. This assumption is not restrictive as the model is not deemed to be applicable to cocrystals.

The subsequent lines define two dictionaries m and n. An entry m[c] contains the list of substructures matching the pattern associated with code c, whereas n[c] contains the number of such substructures. Because the model relies primarily on the occurrences of each substructure, the latter dictionary is usually sufficient to estimate the corrections. The only exception is for compounds with a tetrazole ring, for which the correction is positive ($FS^+ = 1$) if the ring exhibits a *R*–N< group, and negative ($FS^- = 0.8$) otherwise. To test this condition, the present script considers the set of all atoms in tetrazole rings (in_tetrazole_rings) prepared from the list of all such rings (m['tetrazole]), and the set of all nitrogen atoms in *R*–N< groups (inNX3H0) prepared from the list of all such groups (m['NX3H0']). The intersection of both sets determines whether $FS^+ = 1$ or $FS^- = 0.8$.

It should be noted that Eq. (16.5) might misleadingly suggest that the correction for the presence of specific groups to the estimated friction sensitivity generally includes two contributions of opposite signs. In fact, a detailed examination of the model indicates that the two correcting coefficients FS^+ and FS^- are never nonzero for the same compound. This is why the get_correction function in the present code considers a single correcting coefficient cor which is either positive or negative depending on whether it is associated with FS^+ or FS^-. This function is in fact all that is needed to estimate the friction sensitivity in a few lines of code, as shown in the following script.

```
# read SMILES and activation energy
smi = input('SMILES = ')
Eac = eval(input('Activation energy of thermolysis (kJ/mol) = '))
# convert SMILES to RDkit molecules
mol = Chem.MolFromSmiles(smi)
# estimate friction sensitivity from empirical formula: FS0
mf = Counter(a.GetSymbol() for a in mol.GetAtoms())
Omega = 32.67*mf['C'] - 14.50*mf['O'] - 10.21*mf['N']
FS0 = 212 + Omega -85.07*Eac/Descriptors.MolWt(mol)
```

```
# obtain final estimate by adding correcting term
FS = FS0 + get_correction(mol)
# print estimated friction sensitivity in Newton
print(FS)
```

Run from the command line, this script asks for input data, namely the SMILES of the molecule and its activation energy of thermolysis in kJ/mol. The SMILES is converted into an RDkit molecule (mol). The Counter object imported from the standard module collections is used here to count the number of C, O, and N atoms. Note that H atoms would have to be added to mol using Chem.AddHs were their number explicitly required by the model. The final estimate for the friction sensitivity is eventually stored in the FS variable.

Although such a model provides a good real-world example to demonstrate the use of Cheminformatics tools like RDkit, SMILES, and SMARTS, it is well known that extrapolations from such small data sets are bound to fail when the compounds to which they are applied are too different from the reference compounds used to derive them. Unfortunately, in most cases, their limitations are not clear unless one carefully examines the learning panel from which they were developed. As a result, they are commonly applied outside their area of validity. For instance, an empirical equation developed on the basis of various families of molecular explosives being presented as a general correlation [78], it was subsequently applied to bistriazole-derived salts [79], a class of compounds lying clearly beyond its scope.

In general, it is recommended to carry out one's own implementation of empirical models like the one in Ref. [77] so as to better recognize their limitations. For instance, the present implementation of this earlier friction sensitivity model makes its critical dependence on arbitrary rules for the corrections terms FS^+ and FS^- especially clear. The fact that their definition relies on highly specific fragments is especially puzzling. This makes it impossible to know in advance the values of the corrections that might be required as soon as FS is needed for a compound with structural patterns absent from the training set. For instance, although the model is especially devoted to species with $N(NO_2)$ groups (i.e., nitramines) that may also exhibit other energetic structures like 1,3,5-nitrosubstituted benzene rings, a specific correction as large as +115 N is needed for compounds in which a picryl group (or trinitrophenyl: TNP) is directly attached to $N(NO_2)$ in order to account for the specially good resilience to friction of two compounds in the training set. The role of this correction is shown in Fig. 16.4. Because this correction is specific to unsubstituted picryl groups, the model would predict that introducing NH_2 groups as substituents on tetryl would induce a dramatic increase in sensitivity, in sharp contrast with expectations based on the stabilizing influence of this group.

A further drawback of such specific rules is the fact that their application is error prone. In fact, it is necessary to refer to the original publication to decide how they should be understood. For instance, the value originally reported for EDNA implies that the value $FS^- = 1.4$ assigned to compounds with general formula $-RN-(CH_2)_2-NR-$ does not apply to this compound whose formula is $H(NO_2)N-(CH_2)_2-N(NO_2)H$. Otherwise, the predicted sensitivity for this compound would decrease from 73 N down to 5.5 N, as shown in Fig. 16.4. This illustrates the lack of robustness of the model with respect to the composition of the training set, inherent to the ad hoc character of the corrections introduced, which critically depend on the fragments encountered in the training set.

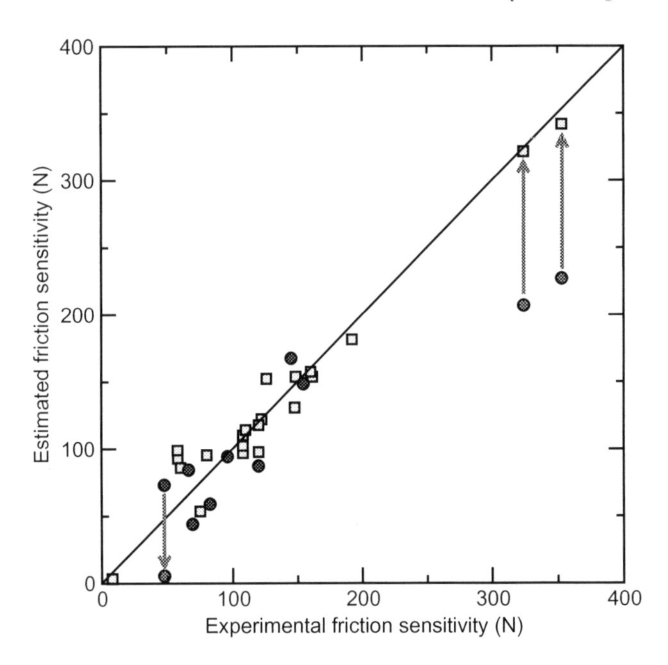

FIG. 16.4 Friction sensitivities evaluated using the present Python lines compared to experiment. The *upward arrows* illustrate the role of the highly specific –N(NO_2)–TNP correction. The *downward arrow* shows the increase in predicted sensitivity for EDNA as the negative correction $FS^- = 1.4$ is applied.

5.2 Taking advantage of previous findings

Notwithstanding the highly empirical approach adopted, one reason for the lack of predictive value of the model described in Section 5.1 arises from the heterogeneous training set considered, including 16 neutral molecules and 5 salts. The only reason why these salts are included in a nitramine database appears to be the fact that they share a common anion including a C=N–NO_2 fragment, along with various other explosophore groups, such as tetrazole, –ONO_2, and >C(NO_2)$_2$ groups. On the other hand, they all exhibit the same anion. Therefore, sensitivity differences between these salts must be ascribed to the cation.

In salts, the strong cohesion due to the magnitude of the Coulomb interactions between ions, and the charges carried by them make the reactivity associated with the functional groups likely to be strongly affected. For such a small dataset, only models with small number of features and description lengths may be considered. Therefore, focusing on nearest neighbors might provide a especially suitable learning bias. In the present context, this implies to consider the six salts present in the data set (including the one in the test set) apart from the neutral molecules. It is obviously not possible to derive a quantitative model from such limited data. What can be done is to ensure that this data is consistent with previously reported general trends, in view of using these as a guide for designing new compounds.

First, the influence of very simple ions on the sensitivity of energetic salts is well documented. In this respect, the especially high sensitivity of the salt with the HO–NH_3^+ cation which exhibits a weak O–N bond is to be expected. From now on, we focus on the five other cations made exclusively from carbon and nitrogen atoms involved in amine/ammonium groups. Among the simple trends reported in literature regarding sensitivities and applicable to energetic salts, one of the best documented (extensively discussed in

Chapter 8) is a tendency for impact sensitivity to increase with the free space per molecule, or equivalently with molecular size. This trend is especially plausible in the present case where the increase in size of the cation on going from NH_4^+ to $NH_2NHC^+(NHNH_2)NHNH_2$ comes with increased flexibility, as more flexible species exhibit enhanced anharmonic coupling between phonons and internal vibrations and thus a more efficient up-pumping leading to more sensitive materials. The theoretical considerations relating impact sensitivities to the efficiency of the up-pumping process apply equally well to friction sensitivity. Therefore, the sensitivity of the five salts considered is theoretically expected to increase (i.e., FS should decrease) as the cation size increases. This is actually the case, as shown in Fig. 16.5 where FS is plotted as a function of the molar volume V_{mol} of the salt estimated using a simple additivity scheme [80]. There are several reasons why such a correlation should be more reliable than Eq. (16.3): (1) it involves less empiricism; (2) it does not depend on a contingent split of the species into substructures, some of which are underrepresented; and (3) it is supported by theoretical considerations. Finally, the results summarized in Fig. 16.5 point to the sensitizing role of NH fragments and illustrates the utmost importance of database consistency.

5.3 Dimensional analysis and scaling behavior

In typical modeling problems in physics, some quantities can often be expressed in terms of others using dimensional analysis. For instance, the drop weight impact height h_{50} has the dimension of a length [L], and the drop weight energy E_{drop} provided by the hammer to the material further depends on the gravitational acceleration g, with the dimension $[LT^{-2}]$, and on the mass m of the hammer [M]. The expression $E_{drop} = mgh_{50}$ can be derived

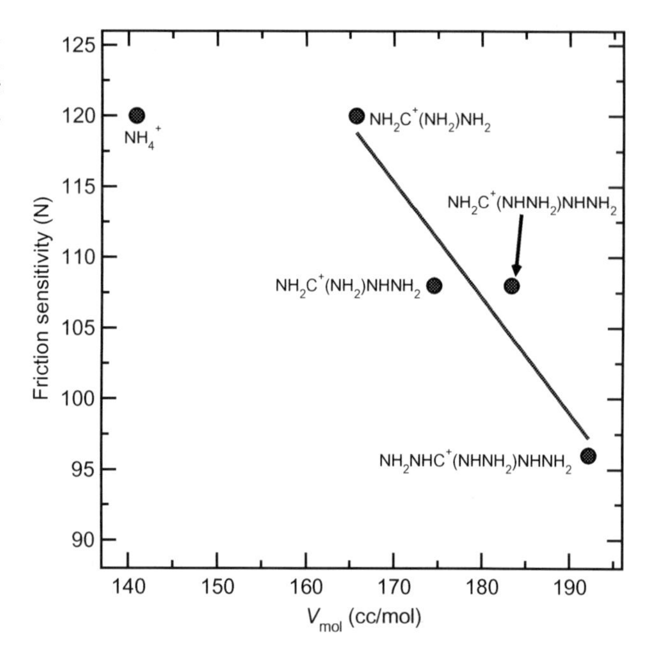

FIG. 16.5 Friction sensitivity (N) of five energetic salts (excluding cation $HO{-}NH_3^+$) sharing the same anion plotted as a function of the theoretical molar volume of the salt. The line stems from a linear regression restricted to salts with organic cations.

from the consistency of this expression with the dimension of E_{drop}, namely $[ML^2T^{-2}]$. However, this approach cannot be applied so easily to the prediction of sensitivities, which are typically characterized by an energy (E_{drop}) or a force (FS). This is because a theoretical expression for such quantities would involve physical quantities on supramolecular scales, some of which are unavailable owing to the complex multiscale character of initiation.

Nevertheless, expressions with unusual dimensions should be considered with caution, especially if they exhibit a spurious scaling behavior. In the above-mentioned model, the fact that FS systematically increases as C atoms are introduced, and decreases with the number of N/O atoms, is acceptable. Such trends are indeed consistent with expectations. However, many published models for sensitivities and decomposition temperatures imply physically inconsistent behaviors, as pointed out in previous articles [3, 71, 72]. For instance, many models for various kinds of sensitivities express them as linear combinations of the number of C, H, N, O atoms, plus some corrections applying only to compounds with specific fragments [81]. This means that except for such specific cases, sensitivities depend only on the compound stoichiometry, which is of course an oversimplified picture.

In addition the six salts studied in Section 5.2, the friction sensitivity data set introduced by Keshavarz et al. [77] exhibits 23 neutral compounds. Although constitutive descriptors allow in principle accurate descriptions of many physical and chemical properties, as mentioned above, this approach is hampered in practice by the large amount of data required to fit the parameters associated with every fragment involved. In this respect, the present data set is too small in view of the structural diversity of the compounds. Indeed, while focused on nitramines, it exhibits many explosophores in addition to the $>N–NO_2$ and $–NH–NO_2$ groups. Therefore, an approach based only on 0D descriptors would require a number of parameters comparable to the number of entries in the data set, which would not allow any serious validation.

In this context, a small number of continuous descriptors taking nonzero values for most compounds in the dataset might be preferred. Before embarking on multiparametric models, it is worth considering special cases. It is clear from the present data set that the four compounds with nitrobenzenic moieties (actually picric groups either substituted or not) are especially resilient to friction. Focusing on the FS differences for the latter, it appears that the activation energy for thermolysis E_a plays a prominent role, as anticipated by Keshavarz et al. [77]. This is clear from the linear correlation observed between FS and E_a/N_{atom} (Fig. 16.6A) where N_{atoms} denotes the total number of atoms per molecule. The need to normalize with N_{atoms} is consistent with the phonon up-pumping model that implies that this number makes the material more sensitive due to the larger number of modes in the doorway region.

While there is no obvious mean to express FS in terms of molecular properties in a dimensionally consistent way, this property might correlate with a dimensionless sensitivity index characterizing its tendency to decompose under friction. Since a higher FS is usually associated with lower performances, such a sensitivity index might be derived from performance criteria. Therefore, we have calculated for the present materials the heat of detonation Q, detonation velocity D_{CJ}, detonation pressure P_{CJ}, and Gurney velocity E_G, using the Kamlet–Jacob [82] and Cγ [83] models with crystal density ρ and formation enthalpy $\Delta_f H^0$ derived from simple procedures [61, 80, 84]. Intriguingly, FS tends to increase with the dimensionless ratio $\rho Q/E_G$ of the heat of detonation to the Gurney energy (Fig. 16.6B). This ratio depends on the material considered only through the ratio ρ/N_g where N_g is the number of moles of gaseous species per gram of explosive in the detonation products [83].

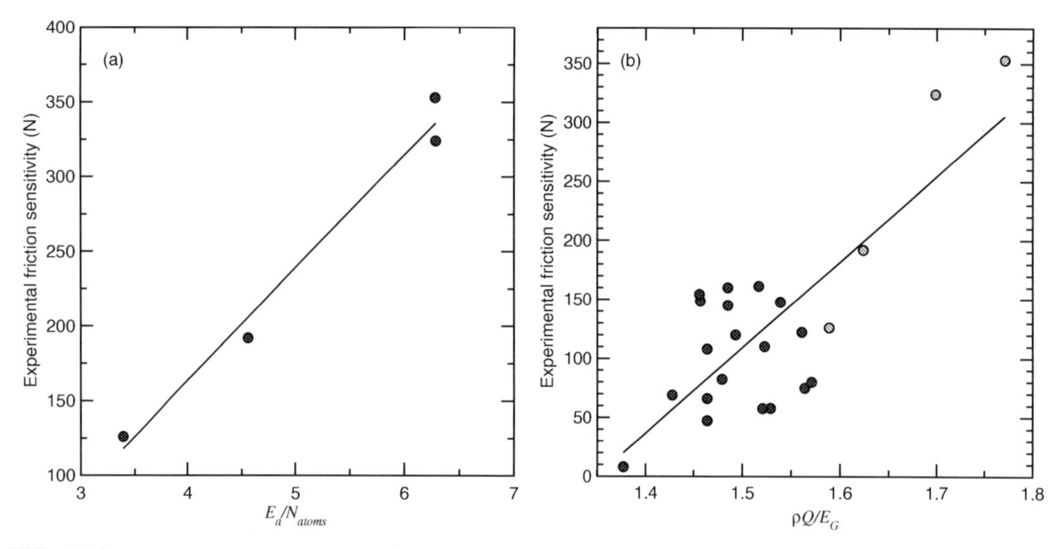

FIG. 16.6 Friction sensitivity data (N) of neutral nitramine compounds reported: (A) as a function of the ratio E_a/N_{atoms} of the activation energy for thermolysis to the number of atoms in the molecule for the four compounds with a nitrobenzenic group; (B) as a function of $\rho Q/E_G$ for all 23 neutral nitramines, with nitrobenzenic ones shown with light symbols.

5.4 Machine learning

To date, QSPR models to evaluate the physicochemical properties of materials are mostly based on a small number of statistical and machine learning techniques, with linear methods (especially ordinary linear regression and partial least squares), neural networks (NN), support vector regression (SVR), and random forest (RF) models being especially popular. Many studies are carried out using integrated software packages like CODESSA [85], which includes all necessary routines to develop standard QSPR models, from the computation of a large pool of descriptors to the development of a model from selected features and its application to new compounds. CODESSA users usually resort to the best multi-linear regression (BMLR) algorithm as feature selection method and end up with relatively simple linear models.

The same functionality is provided by open-source packages, using, e.g., RDkit to handle molecules [86] and Mordred [60] to generate broad pools of descriptors. To handle this data and apply statistical or machine learning (ML) techniques, a wealth of open-source software is available and extensively documented [14]. In particular, we recommend pandas, a very powerful Python library combining the features of spreadsheet software like Excel with those of database management systems. It makes it straightforward to handle very large tables with arbitrary types of data in addition to numerical ones, including strings, molecules, dictionaries, or chemical formulas. Arbitrary operations can be applied to the data and custom partitions of the data set can easily be made. A major advantage of using such open-source software is the fact that it makes it easy to leverage the wealth of specialized modeling techniques and statistical or machine learning packages provided on the Internet through such tools as statsmodels, scikit-learn, waffles, and many others. However, most of these tools require large training sets that are not always available for energetic materials.

In fact, the amount of data available in literature regarding sensitivities and decomposition temperatures of energetic materials is rapidly growing due to the extensive research carried out by various research groups. However, this valuable data is still scattered in research papers. A medium-sized data set of about 300 compounds is only available for impact sensitivities determined using the so-called ERL Type 12 US procedure [87], which is based on the earlier Storm–Stine–Kramer compilation [88]. This data is the standard resource used to train machine learning models for sensitivity prediction. To date, this was done mostly using linear models [24] and neural networks [26, 75, 76, 89, 90].

The application of linear regression methods to the development of QSPR models for sensitivities is motivated by their relative simplicity, compared to the more tedious task of optimizing nonlinear parameters. This enables the application of feature selection techniques to obtain a set of relevant descriptors from the initial pool as model variables. The availability of a very large pool of descriptors proves critical to the identification of significant linear correlations. In other words, the flexibility afforded by the many descriptors in the pool makes up for the constraints inherent to the linearity of the model.

Because the variables of the model are selected so as to yield a satisfactory linear model, it comes as no surprise that little improvement is observed on going from linear (MLR, PLS) to nonlinear (ANN) models [26]. In view of the relatively high cost of fitting nonlinear models, a common practice consists indeed in selecting a subset of descriptors so as to obtain an optimal linear model, and subsequently fitting a nonlinear model using the same set of descriptors.

A drawback of any model based on automatic variable selection techniques is that it is expressed in terms of mathematical (e.g., electrotopological state indices [76]) rather than physical quantities. This hampers its physical interpretation and the determination of its applicability domain (AD). Indeed, due to the high dimensionality of typical descriptor spaces, standard AD definitions based on statistical considerations are only valid for very large datasets [87], which are not available for energetic materials. ADs of models derived from medium-size data sets like those developed for sensitivities should rather be obtained from physical considerations, which is not possible for models built on the basis of a pool of abstract descriptors.

5.5 Symbolic regression

In view of linking material properties to physical properties of the molecules, symbolic regression (SR) is especially attractive. Like other nonlinear regression techniques, it is designed to capture the nonlinearities in the relationship linking the property of interest to the model variables. However, SR is even more versatile as it does not dictate any particular mathematical formalism. Moreover, in contrast to models based on neural networks, obtaining a suitable relationship does not necessarily entail the fit of numerous empirical parameters. Symbolic regression techniques can be divided into two categories that do not have the same advantages.

Conventional SR approaches usually rely on genetic programming (GP). In the QSPR context, they consist in evolving arbitrary mathematical expressions (rather than computer programs) typically represented by graphs, using a fitness function designed to filter out the ones that best approximate the property to be predicted for a given training set. Obviously, such a broad exploration of the space of mathematical relationships between properties and descriptors is extremely costly. This makes it hard to obtain converged results and is a significant obstacle to the development and benchmarking of the technique. However, restricting the search is rather straightforward. In particular, the technique lends itself particularly well to the development of models that are both consistent with theoretical expressions and include empirical aspects.

We tested a number of freely available implementations of SR techniques. Two specially interesting ones are PySR, which uses a neural network as a proxy to facilitate the application of the method to high-dimensional problems by reducing them the conversion of a neural net to an analytic equation [91], and CGP Library [92]. As implies by its name, the latter implements a distinctive variant of GP called cartesian GP, or CGP. It is based on representing a structure to be evolved on a two-dimensional grid and then as a string of integers which is easy to handle by evolutionary algorithms. The CGP Library is especially recommended for anyone interested in applying GP, as it is a very neat, concise, and easy to hack piece of software written in C. In particular, with little programming, the user can introduce any custom fitness function. Since GP algorithms applied to SR are focused on optimizing the mathematical expressions rather than the empirical constants in them, we have introduced a local Levenberg–Marquardt optimization of the latter in the fitness function with the help of the lmfit C library [93]. While this strategy is probably not optimal, it allowed us to discover QSPRs based on new analytic expressions, while avoiding the bloat issue that plagued our earlier experiments with a canonical GP implementation [94]. In particular, we discovered various alternatives to published GIPF models, although not definitely more accurate.

The second group of symbolic regression techniques trade off generality for speed, in an effort to overcome the above-mentioned limitations, as outlined in recent literature reviews [95, 96]. Such methods including built-in linear regression were recently compared [97]. The first of these is the fast function extraction method (FFX) [98]. This is a nonevolutionary and deterministic algorithm with a search space primarily confined to generalized linear space (GLS) so that the property of interest is assumed to depend on the descriptors $\boldsymbol{x}$ according to:

$$f(\boldsymbol{x}) = \beta_0 + \sum_{i=1}^{N} \beta_i \phi_i(\boldsymbol{x}) \tag{16.6}$$

where ϕ_i, $i = 1,2,\ldots,N$ are N basis functions and any β_i is a regression parameter. In principle, FFX also considers rational functions of the bases. However, in our experience, the solutions provided for real-world molecular modeling problems fall systematically into the GLS form given by Eq. (16.6). The FFX algorithm first generates a large pool of basis functions. In a second step, path-regularized learning [99] is used to filter out relevant ones and determine the values of the regression coefficients β_i.

It what follows, we demonstrate the application of the FFX model to impact sensitivity on the basis of the most comprehensive compilation presently available, with over 300 compounds [87]. Following a common procedure, 20% of the data is set aside to be used as an external test set, and the remainder is used as training set. Like for the above modeling of friction sensitivity, the descriptors are simply the counts of the explosophores encountered in the training set: tetrazole rings, nitro groups on aromatic carbons, on sp3 carbons, >N–NO_2, –ONO_2, –NH–NO_2, and –N=NO_2 groups. Explosophores –N_3, –N_2^+, and –NF_2 are simply ignored as they are seldom encountered in the present data set. In addition to these local descriptors, two global ones are introduced to take into account the fact that the amount of energy and the amount of gases released upon decomposition are expected to affect sensitivities. These additional descriptors are heat of explosion Q and Gurney energy E_G.

In a matter of seconds, the FFX algorithm yields a range of 16 models corresponding to an optimal accuracy for a given complexity. These models are ranked in order of increasing complexity and precision. The baseline model #0 simply assigns the average value of 1.62 to $\log(h_{50})$. The second one, model #1, evaluates $\log(h_{50})$ as:

```
log(h50) = 1.43 + 0.000186*max(0,6626-EG)
```

This expression is consistent with the significant correlation observed between impact sensitivity and Gurney energy per volume unit. It suggests that sensitivity tends to increase with performance, but only to a certain extent. Finally, the most complex and best performing model #16 is:

```
log(h50) = 10.5*log10(EG) - 31.6*log10(Q) + 7.84*log10(Q)*log10(EG)
         - 0.00248*EG + 9.19e-8*EG^2 -24.8
         + 2.90*CNO2 + 0.0527*CNO2^2 - 0.0704*cNO2 + 0.00475*cNO2^2
         + 0.481*NNO2 + 0.286*NHNO2 + 3.77*NinsNO2 + 2.70*NinsNO2^2
         + 0.811*ONO2 + 0.0412*ONO2^2 - 0.185*NHNO2^2
         + 0.0250*NNO2*CNO2 + 0.0189*NNO2^2 - 0.00567*ONO2*CNO2
         + 0.0322*NNO2*ONO2 - 1.68*NinsNO2*log10(EG)
         + 0.574*NNO2*log10(Q) - 0.284*NNO2*log10(EG)
         - 0.265*ONO2*log10(Q) - 0.230*ONO2*log10(EG)
         + 2.47*CNO2*log10(Q) - 1.33*CNO2*log10(EG)
```

In the above expressions, Q is the heat of explosion in kJ/g, EG the Gurney energy in MJ/m^3, and NHNO2, NNO2, N=NO2, CNO2, cNO2, and ONO2, the number of nitro groups attached to a trivalent N atom with and without a H neighbor, a divalent N atom, a tetravalent C atom, an aromatic C atom, and an oxygen atom, respectively.

While such expressions are empirical, the present approach has some advantages over commonly used QSPR methodologies, as it does not resort to a pool of purely mathematical descriptors devoid of physical meaning. As a consequence, there is no need to resort to a third-party software to evaluate the descriptors needed. Furthermore, while the purely empirical character of simple models derived from scarce data may lead to overfitting, this issue is mitigated here as the predictive value of the final FFX model is strongly supported by the large number (60) of successfully predicted sensitivities for test set compounds, as shown in Fig. 16.7.

FFX expressions like the one reported above include nonlinear terms, e.g., quadratic terms and logarithms of the input variables. They also include interaction terms involving two input

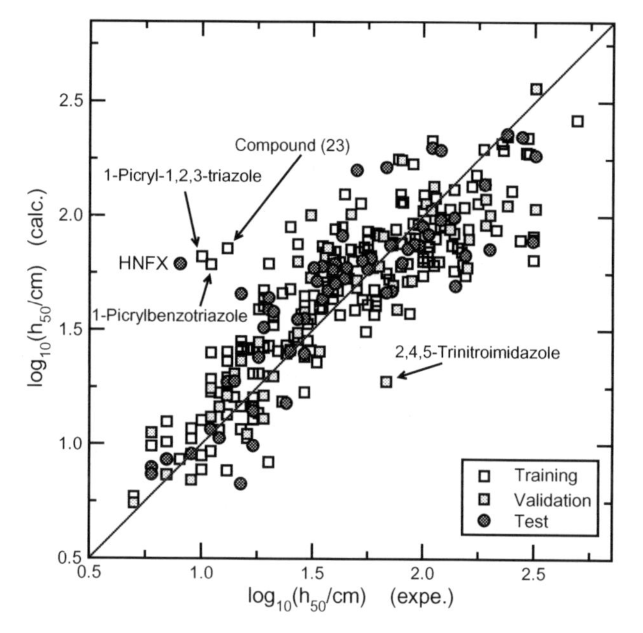

FIG. 16.7 Assessment of the FFX method [98] for impact sensitivity: calculated vs experimental $\log(h_{50})$ values for the training, validation, and test sets. Compound (23) is described in Ref. [100]. HNFX is the only compound bearing the NF_2 moiety, which is not represented in the training set.

variables, like the third term in the above expression that involves the product of the logarithms of Q and E_G. However, notwithstanding the significant number of terms, they remain remarkably simple and easier to interpret than typical machine learned relationships. By varying an input variable and assuming fixed typical values for all others, the role of a specific parameter may be assessed. For instance, the present model is consistent with the empirical observation that materials tend to get more sensitive upon addition of nitro groups, which may be attributed in part to the increased energy content reflected by enhanced heat of explosion (Q) or Gurney energy (E_G). In order to get insight into the role of other factors on the sensitivities of putative compounds with three aromatic nitro groups (e.g., molecules with a picryl group $-C_6H_2N_3O_6$) and a variable number $n(C-NO_2)$ of aliphatic nitro groups, their values of Q and E_G may be arbitrarily fixed to their average values of, respectively, 5.5 kJ/g and 5700 MJ/kg. As shown in Fig. 16.8, the FFX sensitivities thus estimated go through a minimum for $n(C-NO_2) = 3$. This indicate that sensitivity increases faster than anticipated considering only the increase in Q and E_G as the first two aliphatic nitros are introduced, while the reverse is true for $n(C-NO_2) \geq 3$.

Interestingly, compounds that do not fit well into the present FFX model did not fit either in the former kinetic model [87]. These outliers are primarily 1-picryl-1,2,3-triazole, 1-picryl-benzotriazole, and compound (23) in Ref. [100]. The fact that two significantly different models point to a unexpectedly low h_{50} value measured for these compounds should motivate a more in-depth investigation of their sensitivity. For compound (23), the high sensitivity observed might be explained by its especially large size and flexibility, which could hamper the crystallization and lead to an especially large fraction of voids and defects increasing the actual sensitivity.

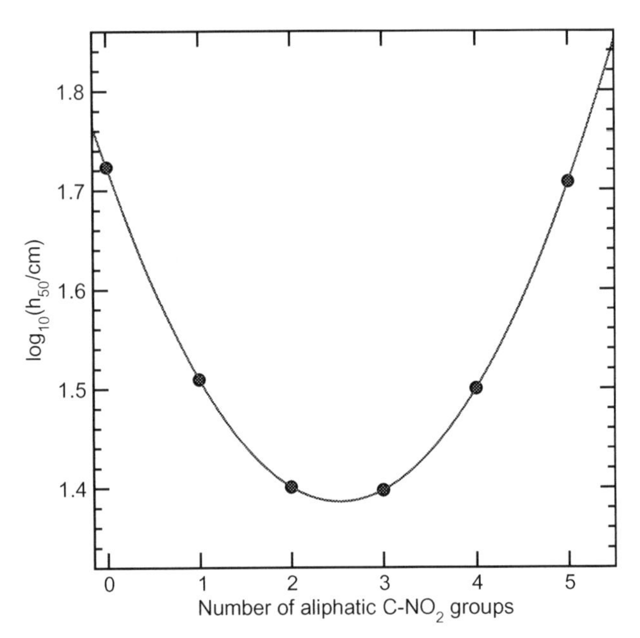

FIG. 16.8 Predicted impact sensitivities for putative compounds bearing three aromatic NO_2 groups and with typical values of the heat of explosion (5.5 kJ/g) and Gurney energy (5700 MJ/kg) as a function of the number of aliphatic C–NO_2 groups.

Finally, among the test set compounds, the widely overestimated h_{50} value for HNFX is to be expected as this molecule is the only one bearing NF_2 groups as explosophores, which is therefore not represented in the training set and thus completely ignored by the present model.

Apart from these outliers, the model tends to account poorly for the large h_{50} values observed for the most insensitives compounds. Nevertheless, it exhibits a surprisingly good average accuracy, as reflected by values of the root mean square deviation (RMSD) from experiment of only 0.23 and 0.22 for the training and test set (excluding HNFX), respectively. It is quite remarkable that this kind of empirical model deprived of physical grounds yields such good predictions for an external test set.

This example demonstrates the interest of efficient regression schemes like FFX. Although the kinetic model reported in Ref. [87] provides more physical insight and might prove equally satisfactory as the present FFX equation for practical applications, it required a stronger effort to be developed. Therefore, FFX and other similar algorithms should prove extremely valuable for material designers, allowing them to instantly develop predictive models without the need for modeling experts, provided that they have large databases at their disposal.

6 Conclusion

Although sensitivity prediction for energetic materials is still in its infancy, valuable models have emerged in recent years for impact sensitivity, and to a lesser extent for decomposition temperatures. This chapter demonstrates that it is surprisingly simple and inexpensive to exploit some of the most advanced of these, in particular through the many software packages freely available on the Internet.

However, the application of the most sophisticated models for impact sensitivities is hampered by their dependence on crystal structures, which are not available for new synthesis targets. Nevertheless, given the relative simplicity of high-energy compounds, the prediction of crystal structures for explosives should be fairly reliable using current state-of-the-art techniques. By comparison with the computational cost of such predictions, the cost of evaluating impact sensitivities from phonons is peanuts, especially if cubic anharmonic coupling coefficients are not calculated explicitly, as usually done. Furthermore, in addition to making possible the application of advanced methods for sensitivity estimation, the availability of crystal structures allows for greater accuracy in the evaluation of other properties. Therefore, the very high cost inherent to crystal structure prediction should be offset if these predictions avoid unnecessary experimentation.

In the lack of crystal structure, quantum chemical calculations allow the determination of a relatively small number of descriptors mirroring the electronic and molecular structure of the compound. Such quantum descriptors appear well suited to the search for correlations or trends that sensitivities follow within a particular set of compounds. This is because a narrow pool of descriptors limits the risk of overlearning. Furthermore, the resulting trends may be easily interpreted on a physical basis.

Finally, models requiring only the structural formula as input are those of most immediate practical interest. This single piece of data is all what is needed to get crude estimates of impact sensitivities using existing models. Developing empirical models is very easy provided sufficient data is available. However, it is much more difficult when data is lacking, as usually the case for the sensitivities to friction or electric discharge. In this case, one must be satisfied with simple correlations within restricted families of compounds and/or simple empirical equations. However, those reported in literature are often overfitted and their applicability domain is rarely made explicit. Therefore, such models should not be used as is. Instead, it is recommended that they should be designed and parameterized on a case-by-case basis, for the specific set of compounds for which predictions are needed. As demonstrated here, this is made easy through the availability of powerful and versatile open-source cheminformatics software.

References

[1] D. Mathieu, Sensitivity of energetic materials: theoretical relationships to detonation performance and molecular structure, Ind. Eng. Chem. Res. 56 (29) (2017) 8191–8201, https://doi.org/10.1021/acs.iecr.7b02021.

[2] M.H. Keshavarz, T.M. Klapötke, M. Sućeska, Energetic materials designing bench (EMDB), version 1.0, Propellants Explos. Pyrotech. 42 (8) (2017) 854–856, https://doi.org/10.1002/prep.201700144.

[3] D. Mathieu, Modeling sensitivities of energetic materials using the Python language and libraries, Propellants Explos. Pyrotech. 45 (6) (2020) 966–973, https://doi.org/10.1002/prep.201900377.

[4] D. Mathieu, Theoretical shock sensitivity index for explosives, J. Phys. Chem. A 116 (2012) 1794–1800.

[5] G. Li, C. Zhang, Review of the molecular and crystal correlations on sensitivities of energetic materials, J. Hazard. Mater. 398 (2020) 122910, https://doi.org/10.1016/j.jhazmat.2020.122910.

[6] A.A.L. Michalchuk, M. Trestman, S. Rudić, P. Portius, P.T. Fincham, C.R. Pulham, C.A. Morrison, Predicting the reactivity of energetic materials: an ab initio multi-phonon approach, J. Mater. Chem. A 7 (33) (2019) 19539–19553.

[7] A.A.L. Michalchuk, Mechanochemical Processes in Energetic Materials: A Computational and Experimental Investigation, Springer International Publishing, 2020, ISBN: 978-3-030-56965-5, https://doi.org/10.1007/978-3-030-56966-2. https://www.springer.com/gp/book/9783030569655.

[8] A. Demenay, Estimation des sensibilités des matériaux énergétiques soumis à divers stimuli. Interprétation de la sensibilité à l'aide de mécanismes cinétiques détaillés, Thèse de l'Université Paris-Saclay, France, 2016. http://www.theses.fr/2016SACLY007.

[9] K.L. McNesby, C.S. Coffey, Spectroscopic determination of impact sensitivities of explosives, J. Phys. Chem. B 101 (1997) 3097–3104.

[10] S. Zeman, M. Jungová, Sensitivity and performance of energetic materials, Propellants Explos. Pyrotech. 41 (3) (2016) 426–451, https://doi.org/10.1002/prep.201500351.

[11] B.M. Rice, S. Sahu, F.J. Owens, Density functional calculations of bond dissociation energies for NO_2 scission in some nitroaromatic molecules, J. Mol. Struct. (Theochem) 583 (2002) 69–72.

[12] S. Majumdar, S.C. Basak, Beware of external validation!–a comparative study of several validation techniques used in QSAR modelling, Current Comput. Aided Drug Design 14 (4) (2018) 284–291, https://doi.org/10.2174/1573409914666180426144304.

[13] D. Mathieu, Physics-based modeling of chemical hazards in a regulatory framework: comparison with quantitative structure-property relationship (QSPR) Methods for impact sensitivities, Ind. Eng. Chem. Res. 55 (27) (2016) 7569–7577, https://doi.org/10.1021/acs.iecr.6b01536.

[14] S. Pirhadi, J. Sunseri, D.R. Koes, Open source molecular modeling, J. Mol. Graph. Model. 69 (2016) 127–143, https://doi.org/10.1016/j.jmgm.2016.07.008.

[15] J. Bernstein, Ab initio study of energy transfer rates and impact sensitivities of crystalline explosives, J. Chem. Phys. 148 (8) (2018) 084502, https://doi.org/10.1063/1.5012989.

[16] A.A.L. Michalchuk, P.T. Fincham, P. Portius, C.R. Pulham, C.A. Morrison, A pathway to the athermal impact initiation of energetic azides, J. Phys. Chem. C 122 (34) (2018) 19395–19408, https://doi.org/10.1021/acs.jpcc.8b05285.

[17] S.D. McGrane, J. Barber, J. Quenneville, Anharmonic vibrational properties of explosives from temperature-dependent Raman, J. Phys. Chem. A 109 (44) (2005) 9919–9927, https://doi.org/10.1021/jp0523219.

[18] A. Togo, I. Tanaka, First principles phonon calculations in materials science, Scr. Mater. 108 (2015) 1–5, https://doi.org/10.1016/j.scriptamat.2015.07.021.

[19] S.J. Clark, M.D. Segall, C.J. Pickard, P.J. Hasnip, M.I.J. Probert, K. Refson, M.C. Payne, First principles methods using CASTEP, Z. Kristallogr. Cryst. Mater. 220 (5-6) (2005) 567–570, https://doi.org/10.1524/zkri.220.5.567.65075. https://www.degruyter.com/view/journals/zkri/220/5-6/article-p567.xml.

[20] J. Hafner, Ab-initio simulations of materials using VASP: density-functional theory and beyond, J. Comput. Chem. 29 (13) (2008) 2044–2078, https://doi.org/10.1002/jcc.21057.

[21] P. Giannozzi, S. Baroni, N. Bonini, M. Calandra, R. Car, C. Cavazzoni, D. Ceresoli, G.L. Chiarotti, M. Cococcioni, I. Dabo, A.D. Corso, S.d. Gironcoli, S. Fabris, G. Fratesi, R. Gebauer, U. Gerstmann, C. Gougoussis, A. Kokalj, M. Lazzeri, L. Martin-Samos, N. Marzari, F. Mauri, R. Mazzarello, S. Paolini, A. Pasquarello, L. Paulatto, C. Sbraccia, S. Scandolo, G. Sclauzero, A.P. Seitsonen, A. Smogunov, P. Umari, R.M. Wentzcovitch, QUANTUM ESPRESSO: a modular and open-source software project for quantum simulations of materials, J. Phys. Condens. Matter 21 (39) (2009) 395502, https://doi.org/10.1088/0953-8984/21/39/395502.

[22] A.H. Larsen, J.J. Mortensen, J. Blomqvist, I.E. Castelli, R. Christensen, M. Du\lak, J. Friis, M.N. Groves, B. Hammer, C. Hargus, E.D. Hermes, P.C. Jennings, P.B. Jensen, J. Kermode, J.R. Kitchin, E.L. Kolsbjerg, J. Kubal, K. Kaasbjerg, S. Lysgaard, J.B. Maronsson, T. Maxson, T. Olsen, L. Pastewka, A. Peterson, C. Rostgaard, J. Schiøtz, O. Schütt, M. Strange, K.S. Thygesen, T. Vegge, L. Vilhelmsen, M. Walter, Z. Zeng, K.W. Jacobsen, The atomic simulation environment–a Python library for working with atoms, J. Phys. Condens. Matter 29 (27) (2017) 273002, https://doi.org/10.1088/1361-648X/aa680e.

[23] A.M. Reilly, R.I. Cooper, C.S. Adjiman, S. Bhattacharya, A.D. Boese, J.G. Brandenburg, P.J. Bygrave, R. Bylsma, J.E. Campbell, R. Car, D.H. Case, R. Chadha, J.C. Cole, K. Cosburn, H.M. Cuppen, F. Curtis, G.M. Day, R.A. DiStasio Jr, A. Dzyabchenko, B.P. van Eijck, D.M. Elking, J.A. van den Ende, J.C. Facelli, M.B. Ferraro, L. Fusti-Molnar, C.-A. Gatsiou, T.S. Gee, R de Gelder, L.M. Ghiringhelli, H. Goto, S. Grimme, R. Guo, D.W.M. Hofmann, J. Hoja, R.K. Hylton, L. Iuzzolino, W. Jankiewicz, D.T de Jong, J. Kendrick, N.J.J de Klerk, H.-Y. Ko, L.N. Kuleshova, X. Li, S. Lohani, F.J.J. Leusen, A.M. Lund, J. Lv, Y. Ma, N. Marom, A.E. Masunov, P. McCabe, D.P. McMahon, H. Meekes, M.P. Metz, A.J. Misquitta, S. Mohamed, B. Monserrat, R.J. Needs, M.A. Neumann, J. Nyman, S. Obata, H. Oberhofer, A.R. Oganov, A.M. Orendt, G.I. Pagola, C.C. Pantelides, C.J. Pickard, R. Podeszwa, L.S. Price, S.L. Price, A. Pulido, M.G. Read, K. Reuter, E. Schneider, C. Schober, G.P. Shields, P. Singh, I.J. Sugden, K. Szalewicz, C.R. Taylor, A. Tkatchenko, M.E.

Tuckerman, F. Vacarro, M. Vasileiadis, A. Vazquez-Mayagoitia, L. Vogt, Y. Wang, R.E. Watson, G.A de Wijs, J. Yang, Q. Zhu, C.R. Groom, Report on the sixth blind test of organic crystal structure prediction methods, Acta Cryst. B 72 (4) (2016) 439–459, https://doi.org/10.1107/S2052520616007447. //scripts.iucr.org/cgi-bin/paper?gp5080.

[24] J.A. Morrill, E.F.C. Byrd, Development of quantitative structure-property relationships for predictive modeling and design of energetic materials, J. Mol. Graph. Model. 27 (2008) 349–355.

[25] J. Xu, L. Zhu, D. Fang, L. Wang, S. Xiao, L. Liu, W. Xu, QSPR Studies of impact sensitivity of nitro energetic compounds using three-dimensional descriptors, J. Mol. Graph. Model. 36 (2012) 10–19.

[26] R. Wang, J. Jiang, Y. Pan, Prediction of impact sensitivity of nonheterocyclic nitroenergetic compounds using genetic algorithm and artificial neural network, Journal of Energetic Materials 30 (2) (2012) 135–155, https://doi.org/10.1080/07370652.2010.550598.

[27] J.S. Murray, T. Brinck, P. Lane, K. Paulsen, P. Politzer, Statistically-based interaction indices derived from molecular surface electrostatic potentials: a general interaction properties function (GIPF), J. Mol. Struct. Theochem. 307 (1994) 55–64, https://doi.org/10.1016/0166-1280(94)80117-7.

[28] C.-K. Kim, S.-G. Cho, J. Li, C.-K. Kim, H.-W. Lee, QSPR studies on impact sensitivities of high energy density molecules, Bull. Korean Chem. Soc. 32 (12) (2011) 4341–4346, https://doi.org/10.5012/bkcs.2011.32.12.4341.

[29] B.M. Rice, J.J. Hare, A quantum mechanical investigation of the relation between impact sensitivity and the charge distribution in energetic molecules, J. Phys. Chem. A 106 (2002) 1770–1783.

[30] N.-O. Friedrich, A. Meyder, C. de Bruyn Kops, K. Sommer, F. Flachsenberg, M. Rarey, J. Kirchmair, High-quality dataset of protein-bound ligand conformations and its application to benchmarking conformer ensemble generators, J. Chem. Inf. Model. 57 (3) (2017) 529–539, https://doi.org/10.1021/acs.jcim.6b00613.

[31] N.-O. Friedrich, C. de Bruyn Kops, F. Flachsenberg, K. Sommer, M. Rarey, J. Kirchmair, Benchmarking Commercial Conformer Ensemble Generators, J. Chem. Inf. Model. 57 (11) (2017) 2719–2728, https://doi.org/10.1021/acs.jcim.7b00505.

[32] S. Riniker, G.A. Landrum, Better informed distance geometry: using what we know to improve conformation generation, J. Chem. Inf. Model. 55 (12) (2015) 2562–2574, https://doi.org/10.1021/acs.jcim.5b00654.

[33] P. Pracht, F. Bohle, S. Grimme, Automated exploration of the low-energy chemical space with fast quantum chemical methods, Phys. Chem. Chem. Phys. 22 (14) (2020) 7169–7192, https://doi.org/10.1039/C9CP06869D. https://pubs.rsc.org/en/content/articlelanding/2020/cp/c9cp06869d.

[34] C. Bannwarth, S. Ehlert, S. Grimme, GFN2-xTB–an accurate and broadly parametrized self-consistent tight-binding quantum chemical method with multipole electrostatics and density-dependent dispersion contributions, J. Chem. Theory Comput. 15 (3) (2019) 1652–1671, https://doi.org/10.1021/acs.jctc.8b01176.

[35] C. Bannwarth, E. Caldeweyher, S. Ehlert, A. Hansen, P. Pracht, J. Seibert, S. Spicher, S. Grimme, Extended tight-binding quantum chemistry methods, WIREs Comput. Mol. Sci. n/a (2021) e01493, https://doi.org/10.1002/wcms.1493.

[36] S. Maeda, Y. Harabuchi, Y. Ono, T. Taketsugu, K. Morokuma, Intrinsic reaction coordinate: calculation, bifurcation, and automated search, Int. J. Quantum Chem. 115 (5) (2015) 258–269, https://doi.org/10.1002/qua.24757.

[37] F. Neese, F. Wennmohs, U. Becker, C. Riplinger, The ORCA quantum chemistry program package, J. Chem. Phys. 152 (22) (2020) 224108, https://doi.org/10.1063/5.0004608.

[38] D.G.A. Smith, L.A. Burns, A.C. Simmonett, R.M. Parrish, M.C. Schieber, R. Galvelis, P. Kraus, H. Kruse, R. Di Remigio, A. Alenaizan, A.M. James, S. Lehtola, J.P. Misiewicz, M. Scheurer, R.A. Shaw, J.B. Schriber, Y. Xie, Z.L. Glick, D.A. Sirianni, J.S. O'Brien, J.M. Waldrop, A. Kumar, E.G. Hohenstein, B.P. Pritchard, B.R. Brooks, H.F. Schaefer, A.Y. Sokolov, K. Patkowski, A.E. DePrince, U. Bozkaya, R.A. King, F.A. Evangelista, J.M. Turney, T.D. Crawford, C.D. Sherrill, PSI4 1.4: open-source software for high-throughput quantum chemistry, J. Chem. Phys. 152 (18) (2020) 184108, https://doi.org/10.1063/5.0006002.

[39] Q. Sun, T.C. Berkelbach, N.S. Blunt, G.H. Booth, S. Guo, Z. Li, J. Liu, J.D. McClain, E.R. Sayfutyarova, S. Sharma, S. Wouters, G.K.-L. Chan, PySCF: the Python-based simulations of chemistry framework, Wiley Interdisciplinary Reviews: Computational Molecular Science 8 (1) (2017) e1340, https://doi.org/10.1002/wcms.1340.

[40] T. Lu, F. Chen, Multiwfn: a multifunctional wavefunction analyzer, J. Comput. Chem. 33 (5) (2012) 580–592, https://doi.org/10.1002/jcc.22885.

[41] B. Hourahine, B. Aradi, V. Blum, F. Bonafé, A. Buccheri, C. Camacho, C. Cevallos, M.Y. Deshaye, T. Dumitrică, A. Dominguez, S. Ehlert, M. Elstner, T. van der Heide, J. Hermann, S. Irle, J.J. Kranz, C. Köhler, T. Kowalczyk,

T. Kubař, I.S. Lee, V. Lutsker, R.J. Maurer, S.K. Min, I. Mitchell, C. Negre, T.A. Niehaus, A.M.N. Niklasson, A.J. Page, A. Pecchia, G. Penazzi, M.P. Persson, J. Řezáč, C.G. Sánchez, M. Sternberg, M. Stöhr, F. Stuckenberg, A. Tkatchenko, V.W.-z. Yu, T. Frauenheim, DFTB+, a software package for efficient approximate density functional theory based atomistic simulations, J. Chem. Phys. 152 (12) (2020) 124101, https://doi.org/10.1063/1.5143190.

[42] F. Bosia, T. Husch, A.C. Vaucher, M. Reiher, qcscine/Sparrow: Release 2.0.1, Zenodo, 2020, https://doi.org/10.5281/zenodo.3907313. https://zenodo.org/record/3907313#.X-jSR_njIXc.

[43] I. Leven, T. Head-Gordon, C-GeM: coarse-grained electron model for predicting the electrostatic potential in molecules, J. Phys. Chem. Lett. 10 (21) (2019) 6820–6826, https://doi.org/10.1021/acs.jpclett.9b02771.

[44] B. Cuevas-Zuviría, L.F. Pacios, Analytical model of electron density and its machine learning inference, J. Chem. Inf. Model. 60 (8) (2020) 3831–3842, https://doi.org/10.1021/acs.jcim.0c00197.

[45] D. Mathieu, E. Germaneau, A fast non-selfconsistent electronegativity equalization method with applications in the field of energetic materials, in: New Trends in Research of Energetic Materials, University of Pardubice, Pardibice, Czech Republic, 2003.

[46] D. Mathieu, P. Bougrat, Model equations for estimating sublimation enthalpies of organic compounds, Chem. Phys. Lett. 303 (1999) 369–375.

[47] D. Weininger, SMILES, a chemical language and information system. 1. Introduction to methodology and encoding rules, J. Chem. Inf. Comput. Sci. 28 (1) (1988) 31–36, https://doi.org/10.1021/ci00057a005.

[48] https://en.wikipedia.org/wiki/Simplified_molecular-input_line-entry_system.

[49] PubChem, https://pubchem.ncbi.nlm.nih.gov, (accessed 13.12.2020).

[50] PubChemPy, https://pubchempy.readthedocs.io/en/latest, (accessed 13.12.2020).

[51] P. Csizmadia, MarvinSketch and marvinview: molecule applets for the world wide web, in: Third International Electronic Conference on Synthetic Organic Chemistry (ECSOC-3), 1999.

[52] D.M. Lowe, P.T. Corbett, P. Murray-Rust, R.C. Glen, Chemical name to structure: OPSIN, an open source solution, J. Chem. Inf. Model. 51 (3) (2011) 739–753, https://doi.org/10.1021/ci100384d.

[53] D. Pavlov, M. Rybalkin, B. Karulin, M. Kozhevnikov, A. Savelyev, A. Churinov, Indigo: universal cheminformatics API, J. Cheminform. 3 (2011) P4.

[54] N.M. O'Boyle, M. Banck, C.A. James, C. Morley, T. Vandermeersch, G.R. Hutchison, Open Babel: an open chemical toolbox, J. Cheminform. 3 (1) (2011) 33, https://doi.org/10.1186/1758-2946-3-33.

[55] G. Landrum, RDKit: open-source cheminformatics, http://www.rdkit.org.

[56] E.K. Brefo-Mensah, M. Palmer, mol2chemfig, a tool for rendering chemical structures from molfile or SMILES format to LATEX code, J. Cheminform. 4 (1) (2012) 24, https://doi.org/10.1186/1758-2946-4-24.

[57] M. Korichi, V. Gerbaud, P. Floquet, A.H. Meniai, S. Nacef, X. Joulia, Computer aided aroma design I-molecular knowledge framework, Chem. Eng. Process. Process Intensification 47 (11) (2008) 1902–1911, https://doi.org/10.1016/j.cep.2008.02.008.

[58] S.C. Basak, Use of graph invariants in quantitative structure-activity relationship studies, Croat. Chem. Acta 89 (4) (2016) 419–429, https://doi.org/10.5562/cca3029. https://hrcak.srce.hr/173872.

[59] C.W. Yap, PaDEL-descriptor: An open source software to calculate molecular descriptors and fingerprints, J. Comput. Chem. 32 (7) (2011) 1466–1474, https://doi.org/10.1002/jcc.21707.

[60] H. Moriwaki, Y.-S. Tian, N. Kawashita, T. Takagi, Mordred: a molecular descriptor calculator, J. Cheminform. 10 (1) (2018) 4, https://doi.org/10.1186/s13321-018-0258-y.

[61] D. Mathieu, Accurate or fast prediction of solid-state formation enthalpies using standard sublimation enthalpies derived from geometrical fragments, Ind. Eng. Chem. Res. 57 (41) (2018) 13856–13865, https://doi.org/10.1021/acs.iecr.8b03001.

[62] D. Mathieu, Pencil and paper estimation of hansen solubility parameters, ACS Omega 3 (12) (2018) 17049–17056, https://doi.org/10.1021/acsomega.8b02601.

[63] R. Bouteloup, D. Mathieu, Predicting dielectric constants of pure liquids: fragment-based Kirkwood-Fröhlich model applicable over a wide range of polarity, Phys. Chem. Chem. Phys. 21 (21) (2019) 11043–11057, https://doi.org/10.1039/C9CP01704F. https://pubs.rsc.org/en/content/articlelanding/2019/cp/c9cp01704f.

[64] https://www.daylight.com/dayhtml/doc/theory/theory.smarts.html, (accessed 19.12.2020).

[65] K.G. Joback, Group contribution techniques: predicting the properties of energetic materials, in: V. Boddu, P. Redner (Eds.), Energetic Materials, CRC Press, 2010, pp. 161–170.

[66] Z. Kolská, M. Zábranský, A. Randova, Group contribution methods for estimation of selected physico-chemical properties of organic compounds, in: R. Morales-Rodriguez (Ed.), Thermodynamics, Fundamentals and its Applications in Science, 2012, pp. 135–162 (Chapter 6), https://doi.org/10.5772/49998. https://www.

intechopen.com/books/thermodynamics-fundamentals-and-its-application-in-science/group-contribution-methods-for-estimation-of-selected-physico-chemical-properties-of-organic-compounds.
[67] B.T. Goodman, W.V. Wilding, J.L. Oscarson, R.L. Rowley, Use of the DIPPR database for development of quantitative structure-property relationship correlations: heat capacity of solid organic compounds, J. Chem. Eng. Data 49 (1) (2004) 24–31, https://doi.org/10.1021/je025656h.
[68] J.A. Lazzús, Hybrid method to predict melting points of organic compounds using group contribution + neural network + particle swarm algorithm, Ind. Eng. Chem. Res. 48 (18) (2009) 8760–8766, https://doi.org/10.1021/ie900431f.
[69] D. Mathieu, Toward a physically based quantitative modeling of impact sensitivities, J. Phys. Chem. A 117 (2013) 2253–2259.
[70] G. Fayet, P. Rotureau, Development of simple QSPR models for the impact sensitivity of nitramines, J. Loss Prev. Process Ind. 30 (2014) 1–8.
[71] D. Mathieu, T. Alaime, J. Beaufrez, From theoretical energy barriers to decomposition temperatures of organic peroxides, J. Therm. Anal. Calorim. 129 (1) (2017) 323–337, https://doi.org/10.1007/s10973-017-6114-x.
[72] D. Mathieu, Alternatives to quantitative structure-property relationships for the evaluation of stability and safety aspects of energetic materials, in: Energetic Materials: Modelling, Simulation and Characterisation of Pyrotechnics, Propellants and Explosives, Fraunhofer Institut für Chemische Technologie ICT, Karlsruhe, Germany, 2011, p. V39.
[73] D. Stumpfe, H. Hu, J. Bajorath, Evolving concept of activity cliffs, ACS Omega 4 (11) (2019) 14360–14368, https://doi.org/10.1021/acsomega.9b02221.
[74] J. Bicerano, Prediction of Polymer Properties, Marcel Dekker, NY, 2002.
[75] H. Nefati, J.-M. Cense, J.-J. Legendre, Prediction of the impact sensitivity by neural networks, J. Chem. Inf. Comput. Sci. 36 (1996) 804–810.
[76] R. Wang, J. Jiang, Y. Pan, H. Cao, Y. Cui, Prediction of impact sensitivity of nitro energetic compounds by neural network based on electrotopological-state indices, J. Hazard. Mater. 166 (2009) 155–186.
[77] M.H. Keshavarz, M. Hayati, S. Ghariban-Lavasani, N. Zohari, Relationship between activation energy of thermolysis and friction sensitivity of cyclic and acyclic nitramines, Z. Anorg. Allg. Chem. 642 (2) (2016) 182–188, https://doi.org/10.1002/zaac.201500706.
[78] M.H. Keshavarz, A new general correlation for predicting impact sensitivity of energetic compounds, Propellants Explos. Pyrotech. 38 (6) (2013) 754–760, https://doi.org/10.1002/prep.201200128.
[79] X.-H. Li, C. Zhang, X.-H. Ju, Theoretical screening of bistriazole-derived energetic salts with high energetic properties and low sensitivity, RSC Advances 9 (45) (2019) 26442–26449, https://doi.org/10.1039/C9RA05141D. https://pubs.rsc.org/en/content/articlelanding/2019/ra/c9ra05141d.
[80] S. Beaucamp, D. Mathieu, V. Agafonov, Optimal partitioning of molecular properties into additive contributions: the case of crystal volumes, Acta Cryst. B 63 (2007) 277–284.
[81] M.H. Keshavarz, S. Damiri, V. Bagheri, Recent advances for prediction of electric spark and shock sensitivities of organic compounds containing energetic functional groups to assess reliable models, Process Saf. Environ. Prot. 131 (2019) 9–15.
[82] M.J. Kamlet, J.E. Ablard, Chemistry of detonations. II. A buffered equilibria, J. Chem. Phys. 48 (1968) 36.
[83] D. Mathieu, Prediction of gurney parameters based on an analytic description of the expanding products, J. Energ. Mater. 33 (2015) 102–115.
[84] D. Mathieu, Atom pair contribution method: fast and general procedure to predict molecular formation enthalpies, J. Chem. Inf. Model. 58 (1) (2018) 12–26, https://doi.org/10.1021/acs.jcim.7b00613.
[85] http://www.codessa-pro.com/index.htm, (accessed 31.01.2021).
[86] G. Landrum, RDKit: open-source cheminformatics, http://www.rdkit.org, (accessed 31.01.2021).
[87] D. Mathieu, T. Alaime, Impact sensitivities of energetic materials: exploring the limitations of a model based only on structural formulas, J. Mol. Graph. Model. 62 (2015) 81–86.
[88] C.B. Storm, J.R. Stine, J.F. Kramer, Sensitivity relationships in energetic materials, in: S.N. Bulusu (Ed.), Chemistry and Physics of Energetic Materials, Kluwer Academic Publishers, Dordrecht, 1990. 605–309.
[89] S.G. Cho, K.T. No, E.M. Goh, J.K. Kim, J.H. Shin, Y.D. Joo, S. Seong, Optimization of neural networks architecture for impact sensitivity of energetic molecules, Bull. Korean Chem. Soc. 26 (2005) 399–408.
[90] Z. Jun, C. Xin-Iu, H. Bi, Y. Xiang-Dong, Neural networks study on the correlation between impact sensitivity and molecular structures for nitramine explosives, Struct. Chem. 17 (2006) 501–507.
[91] M. Cranmer, A. Sanchez-Gonzalez, P. Battaglia, R. Xu, K. Cranmer, D. Spergel, S. Ho, Discovering symbolic models from deep learning with inductive biases, arXiv:2006.11287 [astro-ph, physics:physics, stat] (2020). http://arxiv.org/abs/2006.11287. arXiv: 2006.11287.

[92] A.J. Turner, J.F. Miller, Introducing a cross platform open source Cartesian genetic programming library, Genet. Program Evolvable Mach 16 (1) (2015) 83–91, https://doi.org/10.1007/s10710-014-9233-1.
[93] https://github.com/NSLS-II/lmfit, (accessed 30.12.2020).
[94] R. Poli, W.B. Langdon, N.F. McPhee, A Field Guide to Genetic Programming, Lulu.com, 2008.
[95] F. Olivetti de França, A greedy search tree heuristic for symbolic regression, Inf. Sci. 442-443 (2018) 18–32, https://doi.org/10.1016/j.ins.2018.02.040.
[96] C. Chen, C. Luo, Z. Jiang, Elite bases regression: a real-time algorithm for symbolic regression, arXiv:1704.07313 [cs] (2017), http://arxiv.org/abs/1704.07313. arXiv: 1704.07313.
[97] J. žegklitz, P. Pošík, Symbolic regression algorithms with built-in linear regression, arXiv:1701.03641 [cs] (2017). http://arxiv.org/abs/1701.03641. arXiv: 1701.03641.
[98] T. McConaghy, FFX: fast, scalable, deterministic symbolic regression technology, in: R. Riolo, E. Vladislavleva, J.H. Moore (Eds.), Genetic Programming Theory and Practice IX, Springer, New York, NY, 2011, pp. 235–260, ISBN: 978-1-4614-1770-5, https://doi.org/10.1007/978-1-4614-1770-5_13.
[99] J. Friedman, T. Hastie, R. Tibshirani, Regularization paths for generalized linear models via coordinate descent, J Stat Softw. 33 (2010) 1–22.
[100] M.E. Sitzmann, High-melting aromatic nitrate esters: ethanolamine derivatives of polynitroaromatic compounds, Propellants Explos. Pyrotech. 19 (1994) 249–254.

CHAPTER

17

Molecular and crystal insights into the structural design of low-sensitivity energetic materials

*Yi Wang, Siwei Song, and Qinghua Zhang**

Institute of Chemical Materials, China Academy of Engineering Physics, Mianyang, China

***Corresponding author: E-mail: qinghuazhang@caep.cn**

Energetic materials are a special class of reactive substances with large amounts of stored chemical energy, which are usually consisting of carbon, hydrogen, nitrogen, and oxygen atoms and can release a large quantity of heat and gaseous products by undergoing an extremely rapid self-sustained exothermic decomposition reaction when suitably triggered [1,2]. In the field of energetic materials, energy and safety are the two most important concerns [3]. The energy represents the release efficiency of stored chemical energy within energetic material molecules, which is usually evaluated by reaction heat (heat of detonation or combustion), detonation properties (velocity, pressure, and heat of detonation), power, or working ability. The safety reflects the degree of stability of the energetic materials in response to external stimuli, which is usually evaluated by sensitivity [4]. Both energy and safety are very important for the practical application of energetic materials [5]. Ideally, energetic materials are expected to exhibit both high energy and low sensitivity. However, there is commonly an undesired negative correlation between energy and sensitivity, *i.e.*, high energy of energetic materials is usually accompanied by increased sensitivity (or decreased safety) [6]. Therefore, it is a long-term challenge in this field to balance this energy-sensitivity contradiction in the development of new energetic materials with high-energy and low-sensitivity performances.

In fact, the influence factors on energy and sensitivity are complicated with obvious multi-level characteristics of energetic materials [7], *e.g.*, from chemical composition, molecular structure [8,9] and crystal packing modes [10–12], defects [13–15], to crystal morphology, sizes and interfaces [16–21], to particles and blocks of plastic bonded explosives (PBXs) [22,23], and some others. After decades of efforts, people have come to realize that the

Theoretical and Computational Chemistry, Volume 22
https://doi.org/10.1016/B978-0-12-822971-2.00014-0

micro-scaled molecular and crystal engineering of energetic materials are the most important and inherent factors for determining energy and sensitivity [24]. With the aim of understanding and predicting the energy and sensitivity of energetic materials, numerical molecular- and crystal-level models have been developed [25–34]. However, it is still very difficult to gain insights into the accurate mesoscale (such as crystal defects, morphology, sizes) and macroscale (such as particles and blocks of PBXs) structure–property relationships. Obviously, the development of molecular and crystal-level structure–property correlation models is highly beneficial to guide the molecular design of new energetic materials with well-balanced energy-sensitivity properties.

As mentioned above, the evaluation methods for energy and safety of energetic materials are diverse, but the detonation properties (such as detonation velocity) [35] and mechanical sensitivity (such as impact sensitivity) [36] are widely used to evaluate the energy and safety of energetic materials, respectively. In addition, so far there is no clear definition of high-energy and low-sensitivity energetic materials. Considering that trinitrotoluene (TNT) is a representative modern high explosive, Zhang et al. proposed as a criterion that if an energetic material exhibits energy or safety close or superior to that of TNT, this energetic material can be considered to meet the criteria of high-energy or low-sensitivity [10]. According to this criterion, we employ here the quantifiable detonation velocity (V_D: 7303 m s^{-1}) and impact sensitivity (15 J) of TNT [37] to evaluate the energy and safety of unknown energetic materials. Those energetic materials with a detonation velocity of $>$7303 m s^{-1} and impact sensitivity of $>$15 J are regarded as low-sensitivity and high-energy materials. In this chapter, we will concentrate on typical structure–property correlations from the viewpoints of chemical composition, molecular structure, and crystal packing. We hope these structure–property relationships can provide useful guidance to the structural design of new energetic materials, in particular, low-sensitivity energetic materials with desired high-energy and heat-resistant properties.

1 Molecular concerns in the structural design of low-sensitivity energetic materials

1.1 Molecular composition concerns

Energetic materials are mostly organic molecules consisting of C, H, N, and O elements and their chemical compositions (namely the ratio of C, H, O, N atoms in a molecule) have critical impacts on their mechanical sensitivity [38]. Among them, oxygen balance (OB) is a very important and common composition index to correlate the energy and sensitivity properties, which is usually formulated as follows:

$$\mathrm{OB} = \frac{16(c - 2a - 1/2b)}{Mw} \times 100\% \tag{1}$$

in which a, b, c, and Mw represent the numbers of C, H, and O atoms, and molecular weight of an energetic molecule, respectively [39,40]. This formula (Eq. 1) reflects the ability of oxidizer (O) to reduce (C and H) elements.

Based on in-depth analysis of the crystal densities of more than 1000 organic molecules based on C, H, N, and O elements from the Cambridge Structural Database (CSD), it can be found that the crystal densities of these organic molecules are linearly related to their oxygen balances (Fig. 17.1). Based on the fact that detonation velocity of energetic materials also has a strong linear correlation with their densities, it can be concluded that high oxygen balances usually result in high densities and detonation velocities of energetic materials, indicating the high energy of energetic materials [5]. For example, the results of statistical analysis have demonstrated that a quite high percentage (>30%) of CHNO-based molecules with a positive oxygen balance exhibited densities of >1.90 g cm^{-3} at room temperature (Fig. 17.1, column chart). Undesirably, these molecules with both high density and high oxygen balances are usually very sensitive to external stimuli (*e.g.*, impact sensitivities are worse than 10 J, Fig. 17.1, *red dots*), clearly indicating a contradictory relationship between high energy and low sensitivity. These results are also consistent with the experimental observations from previous studies [42]. By contrast, only 3% of CHNO-based molecules with oxygen balances between −80% and −40% have densities >1.90 g cm^{-3}. Meanwhile, they are all less sensitive to external stimuli with impact sensitivities of >30 J. In addition, some CHNO-based molecules with oxygen balances between −30% and −40% have low impact sensitivities (>20 J). Notably, around 25% (a relatively high ratio) of the CHNO-based molecules with oxygen balances between −20% and −30%, have densities >1.90 g cm^{-3}, including a few ones with low impact sensitivity (superior to that of TNT) (Fig. 17.1). This indicates that it is feasible to design high-energy and low-sensitivity energetic materials in the relatively high oxygen balance regions [41].

Based on the above analyses, a suitable oxygen balance range for the development of high-energy and low-sensitivity energetic materials should be −20%∼−80%. It is worth

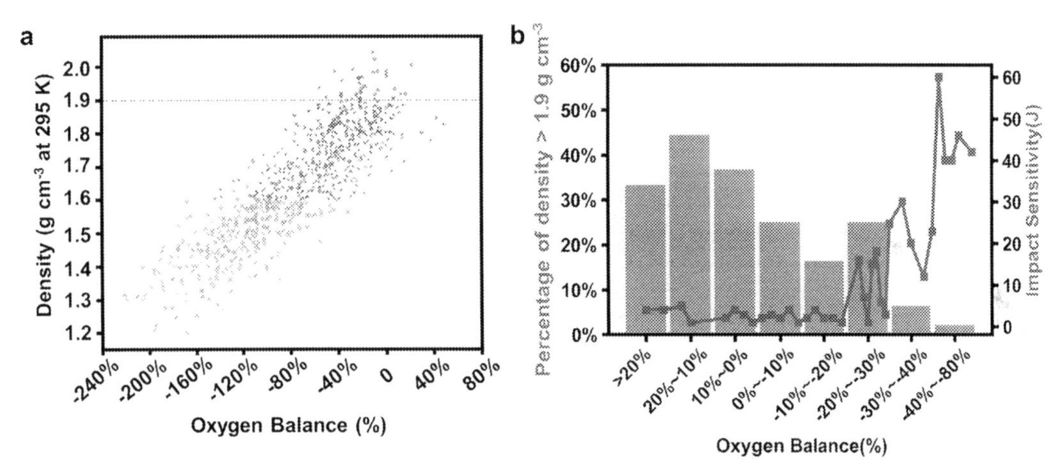

FIG. 17.1 (A) Relationship between crystal densities and oxygen balance of the organic molecules constituted by C, H, N, O atoms. (B) Percentage of the density above 1.90 g cm^{-3} in different oxygen balance ranges (*blue* histogram) and impact insensitivities of some representative energetic molecules with density above 1.90 g cm^{-3} in different oxygen balance ranges (*red dots*). *Reprinted with permission from Y. Wang, Y. Liu, S. Song, Z. Yang, X. Qi, K. Wang, Y. Liu, Q. Zhang, Y. Tian, Accelerating the discovery of insensitive high-energy-density materials by a materials genome approach, Nat. Commun. 9 (2018) 2444.*

noting that most representative high-energy and low-sensitivity energetic materials, such as FOX-7, LLM-105, and TATB, have their oxygen balances (−21.6%, −37.0%, and −55.8%, respectively) falling exactly in this range.

1.2 Molecular skeleton concerns

1.2.1 *Effects of molecular skeletons on density*

Oxygen balance reflects the ratios of C, H, O, N atoms in an energetic molecule with obvious disadvantages in distinguishing isomers. For a given molecular formula, it can generate different types of molecules with the same oxygen balance, but they usually have various physicochemical properties such as different densities, sensitivity and decomposition temperatures, etc. Here, we first classified the above-mentioned more than 1000 CHON-based organic molecules into aliphatic and aromatic and then finely divided them into six categories, including aliphatic chain, aliphatic ring, aliphatic cage, aromatic single-ring, aromatic multiple-ring, and fused heterocycle ring (Fig. 17.2A). After in-depth classifications on the relationships between their oxygen balance, energy, and sensitivity, some interesting results are obtained.

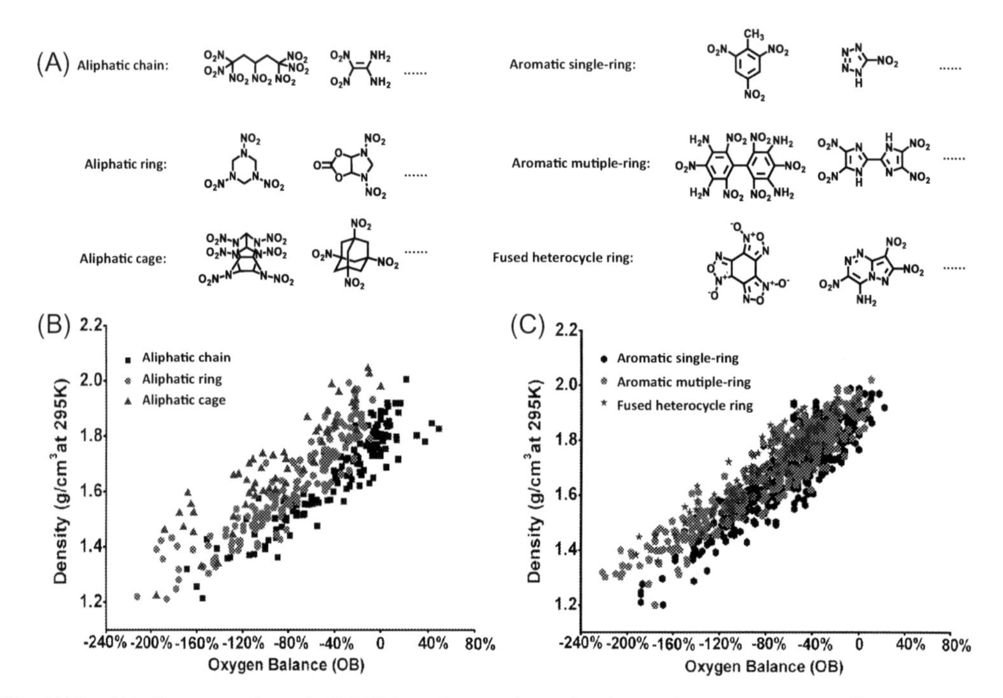

FIG. 17.2 (A) Six categories of CHON-based organic molecules and some corresponding representative energetic molecules. (B) Relationship between crystal densities and oxygen balance of aliphatic CHON-based organic molecules. (C) Relationship between crystal densities and oxygen balance of aromatic CHON-based organic molecules.

Although there is an overall linear correlation between crystal density and oxygen balance for CHNO-based organic molecules, different types of molecules have obviously different trends. As shown in Fig. 17.2B, aliphatic cage, aliphatic ring, and aliphatic chain molecules are commonly distributed in an up-middle-down sequence in the correlation plot between crystal density and oxygen balance. This indicates that, with the same oxygen balance, the aliphatic cage molecules usually have the highest density, while the aliphatic chain molecules usually have the lowest density. For aromatic molecules, the distribution differences are not obvious like the ones of aliphatic molecules, but the fused heterocycle molecules show a higher possibility to possess high density than the other two classes of aromatic molecules (Fig. 17.2C). Through analyzing the distribution regions, we can find that aliphatic cage and fused heterocycle ring molecules should be the most favorable molecular structures to construct higher-energy energetic molecules.

To better understand the relationship between crystal density and oxygen balance for the six categories of molecules, an univariate linear regression has been conducted. As shown in Fig. 17.3A–F, their relationships can be generally formulated as follows:

$$Y = aX + b \tag{2}$$

in which Y and X represent density and oxygen balance of organic molecules, respectively, a and b are two constant terms. The fits are fairly good with the adjusted R-squares (R^2) all above 0.8. Based on these fitting formulas, a quick and rough density prediction of energetic molecules in the period of molecular design can be obtained. Within the oxygen balance range of above −100% (the one where most energetic molecules are located), for the same value of the oxygen balance, aliphatic cage and fused heterocycle ring molecules exhibit obviously higher density than the other four categories of molecules according to the corresponding fitting formulas. Therefore, from the viewpoint of energy, we tend to recommend aliphatic cage and fused heterocycle ring molecules because they show a high possibility to construct high-energy energetic molecules. A famous example is CL-20, which has a typical aliphatic cage structure to exhibit high density (2.04 g cm^{-3}) and high detonation performances (D: 9445 ms^{-1}, P: 46.7 GPa). Unfortunately, the safety of CL-20 is very poor with a very high impact sensitivity (IS: 4 J) [43].

1.2.2 *Effects of a molecular skeleton on impact sensitivity*

Besides above-mentioned differences in density (or energy), different categories of energetic materials show obvious distribution differences in safety (or sensitivity). Because impact sensitivity (IS) is one of the most studied sensitivities, it is chosen here for detailed illustration. The statistical analysis samples are focused on the reported energetic materials (density above 1.70 g cm^{-3}). As shown in Fig. 17.4A and B, the proportion of aliphatic and aromatic energetic molecules located in the high-sensitivity range ($IS \leqq 5$ J) are quite similar (around 27%). In the medium sensitivity ($5\ J < IS \leqq 15$ J) and low-sensitivity ($IS > 15$ J) regions, aliphatic and aromatic energetic molecules exhibit significantly different distribution trends. For aliphatic energetic molecules, the proportion in the medium sensitivity region is a little higher than that in the low-sensitivity region (39.7% vs 32.8%). In contrast, the proportion in the medium sensitivity region is much lower than that in the low-sensitivity region (27.3% vs 46.7%) for aromatic energetic molecules. Obviously, aromatic energetic materials are widely known to exhibit lower sensitivities than aliphatic energetic molecules, hence the respective ratios

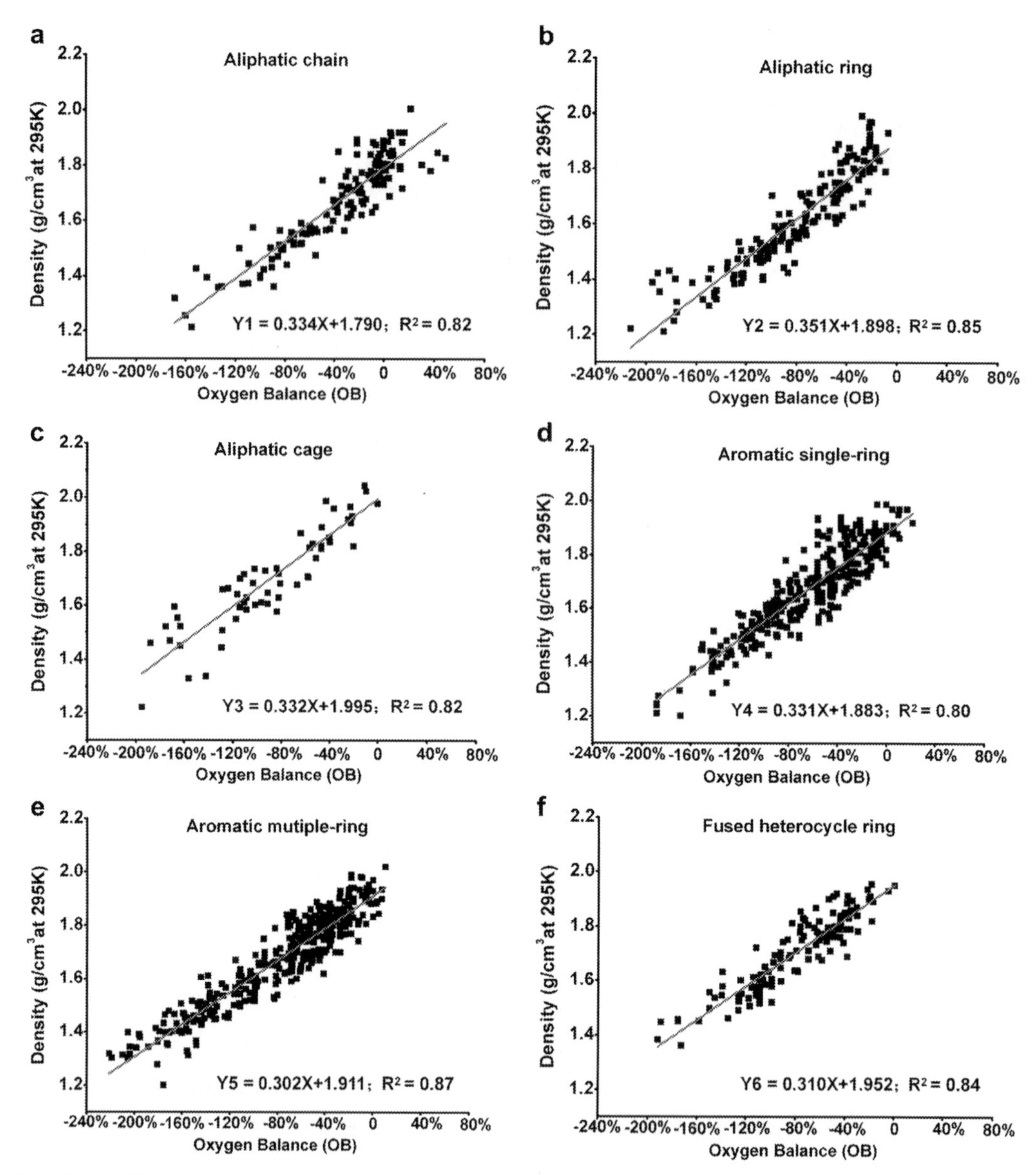

FIG. 17.3 (A) Linear fitting on the relationship between crystal densities and oxygen balance of aliphatic chain molecules. (B) Linear fitting on the relationship between crystal densities and oxygen balance of aliphatic ring molecules. (C) Linear fitting on the relationship between crystal densities and oxygen balance of aliphatic cage molecules. (D) Linear fitting on the relationship between crystal densities and oxygen balance of aromatic single-ring molecules. (E) Linear fitting on the relationship between crystal densities and oxygen balance of aromatic multiple-ring molecules. (F) Linear fitting on the relationship between crystal densities and oxygen balance of fused heterocycle ring molecules.

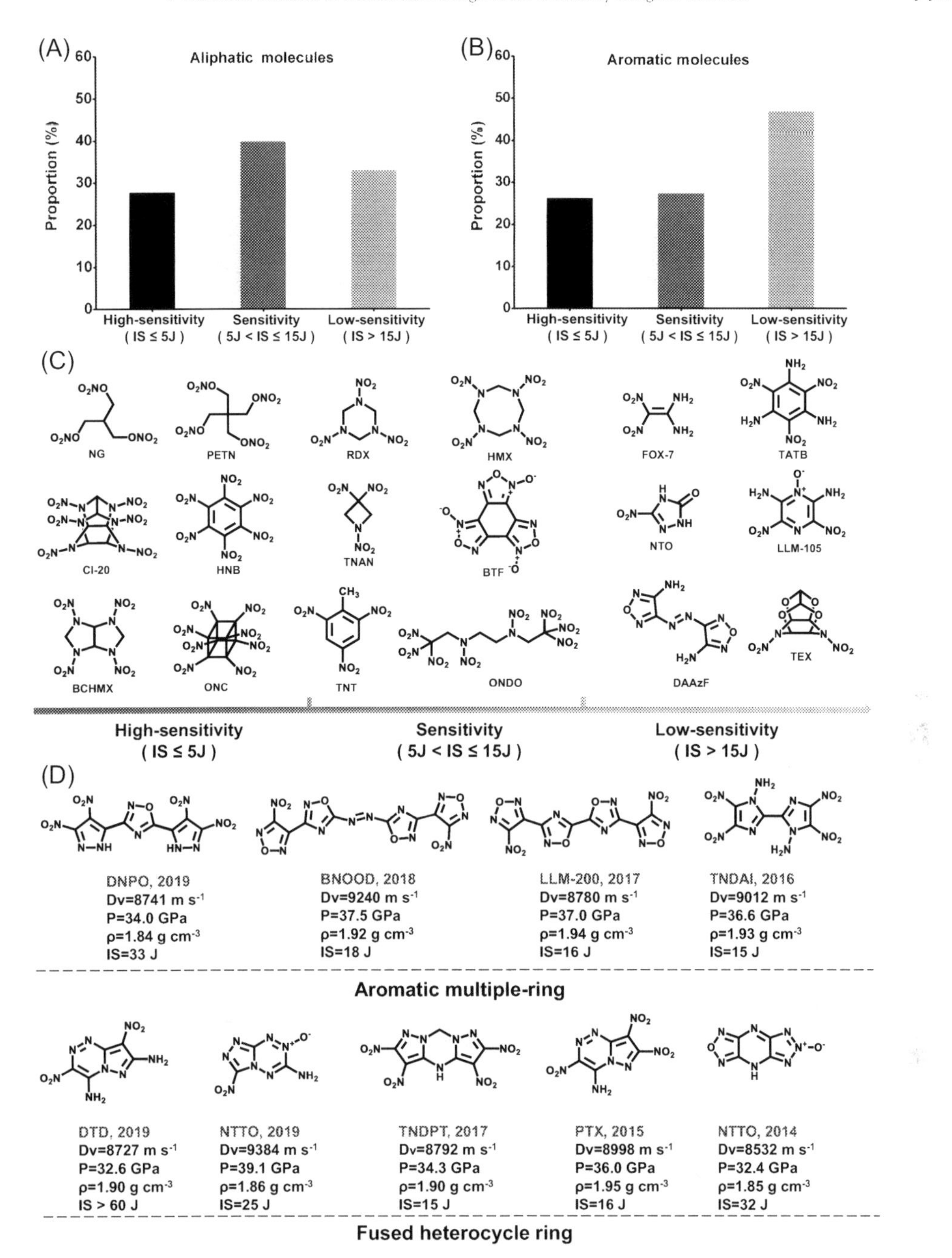

FIG. 17.4 (A) Impact sensitivity distributions for the reported aliphatic energetic molecules (density above 1.70 g cm^{-3}) in recent two decades. (B) Impact sensitivity distributions on impact sensitivities for the reported aromatic energetic molecules (density above 1.70 g cm^{-3}) in recent two decades. (C) Chemical structures and impact sensitivity distributions for present widely used and studied energetic materials. (D) Representative high-energy and low-sensitivity aromatic multiple-ring and fused heterocycle molecules prepared in recent years.

(46.7% vs 32.8%). This result can be further supported by the actual situation in present extensively used and studied energetic materials. As shown in Fig. 17.4C, high-energy but sensitive energetic materials (such as RDX, HMX, PETN, CL-20) are aliphatic organic molecules; conversely, high-energy and low-sensitivity energetic materials (such as TATB, FOX-7, and LLM-105) are aromatic organic molecules. The reason is that the aromatic energetic molecules have large conjugated π-bonded molecular structures efficiently averaging the bond dissociation energies of all bonds, thus leading to increased molecular stability and decreased sensitivity. All in all, aromatic molecular structures should be a good choice for developing new low-sensitivity energetic materials from both statistical analysis and present discussion of actual examples.

Due to the limited number of modifiable reactive sites on aromatic single-ring skeletons and following extensive studies over the past decades, the development potential of aromatic single-ring-based high-energy and low-sensitivity energetic materials has now been fully exploited. In recent years, researchers have shifted their interest to multiple-ring and fused-ring energetic molecules [37]. Some representative examples are listed in Fig. 17.4D, all of which exhibit high energy (density $\geqq 1.84$ g cm^{-3}, detonation velocity: $Dv > 8500$ ms^{-1}) and low impact sensitivities (IS $\geqq 15$ J) [44–52]. Considering that there are more modifiable reactive sites and huge exploration space, we think that aromatic multiple-ring and fused heterocycle ring molecular skeletons should be the primary choices for the construction of high-energy and low-sensitivity energetic materials.

1.3 Substituent group

1.3.1 Effects of explosophores on impact sensitivity and thermostability

CHON-based energetic molecules can usually be divided into two parts, namely molecular skeleton and substituent group. Besides the above discussed molecular skeletons, the substituent groups (including explosophores and stabilizing groups) play a crucial role in the safety of energetic materials. Here, we use a statistical analysis method to illustrate the effects of different explosophores on the impact sensitivities. The analysis samples are still the reported energetic materials (density above 1.70 g cm^{-3}) in recent two decades.

As shown in Fig. 17.5A, the energetic materials with nitric ester (—ONO_2) and azide group (—N_3) as explosophores have quite high percentages (60.0% and 52.9%, respectively) in the high-sensitivity (IS $\leqq 5$ J) region and low percentages in the low-sensitivity (IS > 15 J) region with the values of 4% and 11.8%, respectively (*blue* and *yellow* columns, respectively). In addition, the energetic materials with nitroform group (—$C(NO_2)_3$) as explosophores also exhibit a relatively low percentage (19.4%) in the low-sensitivity (IS > 15 J) region (*pink* column). This indicates that the three explosophores nitric ester (—ONO_2), azide (—N_3) and nitroform (—$C(NO_2)_3$) should not be the first choices for designing new low-sensitivity energetic materials. For the energetic materials with a nitrogen-linked nitro group (N—NO_2) and *gem*-dinitro group (—$C(NO_2)_2$) as explosophores, they have relatively high percentages (37.5% and 35.9%, respectively) in the low-sensitivity (IS > 15 J) region (*red* and *cyan* columns, respectively), indicating that rational combinations between these explosophores and fused-ring molecular skeletons may construct some low-sensitivity energetic materials. Encouragingly, the energetic materials with a carbon-linked nitro group (C—NO_2) as

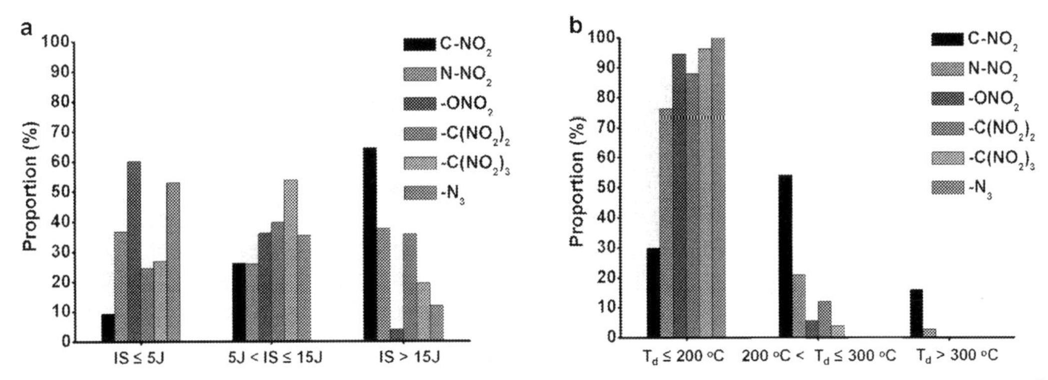

FIG. 17.5 (A) Distributions on impact sensitivities of the reported energetic molecules (density above 1.70 g cm^{-3}) with different explosophores in recent two decades. (B) Distributions on decomposition temperatures of the reported energetic molecules (density above 1.70 g cm^{-3}) with different explosophores in recent two decades.

explosophore show gradually increasing distribution percentages in high-sensitivity, sensitivity, and low-sensitivity regions with values of 9.2%, 26.2%, and 64.5%, respectively (*black* column). Obviously, the ideal explosophore for developing low-sensitivity energetic materials should be the carbon-linked nitro group (C—NO_2), and the next two ones are a nitrogen-linked nitro group (N—NO_2) and *gem*-dinitro group (—$C(NO_2)_2$).

Thermostability is another important property for energetic materials, which is an indicator of resilience to heat stimuli usually characterized by the decomposition temperature (T_d). In general, the decomposition temperatures of widely used energetic materials (such as TNT, RDX, HMX, CL-20, FOX-7, TATB, LLM-105) are all above 200°C. Thus, decomposition temperature should exceed 200°C for an energetic material to be of potential interest in view of applications. Moreover, in some special application fields (such as space and deep-earth explorations), the decomposition temperatures should be higher than 300°C [53,54]. As a result, it is highly desirable to study the effects of different explosophores on the thermostability of energetic materials. Here, statistical analysis can provide some guidance in designing new heat-resistant energetic materials.

As shown in Fig. 17.5B, the energetic materials with the nitric ester (—ONO_2), nitroform (—$C(NO_2)_3$) and azide (—N_3) groups as explosophores have quite high percentages (>90%) in the region below 200°C, and none of them can achieve a high decomposition temperature of above 300°C (*blue*, *pink* and *yellow* columns, respectively). This indicates that three explosophores of nitric ester (—ONO_2), azide group (—N_3) and nitroform group (—$C(NO_2)_3$) should not be good choices for designing new energetic materials with good thermostability. For the energetic materials with nitrogen-linked nitro group (N-NO_2) and *gem*-dinitromethyl group (—$C(NO_2)_2$) as explosophores, they have high percentages (76.3% and 87.9%, respectively) in the region below 200°C, but there are some percentages (21.1% and 12.1%, respectively) in the region between 200°C and 300°C. More importantly, an extremely low percentage (2.6%) of N—NO_2 group-based energetic materials have their decomposition temperatures exceeding 300°C (*red* and *cyan* columns, respectively), which indicates that the N—NO_2 group is an alternative choice for designing new energetic materials with high thermal stability. For the energetic materials with a carbon-linked nitro group (C—NO_2) as explosophore, 54.2% of them are located in the region between 200°C and 300°C, and 15.8% of them are located in the region above 300°C (*black* column), demonstrating that C—NO_2 group should be a good explosophore choice for designing new energetic materials with good thermostability ($T_d > 200$°C).

Based on the above statistical analysis, it can be found that the C—NO_2 group should be the most desirable explosophore to design new energetic material with both low sensitivity and high thermostability, the N—NO_2 group would be the next alternative one. This result is consistent with theoretical studies. It is generally accepted that the chemical decomposition of energetic materials (no matter under mechanical or heat stimuli) is triggered by breaking the weakest bonds (also called the trigger bond) in the energetic molecules [55]. Thus, the bond dissociation energy of the weakest bonds is closely related to the sensitivity and thermostability of energetic materials. High bond dissociation energy of triggered bond will usually result in low sensitivity and high thermostability [56]. In general, the bond dissociation energy of the weakest bonds in C—NO_2 group-based energetic materials are higher than those of the energetic materials with other explosophores (such as N—NO_2 group, nitric ester —ONO_2, and nitroform group —$C(NO_2)_3$) [9] (Fig. 17.6). Consequently, C—NO_2 group-based energetic materials would have the highest possibility to exhibit low sensitivity and high thermal stability.

1.3.2 *Effects of stabilizing groups on density, sensitivity, and thermostability*

The explosophores of energetic materials are usually nitro-based groups (such as N—NO_2 group and nitroform group —$C(NO_2)_3$) with strong electron-withdrawing ability. When they are introduced into the molecular skeletons (especially for aromatic molecular skeletons), their strong electron-withdrawing ability will cause the disequilibrium charge distribution around the entire molecule to form a high molecular polarization degree. As a result, the weakest bonds of energetic molecules are mostly concentrated on the bonds linking nitro and molecule skeleton (Fig. 17.6). To alleviate the adverse effects of strong electron-withdrawing explosophores,

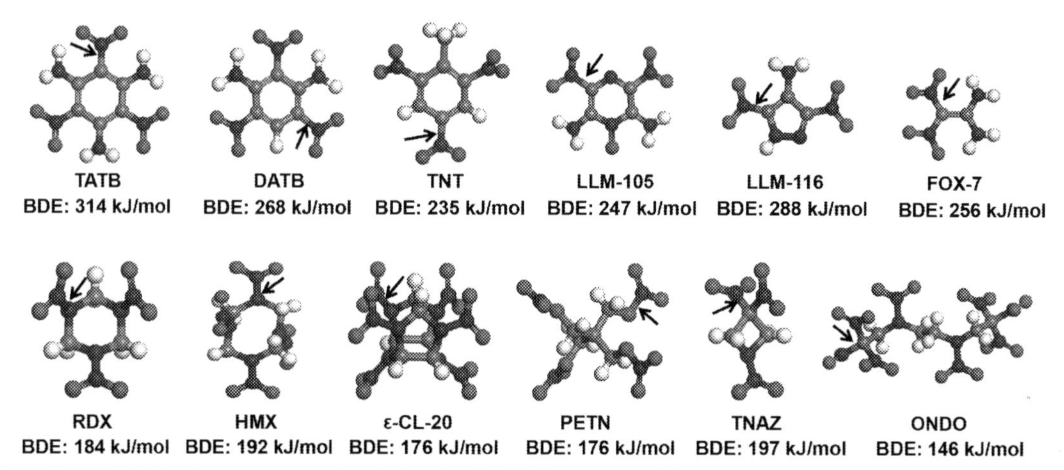

FIG. 17.6 Molecular structures, sites (pointed to by *black arrows*), and dissociation energy of the weakest bond in some typical energetic materials. *Gray*, *white*, *blue*, and *red* balls represent carbon, hydrogen, nitrogen, and oxygen atoms, respectively.

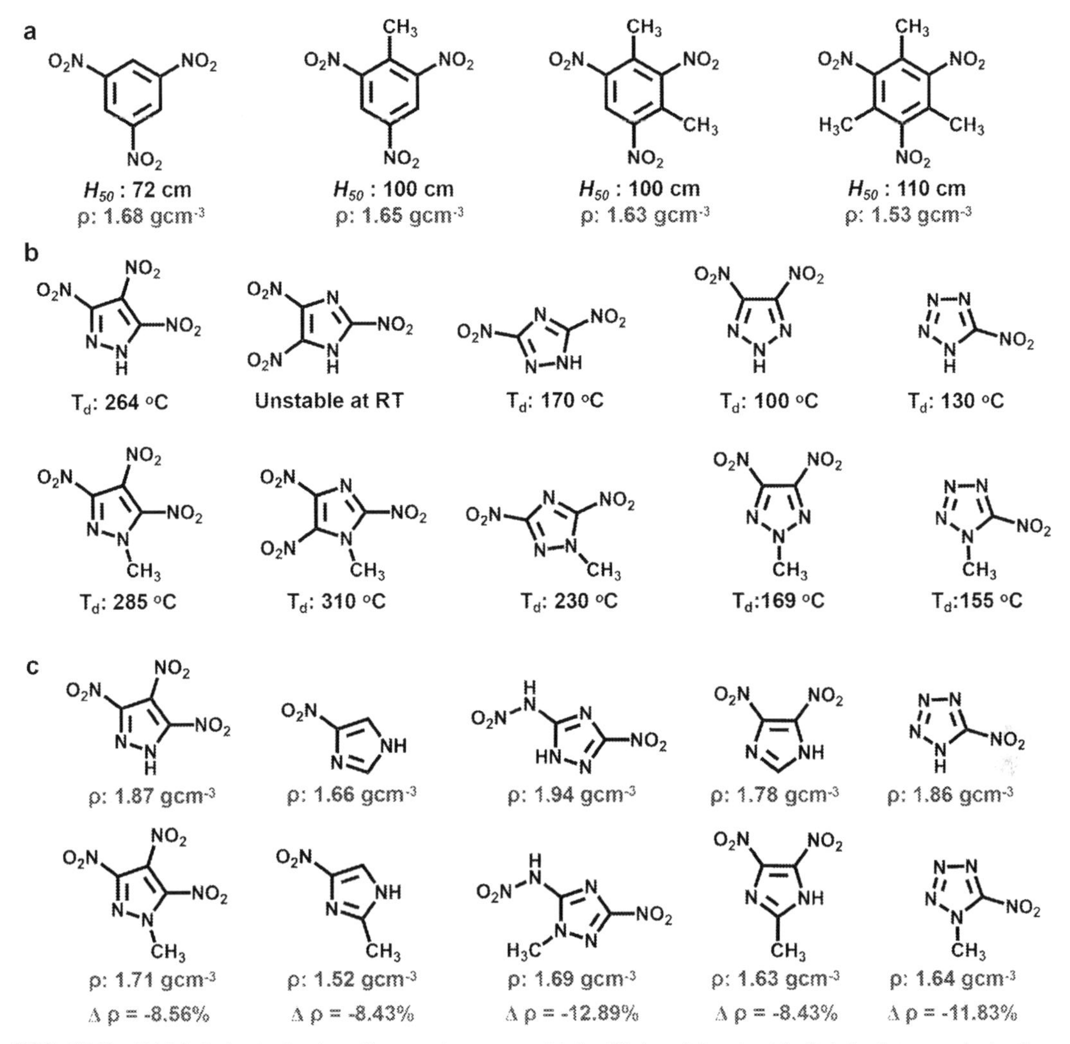

FIG. 17.7 (A) Methyl substitution effects on impact sensitivity (H_{50}) and density (ρ) of trinitrobenzene derivatives. (B) Methyl substitution effects on decomposition temperature (T_d) of nitro-based azole derivatives. (C) Methyl substitution effects on density (ρ) of nitro-based azole derivatives.

an efficient method is to introduce stabilizing groups (such as electron-donating groups methyl and amino groups) into the molecule skeleton of energetic materials [57]. As shown in Fig. 17.7A, when introducing methyl groups into trinitrobenzene, the impact sensitivities of the derivatives are gradually decreasing (H_{50}: changing from 72 to 110 cm) along with the increasing of the number of methyl groups [58]. But the undesired result is that their densities (changing from 1.76 to 1.53 g cm^{-3}) are obviously decreased with increasing the number of methyl groups. In addition, the introduction of methyl groups to nitro-based azole derivatives can also increase their decomposition temperatures (T_d) (Fig. 17.7B), but there is a similar phenomenon that the densities of methyl-substituted derivatives are obviously lower (around 10% decline) than that of the corresponding nitro-based azoles (Fig. 17.7C).

In general, the introduction of electron-donating methyl groups can increase the molecular stability and the resulting energetic materials always exhibit lower sensitivity and higher thermostability than the ones without electron-donating methyl groups. However, these positive changes in impact sensitivities and thermostability are usually the results of sacrificing the energy (obviously decreasing density) of energetic materials. From the viewpoint of high energy, the methyl group should not be a primary choice among stabilizing groups to design new high-energy and low-sensitivity energetic materials.

In contrast to the methyl group, electron-donating amino group usually show relatively positive effects on the energy and density of energetic materials when the amino group is introduced as a stabilizing group. Take trinitrobenzene derivatives as an example, there is a synergistic increase in both energy (D_v: changing from around 7500 m s^{-1} to around 8600 m s^{-1}) and safety (H_{50}: changing from around 80 cm to around 460 cm) with increasing the number of amino groups (Fig. 17.8A). Analyzing the reason for this result, the amino group is not only an electron-donating group to benefit the stabilization of the molecular structure, but also is a good hydrogen-bond donor to form intramolecular and intermolecular hydrogen bonds. As shown in Fig. 17.8B, from TNB to TATB, both intramolecular and intermolecular hydrogen-bond interactions are synchronously strengthened with the increasing number of amino groups. In addition, some typical low-sensitivity energetic materials (LLM-105, TIBMOG, FOX-7, and LLM-116) have extensive intramolecular and intermolecular hydrogen bonds in their crystals (Fig. 17.8B). Obviously, extensive hydrogen-bond networks in energetic molecular crystals can significantly increase packing coefficients and densities of energetic materials to enable them high energy. Moreover, a reasonable hydrogen-bonds combination of energetic molecules can form some unique 3D layered π-π stacking structures (*e.g.*, wave-like and graphite-like for LLM-105 and TIBMOG, respectively) [59] (Fig. 17.8C and D), which are beneficial to low-sensitivity of energetic materials. Therefore, the amino group is an ideal and widely used stabilizing group to develop new low-sensitivity energetic materials.

1.3.3 Effects of substituent arrangement on the density and impact sensitivity

Besides the selection of substituent groups, the substituent arrangement in a molecule also shows a significant impact on the energy and sensitivity properties of energetic materials. Here, two sets of isomers are selected for comparative discussions. As shown in Fig. 17.9A, two energetic molecules of DNTDA-124 and DNTDA-123 share the same molecular composition (molecular formula: $C_4H_4N_{10}O_4$), and their densities and detonation velocity are also relatively close (1.83 g cm^{-3}, 8677 m s^{-1} vs 1.85 g cm^{-3}, 8983 m s^{-1}). However, DNTDA-124 shows much lower impact sensitivity (40 J vs 9 J) and higher decomposition temperature (271°C vs 198°C) than those of DNTDA-123 [60,61]. These significant differences in impact sensitivities and decomposition temperatures could be attributed to their molecular structures with an obviously different arrangement of substituent groups. For DNTDA-124, nitro and amino groups are separately located on the molecular skeleton. Especially for amino groups, they are close to the nitrogen atoms of triazole to facilitate the formation of intramolecular hydrogen bonds. As a result, DNTDA-124 holds a planar configuration with an extended π-conjugated structure, which has a beneficial effect on high molecular stability. In

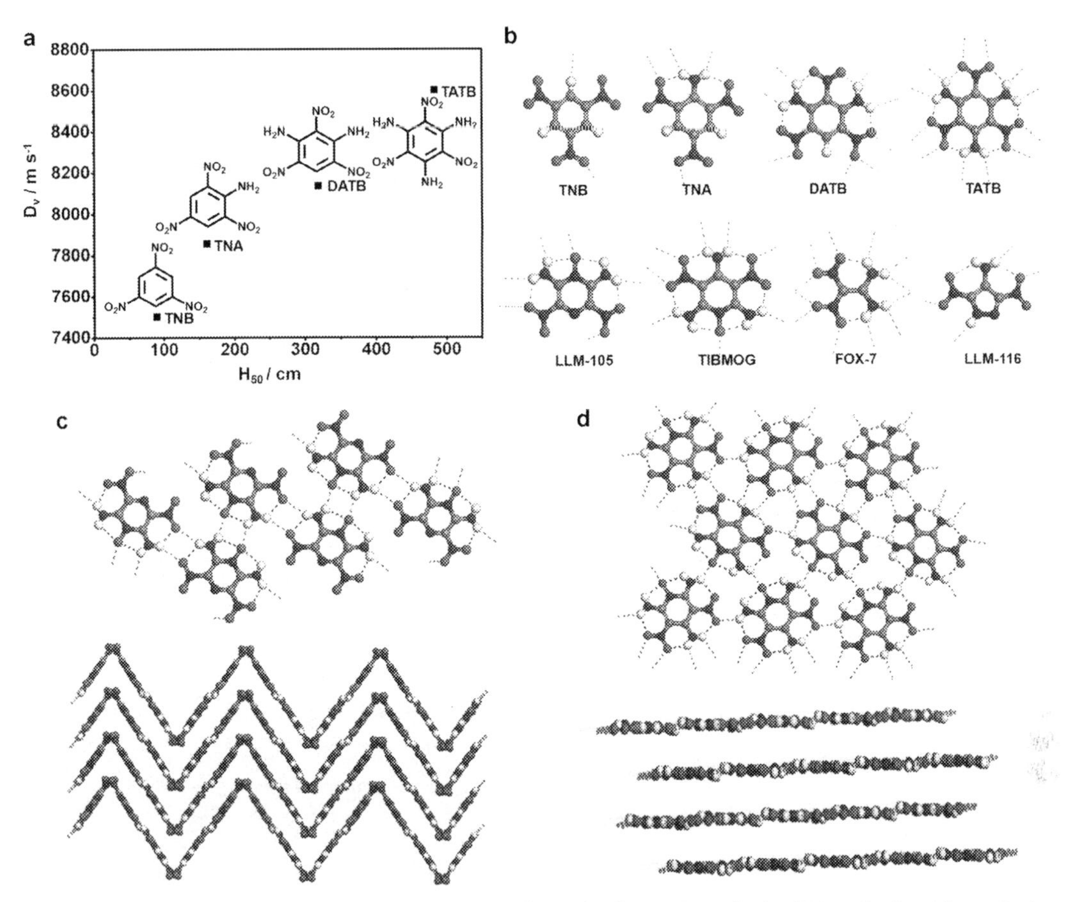

FIG. 17.8 (A) Amino substitution effects on energy (shown by detonation velocity, D_v) and safety (shown by impact sensitivity, H_{50}) of trinitrobenzene derivatives. (B) Molecular structures and intramolecular (*blue dots*) and intermolecular (*cyan dots*) hydrogen bonds of amino-substituted trinitrobenzene derivatives and typical low-sensitivity and high-energy energetic materials. (C) 2D intra-layer hydrogen-bond interactions and 3D wave-like π-π stacking of LLM-105. (D) 2D intra-layer hydrogen-bond interactions and 3D graphite-like π-π stacking of TIBMOG.

contrast, the locations of nitro and amino groups of DNTDA-123 are very close and the strong repulsive effect makes DNTDA-123 a twisted configuration with a limited π-conjugated structure, which is unfavorable to the molecular stability. Similar results are observed in the isomers of LLM-105 and DDPZD-i (Fig. 17.9B). Their energy levels are close (1.92 g cm^{-3}, 8560 m s^{-1} for LLM-105 vs 1.94 g cm^{-3}, 9070 m s^{-1} for DDPZD-i), but the thermostability (T_d: 215°C) and impact sensitivity (IS: 5 J) of DDPZD-i are obviously lower than those of LLM-105 (T_d: 340°C, IS: 23 J) [62]. The relatively low thermostability and high impact sensitivity of DDPZD-i could be ascribed to a more twisted molecular configuration originated from the repulsive effects between N—O bond and two nitro groups. In addition, the alternating arrangement of hydrogen-bond donor (amino group) and hydrogen-bond acceptor

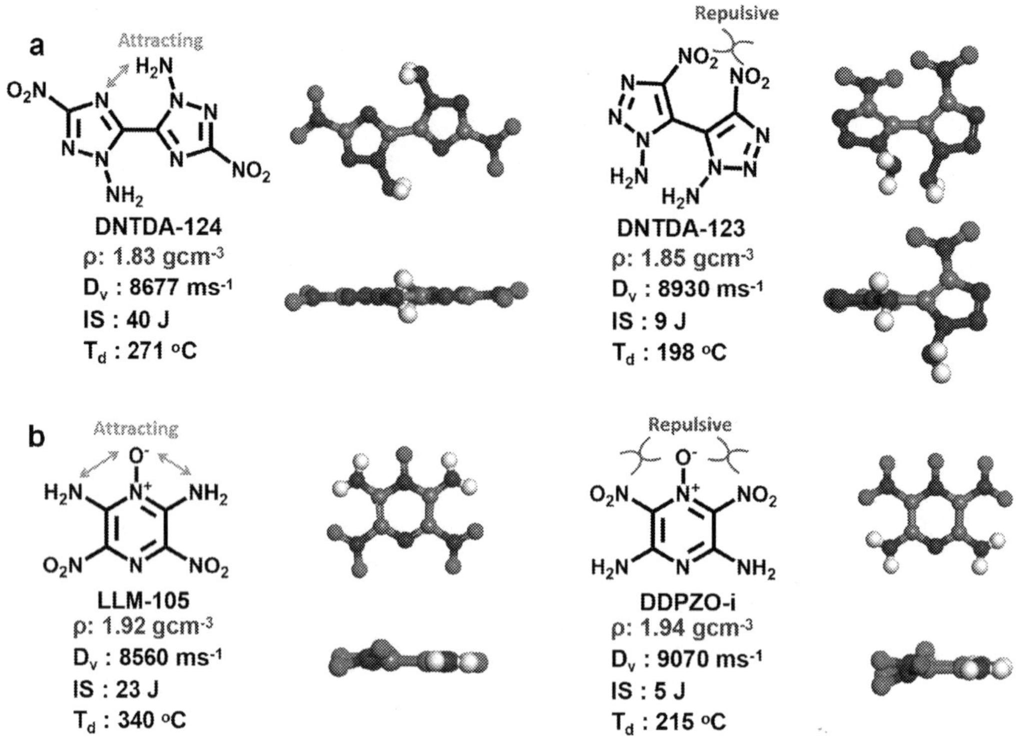

FIG. 17.9 (A) Molecular structure, configuration, and properties of two isomers of DNTDA-124 and DNTDA-123. (B) Molecular structure, configuration, and properties of two isomers of LLM-105 and DDPZD-i.

(N—O bond and nitro group) in LLM-105 are favorable to its high thermostability and low impact sensitivity. First, this alternating arrangement can facilitate to form intramolecular hydrogen bonds and make the energetic molecules an extended π-conjugated configuration with high stability. Second, the alternating arrangement can also facilitate the formation of widely extended intramolecular hydrogen-bond networks to construct a 3D layer-by-layer π-π stacking structure, which is beneficial for low mechanical sensitivities. More importantly, some representative low-sensitivity and high-energy energetic materials (such as TATB, LLM-105, FOX-7) exhibit alternating arrangement of hydrogen-bond donor and hydrogen-bond acceptor in the molecular structure (Fig. 17.8B).

Base on the above analyses, the rational arrangement of substituent groups should be particularly considered in the design of a new low-sensitivity energetic material, in particular, the distribution of explosophore and stabilizing groups on the molecular skeleton without obvious repulsive effects between substituent groups (e.g., alternating arrangement of hydrogen-bond donor and hydrogen-bond acceptor). Ideally, an energetic molecule should hold a large π-conjugated planar (or nearly planar) configuration with a high possibility of constructing intramolecular and intermolecular hydrogen bonds.

2 Crystal concerns in the design of low-sensitivity energetic materials

2.1 Effects of molecular stacking on impact sensitivity of energetic molecules

The above discussions focus on the molecular structure concerns in the design of energetic materials (including molecular composition, molecular skeleton and substituent groups, etc.). But in many cases, these molecular structure concerns could not well guide the design of new energetic materials with well-balanced energy & sensitivity due to the important effects of crystal structures on the sensitivity of energetic materials. For the structures of a given energetic material, its molecular packing is intrinsic and usually remains much lower variability than those extrinsic characteristics (such as crystal defect, size, and shape). Thus, previous studies mainly focused on establishing the relationship between molecular stacking mode and mechanical sensitivity (especially for impact sensitivity) [5,12]. The molecular stacking modes of π-conjugated energetic materials can be divided into four types: face-to-face, wave-like, crossing, and mixing stacking [10]. Among the four types of molecular stacking modes, face-to-face π-π stacking (or graphite-like layered stacking) is the most desirable one since it can well consume external mechanical input by layer-to-layer slipping and buffering and endow energetic materials with low mechanical sensitivities (such as TATB, DAAF, and DAAZF). Although the inter-layer sliding ability of wave-like stacking is relatively weaker than that of graphite-like layered stacking, to a certain extent it is easier to exhibit inter-layer sliding than for crossing and mixing stacking modes. Consequently, several typical low-sensitivity energetic materials such as NTO, FOX-7, and LLM-105 hold the wave-like stacking in crystal structures. Due to relatively high inter-layer sliding inhibition, most energetic materials with crossing and mixing crystal stacking exhibit high mechanical sensitivities. Therefore, from the viewpoint of molecular stacking mode, graphite-like layered stacking is the best choice and wave-like stacking would be an important alternative [63].

2.2 Effect of crystal engineering on the design of graphite-like energetic materials

Based on the above discussions, we can find that the design of new graphite-like energetic molecules is an effective approach for developing new low-sensitivity energetic materials. But up to now, it still remains a great challenge to judge whether the designed energetic molecule holds a desired graphite-like layered stacking or not before experimentally obtaining its crystal structure. Moreover, based on quantum mechanics or molecular mechanics, the existing crystal structure predictions usually suffer from uncertainties regarding the applicability and accuracy of model chemistry or forcefield, and these calculations are time-consuming and costly. Therefore, for graphite-like energetic materials, an in-depth understanding of these self-assembly processes is highly desirable, which will help us to establish a relatively simple and efficient method for predicting the crystal engineering of graphite-like energetic molecules [64].

As shown in Fig. 17.10, the formation of graphite-like layered stacking of energetic molecules can be deduced as a point-chain-plane progressive intermolecular assembly (pcp-PIA). Taking CIWMAW01 (a typical graphite-like energetic molecule) as an example, its 3D graphite-like crystal structure can be progressively exfoliated into a 2D molecular plane, then a one-dimensional (1D) molecular chain, and finally a single planar molecule (Fig. 17.10A).

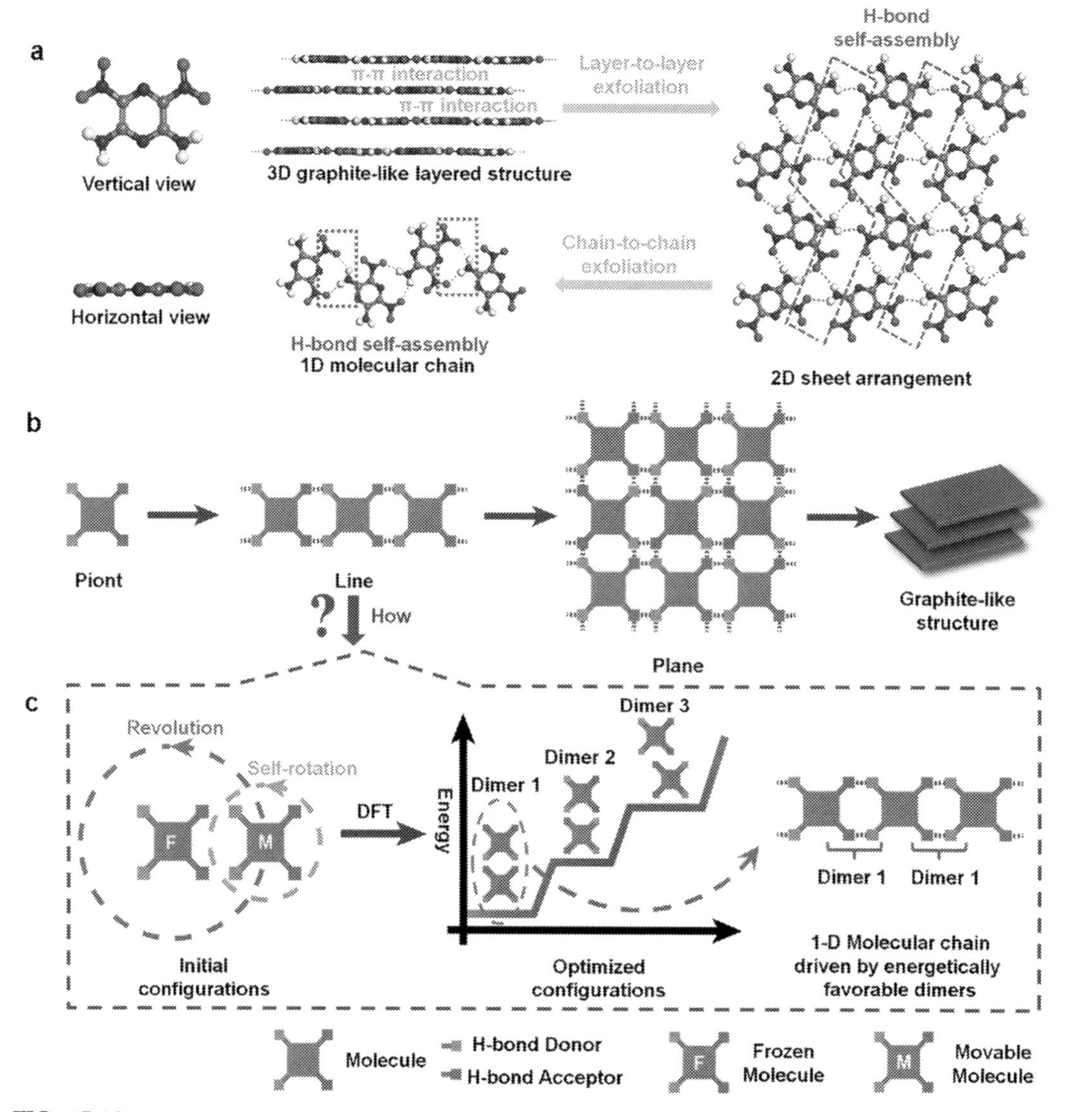

FIG. 17.10 (A) Progressively exfoliated analysis on graphite-like crystal structure using CIWMAW01 (a crystal structure from Cambridge Crystallographic Data Centre, CCDC) as an example. (B) Schematic of progressive intermolecular assembly from point to chain, then to plane, at last to graphite-like layer structure. (C) The searching of energy-favorable dimers for constructing supramolecular "chain." *Reprinted with permission from S. Song, Y. Wang, K. Wang, F. Chen, Q. Zhang, Decoding the crystal engineering of graphite-like energetic materials: from theoretical prediction to experimental verification, J. Mater. Chem. A 8 (2020) 5975.*

Conversely, when we regard a planar molecule as the basic "point", the point-chain-plane progressive intermolecular assembly (pcp-PIA) driven by complex intermolecular interactions can construct a desired 2D supramolecular plane, and the 3D graphite-like crystal structure is finally formed by strong π-π interactions between supramolecular planes (Fig. 17.10B).

With this pcp-PIA rule, we can preliminarily judge whether an organic molecule with planar geometry tends to form a graphite-like crystal structure or not. However, an accurate prediction of graphite-like crystal engineering is far more difficult, because most energetic molecules usually have multiple H-bond donor and acceptor sites with various possibilities of intermolecular assembly. From the molecular assembly process of CIWMAW01 (Fig. 17.10A), it can be found that the most critical step of forming 3D graphite-like crystal structure is the construction of a 1D supramolecular "chain" with planar geometry. From zero-dimensional "point" to 2D "plane," the 1D "chain" is a very important bridge, because it is not only the self-assembled terminal point for low-dimensional "point," but also the self-assembled starting point for high-dimensional "plane." Thus, accurate simulation of the self-assembled process of 1D "chain" is a significant and basic premise for predicting 3D graphite-like crystal structures.

Despite numerous possibilities of intermolecular self-assembly for a given energetic molecule, there is a basic rule that the 1D "chain" is mainly governed by energetically favorable dimers. To well search for the energy-favorable dimers with planar geometry, a "coplanar configuration searching (CCS)" method was developed. As shown in Fig. 17.10C, the CCS method mainly contains three steps: (1) creating initial dimer configurations by revolving and self-rotating the movable molecule around the central of the frozen molecule at a given step-size and distance; (2) optimizing all the generated initial configurations to their nearest energy-minimum configurations by a DFT program and listing possible dimer geometries by an energy sorting; (3) constructing the possible 1D molecular "chain" by energy-favorable dimers with planar configuration (Fig. 17.10C). Thereafter, a 3D graphite-like crystal structure is expected to be formed by the progressive chain-to-chain intermolecular self-assembly and strong π–π interactions between layers (Fig. 17.10B). For four typical energetic molecules (TATNBZ, CIWMAW01, TIBMUM, and TIBMOG) with graphite-like crystal structure, their 1D molecular "chain" are indeed constructed by energy-favorable dimers (located in relatively low-energy regions) (Fig. 17.11), which means that coplanar configuration searching (CCS) method can effectively help to identify the potential coplanar dimers that can realize the formation of 1D molecular chains in the graphite-like energetic crystal structure. Therefore, the combination of coplanar configuration searching (CCS) and point-chain-plane progressive intermolecular assembly (pcp-PIA) is a very effective method to predict and screen new graphite-like energetic materials.

Here, DANAP is selected as an example to illustrate this deduction process of the graphite-like layered stacking by CCS and pcp-PIA method. First, the CCS calculations show that there are six configuration dimers of DANAP in low-energy regions and dimers 1, 2, 3 are thought as the useful dimers of DANAP for possible graphite-like molecular self-assembly except for non-planar configurations (dimers 4 and 5) and structure-similarity (dimer 1 and dimer 6) (Fig. 17.12A). Next, the pcp-PIA analysis shows that although dimers 1 and 2 can self-assemble into a 1D molecular chain, the formation of this molecular chain occupies almost all hydrogen-bond sites. When this molecular chain is considered as a whole, no matter how close these molecular chains are to each other (parallel or mirror flipping approaching), there are not sufficient hydrogen-bond interactions to support the 2D intra-layer spreading of multiple molecular chains to form an ideal 2D molecular plane. As a result, the self-assembly driven by dimer 1 and dimer 2 of DANAP cannot construct a graphite-like crystal structure (Fig. 17.12B). In contrast, the dimer 3 of DANAP can also

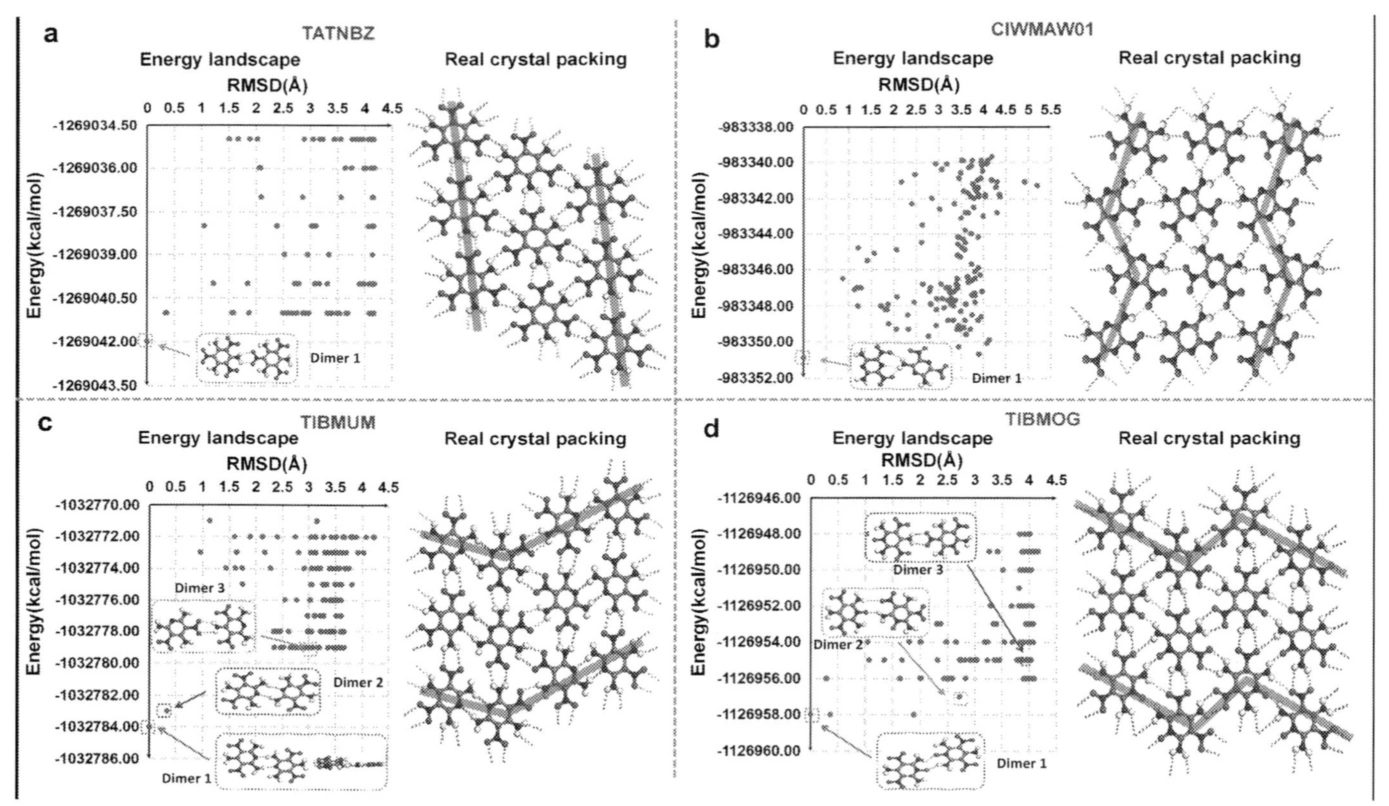

FIG. 17.11 (A) Energy landscape of dimer configurations of TATNBZ (namely TATB) and its real 2D intra-layer crystal packing; (B) energy landscape of dimer configurations of CIWMAW01 and its real 2D intra-layer crystal packing; (C) energy landscape of dimer configurations of TIBMUM and its real 2D intra-layer crystal packing; (D) energy landscape of dimer configurations of TIBMOG and its real 2D intra-layer crystal packing (*white*, *red*, *blue*, and *gray* atoms represent hydrogen, oxygen, nitrogen, and carbon atoms, respectively; the dotted lines represent the hydrogen bond). *Reprinted with permission from S. Song, Y. Wang, K. Wang, F. Chen, Q. Zhang, Decoding the crystal engineering of graphite-like energetic materials: from theoretical prediction to experimental verification, J. Mater. Chem. A 8 (2020) 5975.*

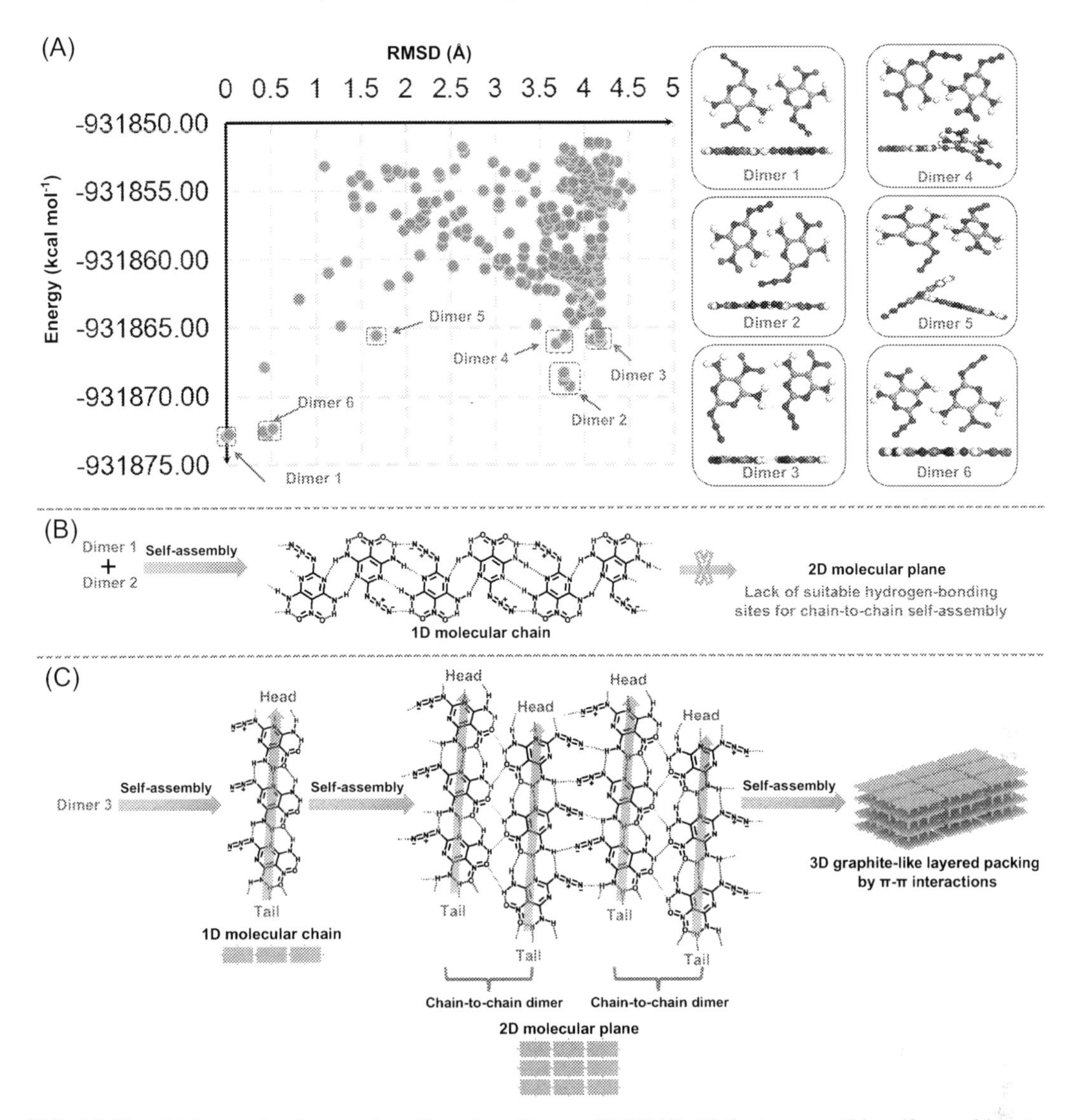

FIG. 17.12 (A) Energy landscape of configuration dimers of DANAP; (B) the incompatible self-assembly into a graphite-like structure by dimer 1 and dimer 2 of DANAP; (C) the theoretically predicted graphite-like structure of DANAP by progressive self-assembly from dimer 3 (the dot lines represent intra/intermolecular hydrogen bonds). *Reprinted with permission from S. Song, Y. Wang, K. Wang, F. Chen, Q. Zhang, Decoding the crystal engineering of graphite-like energetic materials: from theoretical prediction to experimental verification, J. Mater. Chem. A 8 (2020) 5975.*

self-assemble into a 1D molecular chain and two molecular chains further self-assemble into a chain-to-chain dimer by mirror flipping approaching with the aid of hydrogen-bond interactions between nitro and amino groups. The 2D molecular plane can subsequently be fabricated by parallel approaching with the hydrogen bonds between chain-to-chain dimers

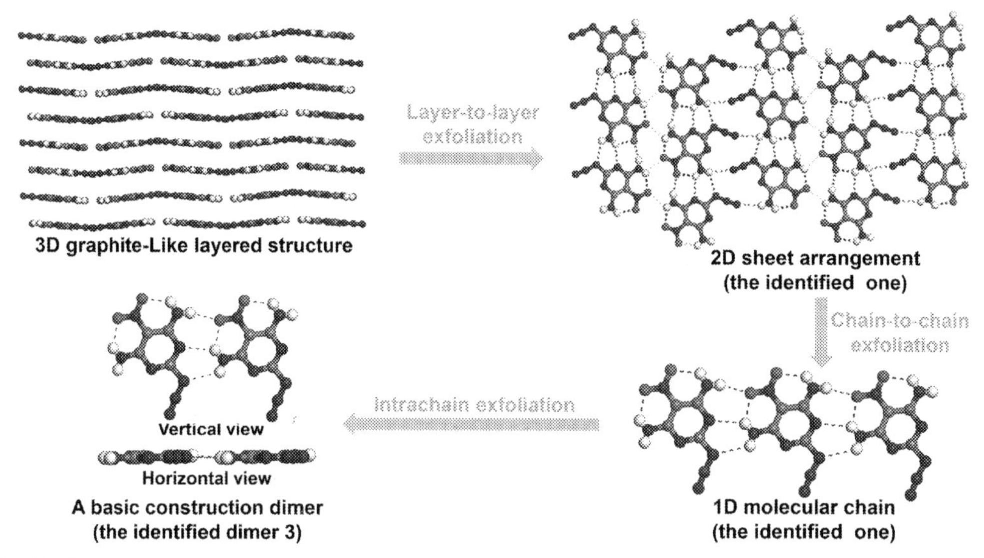

FIG. 17.13 Experimental crystal structure of DANAP (the *left* is the two-dimensional molecular plane and the *right* is three-dimensional graphite-like layered packing, the *dot lines* represent hydrogen bonds). *Reprinted with permission from S. Song, Y. Wang, K. Wang, F. Chen, Q. Zhang, Decoding the crystal engineering of graphite-like energetic materials: from theoretical prediction to experimental verification, J. Mater. Chem. A 8 (2020) 5975.*

(Fig. 17.12C middle). Finally, a 3D graphite-like crystal structure of DANAP is formed through strong π-π interactions between 2D layers (Fig. 17.12C). Thus, dimer 3 can be thought of as an important construction unit for DANAP molecules to form graphite-like crystal structures.

Further experimental verification indicates that DANAP indeed crystallizes into a graphite-like structure (Fig. 17.13 *left*), and the space group of DANAP belongs to -P 2 ac 2ab. Note that the theoretically predicted dimer 3 acts as a basic building unit in the formation of a 1D molecular chain, and the planar layered structure is readily constructed by the remaining hydrogen-bond sites around the molecular chains (Fig. 17.13). The detonation velocity and detonation pressure of DANAP are calculated to be 8100 m s^{-1} and 24.7 GPa, respectively, which are superior to those of TNT (7303 m s^{-1} and 21.3 GPa) [37]. Although there is a sensitive azide group in the molecular structure of DANAP, the impact sensitivity (IS) of DANAP is also better than those of TNT (25 J vs 15 J) due to a graphite-like crystal structure. These results indicate that the strategy of designing graphite-like energetic molecules *via* the combined CCS and pcp-PIA method is very efficient for developing low-sensitivity energetic materials.

3 Conclusion

Although the factors influencing the energy and sensitivity of energetic materials are very complicated, the chemical composition, molecular structure, and crystal packing modes are usually the intrinsic ones to determine the energy and sensitivity of energetic materials. Thus, the understanding of structure–property relationships at the molecular and crystal levels is very important to design new energetic materials. This chapter mainly focuses on the molecular and crystal design tips on new energetic materials, which is expected to provide useful guidance for developing low-sensitivity energetic materials with desired properties.

On the molecular level, oxygen balance (OB) should be first concerned because it is the basic and accessible composition index of correlating energy and sensitivity. A reasonable oxygen balance ranging from −20% to −80% is a key for the development of low-sensitivity energetic materials. With the same oxygen balance, different types of energetic materials exhibit obvious differences in energy and sensitivity. Static statistical analysis shows that reasonable combinations between molecular skeleton, substituent group, and substituent arrangement enable the construction of new low-sensitivity energetic materials with high-energy and heat-resistant properties.

On the crystal level, due to the different dissipation capacities for external mechanical stimuli, the crystal packing modes play a crucial role in the mechanical sensitivity of energetic materials. A large amount of theoretical and experimental studies have demonstrated that graphite-like layered stacking can consume external energy input well to make energetic materials exhibit low sensitivity. Thus, the graphite-like layered stacking is an optimal crystal packing model for developing low-sensitivity energetic materials and the alternative one is wave-like stacking. Nowadays it still remains a big challenge to achieve the efficient and accurate crystal structural prediction. So for the development of new energetic materials with graphite-like layered stacking, an alternative choice is to combine some theoretical calculations on key steps and logic deduction of molecular self-assembly. The combined method of coplanar configuration searching (CCS) and point-chain-plane progressive intermolecular assembly (pcp-PIA) has demonstrated the great potential for predicting graphite-like layered stacking of energetic materials.

Obviously, above-mentioned design tips are mostly discrete information and qualitative cognition. There is still a long way to realize the rapid and efficient design of new energetic materials. The next step will see research in this field explore new technologies (such as machine learning and artificial intelligence) [65,66] that can integrate multi-level information and establish a set of systematic structural-performance prediction models, which will efficiently and accurately guide the development of new-generation energetic materials.

References

[1] K.B. Landenberger, A. Matzger, Cocrystal engineering of a prototype energetic material: supramolecular chemistry of 2, 4, 6-trinitrotoluene, Cryst. Growth Des. 10 (2010) 5341.

[2] Y. Wang, S. Song, C. Huang, X. Qi, K. Wang, Y. Liu, Q. Zhang, Hunting for advanced high-energy-density materials with well-balanced energy and safety through an energetic host-guest inclusion strategy, J. Mater. Chem. A 7 (2019) 19248.

[3] B. Tian, Y. Xiong, L. Chen, C. Zhang, Relationship between the crystal packing and impact sensitivity of energetic materials, CrstEngComm 20 (2018) 837.

[4] G. Li, C. Zhang, Review of the molecular and crystal correlations on sensitivities of energetic materials, J. Hazard. Mater. 398 (2020) 122910.

[5] F. Jiao, Y. Xiong, H. Li, C. Zhang, Alleviating the energy & safety contradiction to construct new low sensitivity and highly energetic materials through crystal engineering, CrstEngComm 20 (2018) 1757.

[6] P. Politzer, J.S. Murray, High performance, low sensitivity: conflicting or compatible, Propellants Explos. Pyrotech. 41 (2016) 414.

[7] C. Zhang, in: L.T. DeLuca, V.P. Sindiskii, T. Shimada, M. Calabro (Eds.), Chemical Rocket Propulsion: A Comprehensive Survey of Energetic Materials, Springer, 2016, pp. 26–29.
[8] J. Mullay, A relationship between impact sensitivity and molecular electronegativity, Propellants Explos. Pyrotech. 12 (1987) 60.
[9] Y. Ma, A. Zhang, X. Xue, D. Jiang, Y. Zhu, C. Zhang, Crystal packing of impact-sensitive high-energy explosives, Cryst. Growth Des. 14 (2014) 6101.
[10] Y. Ma, A. Zhang, C. Zhang, D. Jiang, Y. Zhu, C. Zhang, Crystal packing of low-sensitivity and high-energy explosives, Cryst. Growth Des. 14 (2014) 4703.
[11] C. Zhang, X. Xue, Y. Cao, Y. Zhou, H. Li, J. Zhou, T. Gao, Intermolecular friction symbol derived from crystal information, CrstEngComm 15 (2013) 6837.
[12] C. Zhang, F. Jiao, H. Li, Crystal engineering for creating low sensitivity and highly energetic materials, Cryst. Growth Des. 18 (2018) 5713.
[13] A. Pal, V. Meunier, C.R. Picu, Investigating orientational defects in energetic material RDX using first-principles calculations, J. Phys. Chem. A 120 (2016) 1917.
[14] M.M. Kuklja, E.V. Stefanovich, A.B. Kunz, An excitonic mechanism of detonation initiation in explosives, J. Chem. Phys. 112 (2000) 3417.
[15] M.M. Kuklja, B.P. Aduev, E.D. Aluker, V.I. Krasheninin, A.G. Krechetov, A.Y. Mitrofanov, Role of electronic excitations in explosive decomposition of solids, J. Appl. Phys. 89 (2001) 4156.
[16] H. Chen, L. Li, S. Jin, S. Chen, Q. Jiao, Effects of additives on ε-HNIW crystal morphology and impact sensitivity, Propellants Explos. Pyrotech. 37 (2012) 77.
[17] N. Zohari, M.H. Keshavarz, S.A. Ssyedsadjadi, The advantages and shortcomings of using nano-sized energetic materials, Cent. Eur. J. Energ. Mat. 10 (2013) 135.
[18] M.F. Gogulya, M.A. Brazhnikov, Effect of the dispersity of the components of explosive materials on the detonation velocity and sensitivity to mechanical action, Russ. J. Phys. Chem. B 4 (2010) 286.
[19] X. Song, Y. Wang, C. An, X. Guo, F. Li, Dependence of particle morphology and size on the mechanical sensitivity and thermal stability of octahydro-1, 3, 5, 7-tetranitro-1, 3, 5, 7-tetrazocine, J. Hazard. Mater. 159 (2008) 222.
[20] G. Zhang, H. Sun, J.M. Abbott, B.L. Weeks, Engineering the microstructure of organic energetic materials, ACS Appl. Mater. Interfaces 1 (2009) 1086.
[21] R. Kumar, P.F. Siril, P. Soni, Tuning the particle size and morphology of high energetic material nanocrystals, Def. Technol. 11 (2015) 382.
[22] Y. Li, P. Wu, C. Hua, J. Wang, B. Huang, J. Chen, Z. Qiao, G. Yang, Determination of the mechanical and thermal properties, and impact sensitivity of pressed HMX-based PBX, Cent. Eur. J. Energ. Mater. 16 (2019) 299.
[23] G. He, Z. Yang, X. Zhou, J. Zhang, L. Pan, S. Liu, Polymer bonded explosives (PBXs) with reduced thermal stress and sensitivity by thermal conductivity enhancement with graphene nanoplatelets, Compos. Sci. Technol. 131 (2016) 22.
[24] C. Zhang, Origins of the energy and safety of energetic materials and of the energy & safety contradiction, Propellants Explos. Pyrotech. 43 (2018) 855.
[25] J.S. Murray, P. Lane, P. Politzer, A relationship between impact sensitivity and the electrostatic potentials at the midpoints of C-NO2 bonds in nitroaromatics, Chem. Phys. Lett. 168 (1990) 135.
[26] Q.-L. Yan, S. Zeman, Theoretical evaluation of sensitivity and thermal stability for high explosives based on quantum chemistry methods: a brief review, Int. J. Quantum. 113 (2013) 1049.
[27] M.H. Keshavarz, H.R. Pouretedal, A. Semnani, Novel correlation for predicting impact sensitivity of nitroheterocyclic energetic molecules, J. Hazard. Mater. 141 (2007) 803.
[28] M. Pospíšil, P. Vávra, M.C. Concha, J.S. Murray, P. Politzer, A possible crystal volume factor in the impact sensitivities of some energetic compounds, J. Mol. Model. 16 (2010) 895.
[29] B.M. Rice, J.J. Hare, A quantum mechanical investigation of the relation between impact sensitivity and the charge distribution in energetic molecules, J. Phys. Chem. A 106 (2002) 1770.
[30] C. Zhang, Y. Shu, Y. Huang, X. Zhao, H. Dong, Investigation of correlation between impact sensitivities and nitro group charges in nitro compounds, J. Phys. Chem. B 109 (2005) 8978.
[31] C. Cao, S. Gao, Two dominant factors influencing the impact sensitivities of nitrobenzenes and saturated nitro compounds, J. Phys. Chem. B 111 (2007) 12399.
[32] J.S. Murray, M.C. Concha, P. Politzer, Links between surface electrostatic potentials of energetic molecules, impact sensitivities and C–NO2/N–NO2 bond dissociation energies, Mol. Phys. 107 (2009) 89.

[33] J. Edwards, C. Eybl, B. Johnson, Correlation between sensitivity and approximated heats of detonation of several nitroamines using quantum mechanical methods, J. Quantum. Chem. 100 (2004) 713.
[34] S. Zeman, M. Jungová, Sensitivity and performance of energetic materials, Propellants Explos. Pyrotech. 41 (2016) 426.
[35] T.M. Klapötke, T.G. Witkowski, Covalent and ionic insensitive high-explosives, Propellants Explos. Pyrotech. 41 (2016) 470.
[36] P. Pagoria, A comparison of the structure, synthesis, and properties of insensitive energetic compounds, Propellants Explos. Pyrotech. 41 (2016) 452.
[37] H. Gao, Q. Zhang, J.M. Shreeve, Fused heterocycle-based energetic materials (2012–2019), J. Mater. Chem. A 8 (2020) 4193.
[38] S. Zeman, New dependence of activation energies of nitroesters thermolysis and possibility of its application, Propellants Explos. Pyrotech. 17 (1992) 19.
[39] J. Zhang, P. Yin, L.A. Mitchell, D.A. Parrish, J.M. Shreeve, N-functionalized nitroxy/azido fused-ring azoles as high-performance energetic materials, J. Mater. Chem. A 4 (2016) 7430.
[40] C. He, Y. Tang, L.A. Mitchell, D.A. Parrish, J.M. Shreeve, N-oxides light up energetic performances: synthesis and characterization of dinitraminobisfuroxans and their salts, J. Mater. Chem. A 4 (2016) 8969.
[41] Y. Wang, Y. Liu, S. Song, Z. Yang, X. Qi, K. Wang, Y. Liu, Q. Zhang, Y. Tian, Accelerating the discovery of insensitive high-energy-density materials by a materials genome approach, Nat. Commun. 9 (2018) 2444.
[42] M.J. Kamlet, H.G. Adolph, The relationship of impact sensitivity with structure of organic high explosives. II. Polynitroaromatic explosives, Propellants Explos. Pyrotech. 4 (1979) 30.
[43] W. Zhang, J. Zhang, M. Deng, X. Qi, F. Nie, Q. Zhang, A promising high-energy-density material, Nat. Commun. 8 (2017) 181.
[44] T. Yan, G. Cheng, H. Yang, 1, 2, 4-Oxadiazole-bridged polynitropyrazole energetic materials with enhanced thermal stability and low sensitivity, Chem. Plus. Chem. 84 (2019) 1567.
[45] Q. Wang, Y. Shao, M. Lu, $C_8N_{12}O_8$: a promising insensitive high-energy-density material, Cryst. Growth Des. 18 (2018) 6150.
[46] R. Tsyshevsky, P. Pagoria, M. Zhang, A. Racoveanu, D.A. Parrish, A.S. Smirnov, M.M. Kuklja, Comprehensive end-to-end design of novel high energy density materials: I. synthesis and characterization of oxadiazole based heterocycles, J. Phys. Chem. C 121 (2017) 23853.
[47] P. Yin, C. He, J.M. Shreeve, Fully C/N-polynitro-functionalized 2, 2′-biimidazole derivatives as nitrogen- and oxygen- rich energetic salts, Chem. A Eur. J. 22 (2016) 2108.
[48] Y. Tang, C. He, G.H. Imler, D.A. Parrish, J.M. Shreeve, Aminonitro groups surrounding a fused pyrazolotriazine ring: a superior thermally stable and insensitive energetic material, ACS Appl. Energy Mater. 2 (2019) 2263.
[49] L. Hu, P. Yin, G.H. Imler, D.A. Parrish, H. Gao, J.M. Shreeve, Fused rings with N-oxide and -NH_2: good combination for high density and low sensitivity energetic materials, Chem. Commun. 55 (2019) 8979.
[50] P. Yin, J. Zhang, G.H. Imler, D.A. Parrish, J.M. Shreeve, Polynitro-functionalized dipyrazolo-1, 3, 5-triazinanes: energetic polycyclization toward high density and excellent molecular stability, Angew. Chem. Int. Ed. 56 (2017) 8834.
[51] M.C. Schulze, B.L. Scottb, D.E. Chavez, A high density pyrazolo-triazine explosive (PTX), J. Mater. Chem. A 3 (2015) 17963.
[52] V. Thottempudi, P. Yin, J. Zhang, D.A. Parrish, J.M. Shreeve, 1,2,3-Triazolo[4,5,-e]furazano[3,4,-b]pyrazine 6-oxide - a fused heterocycle with a roving hydrogen forms a new class of insensitive energetic materials, Chem. A Eur. J. 20 (2014) 542.
[53] E.E. Kilmer, Heat-resistant explosives for space applications, J. Spacecr. Rockets 5 (1968) 1216.
[54] C. Li, C. Deng, B. Zhao, M. Wang, M. Zhang, Z. Zhou, A zwitterionic compound with heterocyclic ions as promising heat-resistant explosive, Propellants Explos. Pyrotech. 45 (2020) 531.
[55] F.J. Owens, Calculation of energy barriers for bond rupture in some energetic molecules, J. Mol. Struct. 370 (1996) 11.
[56] V.W. Manner, M.J. Cawkwell, E.M. Kober, T.W. Myers, G.W. Brown, H. Tian, C.J. Snyder, R. Perriot, D.N. Preston, Examining the chemical and structural properties that influence the sensitivity of energetic nitrate esters, Chem. Sci. 9 (2018) 3649.
[57] X. Cao, Y. Wen, B. Xiang, X. Long, C. Zhang, Are amino groups advantageous to insensitive high explosives (IHEs), J. Mol. Model. 18 (2012) 4729.

[58] D.E. Bliss, S.L. Christian, W.S. Wilson, Impact sensitivity of polynitroaromatics, J. Energ. Mater. 9 (1991) 319.
[59] R. Bu, Y. Xiong, X. Wei, H. Li, C. Zhang, Hydrogen bonding in CHON-containing energetic crystals: a review, Cryst. Growth Des. 19 (2019) 5981.
[60] T.M. Klapötke, A. Preimesser, J. Stierstorfer, Energetic derivatives of 4, 4′, 5, 5′-tetranitro-2, 2′-bisimidazole (TNBI), Z. Anorg. Allg. Chem. 638 (2012) 1278.
[61] K.V. Domasevitch, I. Gospodinov, H. Krautscheid, T.M. Klapötke, J. Stierstorfer, Facile and selective polynitrations at the 4-pyrazolyl dual backbone: straightforward access to a series of high-density energetic materials, New J. Chem. 43 (2019) 1305.
[62] Z. Wang, W. Zhang, K. Wang, X. Qi, Q. Zhang, Synthesis and property of 3, 5-diamino-2, 6-dinitropyrazine-1-oxide, Chin. J. Energ. Mater. 24 (2016) 820.
[63] C. Li, H. Li, H.-H. Zong, Y. Huang, M. Gozin, C.Q. Sun, L. Zhang, Strategies for achieving balance between detonation performance and crystal stability of high-energy-density materials, iScience 23 (2020) 100944.
[64] S. Song, Y. Wang, K. Wang, F. Chen, Q. Zhang, Decoding the crystal engineering of graphite-like energetic materials: from theoretical prediction to experimental verification, J. Mater. Chem. A 8 (2020) 5975.
[65] D.C. Elton, Z. Boukouvalas, M.S. Butrico, M.D. Fuge, P.W. Chung, Applying machine learning techniques to predict the properties of energetic materials, Sci. Rep. 8 (2018) 9059.
[66] K.T. Butler, D.W. Davies, H. Cartwright, O. Isayev, A. Walsh, Machine learning for molecular and materials science, Nature 559 (2018) 547.

Index

Note: Page numbers followed by "*f*" indicate figures and "*t*" indicate tables.

D

F

G

T

Printed in the United States
by Baker & Taylor Publisher Services